Advances in

Heterocyclic Chemistry

Volume 46

Editorial Advisory Board

R. A. Abramovitch, *Clemson, South Carolina*
A. Albert, *Canberra, Australia*
A. T. Balaban, *Bucharest, Romania*
A. J. Boulton, *Norwich, England*
H. Dorn, *Berlin, G.D.R.*
J. Elguero, *Madrid, Spain*
S. Gronowitz, *Lund, Sweden*
T. Kametani, *Tokyo, Japan*
O. Meth-Cohn, *South Africa*
C. W. Rees, FRS, *London, England*
E. C. Taylor, *Princeton, New Jersey*
M. Tišler, *Ljubljana, Yugoslavia*
J. A. Zoltewicz, *Gainesville, Florida*

Advances in

HETEROCYCLIC CHEMISTRY

Edited by

ALAN R. KATRITZKY, FRS

Kenan Professor of Chemistry
Department of Chemistry
University of Florida
Gainesville, Florida

Volume 46

ACADEMIC PRESS, INC.
Harcourt Brace Jovanovich, Publishers
San Diego New York Berkeley Boston
London Sydney Tokyo Toronto

COPYRIGHT © 1989 BY ACADEMIC PRESS, INC.
All Rights Reserved.
No part of this publication may be reproduced or transmitted in any form or by any means, electronic or mechanical, including photocopy, recording, or any information storage and retrieval system, without permission in writing from the publisher.

ACADEMIC PRESS, INC.
San Diego, California 92101

United Kingdom Edition published by
ACADEMIC PRESS LIMITED
24-28 Oval Road, London NW1 7DX

LIBRARY OF CONGRESS CATALOG CARD NUMBER: 62-13037

ISBN 0-12-020646-3 (alk. paper)

PRINTED IN THE UNITED STATES OF AMERICA
89 90 91 92   9 8 7 6 5 4 3 2 1

Contents

PREFACE vii

1,5-Diazocines
HOWARD D. PERLMUTTER

I. Scope and Nomenclature	2
II. Preparative Methods	2
III. Theoretical and Structural Studies	42
IV. Reactions	47
V. Applications	58
References	62

Behavior of Monocyclic 1,2,4-Triazines in Reactions with C-, N-, O-, and S-Nucleophiles
V. N. CHARUSHIN, S. G. ALEXEEV, O. N. CHUPAHKIN, AND H. C. VAN DER PLAS

I. Introduction	74
II. Some Physical and Spectroscopic Properties of 1,2,4-Triazines	75
III. Nucleophilic Addition Reactions	84
IV. Nucleophilic Substitution Reactions	97
V. Nucleophilic Substitution of Hydrogen (S_NH Reactions)	119
VI. Transformations of 1,2,4-Triazine Ring	125
VII. Conclusion	134
References	135

Boron-Substituted Heteroaromatic Compounds
MASANAO TERASHIMA AND MINORU ISHIKURA

I. Introduction	143
II. Syntheses	144
III. Reactions	147
IV. Nuclear Magnetic Resonance Spectra	161
References	165

1,2,4-Triazolines
Pankaja K. Kadaba

I. Introduction	170
II. Synthesis of the Triazoline Ring	171
III. Structure and Physical Properties	235
IV. Reactivity of 1,2,4-Triazolines	254
References	274

Cumulative Index of Authors, Volumes 1–45 283

Cumulative Index of Titles, Volumes 1–45 293

Cumulative Subject Index, Volumes 41–45 303

Preface

Volume 46 of *Advances in Heterocyclic Chemistry* is an "Index Volume." It contains a cumulative index of the *titles* of articles which have appeared in the series (Volumes 1–45), and a cumulative index of *authors* who have written these contributions. Additionally it contains a *subject* index covering Volumes 41–45, which can be used in conjunction with the subject index in Volume 40 (which covers Volumes 1–40). It is hoped that the provision of these indexes will help enhance the value of the series.

Apart from these indexes, the present volume contains four chapters spanning a wide range of heterocyclic chemistry. "1,5-Diazocines" by Perlmutter, continues his coverage of important eight-membered heterocycles (cf. "Azocines" in Volume 31, and "1,4-Diazocines" in Volume 45). Charushin, Alexeev, and Chupahkin from the Soviet Union, and Van der Plas from Holland cover reactions of 1,2,4-triazines with nucleophiles, a subject to which they bring much expertise.

There have been enormous developments in organo-boron chemistry during the last few decades, but boron-containing heterocycles are still somewhat rareties. The field is ripe for exploitation, and the chapter on this subject by Terashima and Ishikura should help in this respect.

Finally, Kadaba has contributed the first comprehensive review of the extensive chemistry of 1,2,4-triazolines. This chapter will complement the earlier review (*Advances*, Volume 37) on 1,2,3-triazolines by Kadaba, Stanovnik, and Tisler.

ALAN R. KATRITZKY

1,5-Diazocines

HOWARD D. PERLMUTTER

Department of Chemical Engineering and Chemistry,
New Jersey Institute of Technology,
Newark, New Jersey 07102

I. Scope and Nomenclature 2
II. Preparative Methods 2
   A. Via Ring Closure 2
      1. Nucleophilic Displacement by Amines on Electrophilic Carbon 2
      2. Cyclization of Amino Aldehydes and Amino Ketones 3
      3. Cyclization of Amino Carboxylic Acids and Related Compounds . . . 7
      4. Cyclization Involving Carbon–Carbon Bond Formation 10
   B. Via Ring Expansion 11
      1. Schmidt or Beckmann Rearrangement of Cyclohexanediones or Cycloheptanones or Their Oximes 11
      2. 0-Bridge Opening of Bicyclo [3.3.0] Compounds 12
      3. n-Bridge Opening of Bicyclo[3.3.n]alkanes (n ⩾ 1) 14
      4. Other Ring-Expansion Methods 16
   C. Via Condensation Reactions 20
      1. Condensation of Amines with Saturated Electrophilic Carbon 20
      2. Reaction of Acyclic Diamines with Dicarboxylic Acids or Acid Derivatives 24
      3. Reaction of Acyclic Diamines with Aldehydes and Ketones 29
      4. Mixed and Other Condensation Reactions 32
   D. Homaline and Related Compounds 38
III. Theoretical and Structural Studies 42
IV. Reactions 47
   A. Eight-Membered Ring-Preserved—Annelations 47
   B. Eight-Membered Ring-Preserved—Substitutions, Functionalizations, Defunctionalizations, Oxidations, and Reductions 51
   C. Eight-Membered Ring-Preserved—Other Reactions 51
   D. Diazocine Ring Contractions and Expansions 52
   E. Ring Openings 56
V. Applications 58
   References 62

I. Scope and Nomenclature

This article covers eight-membered rings with two nitrogen atoms in a 1,5-position to each other (**1**), and their fused-ring analogues. Excluded from this chapter are (a) fused-ring analogues in which both nitrogens form part of a fusion bond (e.g., **2**), and (b) structures in which nonadjacent ring atoms are linked by a bridge (e.g., **3** and **4**).

The nomenclature of 1,5-diazocines is similar to that for azocines (82AHC115) and 1,4-diazocines (89AHC185), especially the latter. The completely conjugated compound (**5**) is called 1,5-diazocine. The partially saturated derivatives are prefixed dihydro-, tetrahydro-, etc. The totally saturated compound is called perhydro-1,5-diazocine, or 1,5-diazacyclooctane.

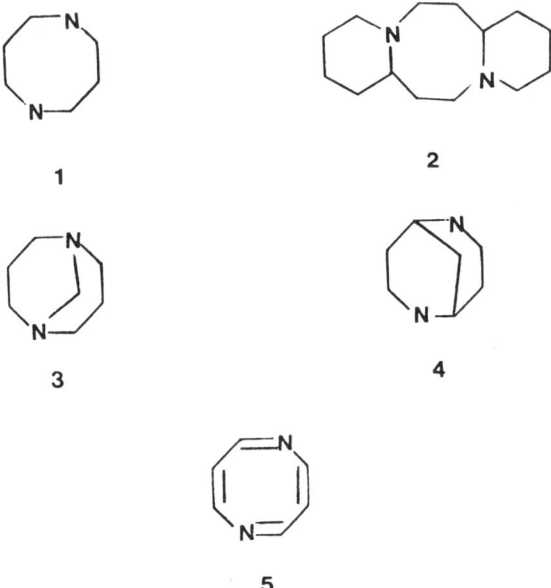

II. Preparative Methods

A. Via Ring Closure

1. Nucleophilic Displacement by Amines on Electrophilic Carbon

Yost and Margerison treated the halocyanoamide **6** with three equivalents of lithium aluminum hydride and obtained 1,5-diazocine **7** (R = H,

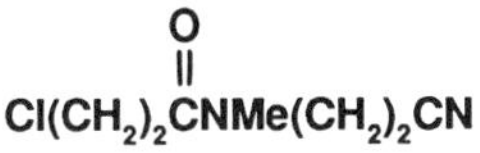

R^1 = Me) (64FRP1378964; 66USP3247206). Gatta and Landi-Vittory reacted N-phenyl- and N-benzyl,N-(3-chlorophenyl)anthranilamides with potassium carbonate and isolated 1,5-benzodiazocinones **8** (70FES830). The bisamide **9** was reductively cyclized using diborane to afford the diazocine **10** (70USP3488345).

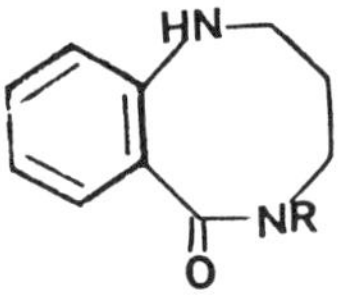

2. Cyclization of Amino Aldehydes and Amino Ketones

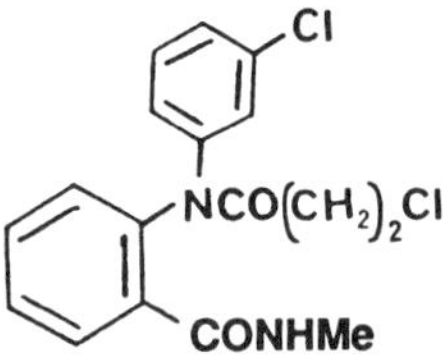

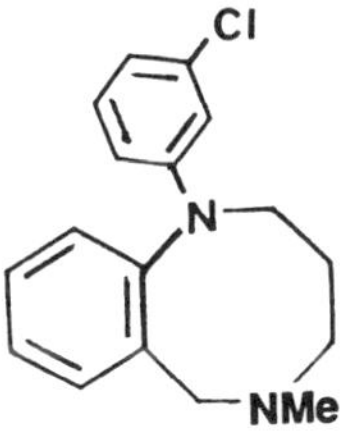

Several groups have synthesized 1,5-benzodiazocines by this route. Some of these compounds have been pharmacologically useful (see Section V). Sulkowski (66USP3294782), Fryer and co-workers (69JOC179; 70GEP1920908), and Steinman and Topliss (69JPS830) reported that the cyclization of protected aminoketoamide **11** (R = Me, R^1 = Cl, R^2 = NHCOOCH$_2$Ph) with hydrobromic acid afforded diazocinone **12** (R = Me, R^1 = Cl). Sulkowski also reported that **12** (R = H, R^1 = Cl) was prepared by this route. However, Fryer and co-workers (69JOC179; 70GEP1920908) and Denzer and Ott (69JOC183) tried to repeat the preparation of the latter unmethylated compound, but found that the product was in fact a dimer. Fryer and co-workers postulated a 16-membered ring as the dimer structure, but could not exclude structure **13** as a possibility (69JOC179; 70GEP1920908). Yamamoto and co-workers treated the unprotected analogue of **11** (R = Me, R^1 = Cl, R^2 = NH_2) with hydrogen chloride in pyridine and also obtained **12** (R = Me, R^1 = Cl) (71JAP71/04176). This same group oxidized indole **14** with chromic oxide, and, without isolation, heated the resulting solid, presumably **11** (R = Me, R^1 = Cl, R^2 = NH_2), to produce **12** (R = Me, R^1 = Cl) (71JAP71/04177). Bogatskii *et al.* cyclized the ketochloropropionamide **11** (R = H, R^1 = Cl, Br; R^2 = Cl) with ammonia to produce **12** (R = H, R^1 = Cl, Br), presumably via the aminoamide **11** (R = H; R^1 = Cl, Br; R^2 = NH_2) (72KGS1705).

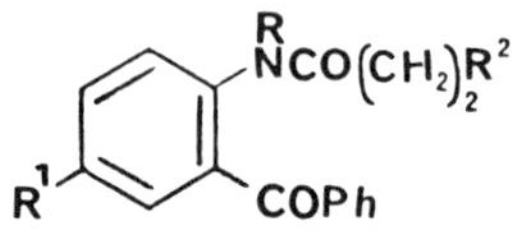

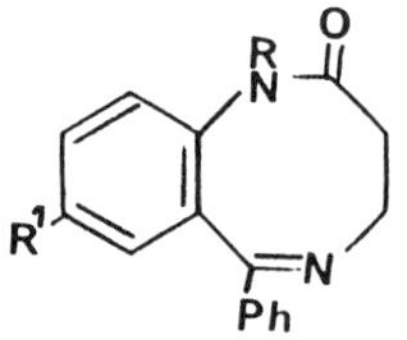

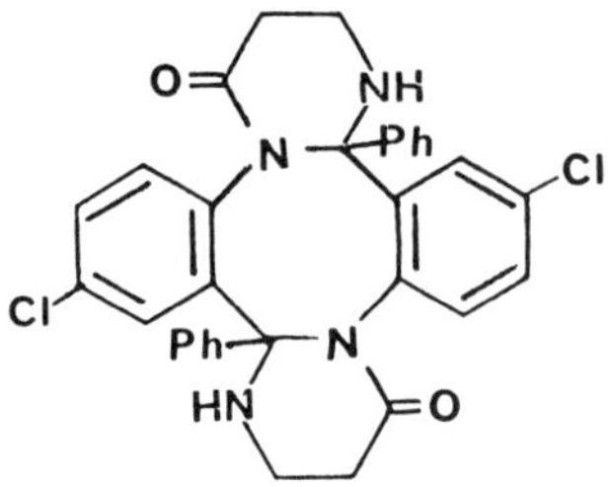

Steinman and Topliss reductively cyclized the cyanoketoamide **15** (R = H, Me) to give **16** (R = H, Me). The reduction of **15** (R = H) afforded the intermediate amino compound, which was then cyclized; but **15** (R = Me) spontaneously cyclized to product upon reduction (69JPS830). Derieg *et al.* also cyclized **17** (R = H, R^1 = R^2 = Cl) to obtain **18** (R = H, R^1 = Cl). The N-methyl compound **17** (R = Me, R^1 = R^2 = Cl) was converted directly into the diazocine **18** (R = Me, R^1 = Cl) without isolating intermediates, by treatment with potassium iodide followed by ammonia (69JOC179; 70GEP1920908). Gatta and Chiavarelli cyclized **17** (R

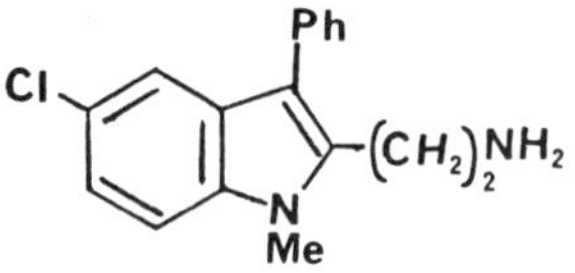

= Ac, R^1 = H, R^2 = NH_2) to yield **18** (R = Ac, R^1 = H). Deacylation afforded **18** (R = R^1 = H) (77FES33). Interestingly, these workers obtained precursor **17** (R = Ac, R^1 = H, R^2 = NH_2) by ring opening an indole, whereas Derieg *et al.* started with a benzodiazepine to obtain the precursor **17** (R = H, R^1 = Cl, R^2 = NH_2) (69JOC179; 70GEP1920908).

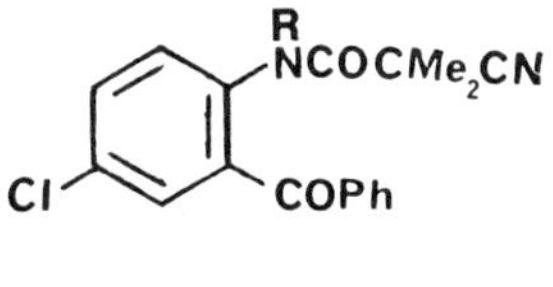

15

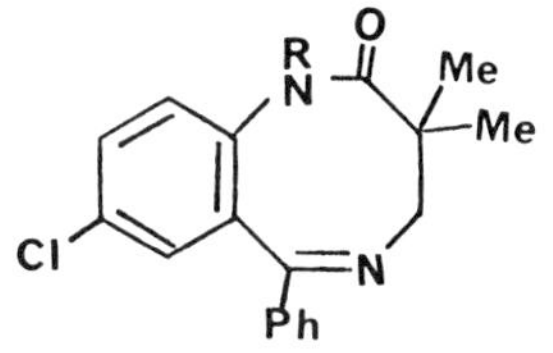

16

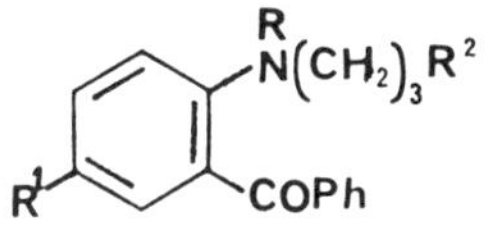

17

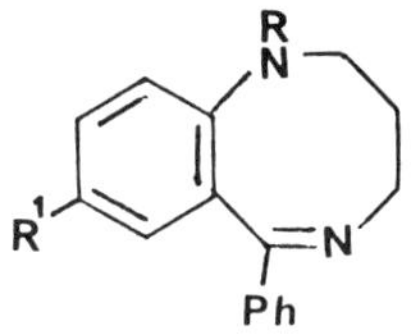

18

Singh and Mehta heated an enamine postulated as either structure **19** or **20** with concentrated sulfuric acid and obtained the cyclopentenodiazocinethione **21** (77IJC(B)786). Topliss and co-workers reductively cy-

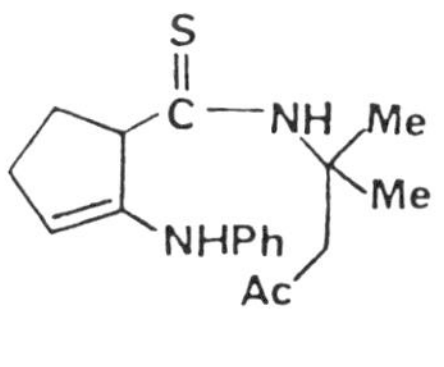

19

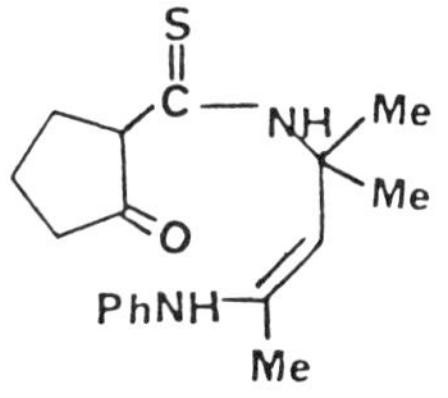

20

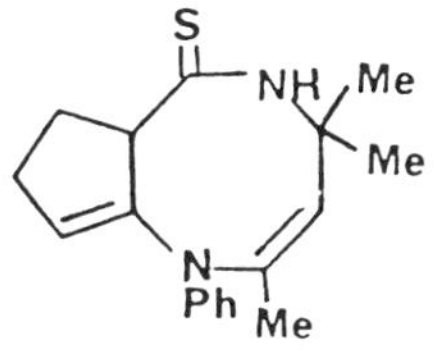

21

clized nitro ketone **22** (R = Cl, R^1 = R^2 = H, R^3 = NO_2) to afford the dibenzo[b,f]-1,5-diazocine **23** (R = Cl, R^1 = R^2 = H) (67JMC642; 68USP3409608). The same reactions were carried out by Bogatskii *et al.*, who cyclized molecules of type **22** directly (R^3 = NH_2) or reductively (R^3 = NO_2) to obtain **23** (R = H, Me, Cl Br; R^1 = H, NO_2; R^2 = H, *o*-Cl, *p*-Cl) (80MI3; 82UKZ1077).

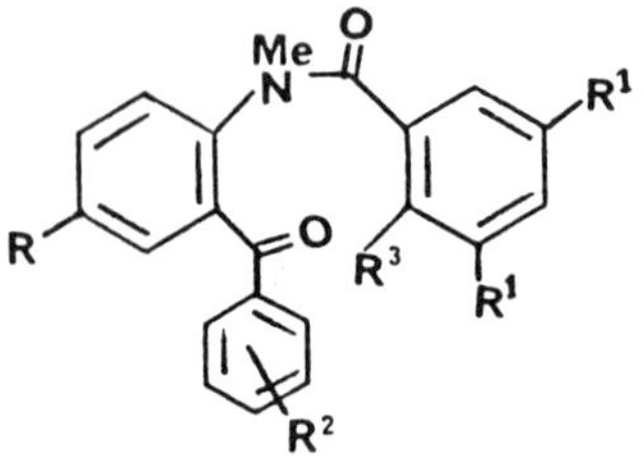

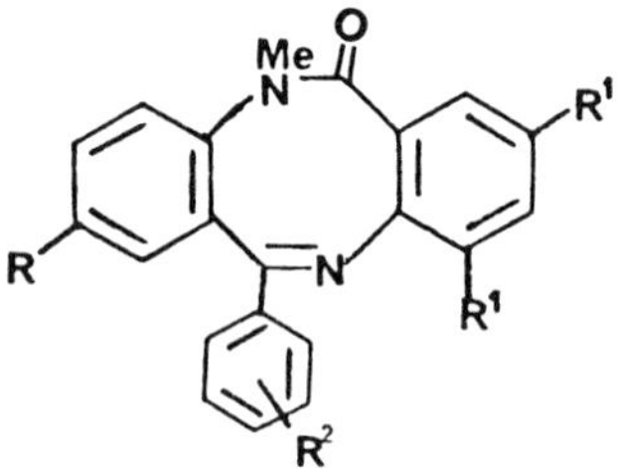

Interesting heterocycle-fused 1,5-benzodiazocines have been prepared. Two different groups have synthesized triazolo-1,5-benzodiazocines **24** by cyclization of triazolobenzophenones **25** (74JAP74/85095; 80JMC392), while pyrazolo-1,5-benzodiazocines **26** (R = Me, Ph) have been isolated by cyclization of **27** (R = Me, R^1 = NH_2) and by reductive cyclization of **27** (R = Ph, R^1 = NO_2) (79JHC935).

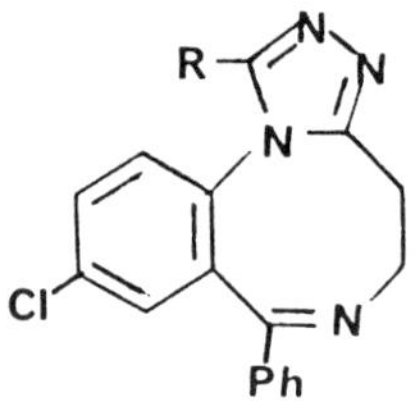

R=Me, CH_2NMe_2

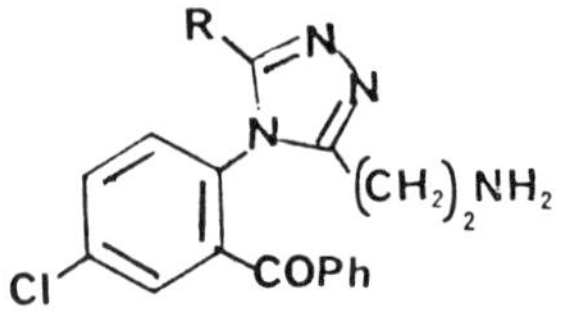

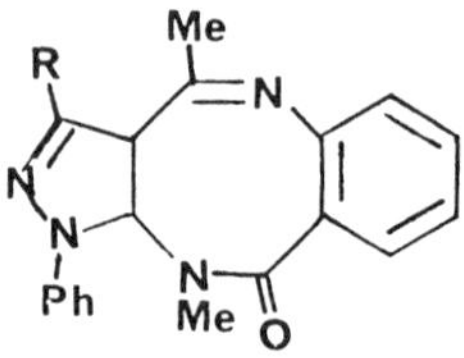

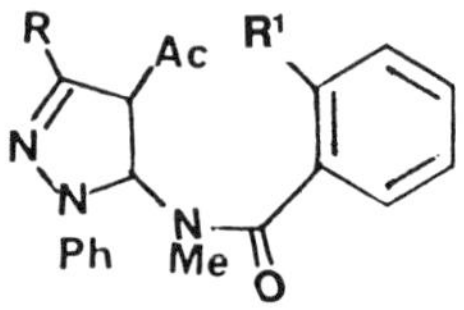

3. Cyclization of Amino Carboxylic Acids and Related Compounds

Cyclization of L,L-**28** in methanol afforded the optically active bislactamdicarboxylate **29** (74MI1). Reductive cyclization of 3-(*N*-cyanoethylamino)-1-propylamine using sodium in butanol yielded 1,5-diazocine **7** (R = R^1 = H) (51ZOB268). A number of 1,5-diazacyclooctanes **7** (R = alk, R^1 = H) were prepared by similar reductive cyclization of *N*-alkyl,*N*,*N*-bis(cyanoethyl)amines using hydrogen and Raney nickel catalyst (54ZOB163; 66MI1).

$H_2NCH(COOMe)CH_2CONHCH(COOMe)CH_2COOC_6Cl_5\text{-HBr}$

28

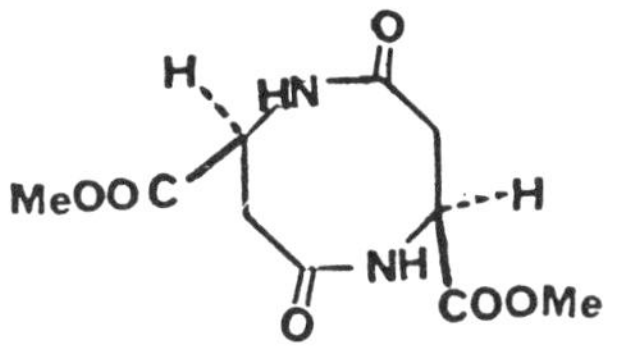

29

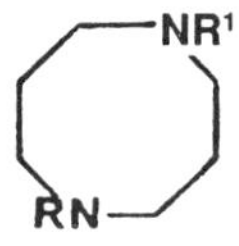

7

Treatment of amino acids **30** with ethyl chloroformate in trimethylamine yielded the 1,5-benzodiazocinediones **31** (70FES991). Two groups cy-

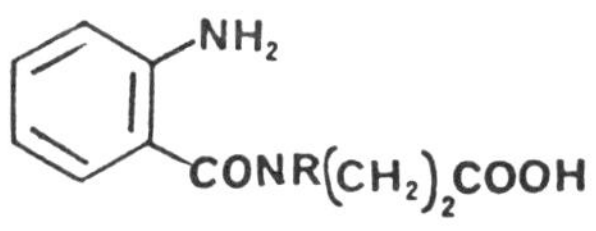

R = Ph, CH_2Ph

30

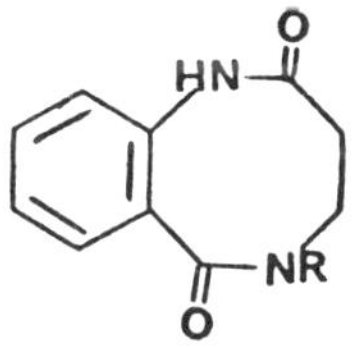

31

clized 2-aminobenzophenonecyanoethylimines **32** with acid to obtain the amidines **33** (R = H) or **34** (71GEP2024472; 79CPB2589). Greve *et al.* treated **32** (R = Me; R^1 = H, Me; R^2 = R^3 = R^4 = R^6 = R^7 = H; R^5 = NO_2, Cl, CF_3) with hydrogen chloride in benzene and obtained the corresponding compound **33** (71GEP2024472). Natsugari *et al.* reacted **32** (R = H; R^1 = R^2 = H, alkyl; R^3–R^7 = H, Cl, OR, R, CF_3, NO_2) with methanolic hydrogen chloride to get the appropriately substituted compound **33** (79CPB2589). The latter group reported some interesting reactions of N-oxides of **33** with phosgene (see Section IV,A). Thomae

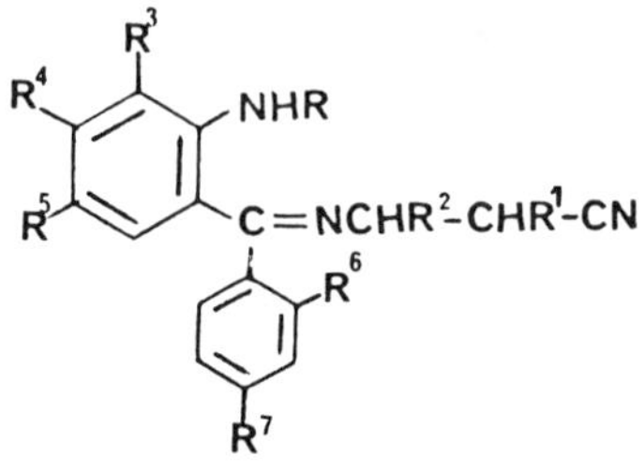

32

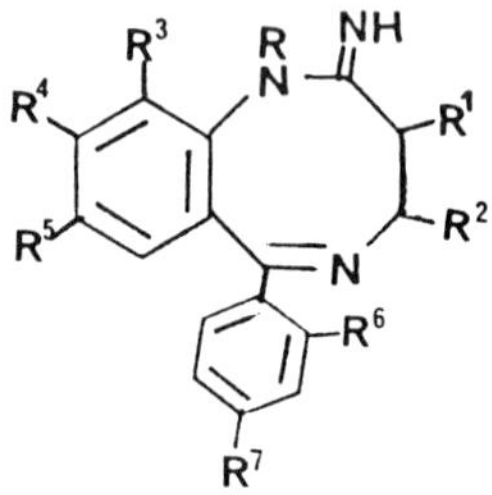

33 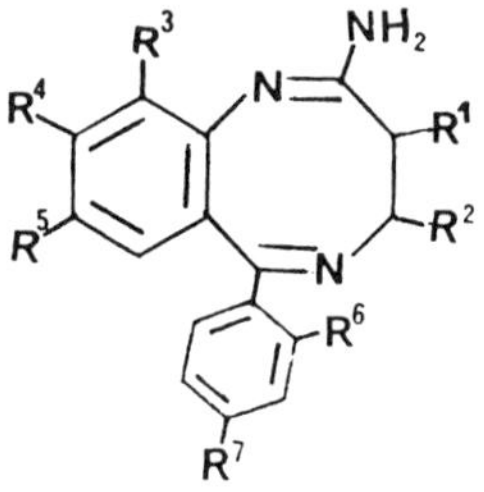   **34**

prepared dibenzodiazocine **35** [R = Et$_2$N(CH$_2$)$_2$, R^1 = H] by cyclization of *N*-[*N*-(2-diethylaminoethyl)anthranoyl]anthranilic acid with dicyclohexylcarbodiimide (DOC) (67FRP1497272). β-[*N*-(*o*-Aminophenyl)pyrrolyl]propionic acids were cyclized in thionyl chloride to afford pyrrolo-1,5-benzodiazocines **36** (81JHC1153). Gatta and Ponti cyanoethylated the substituted indole **37** at the indole nitrogen. Hydrolysis of the resulting nitrile to the carboxylic acid was followed by cyclization to the indolobenzodiazocine **38** (81MI1).

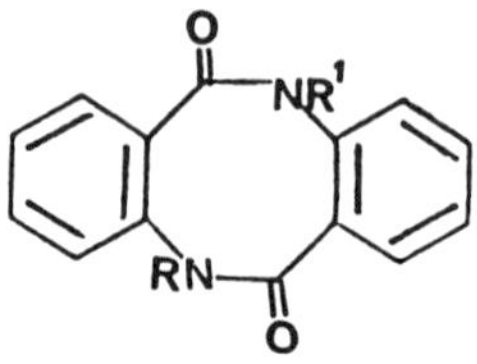

35 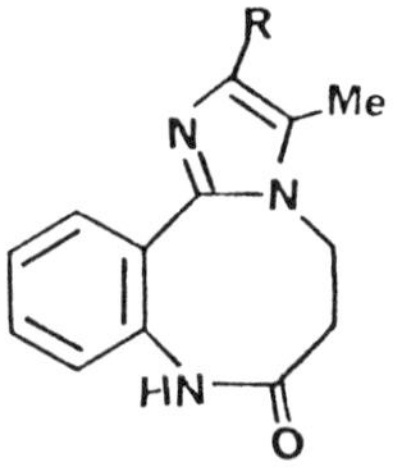   **36**

R = Ac, CoEt

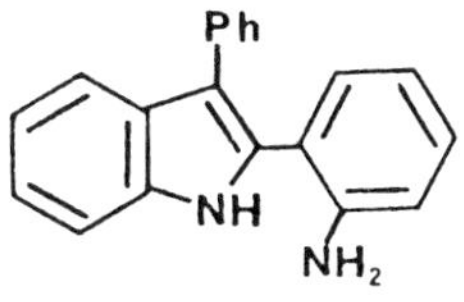

37

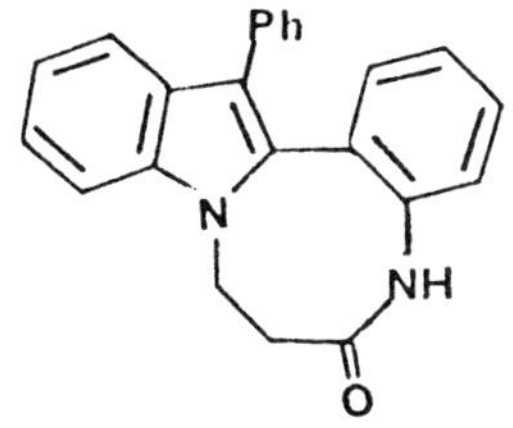

38

Heating the hydroxybenzotriazinone **39** for prolonged periods in diglyme yielded a mixture, which, according to mass spectral analysis, contained diazocine **40** and a series of polymeric diazocines **41**. These compounds were said to arise from the anthranoyl derivative of **39** (**42**) which is formed by shorter duration heating of **39**. Compound **42** was postulated to undergo self-anthranoylation by another molecule of **42** to produce amide **43**, which can be cyclodehydrated to **40**, or react repeatedly with **42** to yield polymer **41** (77CJC630).

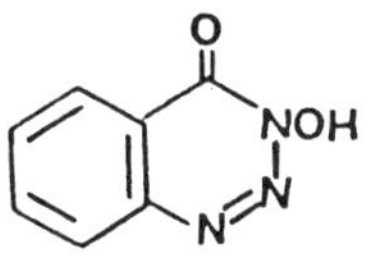

39

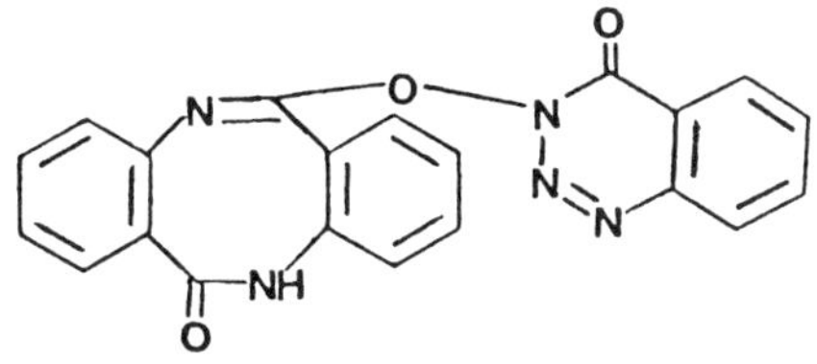

40

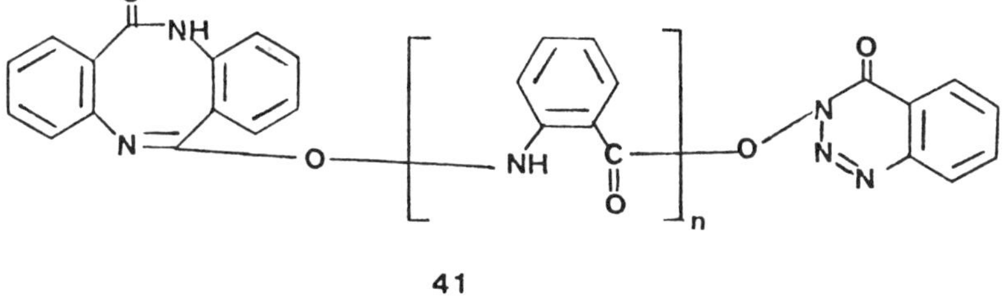

41

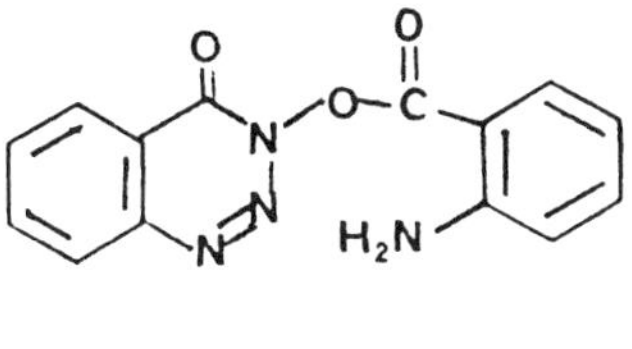

42

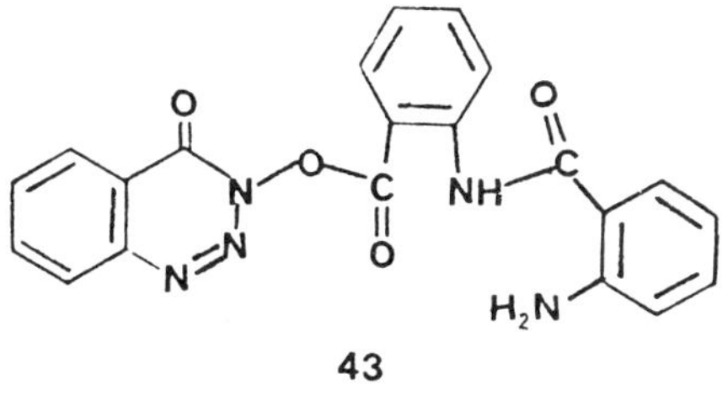

43

4. Cyclization Involving Carbon–Carbon Bond Formation

Cyanoaminoamides **44**, when treated with potassium *tert*-butoxide, afforded 1,5-diazacyclooctan-2-ones **45** (85HCA750). Milkowski and co-workers, in a series of papers, cyclized a number of 2-substituted *N*-aryl-*N'*-aroyl-1,3-propandiamines with phosphorus oxychloride to obtain the 1,5-benzodiazocines **46**, along with the pharmacologically more useful

RCH$_2$CONR^1(CH$_2$)$_3$NH(CH$_2$)$_3$CN

R = R^1 = Me
R = nC$_5$H$_{11}$, R^1 = nC$_6$H$_{13}$

44

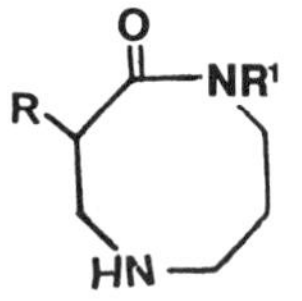

45

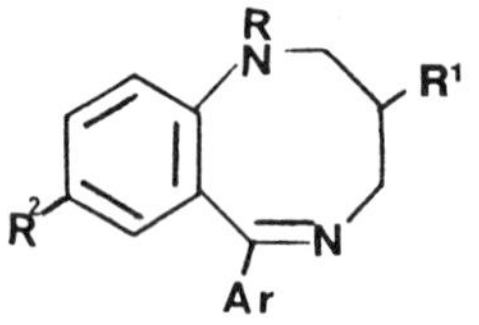

46

R=H; Me; PhCH$_2$
R^1=OH; OAc; OCOPh; Cl
Ar=Ph; 2halC$_6$H$_4$; 2, 4Cl$_2$C$_6$H$_3$; 3, 4 Cl$_2$ C$_6$H$_3$; 3CF$_3$C$_6$H$_4$; 2CF$_3$C$_6$H$_4$; 3, 4, 5 (OMe)$_3$C$_6$H$_2$; 2-furyl; 2, 6 F$_2$ C$_6$H$_3$
R^2=H; Cl; Br; NO$_2$

ring-contracted 1,4-benzodiazepines (73GEP2221558; 74GEP2314993, 74GEP2353165; 76GEP2520937, 76MI1, 76MI2; 85MI1). The diazocines could be rearranged into the benzodiazepines (see Section IV,D). The *N*-pyrazolylbenzamide **47** was treated with phosphorus oxychloride to yield the benzopyrazolodiazocine **48** (82MI1).

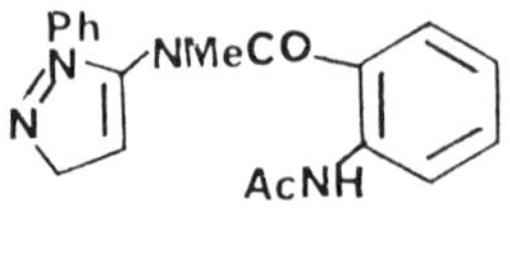

47

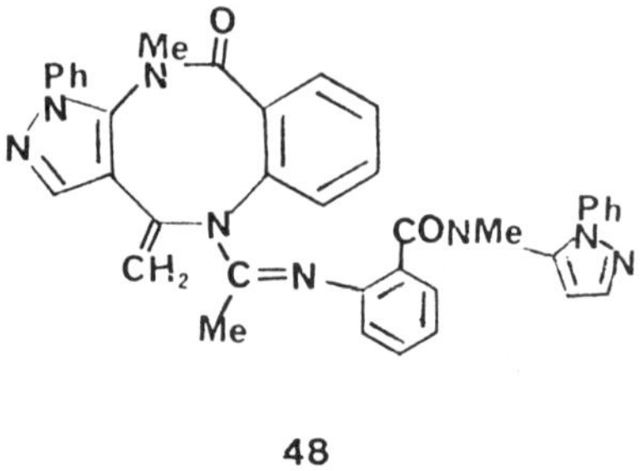

48

B. Via Ring Expansion

1. Schmidt or Beckmann Rearrangement of Cyclohexanediones or Cycloheptanones or Their Oximes

Weygand and Dietrich treated diethyl cyclohexane-1,4-dione-2,5-dicarboxylate with hydrazoic acid and obtained diazocine **49** (R = COOMe) (54CB482) (For a discussion of the chemistry of **49**, see Section V.) Koyama and co-workers attempted the Beckmann rearrangement of the ditosylate of cyclohexane-1,4-dionedioxime, but obtained no diazocine (61CPB834). Schmidt reaction of the dione itself using hydrazoic acid afforded a crude mixture believed to be mainly the 1,5-diazocine **49** (R = H) and a small amount of the isomeric 1,4-diazocine (89AHC185); however, purification yielded only **49** (R = H), as evidenced by its hydrolysis to β-alanine (61CPB834). About the same time, Rothe and Timler reported obtaining **49** (R = H) by both Schmidt and Beckmann rearrangement of cyclohexane-1,4-dione and its dioxime, respectively (62CB783). Meanwhile, Sekiguchi subjected cyclohexane-1,3-dione to typical Schmidt rearrangement conditions and obtained 1,5-diazocinedione **50** (R = R^1 = H) along with triazolodiazocine **51** (65BSF691), [Very small amounts of the corresponding 1,4-diazocines were also isolated (65BSF691; 89AHC185).] Iwakura *et al.* obtained **50** (R = R^1 = H) by

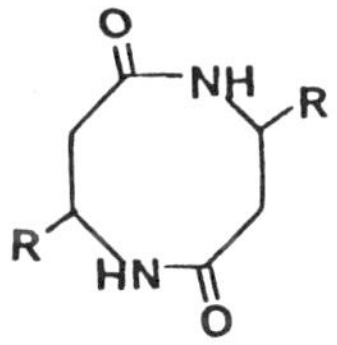

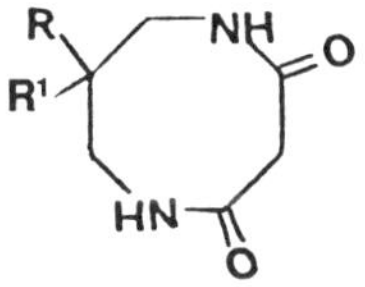

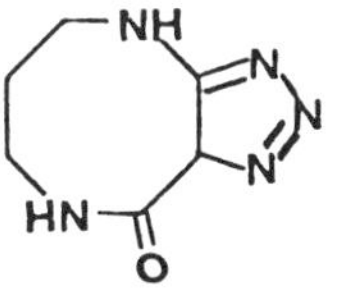

Beckmann rearrangement of cyclohexane-1,3-dione. Methyl (**50**, R = Me, R^1 = H) and dimethyl **50** (R = R^1 = Me) derivatives were also prepared from 5-methyl- and 5,5-dimethylcyclohexane-1,3-dione, respectively. [Polymerization of these diazocines was also reported; see Section IV,E) (73MI1).] Misiti *et al.* treated benzazepinone **52** with sodium azide in acid and obtained a small amount of 1,5-benzodiazocinone **53**, in addition to a 1,4-diazocine (73JHC689; 89AHC185).

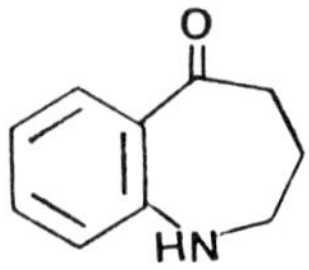

52

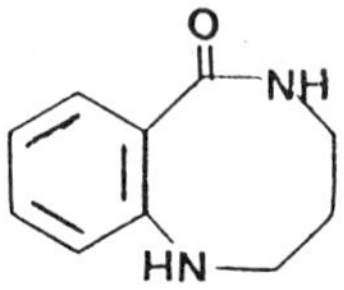

53

Rydon *et al.* reacted 9,10-anthraquinonedioxime with hot polyphosphoric acid and achieved a Beckmann rearrangement that afforded dibenz[*b*,*f*]-1,5-diazocine-6,12-dione (**54,** R = H). Similar reaction of 1,5-dichloroanthraquinone yielded mainly the dichloro derivative **54** (R = Cl) (plus a benzimidazoisoindolone). These results showed that the dioximes must have been *trans,trans* (or mostly *trans,trans*, in the case of the dichloro- compound), or else isomeric 1,4-diazocine **55** (89AHC185) would have formed instead (57JCS1900). The aforementioned Beckmann rearrangement to form **54** (R = H) was repeated by Aubagnac *et al.*, who also reported some chemistry of the N,N'-dimethyl derivative of **54** (R = H; also **35** R = R^1 = Me) (see Section IV,E) (72BSF2868). Costa *et al.* also reported this reaction, along with an interesting rearrangement of **54** (R = H) (also **35** R = R^1 = H) (see Section IV,C) (79MI1).

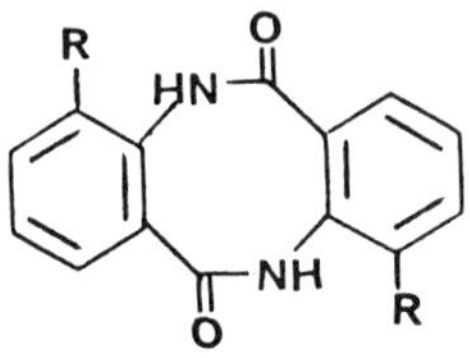

54

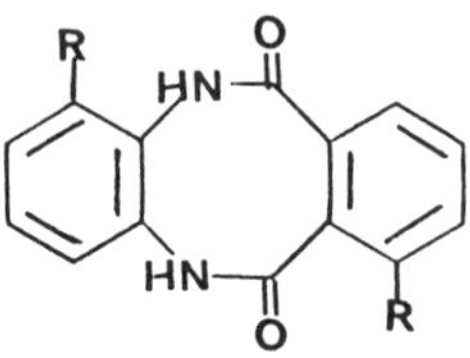

55

2. *0-Bridge Opening of Bicyclo [3.3.0] Compounds*

One of the more frequently reported routes to 1,5-diazocines involves reductive nitrogen–nitrogen bond cleavage of N,N'-trimethylenepyrazolidines (**56**) to yield **57**. This reaction was first reported by Wiselogle and co-workers, who reacted 2 mol of trimethylene bromide or chloride with 1 mol of hydrazine. They obtained, in addition to **56** (R = R^1 = H) and pyrazolidine, the corresponding ring-opened compound **57** (R = R^1 = H)

(43JA29). The same product was obtained by Stetter and Spangenberger by Raney nickle-catalyzed hydrogenation of **56** (R = R^1 = H) (58CB1982).

Two groups reported cleaving the N-methyl derivative of **56** (R = H, R^1 = Me) with aluminum amalgam to isolate **57** (R = H, R^1 = Me). [This compound underwent reaction at the unsubstituted nitrogen (see Section IV,B) to give useful products (see Section IV,B) (62BEP610039; 63BRP926624).] Grob and Schier, using this N—N cleavage method, prepared a series of N-substituted derivatives of **57** [R = H; R^1 = Et, Me, (CH$_2$)$_3$OH,(CH$_2$)$_3$OPh,(CH$_2$)$_2$OH, n-C$_{12}$H$_{25}$). They used either aluminum amalgam or lithium aluminum hydride to reduce **56** (61GEP1104516). The N-hydroxyethyl derivative [**57**, R = H, R^1 = (CH$_2$)$_2$OH] was prepared by reacting β-hydroxyethylhydrazine with 2 mol of trimethylene bromide, followed by Al(Hg) reduction of the resulting **56** [R = H, R^1 = (CH$_2$)$_2$OH] (65SZP393342). However, these same workers N-alkylated **56** (R = R^1 = H) with α-chloroacetamide and then reduced the product to obtain **57** (R = H, R^1 = CH$_2$CONH$_2$) (67SZP416654). Weigert reduced **56** (R = CH$_2$CN, R^1 = H) with Raney nickel to afford **57** [R = (CH$_2$)$_2$NH$_2$, R^1 = H] (78JOC622).

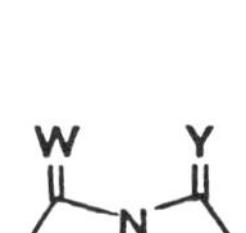

56

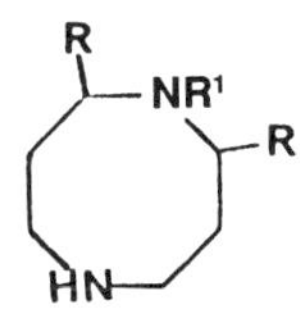

57

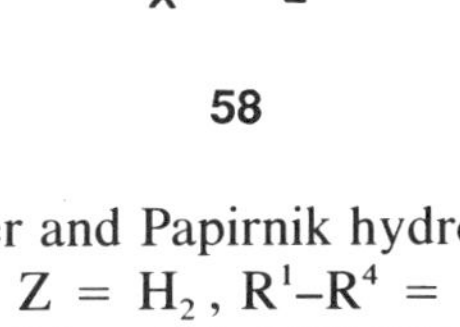

58

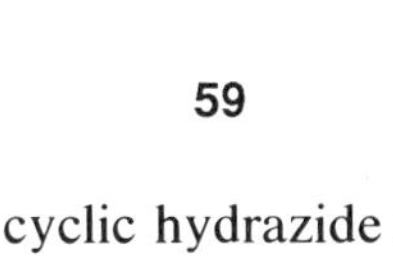

59

Moldaver and Papirnik hydrogenated the cyclic hydrazide **58** (W = X = O, Y = Z = H$_2$, R^1–R^4 = Et) with Raney nickel and obtained **59** (X = H$_2$, R^1–R^4 = Et) (71KGS1097). Kemp *et al.* reduced tetraoxo compound **58** (W = X = Y = Z = O, R^1–R^4 = Me; R^1–R^3 = Me, R^2 = CH$_2$OH; R^1 = R^3 = Me, R^2 = R^4 = CH$_2$OCH$_2$Ph) with diborane to afford the correspondingly alkylated diazocines **59** (X = H$_2$) (68TL513). The bis(benzyloxymethyl)diazocine **59** (X = H_2, R^1 = R^3 = H, R^2 = R^4 = CH_2OCH_2Ph) was used to make novel caged compounds containing the 1,5-diazocine ring (see Section IV,A) (68TL547). In related work, Kemp *et al.* devised an alternate route to 1,5-diazocines by sodium and ammonia reduction of hydrazide **58** (W = Z = H_2; X = Y = O; R^1–R^3 = Me; R^4 = Me, CH_2OCH_2Ph, H, Ac, THF) to yield the corresponding diazocinediones **59** (X = O). However, under the reaction conditions, the benzyloxymethyl- and acetyl-substituted **59** underwent debenzylation and deacylation, respectively (79JOC4473). Kosower and Pazhenchevsky hydrogenated **60** to afford 1,5-diazocinedione **61** (80JA4983).

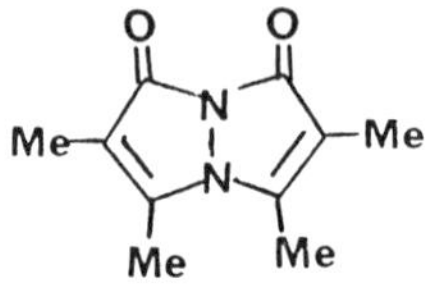

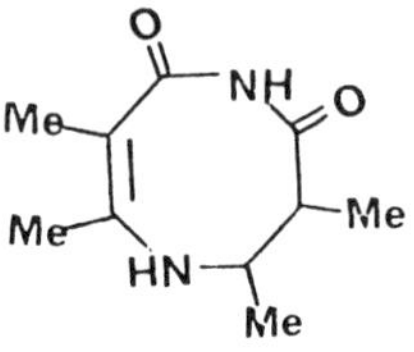

An interesting reaction was reported by Evnin *et al.*, who pyrolyzed the azo-bridged pyrazolidine **62** and obtained 1,5-diazocine **63** (69TL4497).

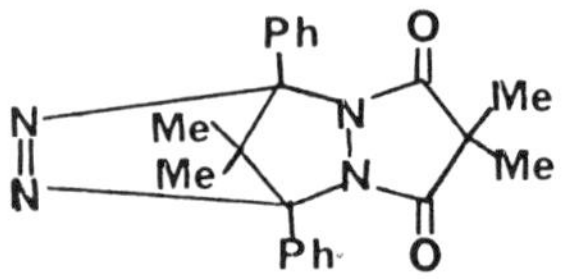

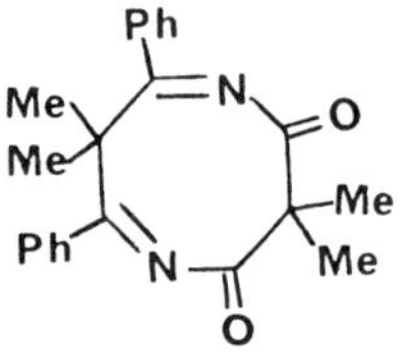

3. *n-Bridge Opening of Bicyclo[3.3.n]alkanes (n ⩾ 1)*

Denzer and Ott prepared the benzodiazabicyclo[3.3.1]nonanes **64** by reactions of 1,2,3,4-tetrahydroquinazolines with acrylates. The methylene bridge in **64** was removed with acid to yield 1,5-diazocines **65** (R = H) (69JOC183; 71USP3577557; 73USP3741969). In addition, **64** (X = Cl) was hydrogenolyzed to give **65** (X = Cl, R = Me) (69JOC183).

Spielman corrected the structure originally proposed by Tröger for the compound formed by acid-catalyzed condensation of formaldehyde with *p*-toluidine. Spielman's proposed structure (**66**, R^1 = Me, R^2 = R^3 = H) was supported by the reaction of **66** (R^1 = Me, R^2 = R^3 = H) with acetic

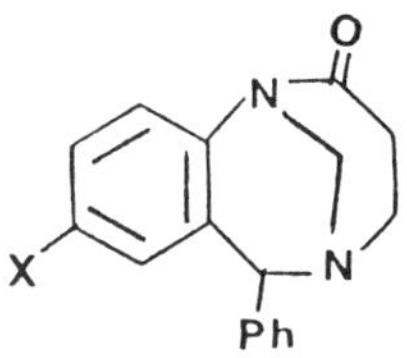

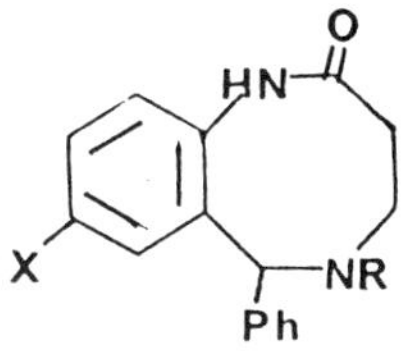

64 X = H, Cl, Br, NO₂, CF₃ **65**

anhydride, benzoyl chloride, or nitrous acid to give 1,5-diazocines **67** (R^1 = Me, R^2 = R^3 = H, R^4 = R^5 = Ac), **67** (R^1 = Me, R^2 = R^3 = H, R^4 = R^5 = PhCO), or **67** (R^1 = Me, R^2 = R^3 = H, R^4 = R^5 = NO), respectively (35JA583). [Note that compounds of type **67** were often called *tetrahydrophenhomazines* (e.g., see 63HCA2970).] Iwao and Tomio similarly treated **66** (R^1 = R^2 = CH₂Ph, R^3 = H) with acetic anyhdride and obtained **67** (R^1 = R^2 = CH₂Ph, R^3 = H, R^4 = R^5 = Ac) (56MI2).

Various N-alkylated dibenzo-1,5-diazocines were synthesized by reaction of Tröger's bases with various alkylating agents. For example, reaction of **66** (R^1 = Me, R^2 = R^3 = H), **66** (R^1 = MeS, R^2 = R^3 = H), **66** (R^1 = R^3 = Me, R^2 = H), and **66** (R^1 = OMe, R^2 = R^3 = H) with dimethyl sulfate and alkali afforded the corresponding diazocines **67** (R^4 = R^5 = Me, R^2 = R^3 = H, R^1 = various). Reaction of **66** (R^1 = Me, R^2 = R^3 = H) with allyl and benzyl bromide gave **67** R^1 = R^4 = Me, R^2 = R^3 = H, R^5 = allyl) and **67** (R^1 = R^4 = Me, R^2 = R^3 = H, R^5 = PhCH₂), respectively. Reaction of **66** (R^1 = Me, R^2 = R^3 = H) with benzyl bromide followed by methanol afforded **67** (R^1 = Me, R^2 = R^3 = H, R^4 = CH₂Ph, R^5 = CH₂OMe) (63HCA2970). Greenberg *et al.* reported that the dimethylated Tröger's base **68** (R = Me) exists as the "open" isopropylidene iminium ion **69** (R = Me) in concentrated acid (84JOC1127).

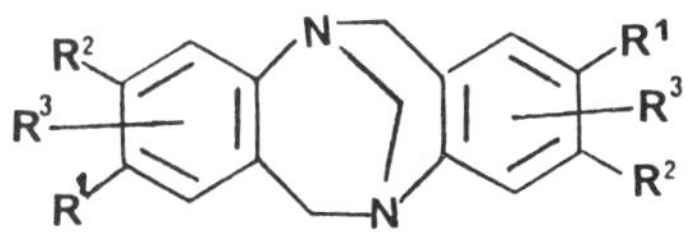

66

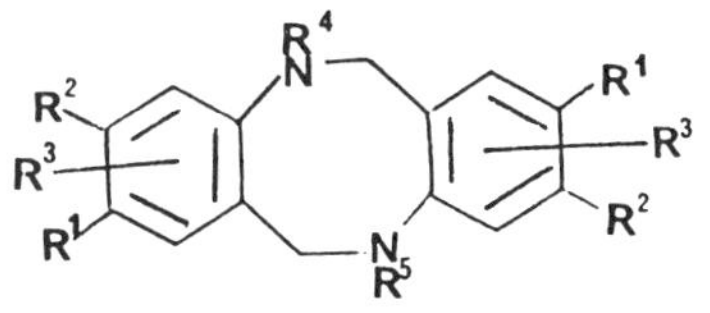

67

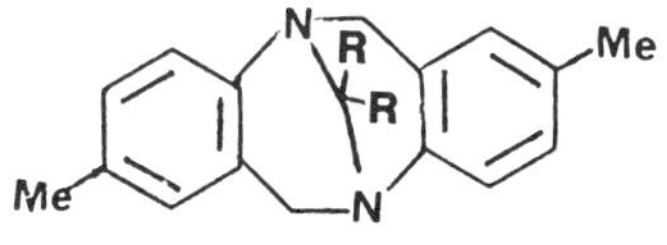

68

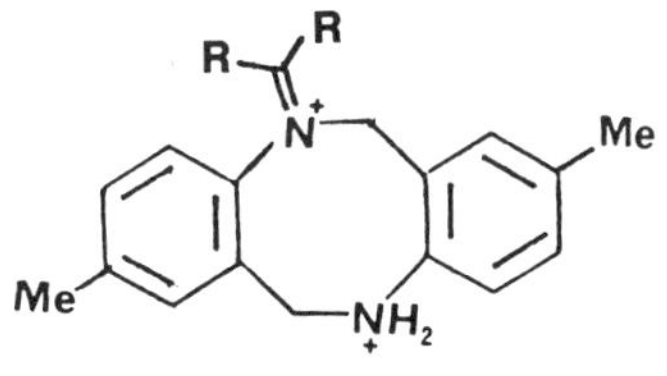

69

Jones et al. found that the diazabicyclo[3.3.3] undecane **70** (prepared from 1-nitropropane, formaldehyde, and ammonia) is cleaved in acid to give formaldehyde, 1-nitropropane, and 1,5-diazocine **71** (56MI1; 57RC101). Williams prepared the oxadiazabicyclo[3.3.1]nonanes **72** by acid-catalyzed dimerization of N-substituted β-amino-α,α-dimethylpropionaldehydes. Lithium aluminum hydride deoxygenation of **72** (R = Me, CH_2Ph) afforded the diazocines **73** ($R^1 = R^2$ = Me, R = Me, CH_2Ph). Compound **73** ($R^1 = R^2$ = Me, R = CH_2Ph) is further hydrogenolyzed to **73** ($R^1 = R^2$ = Me, R = H). Use of appropriately substituted β-aminopropionaldehydes afforded variously substituted compounds **72**, which, on reduction, gave **73** (R = H; $R^1 = R^2$ = Et, n-Pr, n-Bu), **73** (R = H; R^1 = Et; R^2 = n-Bu, n-Pr), and **73** (R = H, R^1 = Me, R^2 = Ph) (68JOC3946; 70USP3503939). Diazocine **73** (R = H, $R^1 = R^2$ = Me) was polymerized (see Section V).

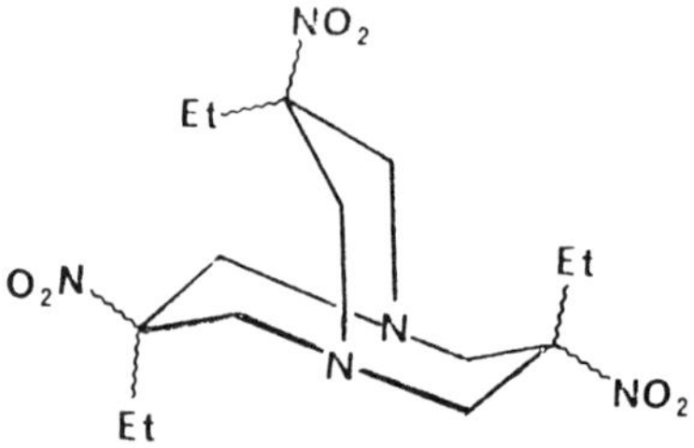

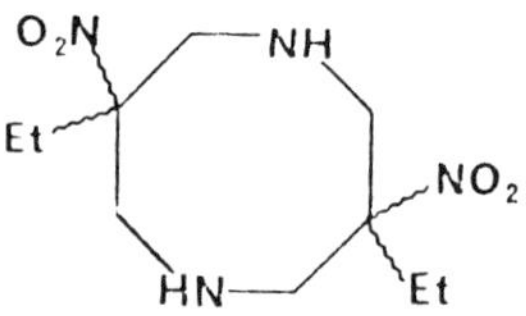

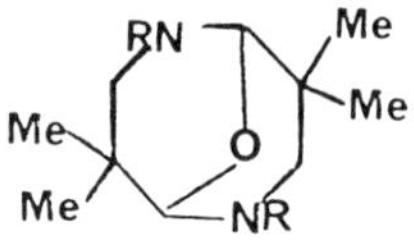

4. *Other Ring-Expansion Methods*

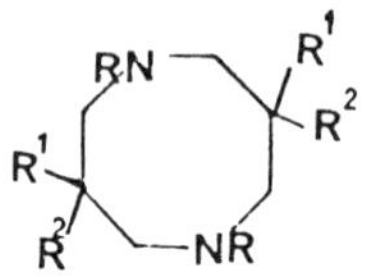

Shenoy treated 1,4-benzodiazepine N-oxides **74** (or related quinazoline 3-oxides) with either acid (**74**, R = alkyl, R^1 = H) or base (**74**, R = alkyl, aryl; R^1 = H, CH_2OH) and obtained pharmacologically useful (Section V) 1,5-benzodiazocine N-oxides **75** (75BRP1413599; 76GEP2525094, 76GEP2627461). When the seven-membered lactone **76** was treated with sodium hydride, a ring-expansion/dimerization occurred, affording 1,5-benzodiazocinedione **35** (R = R^1 = o-HOC_6H_4) (66JHC527).

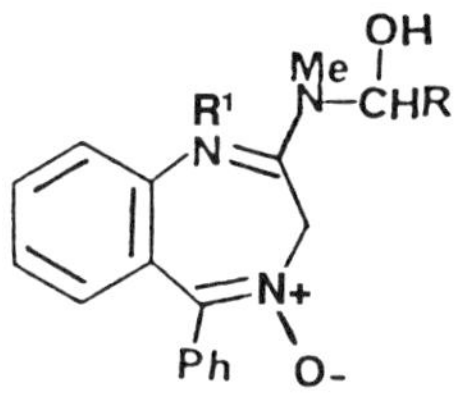

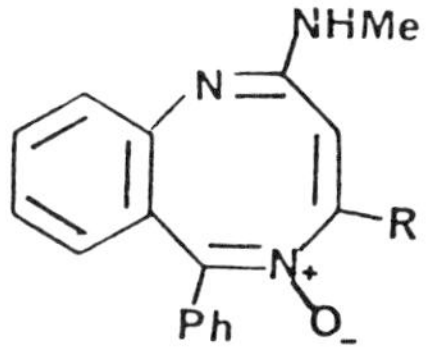

74

R¹ = H, Me, Et, n-Pr, Ph, 2-furyl
R¹ = H, CH₂OH

75

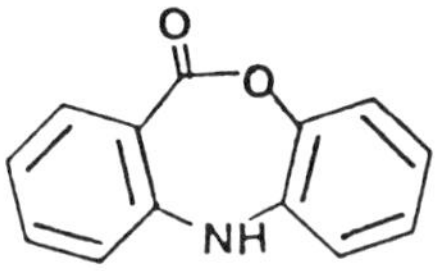

76

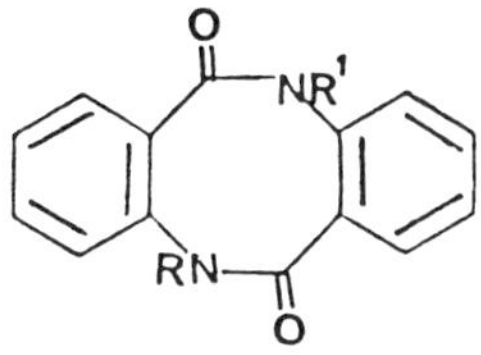

35

A number of ring-opening reactions have resulted in 1,5-diazocine formation. De Diesbach and Frossard treated the naphthindole derivative **77** with ammonia in methanol and obtained diazocine **35** (R = R¹ = COPh) (54HCA701). Ege *et al.* photolyzed the quinazolinobenzotriazine **78** in inert solvent and obtained imide **79**, assertedly via azetoquinazoline **80**. If **78** is photolyzed in nucleophilic solvent, diazocines **81** (R = Nuc) are formed along with **79** (72CB2898). (For another interesting route to **79**, see Section IV,C.)

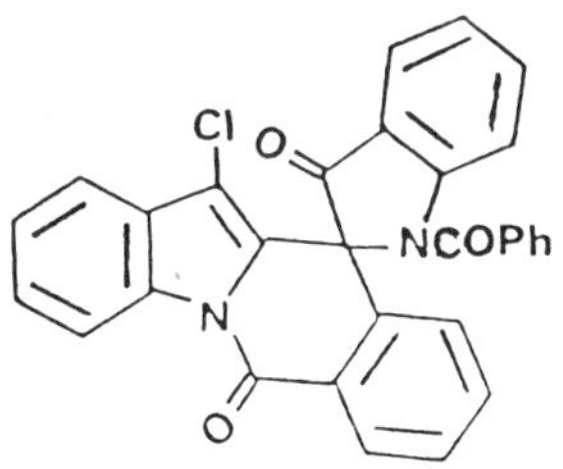

77

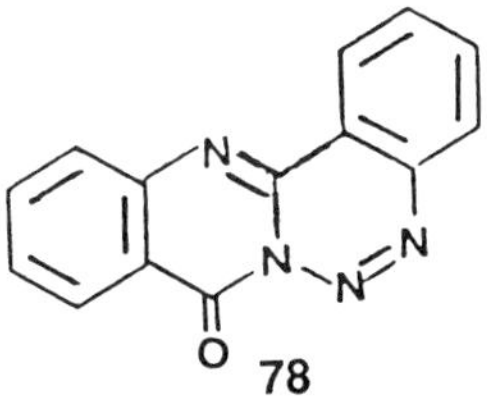

78

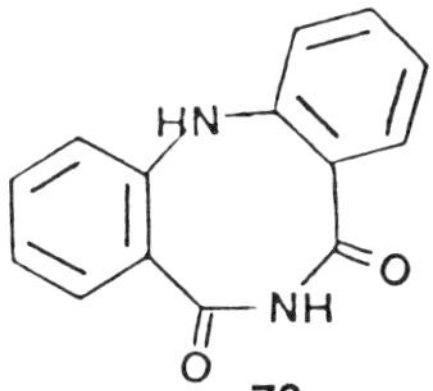

79

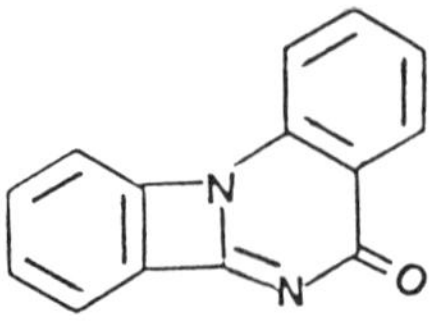

80

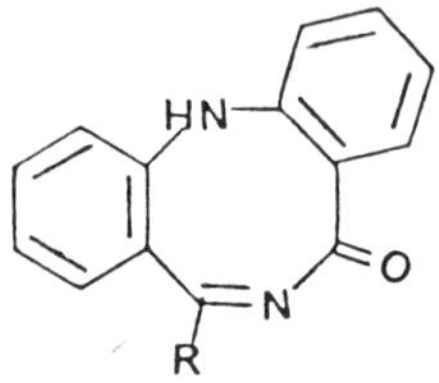

81

Another interesting photochemical rearrangement was reported by Katsuyama and co-workers, who prepared spiro(indolineisoxazolines) **82** and **83** by 1,3-dipolar cycloaddition of 2-methyleneindolene derivatives **84** with mono- and bis(hydroxamic chlorides), respectively. Photolysis of **82** and **83** (R^1 in both = H, Cl, OCOEt, NO_2; R^2 = H, Cl) resulted in formation of correspondingly substituted diazocines **85** and **86**, respectively (72GEP2207291; 78MI1). Photolysis of the polyspiro(indoleneisoxazolines) **87** resulted in isolation of novel poly(1,5-benzodiazocines) **88**.

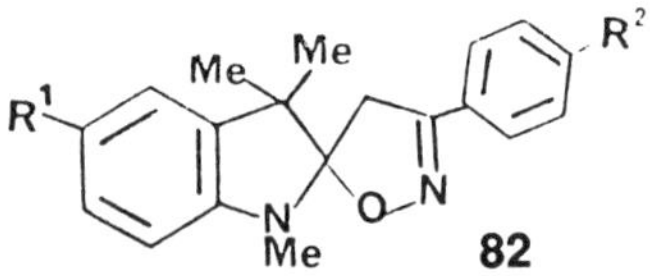

82

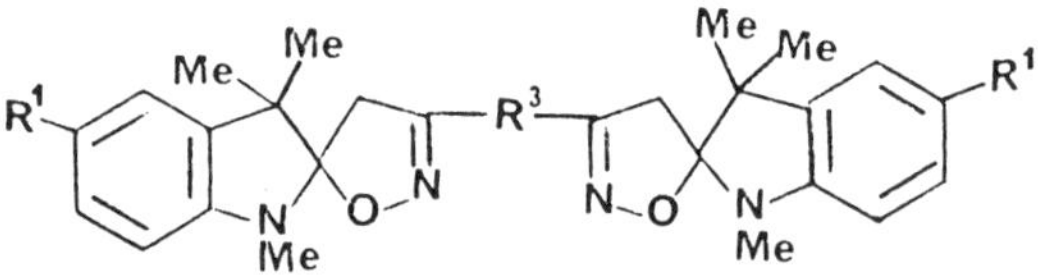

83

R^3 = bond, 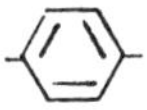

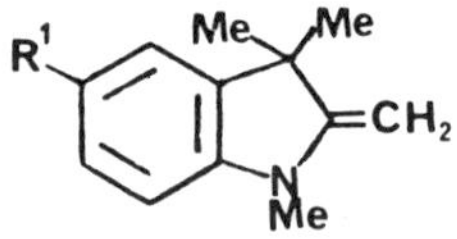

84

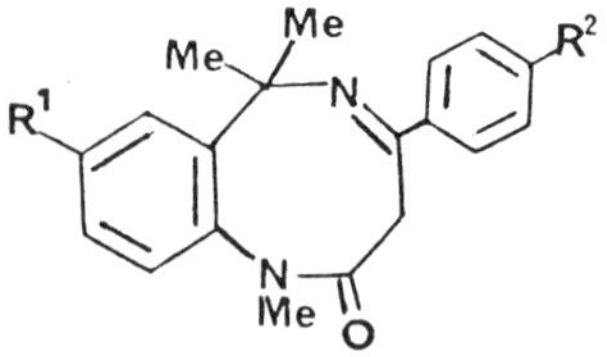

85

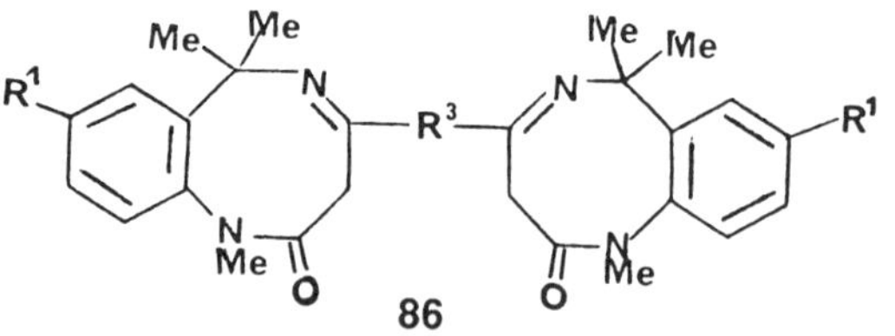

86

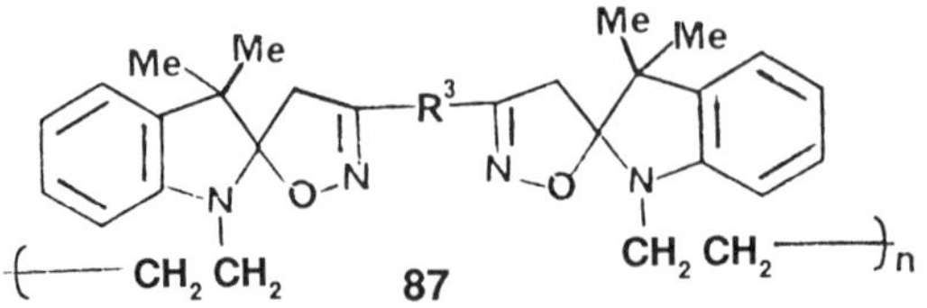

87

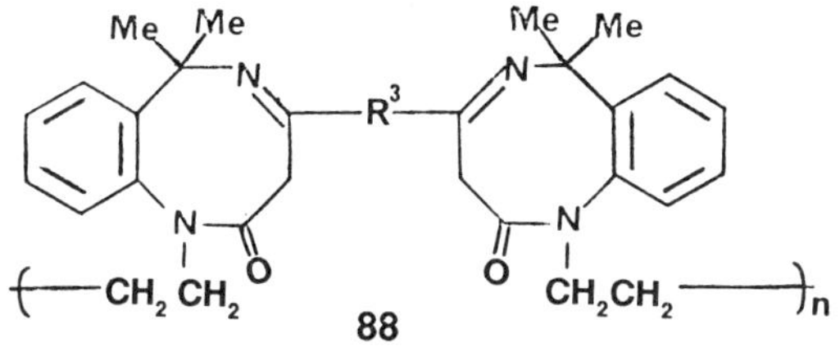

These compounds were investigated for their photoconductivities (see Section V) (78MI1). More recently, Schneiders et al. prepared the completely conjugated 1,5-diazocines **89** and **90** by thermolysis of diazasemibullvalenes **91**. Photolysis of **91** ($R = R^1 = H$) afforded both double bond isomers **89** ($R = R^1 = H$) and **90** ($R = R^1 = H$), while reaction of **91** ($R = Br, R^1 = H$, or Br) resulted in the formation of the single isomer **89** ($R = Br, R^1 = H$ or Br). The isomers **89** ($R = R^1 = H$) and **90** ($R = R^1 = H$) could not be separated, but if the dianion of this mixture was treated with iodine or oxygen at low temperatures, pure **89** ($R = R^1 = H$) was isolated. Warming of a solution of this material to room temperature reestablished the equilibrium mixture (85AG579).

1,5-Diazocines have been prepared by reaction of small rings with unsaturated compounds. Wegener reacted β-lactams **92** [$R = Me, H; R^1 = H; RR^1 = (CH_2)_4; R^2 = H$] with acrylonitrile and alkali. Subsequent cata-

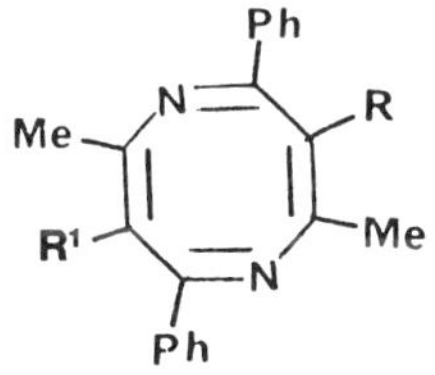

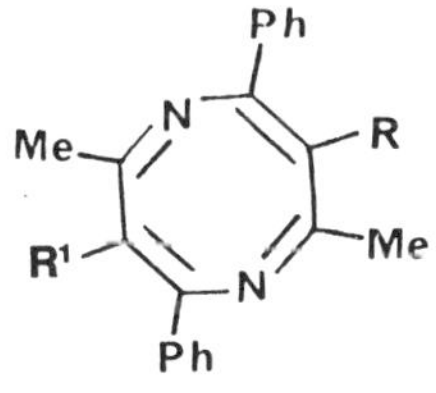

lytic hydrogenation yielded 1,5-diazocinones **93** [$RR^1 = -(CH_2)_4-, R^2 = R^3 = H$] (70GEP1802207). Crombie et al. reported that N-haloalkyl-4-substituted azetidin-2-ones **92** [$R = Ph, n-C_5H_{11}, n-C_7H_{15}; R^1 = H; R^2 = (CH_2)_3hal$] underwent transamidation in the presence of liquid ammonia to afford the ring-expanded 1,5-diazocinones **93** ($R = Ph, n-C_5H_{11}, n-C_7H_{15}; R^1 = R^2 = R^3 = H$). Similar reaction of **92** ($R = Ph, R^1 = H, R^2 = CH_2CH_2CHBrMe$) yielded **93** ($R = Ph, R^1 = R^3 = H, R^2 = Me$).

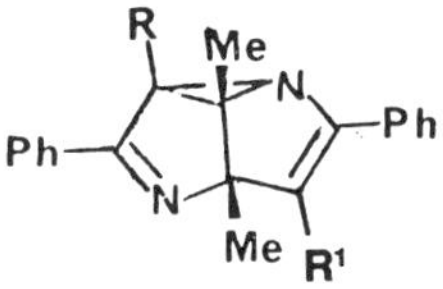

When 92 [R = Ph, R^1 = H, R^2 = $(CH_2)_3Cl$]) was treated with primary amines such as ethyl- or allylamine instead of ammonia, diazocines 93 (R = Ph, R^1 = R^2 = H, R^3 = Et, allyl) resulted. These reactions were found to proceed via aminoalkyl intermediates 92 (R = Ph, R^1 = H, R^2 = aminoalkyl), some of which, when isolated, underwent spontaneous transamidation to afford 93 (83CC959). [This interesting reaction (82TL465) was used in the preparation of the alkaloid homaline (see Section II,D).]

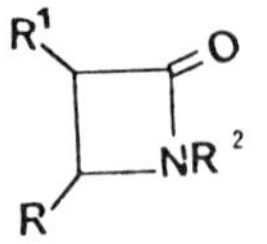

92

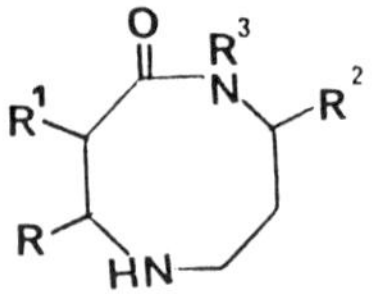

93

Barluenga et al. reacted 1-alkyl-3-phenyl-3-(N-alkylamino)-2-propen-1-oneimines with dichlorodiphenylsilane and triethylamine and obtained diazasilacyclohexadienes 94. The sila compounds underwent ring expansion and rearrangement with dialkylacetylenedicarboxylates to give silyl-substituted 1,5-diazocines 95 (86AG190).

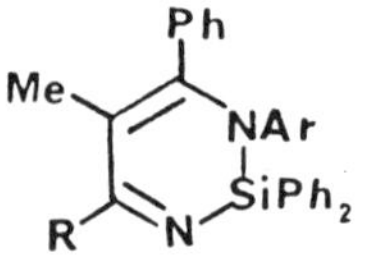

94

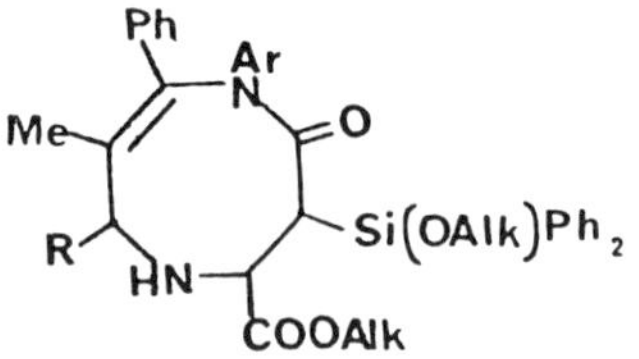

95

R = Ph, p-tolyl, cyclohexyl
Ar = Ph, p-tolyl

C. Via Condensation Reactions

1. *Condensation of Amines with Saturated Electrophilic Carbon*

Marckwald and von Droste-Huelshoff treated the bis(*p*-toluenesulfonamide) of 1,3-propanediamine with alkali, followed by trimethylene bromide, and obtained diazocine 7 (R = R^1 = Ts) (1898CB3264). Shortly thereafter, Howard and Marckwald reported obtaining the free base 7 (R = R^1 = H) by reacting the bistosyl derivative of 7 with hydrochloric acid (1899CB2038). Earlier, Gabriel and Weiner reported isolating a base, $C_6H_{14}N_2$, by heating γ-bromopropylamine with alkali (1888CB2669), but this was described as a primary amine (1896CB2381), and was found by Howard and Marckwald (1899CB2031) not to be the same as **7** (R = R¹ = H) reported by them (1899CB2038). Gabriel and Stelzner claimed **96** as the structure of the salt obtained by heating N-(γ-bromopropyl)piperidine (1896CB2381). However, Knorr and Roth heated N-(γ-chloropropyl)dimethylamine in alkali and reported obtaining diazocine **97** (R = Me) (06CB1420). Moreover, Hörlein and Kneisel treated N-(γ-chloropropyl)-piperidine with alkali and isolated a product having structure **97** [RR = $(CH_2)_5$] (06CB1429), not **96** as postulated earlier by Gabriel and Stelzner (1896CB2381). In later work, von Braun and Goll prepared salt **98** by reaction of tetrahydroisoquinoline with trimethylene bromide to give N,N-trimethylenebis(tetrahydroisoquinoline), followed by reaction of this product with more trimethylene bromide (27CB339). (For ring-opening reactions of these salts, see Section IV,E.)

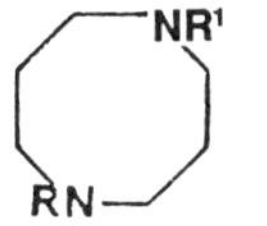

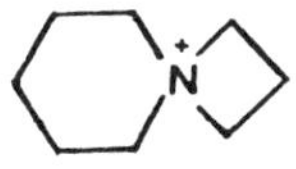

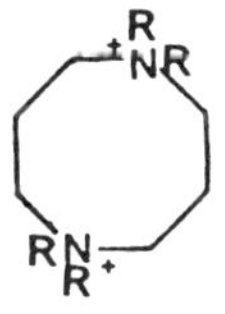

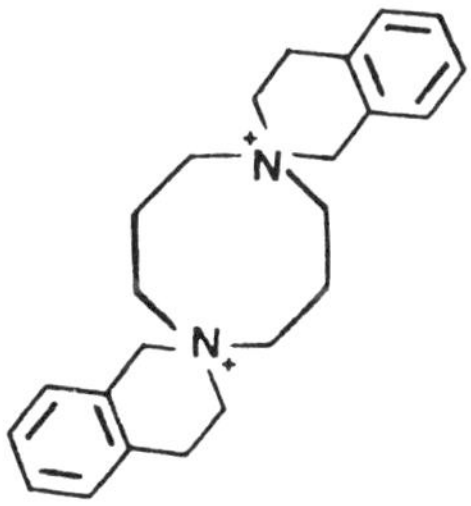

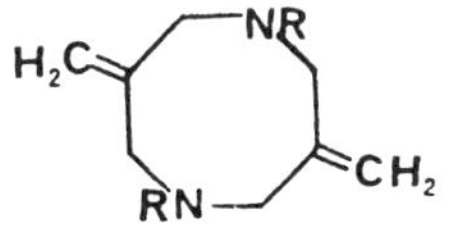

R = Me, Et, i-Pr, Ph

Mikolajewska heated alkylbis(γ-chloropropyl)amines with ethanolamine in the presence of sodium bicarbonate and obtained the hydroxyethyl-1,5-diazocines **7** (R = CH_2CH_2OH; R¹ = Me, Et, i-Pr, $PhCH_2$)

(71MI1). N,N'-dialkyl-1,3-propanediamines were condensed with trimethylene bromide to afford substituted diazocines 7 (R = CH_2Ph; R^1 = Et, n-Bu, CH_2Ph) (75MI2). Bismethylenediazocines 99 were prepared by treating 3-halo-2-halomethyl-1-propenes with primary amines, or by condensing 3-iodo-2-iodomethyl-1-propene with N,N'-substituted 3-amino-2-aminomethyl-1-propenes (77ZC174).

A Mannich-type condensation (see also Section II, C, 4) of nitroalkanes with formaldehyde and ammonia provided an entry into 3,7-dinitro-1,5-diazocines. Urbanski and co-workers mixed 1 mol of nitroethane with 1 mol of ammonia and 1 mol of formaldehyde and obtained the dinitrodihydroxyamine 100. This compound reacted with ammonia to give substituted 1,5-diazocine 101 (R = Me, R^1 = H) (52RC182). [For a review containing the synthesis, properties, and synthetic use of nitroparaffins, including diazocines, see (54MI1).] When 2-nitro-2-ethyl-1,3-propanediol was treated with ammonium hydroxide, the diazocine 101 (R = Et, R^1 = H) resulted (58JCS2319). In addition, condensation of 2-nitro-2-n-propyl-1,3-propanediol with ammonia afforded 101 (R = n-Pr, R^1 = H) (55RC379). The diethyl derivative (101, R = Et, R^1 = H) was also prepared by ring opening of a diazabicycloundecane (see Section II,B,3).

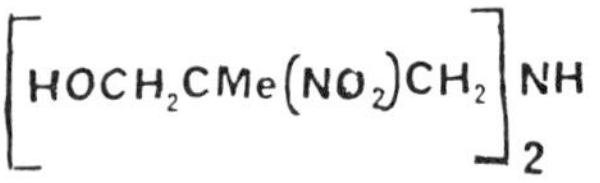

$$\left[HOCH_2CMe(NO_2)CH_2 \right]_2 NH$$

100

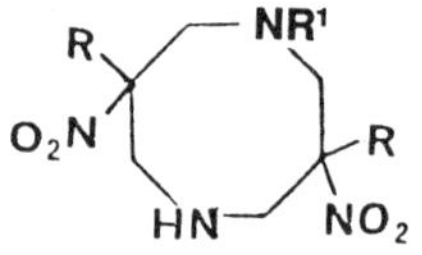

101

A number of diazocinediols have been prepared by reaction of amines with epoxides or haloalcohols. Paudler et al. condensed 1,3-dibromo-2-propanol with p-toluenesulfonamide anion and obtained trans-diol 102 (R = R^1 = Ts, R^2 = H), whereas reaction of 1,3-bis(p-toluenesulfonamido)-2-propanol with epichlorohydrin afforded both trans and cis isomers of 102 (66JOC277). Reaction of methallyl chloride oxide with p-toluenesulfonamide in base yielded the diastereomeric diols 102 (R = R^1 = Ts, R^2 = Me) (67JOC2425). (For reactions of these compounds, see Section IV,A.) Piper et al. condensed N,N'-trimethylenebis (p-toluenesulfonamide) dianion with epichlorohydrin and obtained diazocinol 103 (75JMC803). Gaertner cyclodimerized N-substituted 2,3-epoxypropylamines and obtained a number of diols 102 (R = R^1 = t-Bu, t-Oct, Ph; R^2 = H). The tert-butyl compound was shown to be the trans isomer.

Support for these structural assignments was obtained when reaction of bis(2,3-epoxypropyl)-tert-butylamine with tert-butylamine and reaction of 1,3-bis (*tert*-butylamino)-2-propanol both afforded diazocine **102** (R = R^1 = *t*-Bu, R^2 = H), probably via acyclic intermediate **104** (which was never isolated as a pure compound). This bis(halohydrin)amine **105**, when reacted with *tert*-butylamine, gave **102** (R = R^1 = *t*-Bu, R^2 = H) directly. Also, *N,N*-bis(2,3-diepoxypropyl)aniline, when reacted with aniline, yielded **102** (R = R^1 = Ph, R^2 = H). Mixed condensations were also performed by Gaertner. Although 2,3-epoxypropyl-*tert*-butylamine reacted with *N*-(2,3-epoxypropyl)aniline to afford only a trace of **102** (R = *t*-Bu, R^1 = Ph, R^2 = H), this compound was readily obtained by reaction of bis(2,3-epoxypropyl)aniline with *tert*-butylamine or by reaction of bis(2,3-epoxypropyl)-*tert*-butylamine with aniline. In a similar way, diazocine **102** (R = *n*-Bu, R^1 = Ph, R^2 = H) was obtained (66USP3236837; 67T2123). In later work, Sorokin *et al.* condensed bis(2,3-epoxypropyl)-*n*-butylamine with *n*-butylamine and isolated **102** (R = R^1 = *n*-Bu, R^2 = H) (72MI2).

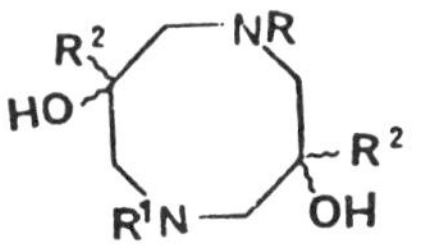

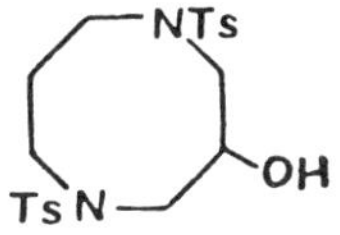

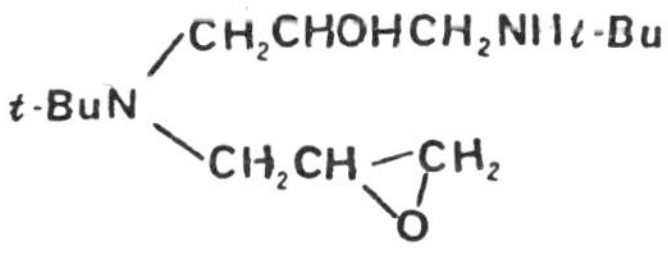

Shiotani and Mitsuhashi condensed the bis-*p*-toluenesufonamide of *o*-aminobenzylamine with trimethylene bromide and obtained, after detosylation, 2,6-benzodiazocine **106** (R = R^1 = H) [Reaction of **106** with formaldehyde afforded *N,N'*-methanobenzodiazocines of some pharmacological interest; see Sections IV,A and V) (64YZ656; 74H349).] The benzodiazocine **107** was prepared by heating the epoxy amidoester **108** with methylamine [Elimination of alcohol from **107** afforded an enedione) (68JCS(C)2862.] Kato and Mori reacted the bis(dihaloalkyl)piperidine **109** with various primary amines and obtained the medicinally useful N-substituted piperidinodiazocines **110** (72GEP2141464) (see Section V).

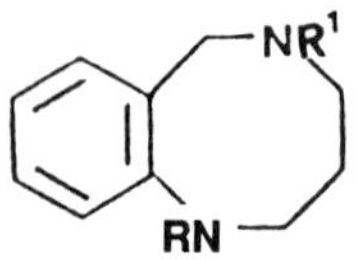

106

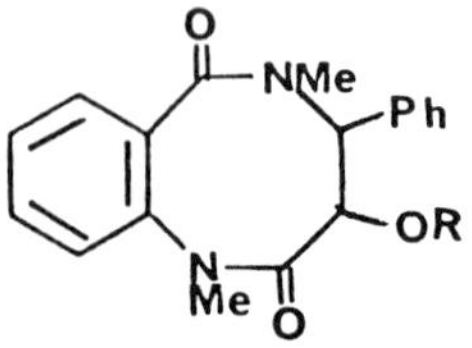

107

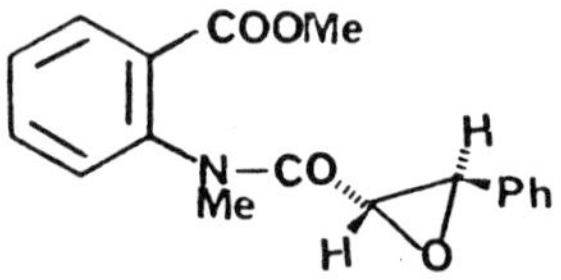

108

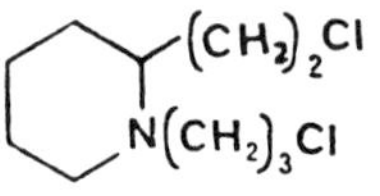

109

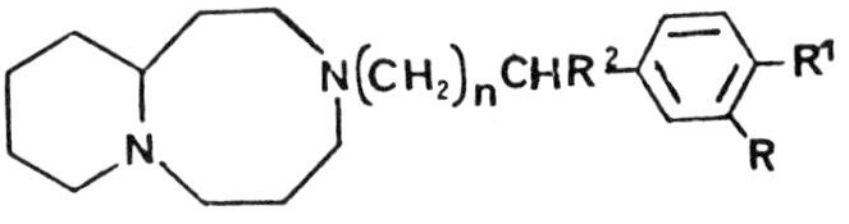

110
n = 0, 1, 2
R = H, OMe
R^1 = H, Cl, F, OMe
RR^1 = –OCH$_2$O–
R^2 = H, Ph, CH$_2$Ph

2. Reaction of Acyclic Diamines with Dicarboxylic Acids or Acid Derivatives

Rothe and Timler synthesized the diazocinedione **49** (R = H) (the cyclic dipeptide of β-alanine) by reacting the β-alanine derivative **111** with triethylamine under high dilution (62CB783). Yamazaki and Nagata heated β-(3,4-dimethoxyphenyl)ethylamine with diethyl methylaminedipropionate and obtained imide **112** (59YZ1219). The N,N'-dicyclohexyl-1,5-diazocines **113**, prepared by condensing dialkylmalonyl chlorides with N,N'-dicyclohexyltrimethylenediamine, demonstrated some pharmacological activity (see Section V) (81FES135).

Schroeter and Eisleb reported the first preparation of a dibenzo-1,5-diazocine when they warmed various N-substituted anthranilic acid chlorides with pyridine. In this manner, diazocines **35** were prepared (called, at the time, "dianthranilides"; cf. structure **54**, Section II,B,1). Acetolysis of **35** (R = R^1 = H) afforded **35** (R = H, R^1 = Ac). Hydrolysis of this latter compound yielded the original unsubstituted compound **35** (R = R^1 = H). Also prepared was **35** (R = H, R^1 = Me) (09LA101). In a

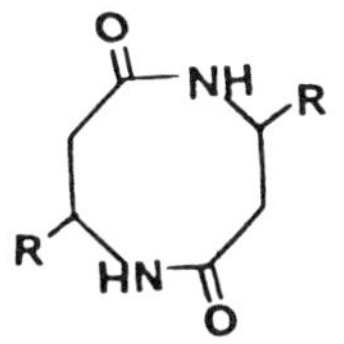

49

N$_3\overset{+}{\text{N}}$(CH$_2$)$_2$CONH(CH$_2$)$_2$COSPh

111

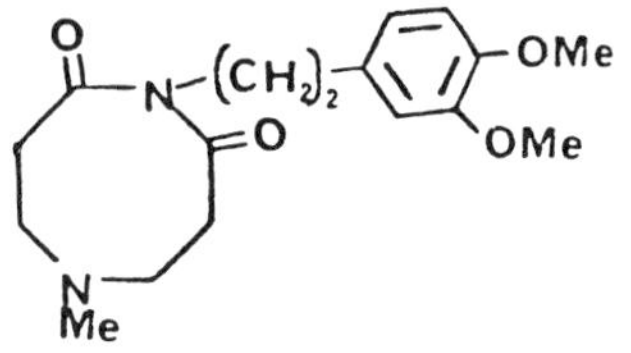

112

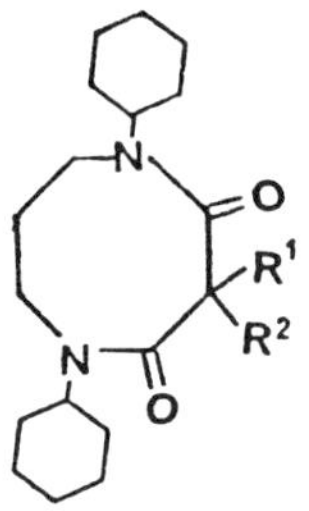

113
R^1 = Ar, Et
R^2 = allyl, Et, Ph

later report, Schroeter treated **35** (R = R^1 = H) with phosphorus pentachloride and obtained the unsaturated dichloride **114** (R = Cl). However, attempts to reduce the dichloride to the parent (**114**, R = H) were unsuccessful (19CB2224).

Since this early work, numerous similar condensations have been reported. Mazza and Ferrajolo, by allowing the ethyl esters of hexahydroanthranilic acid and tetrahydroanthranilic acid to stand, obtained diazocines **115** and **116**, respectively (27MI1). Partridge et al. published several papers in this area. When o-cyanoanilinium tosylate was heated, the amidine **114** (R = NH$_2$) was formed. This compound was also obtained by heating the dichloride **114** (R = Cl) with ammonia. Condensation of o-cyanoaniline with the p-toluenesulfonamide of methyl anthranilate gave diazocine **117** (62JCS2545). Also, an improved synthesis of the dianthranilide **35** (R = R^1 = H) by condensation of 2 mol of methyl anthranilate with a nitrile and sodium was reported (54JCS3429). The 2,8-dimethyl derivative of **35** (R = R^1 = H) was prepared analogously from methyl 5-methylanthranilate. This compound was then converted into the 2,8-dimethyl derivative of **114** (R = Cl). However, attempts to reduce this and the nonmethylated compound to the parent (**114**, R = H) were unsuccessful (55JCS991). [For reactions of **114** (R = Cl) and its dimethyl derivative to make other diazocine derivatives, see Section IV,B.]

The diamine derivative (**114**, R = NH$_2$) has also been obtained, among other products, by reaction of anthranilamide with phenylacetic acid in

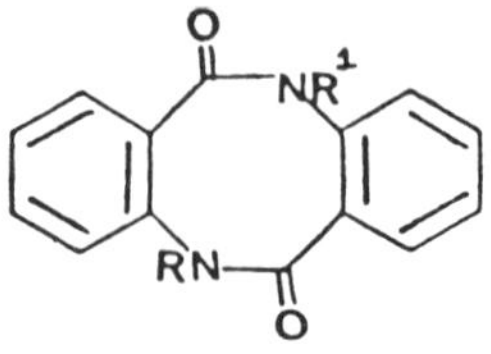

35 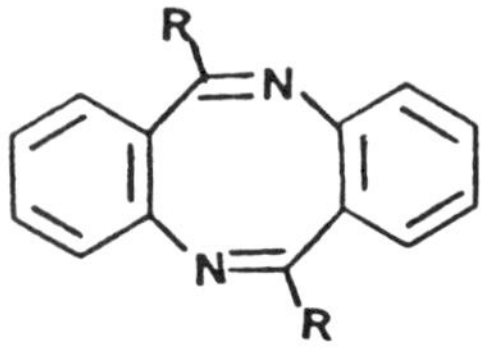   **114**

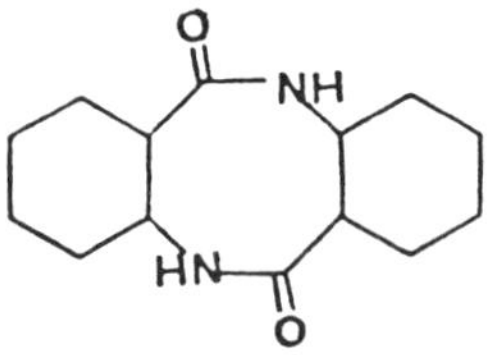

115 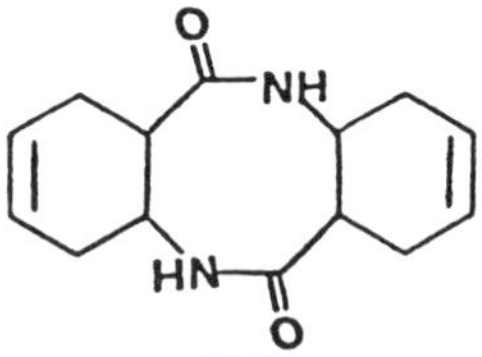   **116**

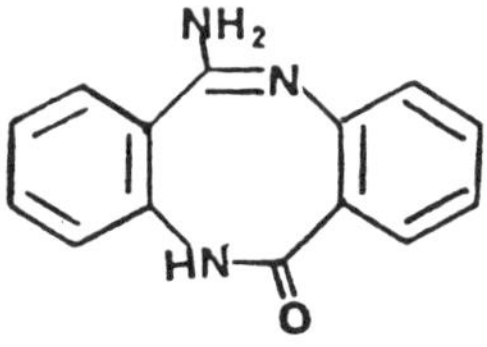

117

the presence of phosphorus pentoxide (71JOC642). Osbirk and Pedersen isolated the bis(dimethylamino)diazocine **114** (R = NMe$_2$) as a byproduct of the reaction of methyl anthranilate with cycloalkanones in the presence of hexamethylphosphorictriamide to give polymethylenequinolines (79ACS(B)313). Synthesis of the parent dibenzo[b,f]-1,5-diazocine (**114**, R = H) was finally achieved by Paudler and Zeiler, who found that the dihydrazine derivative **114** (R = NHNH$_2$) [prepared from **114** (R = Cl) by reaction with hydrazine] reacted with copper(II) acetate to give **114** (R = H) (67CC1077). [Interestingly, the 1,4-isomer of **114** (R = H) has not yet been prepared (89AHC185).]

Reaction of the benzoxazine **118** with ammonia afforded a compound with formula $C_{28}H_{23}O_4N_5$. Structure **119** was suggested, since, when the compound was heated above its melting point, diazocine **35** (R = R^1 =

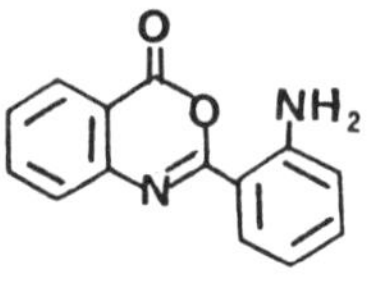

118 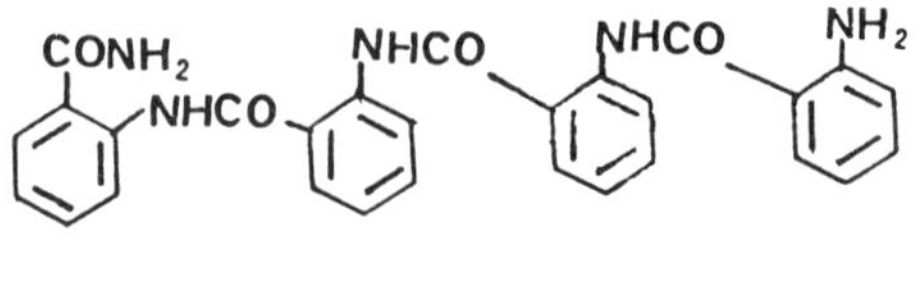   **119**

H) resulted (59JCS2396). Other related condensation reactions have been reported. The reaction of N-benzylanthranilic acid with phosphorus oxychloride gave **35** (R = R^1 = CH$_2$Ph) (65CCC445). Condensation of anthranilic acid with methanesulfonyl chloride in the presence of pyridine afforded **35** (R = H, R^2 = SO$_2$Me) (66YZ107). The reaction of sulfinamide anhydride **120** with formamide gave, in addition to a quinoxaline derivative, the dimethyldiazocine **35** (R = R^1 = Me). This compound was also formed quantitatively when moist **120** was allowed to stand at room temperature (77JCS(Pl)2347). Bitter *et al.* also prepared the dimethyldiazocine **35** (R = R^1 = Me) by reaction of methyl anthranilate with dimethyl(dichloromethyl) ammonium chloride (81ACH171). Bolotin *et al.* reacted ring-substituted anthranilic acid derivatives with benzenesulfonyl chlorides and obtained, in addition to benzoxazine derivatives and N-arylsulfonylanthranilic acids, a variety of substituted diazocines **121** (71M13; 72KGS1341). Interestingly, reaction of nitro-substituted anthranilic acids failed to produced diazocines (72KGS1341). Reaction of N-benzenesulfonylanthranilic acid with dicyclohexylcarbodiimide yielded **35** (R = R^1 = SO$_2$Ph) (78IJC(B)393).

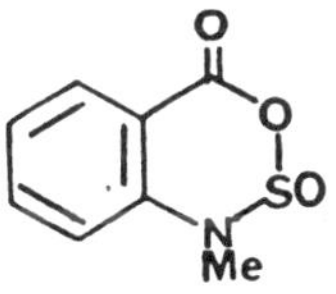

120

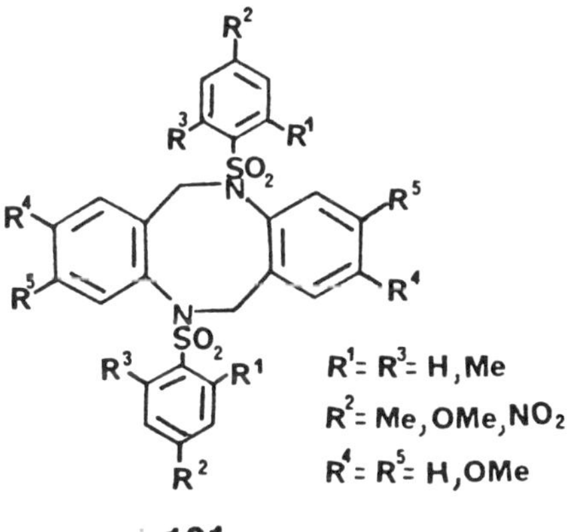

121

R^1 = R^3 = H, Me
R^2 = Me, OMe, NO$_2$
R^4 = R^5 = H, OMe

Katritzky and co-workers found that pyrolysis of imidazole derivative **122** afforded the diazocine **123**. However, similar treatment of corresponding benzimidazole and 5,6-dimethylbenzimidazole derivatives gave, instead, a novel hexacyclic triazadibenzacephenanthrylenedione (76CC48; 77JCS(Pl)1162). When N-hydroxyphthalimide was treated with benzenesulfonyl chloride in aqueous alkaline solution, a novel rearrangement occurred via hydroxamic acid **124** to yield anthranilic acid derivatives, among them the diazocine **121** (R^1–R^5 = H) (57JA1983). Treatment of trimethyl orthoanthranilate with glacial acetic acid afforded the dimethoxydiazocine **114** (R = OMe) (72JHC947). The isoxazoline **125** (R = OMe), when heated, rearranged to yield methyl *N*-ethylanthranilate and diazocine **35** (R = R^1 = Et) (71JA1543).

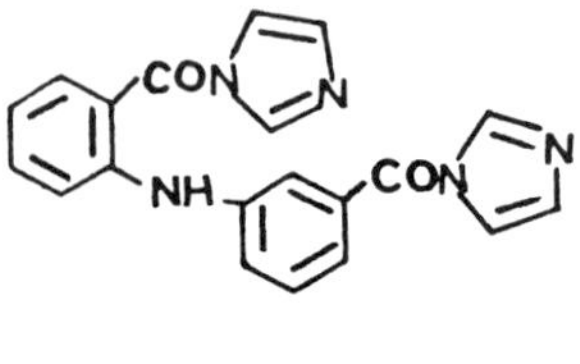

122

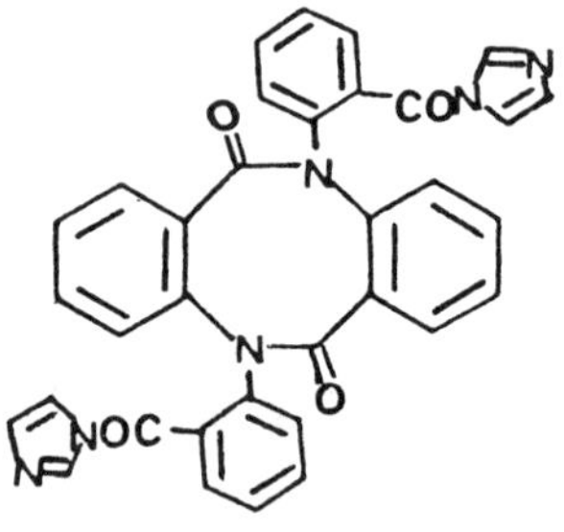

123

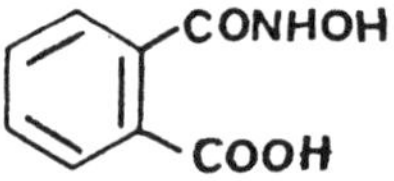

124

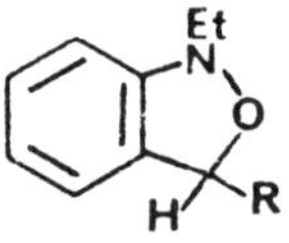

125

A number of heterocycle-fused 1,5-diazocines have been reported. Klisiecki and Sucharda treated an ethyl acetate solution of methyl nicotinate with sodium and obtained diazocine **126** (23RC251). In a more recent, but similar, report, Oakes and Rydon attempted a Claisen condensation of ethyl 3-aminopicolinate with ethyl acetate, but obtained instead the intermolecular amidation product **127** (58JCS205). A similar result was reported by Brederick *et al.*, who found that reaction of 4-amino-5-carbethoxypyrimidine with ethyl acetate in toluene in the presence of sodium

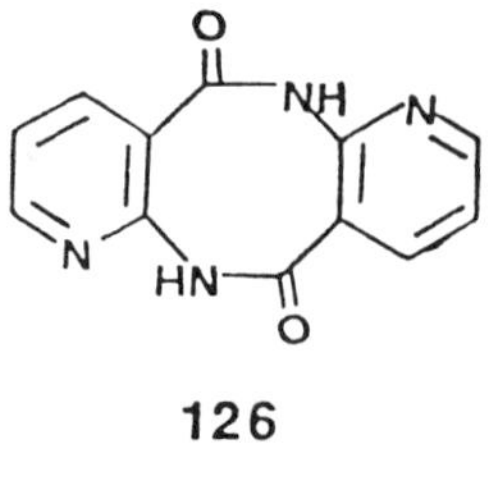

126

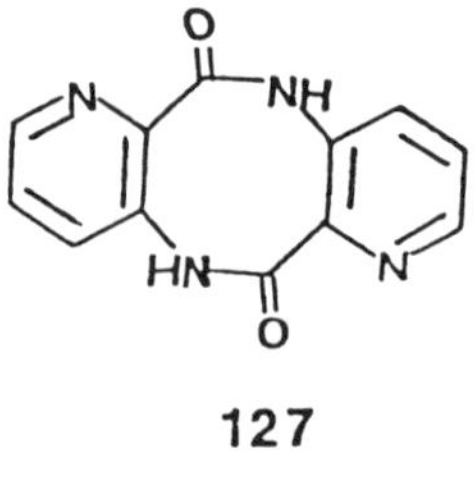

127

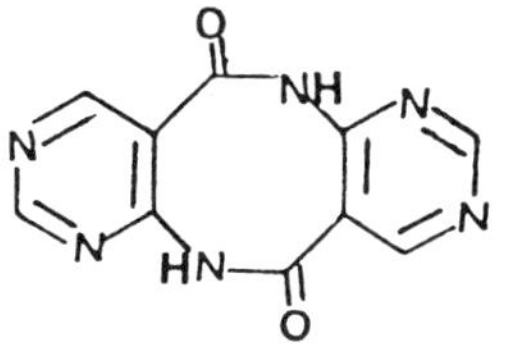

128

afforded diazocine **128**. [The authors proposed an enol structure related to **114** (R = OH), but **128** is more likely.] Use of alcohol as solvent, however, resulted in formation of pyridopyrimidine, as did replacement of ethyl acetate with higher esters, nitriles, or amides (63CB1868).

Reaction of thiophene derivative **129** with hexamethylphosphorictriamide yielded, in addition to a thienopyridine derivative, the diazocine **130** (77T2089). Treatment of the aminotriazepinecarboxylate **131** with alkali resulted in formation of bis(triazepino)diazocine **132** (77BCJ957).

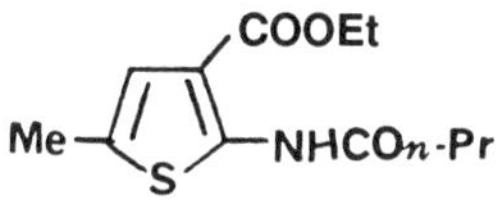

129 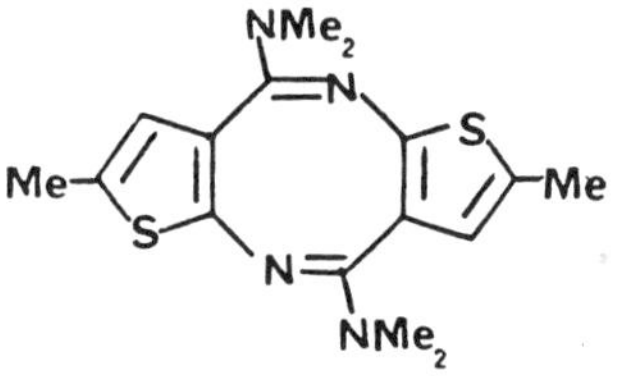   **130**

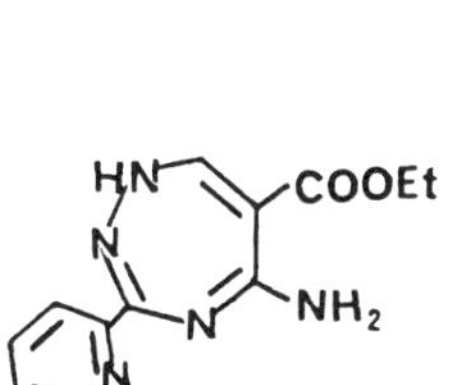

131 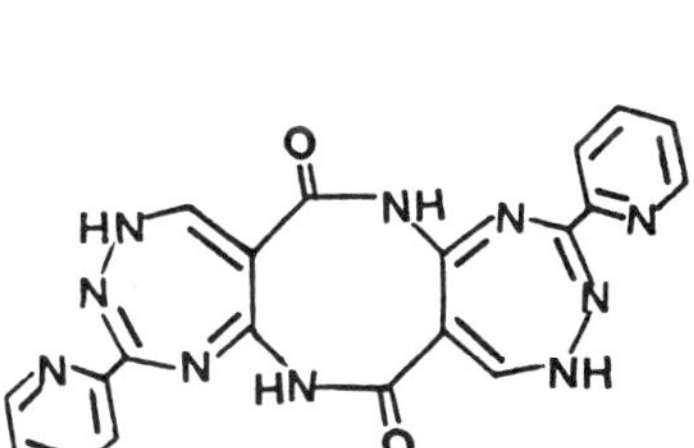   **132**

3. Reaction of Acyclic Diamines with Aldehydes and Ketones

Dietsche treated 2,4-pentanedione with methanol and ammonia and obtained diazocine **133** (77USP4001215). Giacalone condensed benzaldehyde with benzal p-tolylhydrazone (**134**) in the presence of zinc chloride and obtained a substance of formula $C_{42}H_{36}N_4$ that was postulated to be either 1,3-diazocine **135** or the 1,5-diazocine **136** (33G20). Independent synthesis of **135** showed it to be different from the benzaldehyde–benzal p-tolylhydrazone condensation product, and the latter compound was therefore confirmed as **136** (33G764).

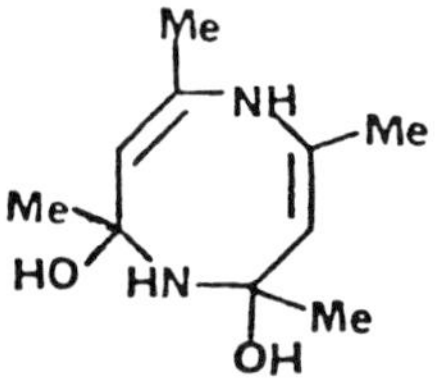

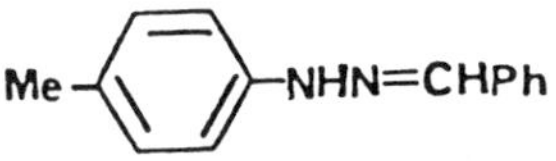

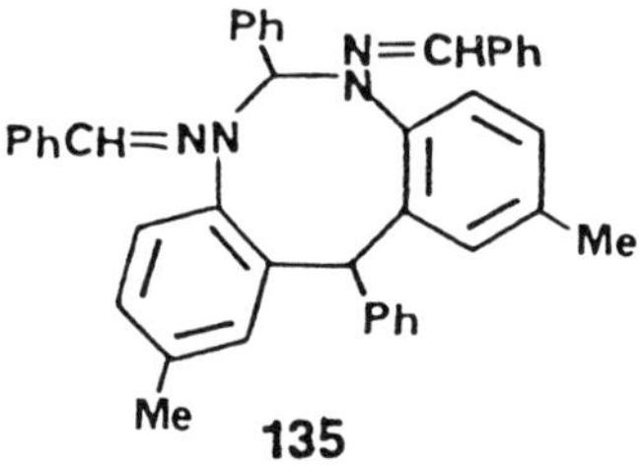

Nearly a century ago, Posner obtained a material of empirical formula C_7H_5N from zinc chloride-catalyzed reduction of o-nitrobenzaldehyde. The compound, probably arising from intermediary o-aminobenzaldehyde, was assigned either structure **137** or **114** (R = H) (1898CB658). [Compound **114** (R = H) has never been accessible from o-aminobenzaldehyde; for its synthesis, see Section II,C,2.] At about that time, Sondheimer reacted o-aminobenzophenone with hydrochloric acid and obtained 6,12-diphenyldizenzo[b,f]-1,5-diazocine (**138**, R–R⁶ = H)

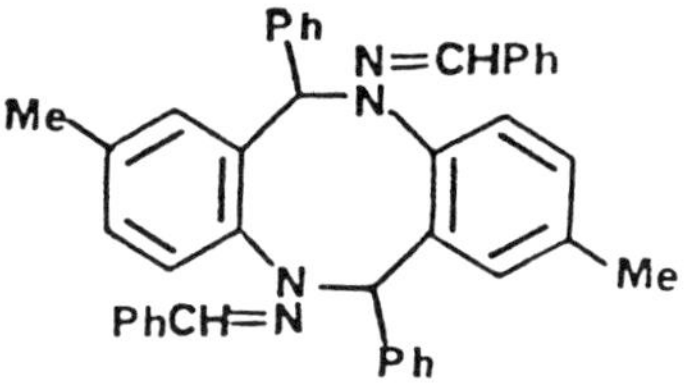

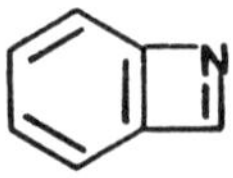

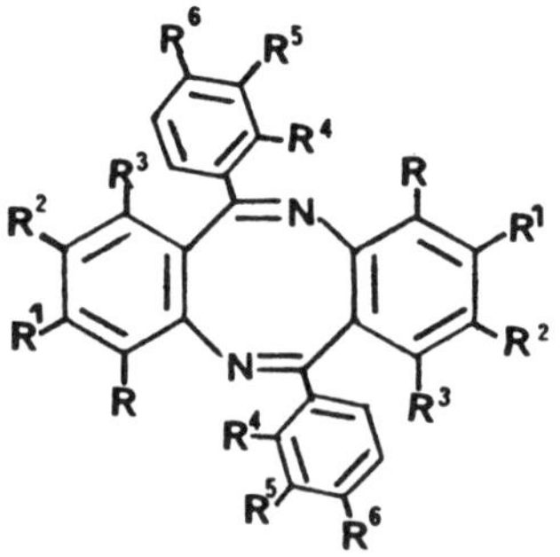

(1896CB1272). Almost 40 years later, Giacalone similarly prepared the dimethyl derivative **138** (R^2 = Me; R, R^1, R^3–R^6 = H) starting with 5-methyl-2-aminobenzophenone (35G120).

More recently, a number of groups have prepared 6,12-diaryldibenzo[b,f]-1,5-diazocines **138** (R–R^6 = various), which have some medicinal use (see Section V), from the corresponding o-aminobenzophenones in the presence of various acid and base catalysts. A group at Upjohn prepared compounds **138** (R^2 = H, Cl; R^6 = H, OMe; R = R^1 = R^3 = R^4 = R^5 = H) using p-toluenesulfonic acid (66NEP6602819). Sternbach and his group at Hoffmann La Roche similarly prepared **138** (R–R^6 = H) and a number of derivatives (**138**, R^2 = Cl, F, Br, CF_3, OMe, EtS, NH_2, NO_2; R = H, Cl, Br; R^1 = H, Cl; R^3 = H, Cl; R^4 = H, F, Me, OMe; R^5 = H, Cl; R^6 = H, F) using boron trifluoride etherate, aluminum and titanium chlorides, polyphosphoric acid, and sodium hydride as catalysts (66FRP1463527, 66USP3243430). The dichloro derivatives **138** (R, R^1, R^3–R^6 = H; R^2 = Cl) were synthesized by heating the chloroaminobenzophenone with either zinc chloride, hydrochloric acid, or without any catalyst (67NKZ102; 69BCJ2066). Compounds **138** (R = R^1 = R^3 = H; R^2 = H, Cl, Br, NO_2; R^4 = H, Br, Cl; R^5 = R^6 = H, Br, NO_2) were made using phosphorus oxychloride and p-toluenesulfonic acid (76ZOB1893), whereas polyphosphate ester was used to prepare **138** (R = R^1 = R^3 = R^4 = H; R^2 = Cl, NO_2; R^5 = R^6 = OMe) (80IJC(B)703). Moriyama et al. heated the appropriate aminobenzophenone without catalyst, affording **138** (R = R^1 = R^3 = R^5 = R^6 = H; R^2 = Cl, NO_2, CF_3; R^4 = H, Me) (70JAP70/06271). In a different report, Moriyama et al., instead of

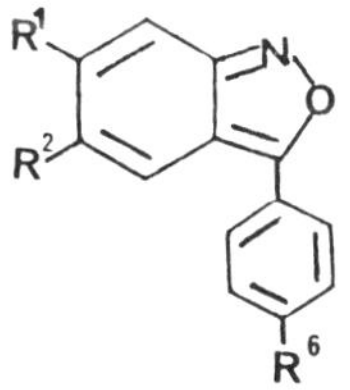

139

R^1 = H, Me
R^2 = Cl, Br, NO_2
R^6 = H, OMe

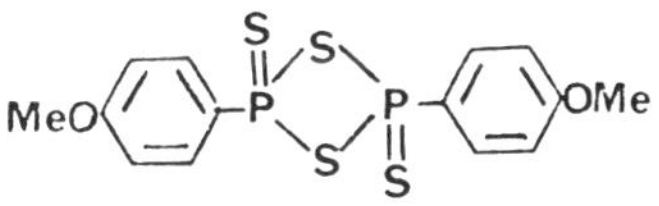

140

using o-aminobenzophenone as a starting material, employed o-aminobenzophenoneimines. When these compounds were heated under nitrogen, compounds **138** (R = R^1 = R^3 = R^5 = R^6 = H; R^2 = Cl, NO_2, CF_3, Cl; R^4 = H, Me) were obtained (70JAP70/06270). The same compounds were obtained by heating an equimolar mixture of substituted o-aminobenzophenone and substituted *o*-aminobenzophenoneimine under nitrogen (70JAP70/06272). The 2,1-benzisoxazoles **139**, upon heating in the presence of Lawesson's reagent (**140**), afforded substituted **138** (82IJC(B)899).

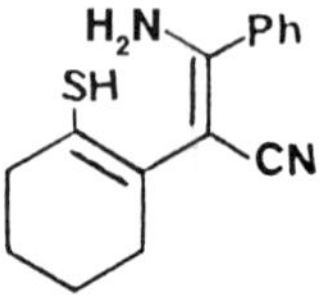

141

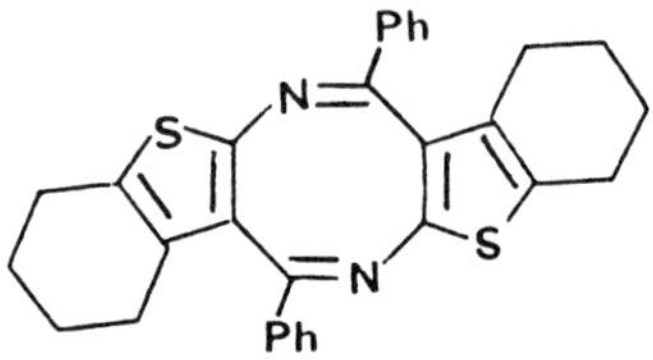

142

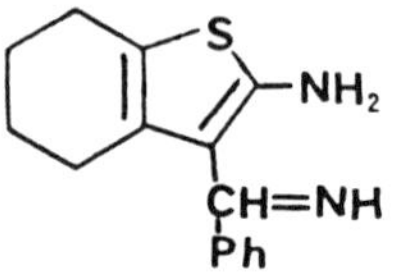

143

There have been additional interesting reports of heterocycle-fused 1,5-diazocine syntheses. Heckendorn and Gagneux thermolyzed the cyanoaminothiol **141** and obtained the dithienodiazocine **142**. The reaction most likely proceeded via iminothiophene **143**, which could be isolated when **141** was allowed to stand at room temperature (73TL2279). When Roma *et al.* reacted the naphthopyran **144** with various amidines in pyridine, napthopyranopyrimidines were obtained. However, reaction of **144** with *O*-methylisourea in pyridine afforded, instead, the dinaphthopyranodiazocine **145** (76JHC761).

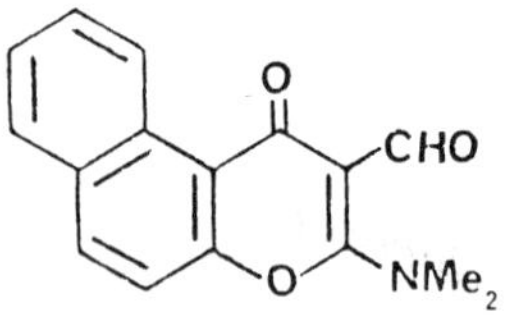

144

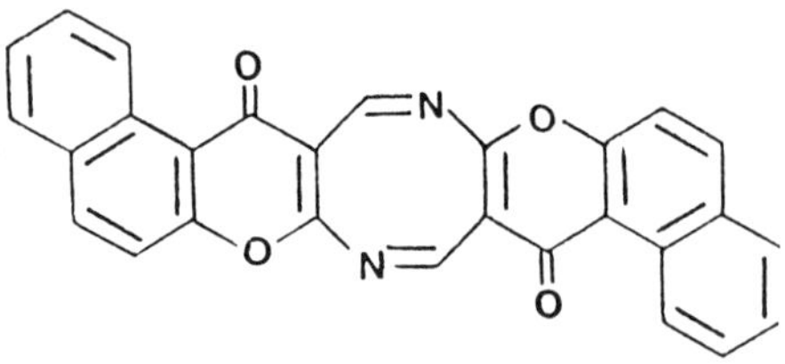

145

4. *Mixed and Other Condensation Reactions*

A number of 1,5-diazocines and fused-ring analogues have been prepared by condensation reactions forming both amide and imine bonds concurrently. Shinano *et al.* condensed N-substituted 1,3-propanediamines with 3-formyl propionates and acetoacetates and obtained diazocines **146** (1,4-diazocines were also prepared) (89AHC185). (For applications of these compounds, see Section V.) Bogatskii *et al.* reacted various N- and ring-substituted *o*-aminobenzophenones with β-alanine or β-alanyl chloride and isolated 1,5-benzodiazocines **147** (75MI1; 80KFZ45). Murakami *et al.* condensed 5-chloro-2-aminobenzophenone with 2-chloro-5-aminobenzoic acid and isolated dibenzo-1,5-diazocine **148** (71JAP71/10861). Barchet and Merz treated 1,8-diaminonaphthalene with ethyl benzoylacetate and ethyl α-benzyl-β-phenylpropionate and ob-

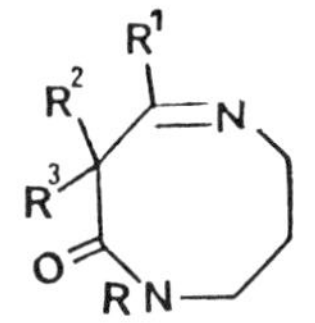

146
R = long chain
R¹, R², R³, = H, Me

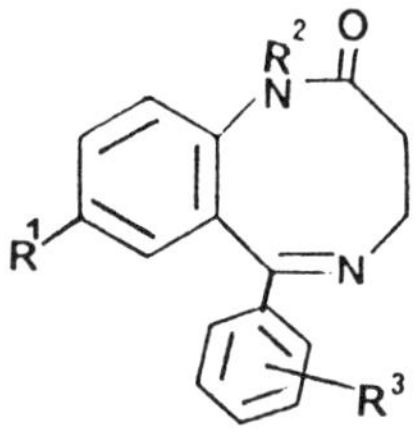

147
R¹ = Cl, Br, Me, H, OCF$_3$, SCHF$_2$,
SO$_2$CHF$_2$, NO$_2$, NH$_2$, NHAc
R² = Me, Et, *n*-Pr, *n*-Bu, allyl, CH$_2$Ph
R³ = H, o-Cl, p-Cl, o-Br, p-F, m-Br,
m-NO$_2$, 3, 5-(NO$_2$)$_2$, p-NO$_2$

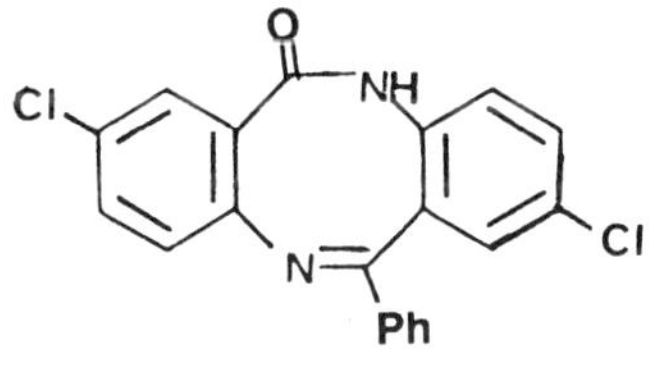

148

tained naphthodiazocines **149** (R = H) and **149** (R = CH$_2$Ph), respectively (64TL33). An interesting reaction was reported by Singh and Singh, who condensed *N*-cyclohex-1-enylaniline with the ketoisothiocyanate **150** and isolated the diazocinethione **151** (75AJC143).

Other Mannich and Mannich-type condensations (also see Section II, C, 1) have been used to synthesize 1,5-diazocines. Two groups reported reacting the tetranitro-*N*-nitroso compound **152** (R = NO) with alkylamines and formaldehyde, affording tetranitro-*N*-nitrosodiazocines **153** (R = NO, R¹ = Me) (81IZV1676; 83S830) and **153** (R = NO; R¹ = H, *i*-Pr)

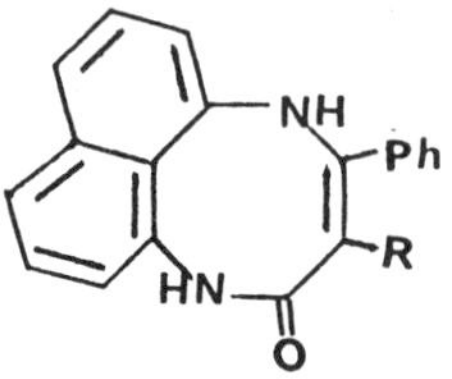

149

150 SCNC(Me)$_2$CH$_2$COCH$_3$

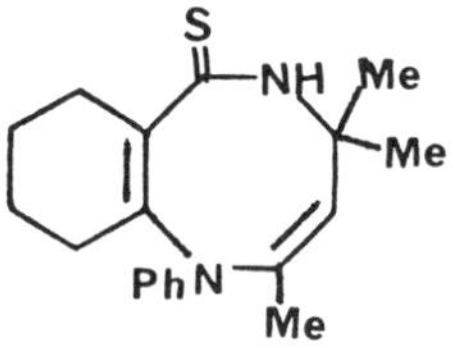

151

(83S830). Use of the N-nitro derivative **152** (R = NO$_2$) gave the N-nitrodiazocines **153** (R = NO$_2$; R^1 = H, Me) (83S830). There have been two reports of reaction of N-methyl-p-toluidine with acidic formaldehyde solution to yield the diazocine **67** (R^1–R^3 = H, R^4 = R^5 = Me) (76CI(L)1644). Similarly, reaction of formaldehyde with diphenylamine afforded

152 RN[CH$_2$CH(NO$_2$)$_2$]$_2$

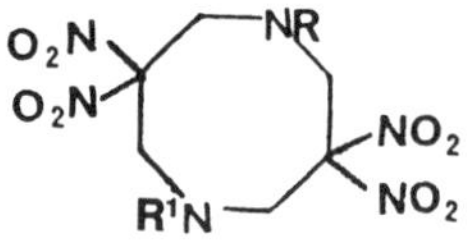

153

diazocine **67** (R^1–R^3 = H, R^4 = R^5 = Ph) (78GEP2729363). Kutlu condensed saccharin and paraformaldehyde with o-, m-, and p-aminophenols and obtained diazocines **154**, **155**, and **156**, respectively (70MI2).

Oxidative cyclodimerization of N,N-dimethylaniline and N,N-dimethyl-p-toluidine has been investigated by a number of groups. Roy and Swan found that these two compounds, when heated with benzoyl peroxide, afforded diazocines **67** (R^1–R^3 = H, R^4 = R^5 = Me) and **67** (R^1 = R^4 = R^5 = Me, R^2 = R^3 = H), respectively (66CC427; 68CCl376). Zelent and co-workers photolyzed the two aforementioned anilines in dichloromethanes and obtained identical products (79JOC3559; 80MI2). Photolysis of N-methylaniline afforded **67** (R^1–R^3 = H, R^4 = R^5 = Me) through incorporation of solvent molecules (80MI2). Interestingly, the same reaction using chloroform or carbon tetrachloride as solvent failed to produce diazocines. The reaction of N,N-dimethylanilines para-substituted by electron-donating groups with palladium(II) acetate in benzene–acetic acid produced similarly substituted diazocines **67** (80CL1331; 82CL1823).

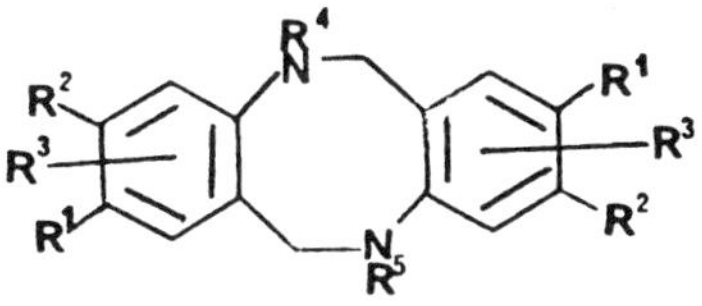

Sec. II.C] 1,5-DIAZOCINES 35

67

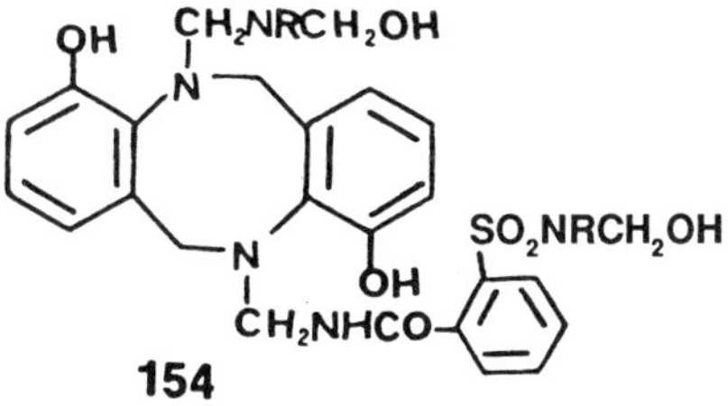

154 R = o-HOC$_6$H$_4$

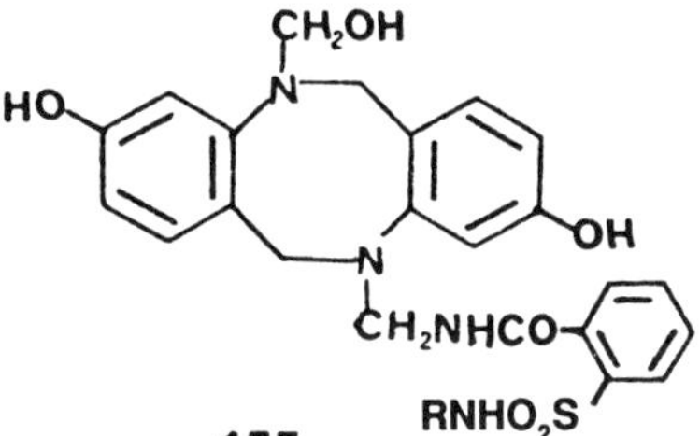

155 R = m-HOC$_6$H$_4$

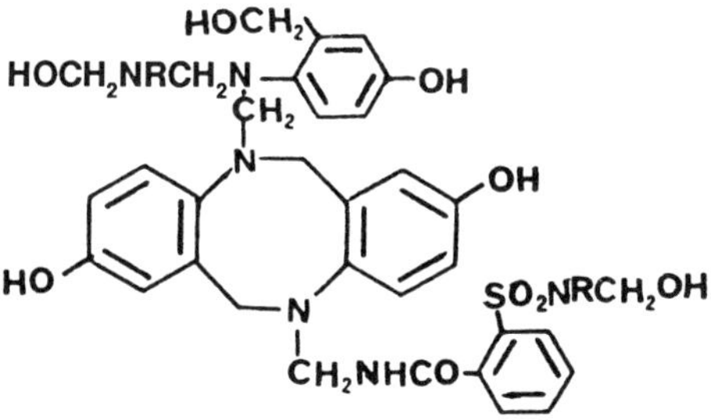

156 R = p-HOC$_6$H$_4$

Reaction of two different anilines, e.g., **157** (R = OMe, R^1 = H) and **157** (R = Me, R^1 = H), afforded both homo-coupling diazocines **158** (R = R^2 = OMe, R^1 = R^3 = H) and **158** (R = R^2 = Me, R^1 = R^3 = H) and the cross-coupling product **158** (R = OMe, R^2 = Me, R^1 = R^3 = H) (82CL1823).

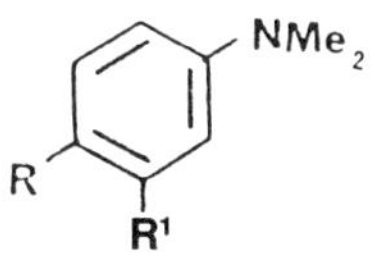

157 R = OMe, Me, Cl; R^1 = H, Me

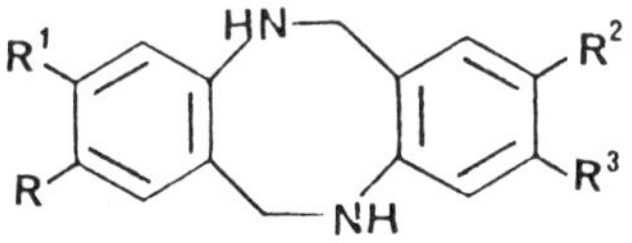

158

Swan found that when N,N'-disubstituted methanedianilines **159** were treated with weak acids, the diazocines **67** (R^1 = H, Me; R^2 = R^3 = H; R^4 = R^5 = Me, Et, Ph) resulted, primarily via decomposition to an iminium ion, which attacks a second molecule of **159**. Interestingly, quaternization of tetrahydroquinazoline **160** with methyl bromide yields diazocine **67** ($R^1 = R^4 = R^5 = Me$, $R^2 = R^3 = H$) (71JCS(C)2880). This reaction may proceed via iminium ion intermediates similar to those postulated for the formation of **67** and α-diamines **159**, and may be related to the aforementioned formation of **67** by reaction of *N*-methyl-*p*-toluidine with acidic formaldehyde *(vide supra)* (76CI(L)112).

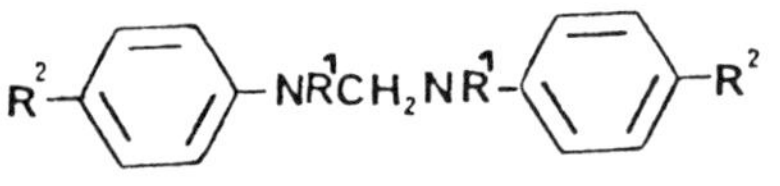

159
R^1 = H, Et, Ph
R^2 = H, Me

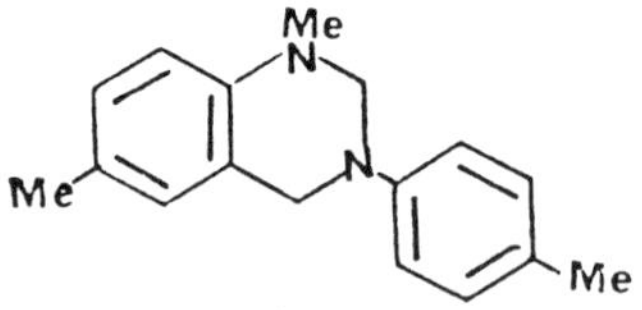

160

Diazocines **54** (R = H, Me) were synthesized by treating phenyl and *o*-tolyl isocyanate with aluminum chloride (51DOK1073). A Michael addition was involved in the reaction of benzylideneacetone with 1,3-diaminopropane to yield diazocine **161** (71JCS(D)372). Photolysis of diphenylaryl-

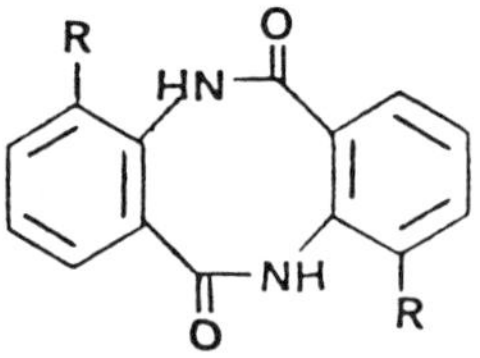

54

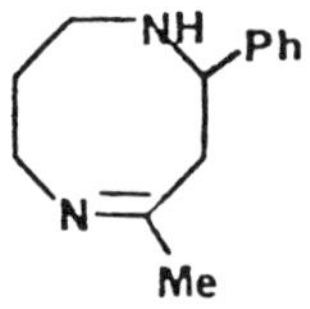

161

triazafulvenes **162** afforded diazocine **163** (R = Ar, $R^1 = R^2$ = Ph), presumably via dimerization of an intermediate azacyclobutadiene (73JOC176). A similar intermediate was postulated in the photolysis of pyridazine **164** to produce diazocine **163** [R = R^2 = F, R^1 = $CF(CF_3)_2$] (76CC1005).

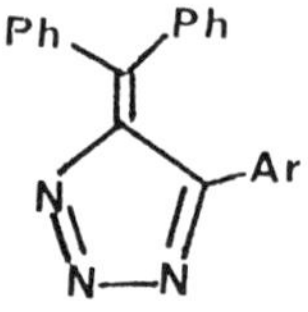

162
Ar = Ph, p - ClC_6H_4

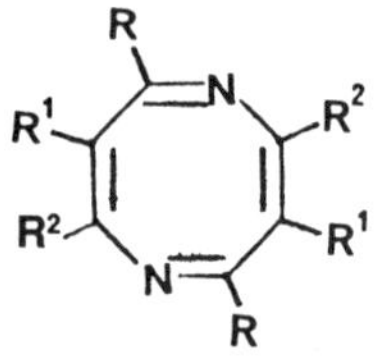

163

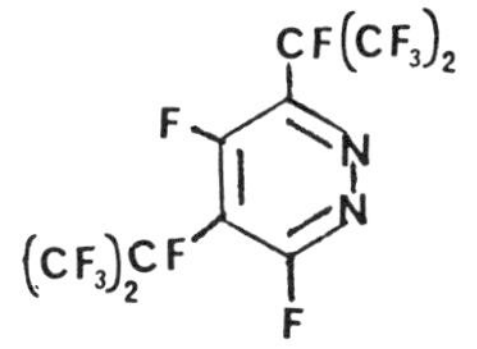

164

Heterocycle-fused, 1,5-diazocines have been prepared by nucleophilic attack by nitrogen at the carbon alpha to the hetero atom. Eberle and Houlihan reacted 1,3-propanediamine with the pyrroline **165** and obtained the pyrrolidinodiazocine **166** (71GEP2050344). Wamhoff and Materne treated the dihydrofurans **167** with 1,3-diaminopropane and isolated the

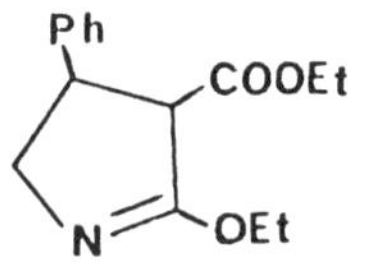

165 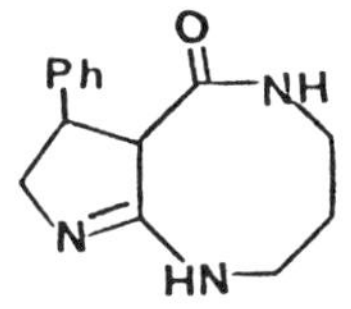   **166**

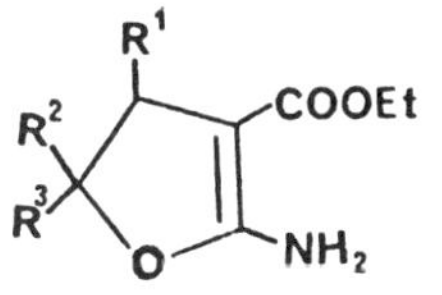

167 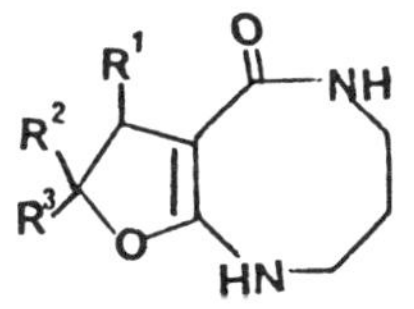   **168**

$R^1 - R^3 = H, Me$

furanodiazocine **168** (74CB1784). DeWald and L'Italien found that cyclizing the aroylpyrazoles **169** with 1,3-diamines afforded the pyrazolodiazocines **170**. Various N-substituted derivatives **170** and their imine N-oxides were prepared (74GEP2423642).

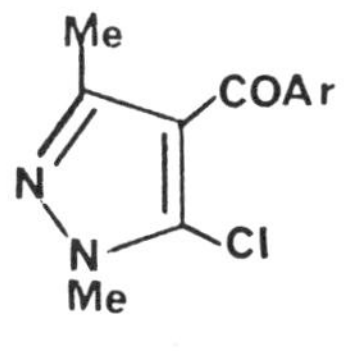

169

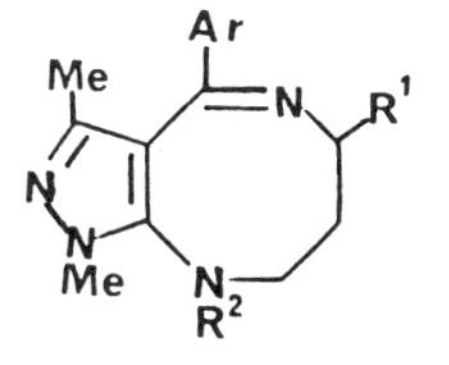

170
R^1 = various; R^2=H
Ar = m-ClC$_6$H$_4$, m-FC$_6$H$_4$,
p-ClC$_6$H$_4$, m-CF$_3$C$_6$H$_4$, m-BrC$_6$H$_4$

D. HOMALINE AND RELATED COMPOUNDS

The structure of homaline, a plant alkaloid related to spermine and isolated from species of *Homalium*, was studied by Pais *et al.* Homaline was shown to have the 1,4-bis(1,5-diazocinyl)butane structure **171** (R = Me), and not structures involving diazepines (68CR(C)37) or 1,5,9,13-tetraazabicyclo[7.7.4]eicosanes (68CR(C)82). Structural determination is based on degradative studies (68CR(C)37; 73T1001) and spectroscopic properties (68CR(C)82; 73T1001) of homaline, and by the synthesis of bisdeoxohomaline by the reactions shown in Eq. (1). β-Alanyl-β-alanine (**172**) was prepared by condensation of amino-protected β-alanine with the methyl ester of *N*-methyl-β-phenyl-β-alanine. Compound **172** was then cyclized with hydrogen bromide in acetic acid to afford bislactam **173**, which was reduced to give the diazocine **174**. Condensation of 2 mol of **174** with succinoyl chloride, followed by reduction of the product with lithium aluminum hydride, gave deoxohomaline (**175**). This latter compound was shown to be identical to deoxohomaline obtained from naturally occurring homaline (**171**, R = Me) (71BSF255, 71CR(C)1728; 73BSF331). The detailed structure of homaline was later elucidated by X-ray diffraction (76AX(B)1305).

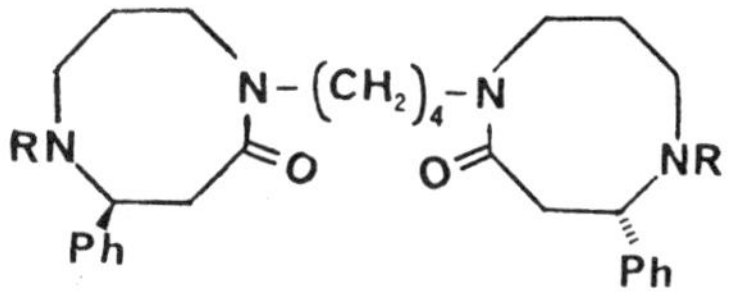

171
homaline (R=Me)

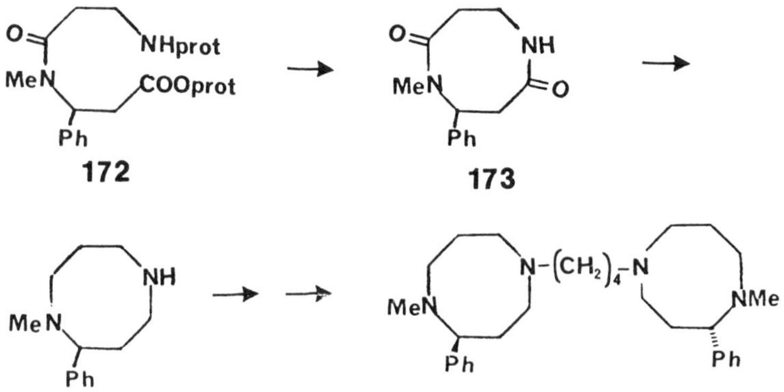

172 → **173** →

174 → → **175**

(1)

The total synthesis of homaline (**171**, R = Me) was accomplished by Wasserman *et al.*, who began with 1,4-diaminobutane and in six steps produced bistosylate **176** (BOC = *tert*-butoxycarbonyl). This compound was reacted with β-lactam **177** (R = H) (derived from β-phenyl-β-alanine) to afford the bis(β-lactam) **178**. This compound underwent thermal transamidation to yield the bis(*N*,*N*'-demethyl) homaline (**171**, R = H), which could be methylated to yield homaline (**171**, R = Me) [Eq. (2)] (82TL465; 83T2459).

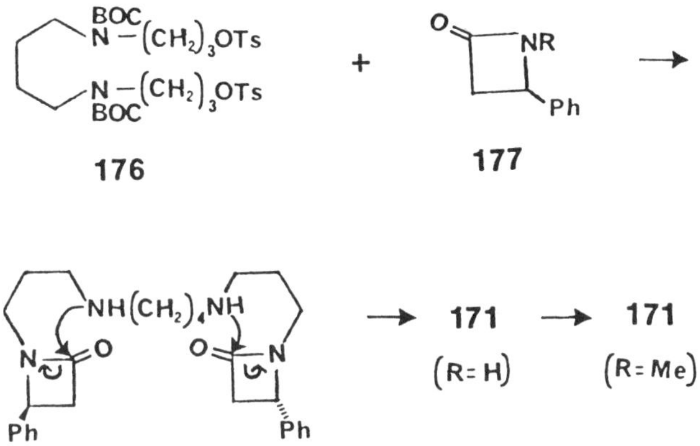

$$\textbf{176} + \textbf{177} \longrightarrow$$

$$\textbf{178} \longrightarrow \textbf{171}\,(R=H) \longrightarrow \textbf{171}\,(R=Me) \quad (2)$$

In an approach related to the synthesis of a diastereomeric mixture of homaline (**171**, R = Me) and epihomaline (**179**, R = Me) [Eq. (3)], Crombie *et al.* first prepared 1,5-diazocine **180** (R = R^1 = H) by transamidative

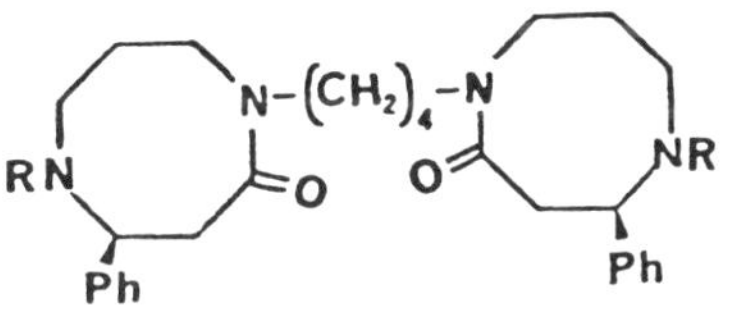

179
epihomaline (R=Me)

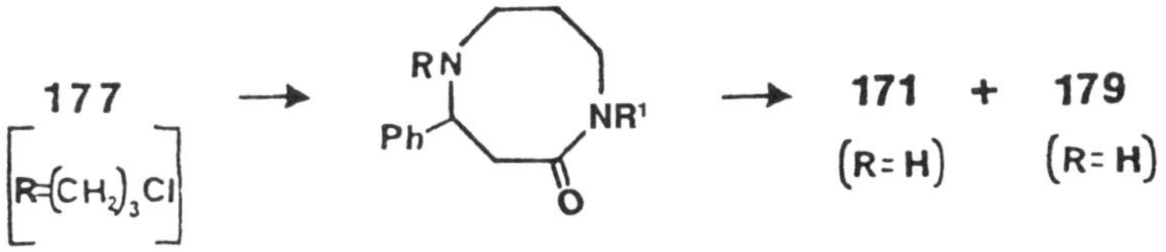

$$\textbf{177} \xrightarrow{[R\text{-}(CH_2)_3Cl]} \textbf{180}\,(R=R^1=H) \longrightarrow \textbf{171}\,(R=H) + \textbf{179}\,(R=H) \quad (3)$$

ring expansion of racemic **177** [R = $(CH_2)_3Cl$] (see Section II,B,4). Two moles of **180** (R = R^1 = H) were treated with 1 mol of 1,4-dibromobutane to afford a mixture of N,N'-demethylhomaline (**171**, R = H) and its epimer **179** (R = H). Reductive methylation of this mixture with formaldehyde and sodium cyanoborohydride gave a mixture of homaline (**171**, R = Me) and epihomaline (**179**, R = Me), from which the latter could be isolated in a pure state by recrystallization.

Alternatively [Eq. (4)], the N-methylated derivative of **180** (R = Me, R^1 = H) [prepared from **180** (R = R^1 = H) by reductive methylation] was treated with 1,4-dibromobutane to give **171** (R = Me) and **179** (R = Me). The naturally occurring (S,S)-($-$)-homaline (**171**, R = Me) was similarly prepared, starting with (S)-($-$)-(**177**) [R = $(CH_2)_3Cl$] in a sequential approach [Eq. (5)]. Compound **180** (R = Me, R^1 = H) can be reacted with 1-bromo-4-chlorobutane (mole ratio of 1 : 1) to afford the N-haloalkyl compound **180** [R = Me, R^1 = $(CH_2)_4Cl$] (see Section II,B,4). This compound was then condensed with **180** (R = Me, R^1 = H) to yield the mixture of homaline (**171**, R = Me) and epihomaline (**179**, R = Me)

$$\underset{(R = Me, R^1 = H)}{\textbf{180}} \longrightarrow \underset{(R = Me)}{\textbf{171}} + \underset{(R = Me)}{\textbf{179}} \quad (4)$$

$$\underset{(R=Me, R^1=H)}{\textbf{180}} \longrightarrow \left[\underset{R=Me, R^1=(CH_2)_4Cl}{\textbf{180}}\right] \quad (5)$$

$$\left[\underset{R=Me, R^1=(CH_2)_4Cl}{\textbf{180}}\right] + \underset{(R = Me, R^1 = H)}{\textbf{180}} \longrightarrow \underset{(R=Me)}{\textbf{171}} + \underset{(R=Me)}{\textbf{179}}$$

(83CC960). Still another synthesis [Eq. (6)] involved reacting **180** (R = Me, R^1 = H) with 1,4-dichloro-2-butene to afford **180** (R = Me, R^1 = $CH_2CH=CHCH_2Cl$). This compound was condensed with **180** (R = Me, R^1 = H) to give 1,4-bis(diazocinyl)-2-butene **181**, hydrogenation of which resulted in the mixture of homaline (**171**, R = Me) and epihomaline (**179**, R = Me) (86TL5147).

Three spermine-type alkaloids closely related to homaline—hopromine, hoprominol, and hopromalinol—had been shown, by degradative and spectroscopic studies, to have structures **182** (R = n-C_5H_{11}, R^1 = n-C_7H_{15}), **182** [R = n-C_5H_{11}, R^1 = CH_2—CHOH—$(CH_2)_4Me$], and **182** [R = Ph, R^1 = CH_2—CHOH—$(CH_2)_4Me$], respectively (73T1001). Crombie

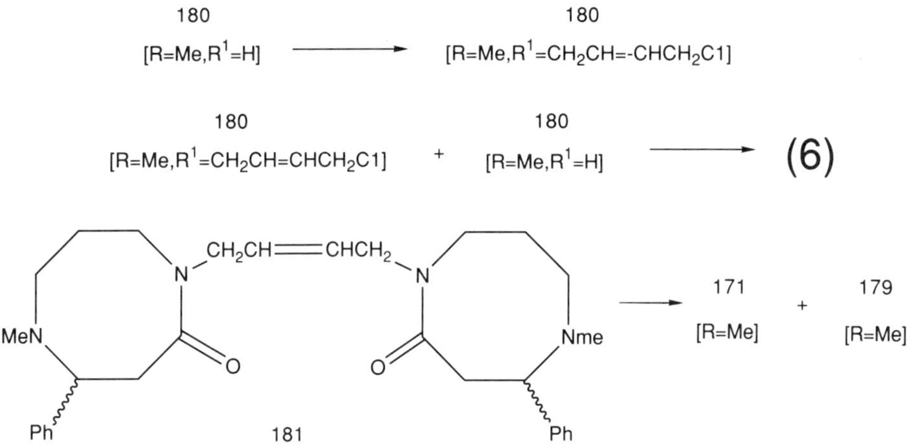

180
[R=Me,R¹=H] ⟶ **180**
[R=Me,R¹=CH$_2$CH=-CHCH$_2$Cl]

180
[R=Me,R¹=CH$_2$CH=CHCH$_2$Cl] + **180**
[R=Me,R¹=H] ⟶ **(6)**

⟶ **171** + **179**
[R=Me]   [R=Me]

181

et al., in an extension of their synthesis of homaline (vide supra), prepared hopramine, hoprominol, and hopromalinol, each as a mixture of stereoisomers. They used the same sequence of reactions used for homaline, i.e., transamidative ring expansion of β-lactams; condensation of the resulting 1,5-diazocines with (E)-1,4-dibromo-2-butene to afford compounds related to **180** (R = Me, R = CH$_2$CH=CHCH$_2$Br); coupling of these latter compounds with the appropriate diazocines related to **180** (R = Me, R¹ = H); and hydrogenation of the resulting compound related to **181**. The β-lactam synthons required for these preparations were **92** [R = Ph, (CH$_2$)$_4$Me, (CH$_2$)$_6$Me, CH$_2$CH(OSiMe$_2$-t-Bu)(CH$_2$)$_4$Me; R¹ = H; R² = (CH$_2$)$_3$Cl] (86TL5147) (see Section II,B,4).

Finally, in an accompanying communication, Crombie et al. described an innovative diazocine ring expansion, in which intramolecular transamidation was used to convert the *Homalium* alkaloid precursor **180** (R = R¹ = H) into the alkaloid dihydroperiphylline (for details, see Section IV,D) (86TL5151).

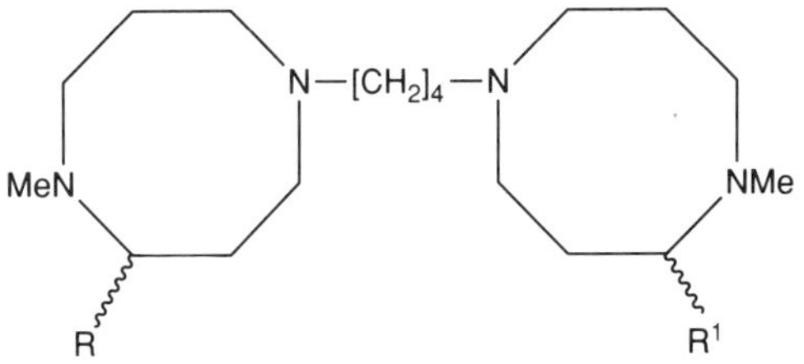

182 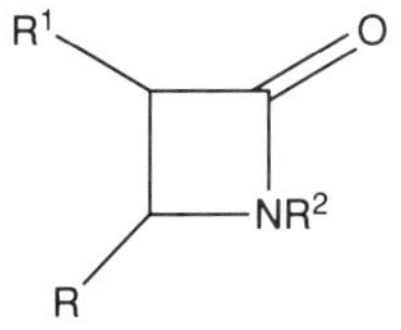   **92**

III. Theoretical and Structural Studies

Hückel calculations on the 8π 1,5-diazocine (5) gave a resonance energy almost identical to the isomeric 1,3-diazocine, and about half as negative as the isomeric 1,4-diazocine (75T295; 83MI1; 89AHC185). Molecular orbital calculations using the PPP method demonstrated that, 1,5-diazocine (5), as well as its 1,3-isomer, have decreased antiaromatic character compared to cyclooctatetraene; these results contradicted those obtained for 1,4-diazocine (89AHC185). It was also claimed that in 5 and its 1,3-diazocine, isomerization via bond alternation may compete with ring inversion (83MI1).

MINDO calculations of the molecular geometries of diazaheterocyclic 1,2-dienes indicated that diazocine 183 has more allenic character than the seven- and six-membered ring homologues (although all three have largely allenic structures) (82JCR(S)80).

Urbanski and co-workers prepared a number of 1,5-diazocines (see Section II,C,l) and investigated their conformations. In one study, molecular refractions were determined for compounds 101 (cis and trans, R = Me, Et, n- Pr; R^1 = H). Infrared spectra of these compounds showed both bonded NH_2 and NO_2 groups; the compounds formed only monohydrochlorides and mononitroso derivatives. These facts suggested intramolecular hydrogen bonding between the two ring nitrogen atoms. The experimental dipole moment of *trans*-101 (R = Et, R^1 = H) was consistent with that calculated for a crown conformation (184), with the two

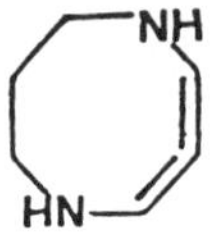

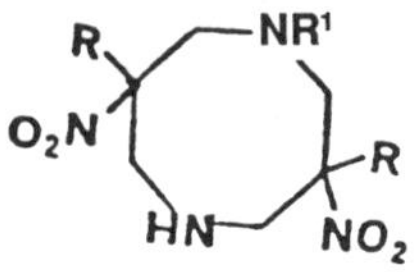

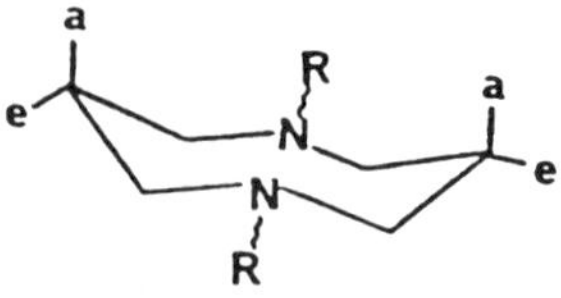

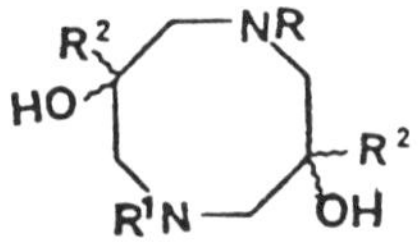

nitro groups in axial (a) and equatorial (e) positions. The *cis* isomer was in the crown form with the equatorial nitro group considered more probable (58JCS2319). In another report, the assignment of the crown form 184 was extended to other derivatives (**101**, R = Me and *n*-Pr, R^1 = H). In the case of **101** (R = *n*-Pr, R^1 = Me), the measured dipole moments favored a crown with the N-methyl groups in equatorial positions (64T207). X-Ray diffraction studies of diazocines **102** (R = R^1 = Ts, *p*-BrC$_6$H$_4$; R^2 = H) indicated that the tosyl derivative had an extended crown conformation (**184**) and exhibited strain (75AX(B)2571), while the *p*-bromophenyl derivative (as the ethanol solvate) existed in a twisted crown form (76CSC905).

The molecular structures of benzodiazocines **12** (R = Me, R^1 = Br, Cl) were determined. The conformations, bond distances, and valency angles of the two compounds were in accord. The bromo derivative has a deep pseudo-boat diazocine ring (79DOK1140), whereas the chloro compound was a boat (82AX(B)638). The main difference between the two is the orientation of the 6-phenyl ring, this difference being reflected in major differences in molecular packing (82AX(B)638).

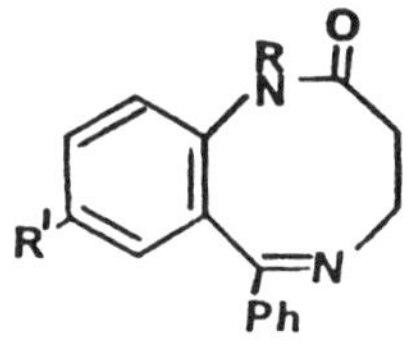

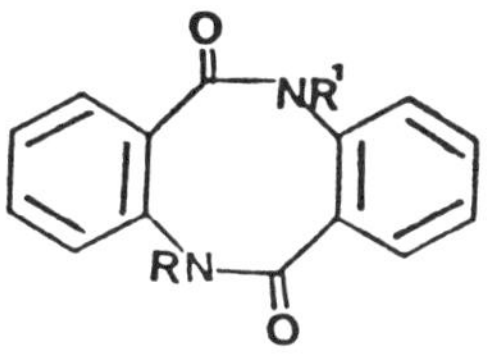

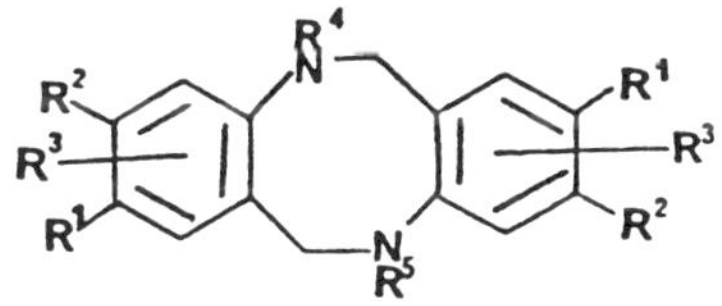

The 1,5-dibenzodiazocine **35** (R = R^1 = CH$_2$Ph) (like its 1,4-isomer) (89AHC185) showed a high degree of conformational stability as demonstrated by its nuclear magnetic resonance (NMR) spectrum, which was unchanged up to 180°C (73CC571; 74T1903). Ollis *et al.* also studied the ^{1}H NMR temperature-dependence of 1,5-diazocines **67** (as well as the 1,4-isomers) (89AHC185). The behavior of **67** was consistent with the presence of a mobile conformation, probably a boat, which inverts by way of folded-boat transition states (73JCS(Pl)205). As with the 1,4-diazocines (89AHC185), the NMR results for **67** were corroborated using molecular mechanics (74T1903). Preliminary X-ray crystallographic data were obtained for diazocine **67** (R^1 = R^4 = R^5 = Me, R^2 = R^3 = H) (70MII).

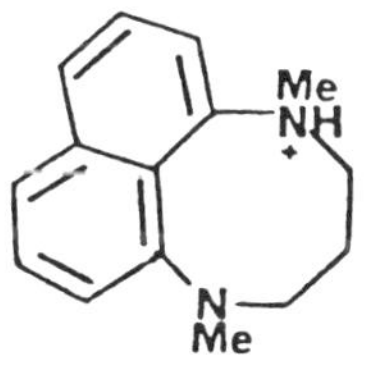

The naphthodiazocine monocation **185** was found to undergo 100% equilibrium N-protonation. This was in contrast to the 9- and 10-membered ring homologues, which underwent 100% C-protonation on the naphthalene nucleus to give an sp^2-hybridized imminium ion $C=N^+$. This phenomenon is believed to be favored in medium rings. However, models of C-protonated **185** did seem strained (82JCS(P2)477).

In a study of Tröger's base derivatives **68** [R = H, Me; RR = $(CH_2)_3$] in acid media (see Section II,B,3), Greenberg *et al.* found the dication of the dimethyl derivative (**186**, R = Me) to have nonidentical benzylic methylene groups and nonequivalent methyl NMR resonances. These results are consistent with the open iminium ion structure **69** (R = Me), and contrast with the results with Tröger's base itself (**68**, R = H) and its spiro (cyclobutyl) derivatives **68** [RR = $(CH_2)_3$], in which the protonated closed structure **186** [RR = $(CH_2)_3M$] is favored (84JOC1127).

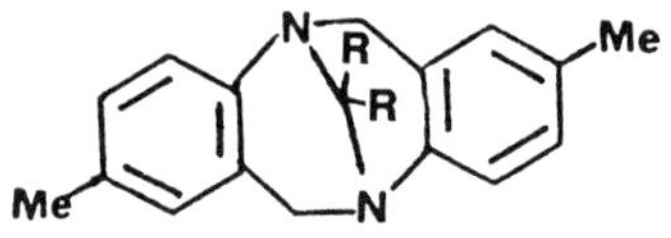

68 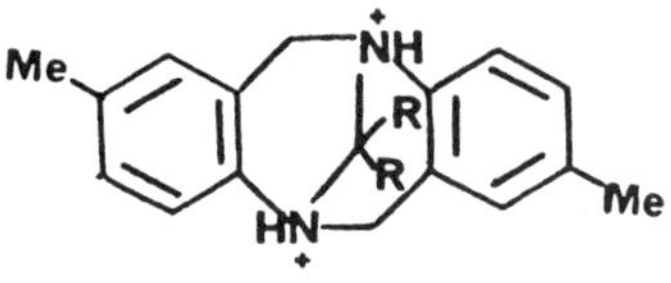   **186**

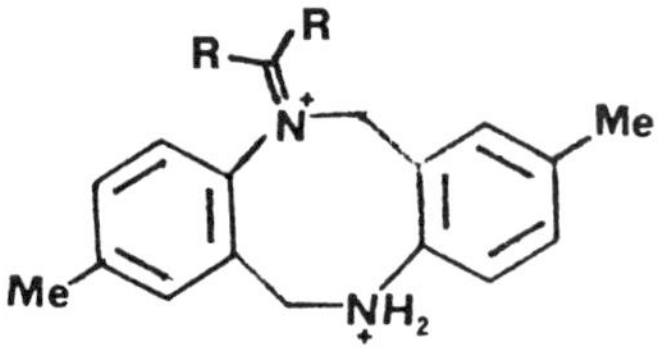

69

In an interesting early paper, von Braun and Goll reported that the reaction of bis(spiro)diazocinium ion **97** [RR = $(CH_2)_5$] (see Section II,C,1) with base resulted in opening of the diazocine ring, thus establishing the relative stabilities of six- and eight-membered cyclic amines (27CB339). The one-electron oxidation of N,N'-dimethyl-1,5-diazocine **7** (R = R^1 = Me) with potassium hexacyanoferrate(III) was found to proceed more rapidly than that of lower ring homologues and bi- and tricyclic amines. The reaction is the only one to develop a transient red color, indicative of an iminium radical (**187**) that is transannularly stabilized (76JCS(P2)1172). Similar results were obtained from the one-electron electrochemical oxidation of **7** (R = R^1 = Me), the peak potential for this compound being lower than that of *N*-methylazacyclooctane and other lower ring diaza and azahomologues (77JCS(P2)1732). Earlier, a similar type of transannular effect was invoked to explain why 7 (R = R^1 = H) reacted with *p*-fluoronitrobenzene almost 10^2 times more rapidly than did azacyclooctane (69JCS(B)544; 82AHC115).

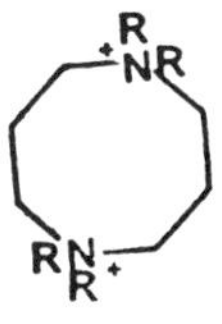

97

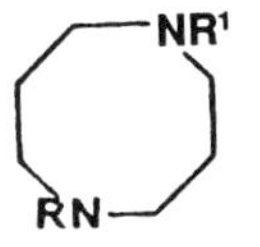

7

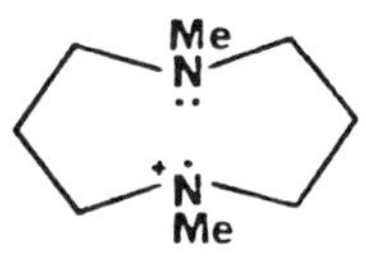

187

The mass spectrum of diazocine 35 (R = R^1 = H) (along with other cyclic *o*-phenylene molecules) showed a molecular ion peak and fragmentation characterized by a pronounced *ortho* effect giving rise to a base peak. In particular, 35 (R = R^1 = H), along with analogous eight-ring bislactams and bisthiolactones, showed, in addition to the *ortho* effect, a loss of CO and CO_2 fragments (77IJC(B)36).

There have been a number of investigations of metal complexes of 1,5-diazocine 7 (R = R^1 = H) and fused-ring analogues. Musker and Hussain found that 7 (R = R^1 = H) or "daco", formed very stable square-planar tetracoordinated bis complexes with Ni(II) and Cu(II). The best conformation for the complex is the chair–boat (**188**), which effectively shields the axial metal positions from further coordination (66IC1416; 69IC528). This insensitivity to further solvent coordination contrasted with the diazacycloheptane homologue, which was transformed from a tetra- to a pentacoordinated complex (69IC528). Legg *et al.* found that 1,5-diazacyclooctane-*N,N'*-diacetic acid 7 (R = R^1 = CH_2CO_2H) ("dacoda") forms a square-pyramidal complex with Ni(II) (with an additional ligand) (**189**), as indicated by X-ray crystallography of [Ni(dacoda)(H_2O)]$2H_2O$. Chemical and spectroscopic evidence suggested that this structure was maintained in solution. Evidently, coordination of the two tertiary amines in the eight-membered ring restricts the CH_2CO_2H groups to such an extent that the two five-membered chelate rings can form only in a plane defined by the metal ion and the two nitrogen atoms (68JA5030;

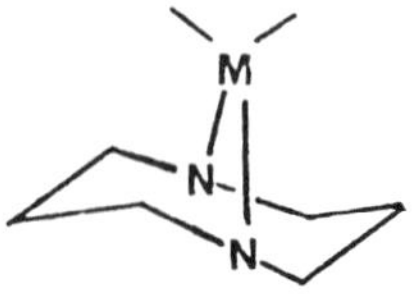

188

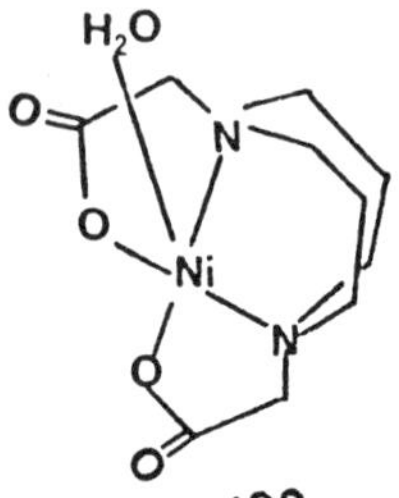

189

71JA5079). The pentacoordinated nature of M[(dacoda)(H_2O)] [M = Co(II), Ni(II)] was established by comparison of X-ray powder patterns and solid and solution spectra. The Zn(II) complex has the same structure as the Cu(II) (68JA5030; 71JA5079), Co(II), and Ni(II) complexes (72IC2344). All of these complexes are resistant to expansion of coordination in solution, due to blocking of a coordination site on the metal ion by an alkyl proton (68JA5030; 71JA5079; 72IC2344). Billo studied the solution equilibrium of M[(dacoda)(H_2O)] [M = Co(II), Ni(II), Cu(II), and Zn(II)] and suggested that coordination geometry, as well as possible hydrophobic interactions, resulted in increased acidity of coordinated water, compared with hexacoordinated complexes (75MI4). West and Legg prepared and characterized the Ni(II), Cu(II), and Zn(II) complexes of 1,5-diazacyclooctane-*N*-monoacetic acid ("dacoma") (**7**, R = H, R^1 = CH_2CO_2H). By comparison of their spectra and chemical and magnetic properties with those of M(dacoda) (**7**, R = R^1 = CH_2CO_2H) complexes (*vide supra*), it was concluded that these complexes also had a square-pyramidal, pentacoordinate structure (76JA6945). Nanosecond laser flash photolysis of the paramagnetic Ni(dacoda)(H_2O) resulted in formation of diamagnetic planar Ni(dacoda), reflecting a pentacoordinate–tetracoordinate equilibrium (78IC3378). The intended preparation of a Cu(II) complex of a 16-membered tetradentate macrocycle turned out instead to be the synthesis of a dinitrato complex of daco (**7**), the structure of which was shown by X-ray crystallography to be chair-boat (**188**) (80AX(B)452).

The dibenzodiazocines **67** (R^1–R^5 = H) were found to form square-planar tetracoordinate complex anions ML_2^{2+} [M = Cd(II), Co(II), Ni(II), Zn(II), Pd(II), Pt(II), Pt(IV)]. The diazocine **67** (R^1–R^4 = H, R^5 = Me) did not form bis complexes with first-row metals. Although the nearly planar boat conformation of the diazocine rings would seem to leave the off-planar protons free for further coordination, detailed X-ray crystallographic analysis showed a slight skewing of two of the carbon atoms alpha to nitrogen, thus causing the hydrogens to block the axial positions, and preventing five- and/or sixfold coordination (82JCS(D)2545, 82MI2).

IV. Reactions

A. EIGHT-MEMBERED RING-PRESERVED—ANNELATIONS

Reaction of diazocine **190** (see Section II,A,3) with phosgene afforded the oxadiazolone derivative **191** (79CPB2608). As mentioned earlier (see Section II,B,1), annelated diazocine **51** accompanied **50** ($R = R^1 = H$) as a product of the Schmidt reaction of 1,3-cyclohexanedione. It was postulated, on the basis of ring-opened hydrolysis products (see Section IV,D), that **51** arose from annelation of **50** ($R = R^1 = H$). [A 1,4-diazocinedione and its annelation product were also obtained (89AHC185), but it was determined that in this case annelation occurred at the intermediate seven-membered β-ketolactam stage, not on the diazocine that followed.]

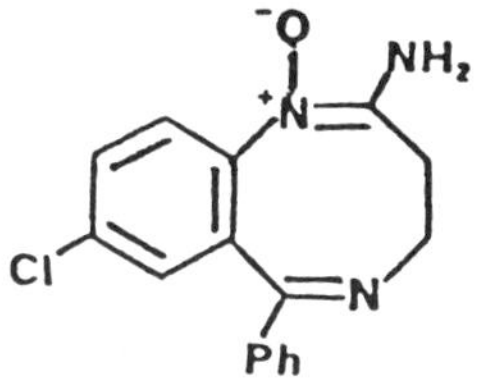

190

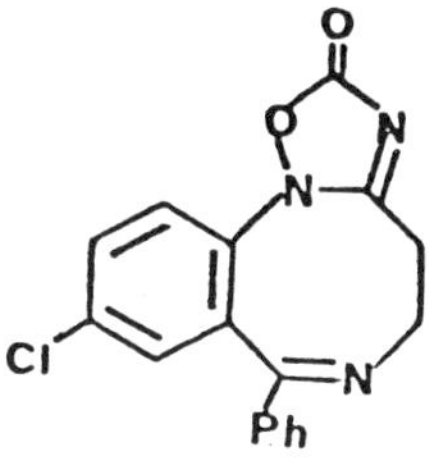

191

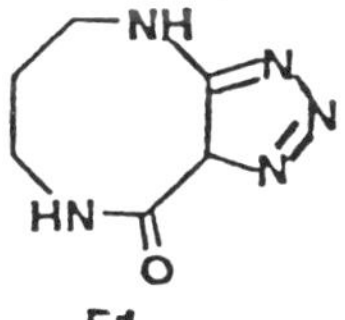

51

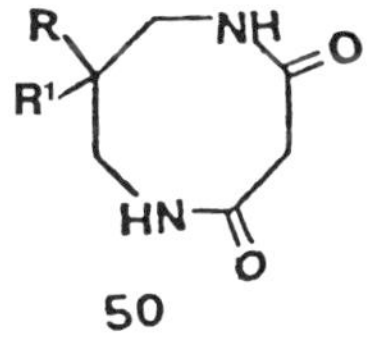

50

Diazocine **192**, when heated in pyridine, gave the dehydrated ring-opened product (**194**) [Eq. (7)]. Formation of **194** was rationalized in terms of annelation of **192** to indolodiazocine **193** (not isolated), followed by ring opening to **194** [Eq. (7)] (79CPB2618). 1,5-Benzodiazocines **106** ($R^1 = CH_2Ph$, $R^2 = H$), when treated with propiolactone and cyclized with polyphosphoric acid, afforded the tricyclic compound **195** ($R = H$). When the cyclization was carried out in the presence of acetic acid, **195** ($R = Ac$) was obtained (72MI1). Treatment of 1,5-diazocine **34** ($R^1 = R^2 = R^3 = R^4 = R^6 = R^7 = H$, $R^5 = Cl$) (see Section II,A,3) with diketene gave two products: an acetoacetyl derivative **196** (see Section IV,B) and annelated product **197**. Treatment of **196** with methanolic hydrogen chlo-

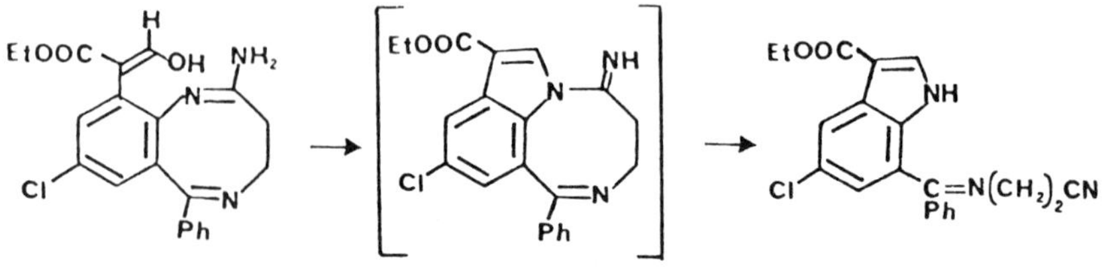

192 → [**193**] → **194** (7)

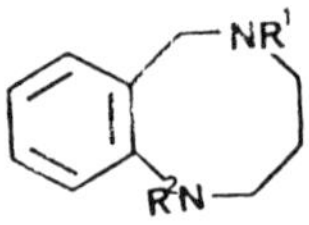

106

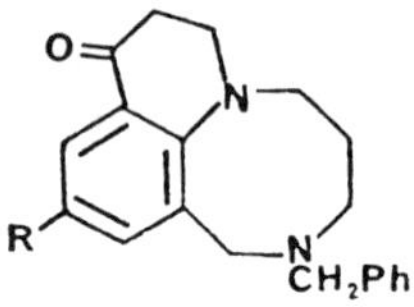

195

ride or thionyl chloride afforded annelated products **198** and **199**. Thermal dehydration of both **196** and **197** gave **198**, indicating that acyl migration took place in the reaction of **196** (79CPB2927).

There have been a number of reactions of 1,5-diazocines leading to N—N or C—C bridging. Examples of the former include the reaction of benzo-1,5-diazocines **106** ($R^1 = R^2 = H$) with aldehydes to give 1,5-meth-

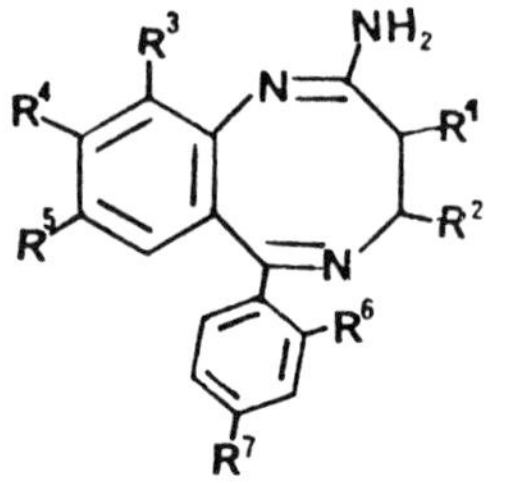

34

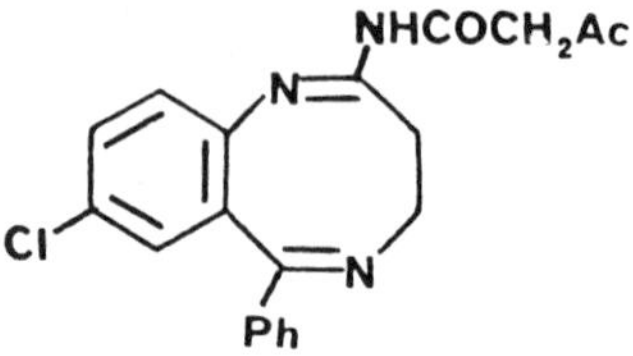

196

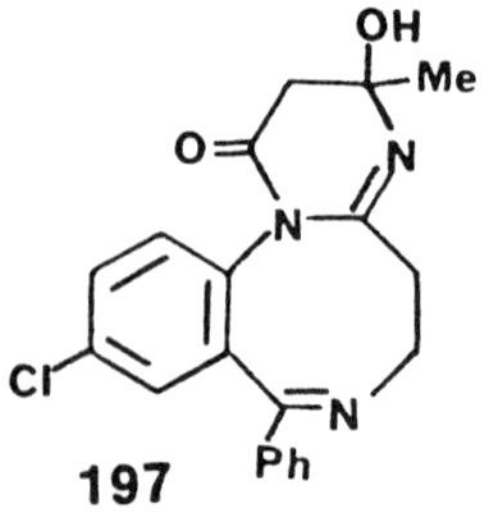

197

198

199

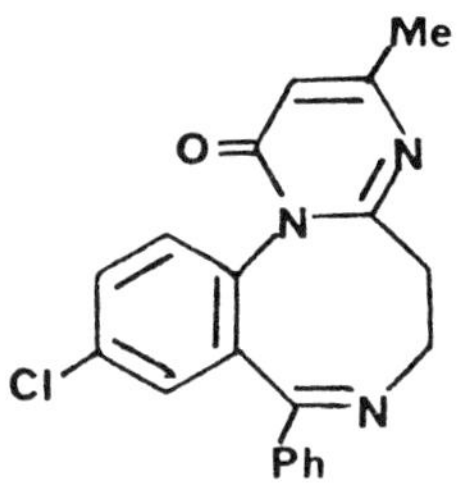

anodiazocines **200** (R = H; R^1 = H, Me, Ph) (64YZ656; 74H349); the conversion of diazocinediol **102** (R = R^1 = R^2 = H) into 1,5-diazabicyclo[3.3.1]nonane (**201**) with formaldehyde (66DIS406B); the condensation of *cis*- and *trans*-dibenzodiazocines **202** to yield the diastereomeric methano compounds **203** (66JOC3356); and imine reduction of **18** (R = R^1 = H) (see Section II,A,2), followed by reaction with formaldehyde, to afford **200** (R = Ph, R^1 = H) (77FES33).

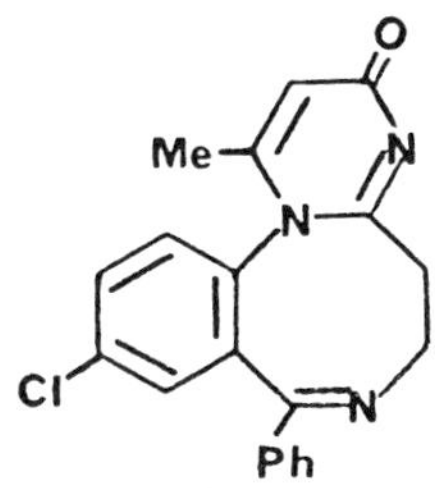

Diazocine C—C bridging was accomplished when dibenzodiazocinedii-

200

102

201

202

203

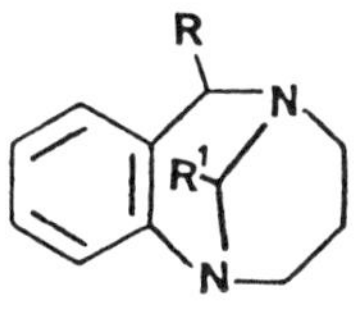

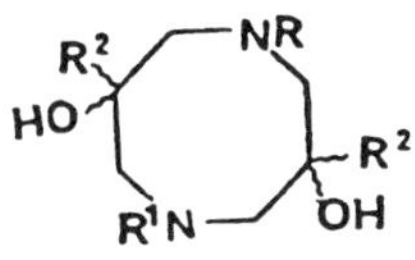

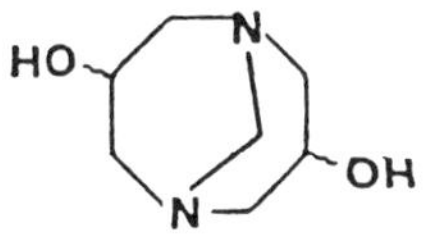

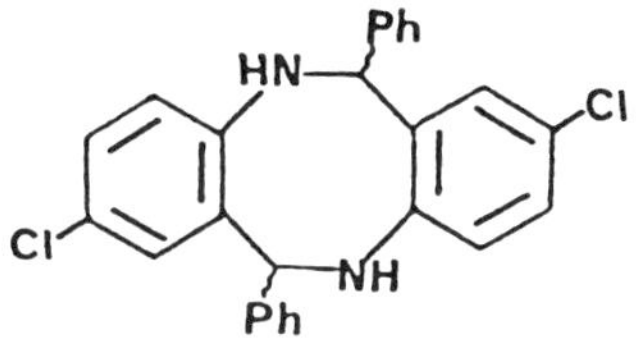

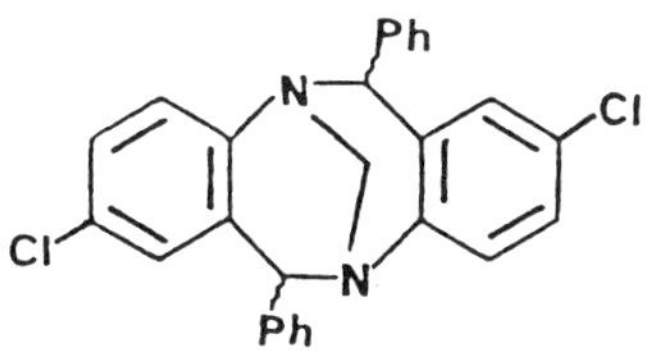

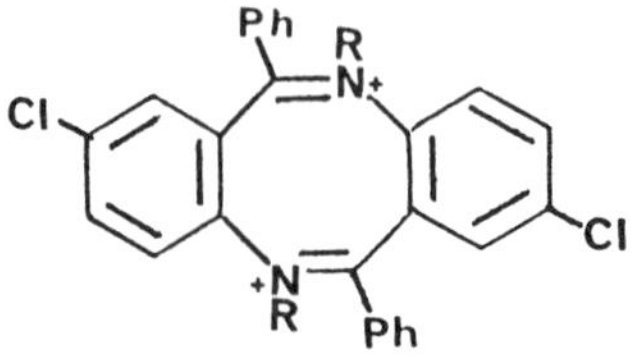

204

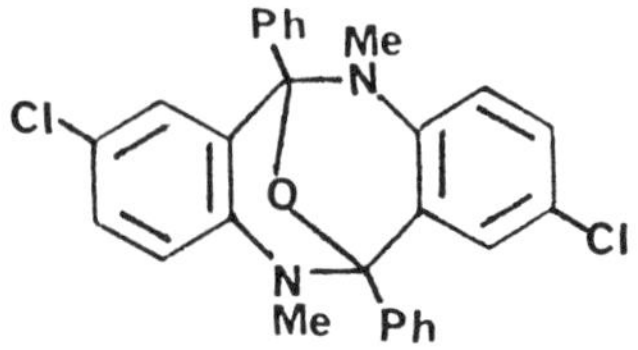

205

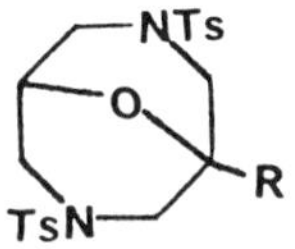

206

minium salt **204** (R = Me) was hydrolyzed in aqueous methanol to afford the 6,12-epoxy compound **205** (67JHC435). Also, a by-product in the acetylation and sulfonation of **102** (R = R^1 = Ts, R^2 = H) is the oxide **206** (R = H) (66DIS406B), whereas oxidation of **102** (R = R^1 = Ts, R^2 = H) led to formation of the hemiketal **206** (R = OH) (66DIS406B, 66JOC277).

The diazocine **59** (R^1 = R^3 = Me, R^2 = R^4 = CH$_2$OCH$_2$Ph) (see Section II,B,2) was used as the starting material for the synthesis of the interesting cage structure **207** (78TL547). The reaction of 1,5-diazacyclooctane

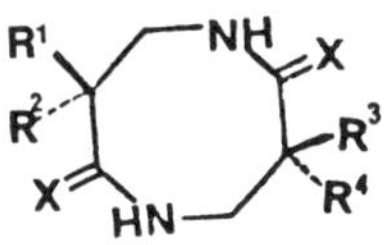

59

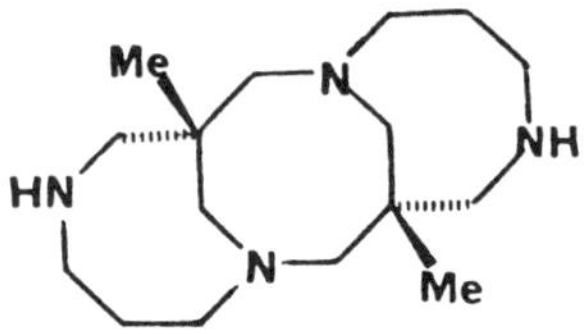

207

(**7**, R = R^1 = H) with 4-chlorobutanal and 5-chloropentanal yielded the azaazoniatricycloalkane chlorides **208** and **209** respectively. The salts were cleaved with lithium aluminum hydride to give bicyclic diamines **210** and **211**, respectively (82TL4181).

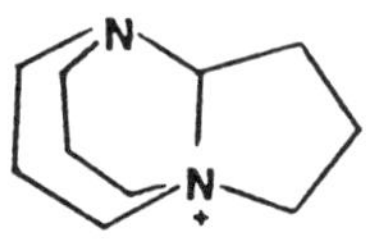

208

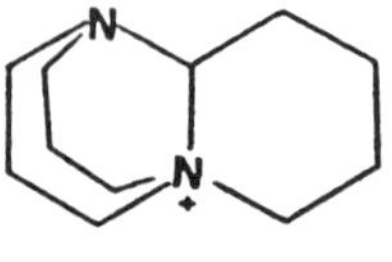

209

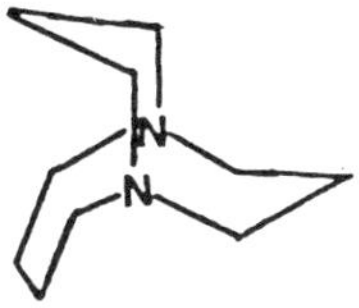

210

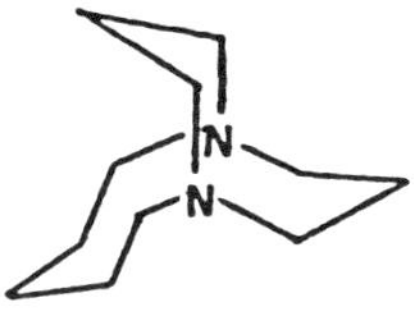

211

B. Eight-Membered Ring-Preserved—Substitutions, Functionalizations, Defunctionalizations, Oxidations, and Reductions

Substitutions include substitutions at carbon (19CB2224; 54JCS3429; 55JCS991; 67CC1077, 67JOC2425; 74ACS(B)313); N-functionalizations include alkylation (63BRP926624; 66JOC3356; 67FRP1497272; 67SZP416654; 81MI1), acylation (67JMC642, 67SZP436288; 69JCS(C)882; 77FES33; 79CPB2927), sulfonation (70FES830), and nitrosation (62BEP610039; 65SZP383392; 66JOC3356; 69JCS(C)882). Defunctionalizations include N-deacylations (54JCS3429; 77FES33), N-desulfonations (72JOC2208), and C—N to C—H transformations [i.e., hydrazide **114** (R = NHNH$_2$) to parent diazocine **114** (R = H)] (67CC1077). An example of a dehydration is the conversion of a saturated hydroxybenzodiazocine to an unsaturated one (see Section II,C,1) (68JCS(C)2862). Oxidations include formation of N-oxides (79CPB2608), conversion of a C—N bond to a C=N bond (71USP3577557), and oxidation of alcohols (66DIS406B, 66JOC277; 66JOC277). Reductions include C=N to C—N transformations (66JA1077, 66JOC3356; 67JHC435; 75M13; 79CPB2618) and carbonyl reductions (66USP3247206; 70FES830; 79JOC4473; 81FES425; 81MI1).

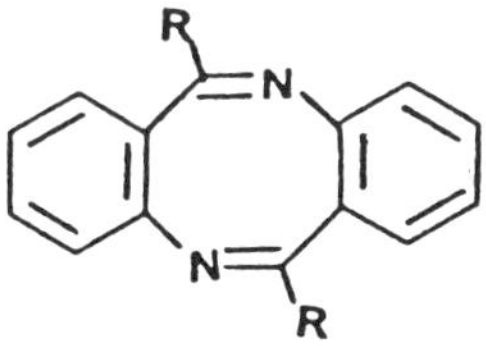

114

C. Eight-Membered Ring-Preserved—Other Reactions

In an interesting rearrangement, the bishydrobromide **212** was converted in dimethyl sulfoxide to the bis(*p*-bromophenyl) compound **213** (72JCS(P1)1209). A unique 1,5-diazocine-to-1,5-diazocine transformation has been reported in which **35** (R = R¹ = H), formed by a polyphosphoric acid-catalyzed Beckmann rearrangement (see Section II,B,1), underwent further reaction with polyphosphoric acid to give imide **79** (79MI1). (See Section II,B,4 for another interesting route to **79**.)

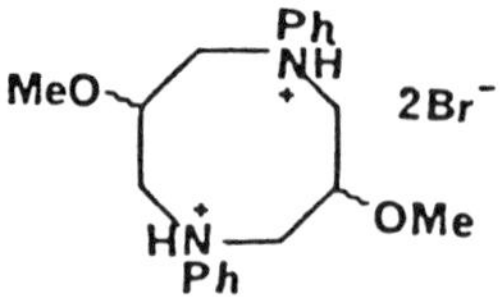

212

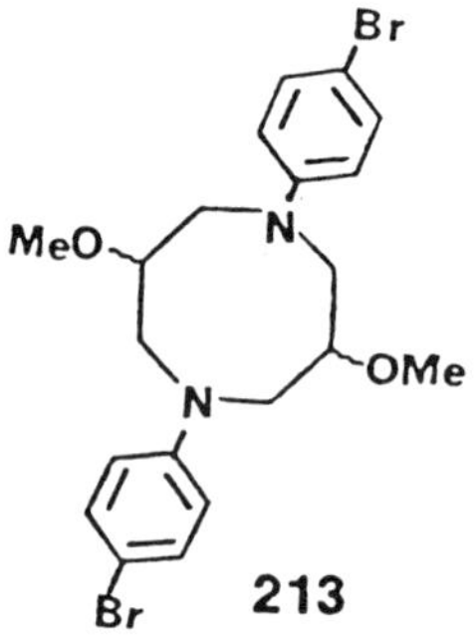

213

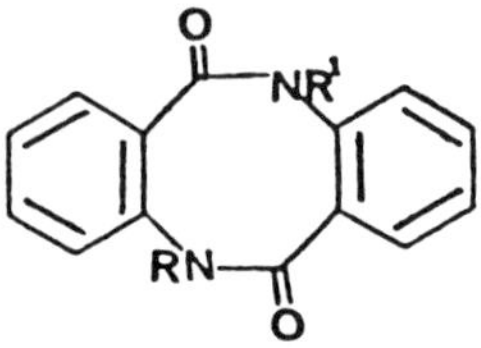

35

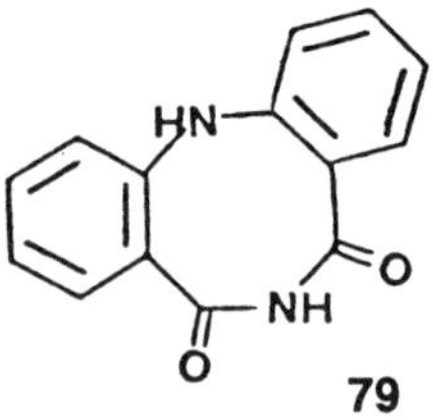

79

D. DIAZOCINE RING CONTRACTIONS AND EXPANSIONS

Milkowski and co-workers, in a series of reports, found that the acyclic benzamidoaniline **214** could be cyclized with phosphorus oxychloride to afford benzo-1,4-diazepines **215** and benzo-1,5-diazocines **46** (see Section II,A,4) (73GEP2221558; 74GEP2314993; 76GEP2520937, 76MI1,76MI2). The pharmacologically important **215** arose by ring contraction of the corresponding **46** by heating the latter with phosphorus oxychloride in refluxing 1,1,2,2-tetrachloroethane (73GEP2221558; 74GEP2314993), or by treating **46** with alkoxide or hydroxide (76GEP2520937; 86USP4595531). In another example of a ring contraction, Sternbach and co-workers treated dibenzodiazocines **138** with quaternizing agents such as dimethyl

ArNRCH₂CHOHCH₂NCOAr¹

214

R = H, Me, CH₂Ph
Ar, Ar¹ = various

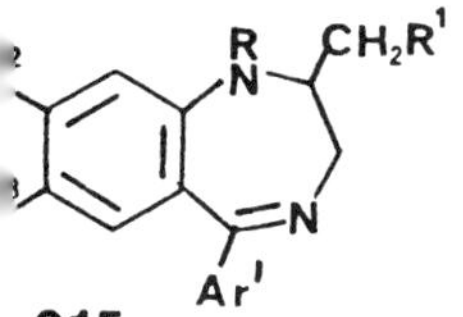

215

R = H, ME, CH₂Ph
R¹ = OR, OH, OAc, OCH₂Ph, Cl, CN, morpholinyl, NH₂, COOEt, etc.

R², R³=H, hal, NO₂, CF₃, R, OR
R², R³=OCH₂CH₂O
Ar¹= Ph; o-halC₆H₄; o-CF₃C₆H₄;
2 - furyl; 2 - thienyl; 2,4, - Cl₂C₆H₃,
3, 4 - Cl₂C₆H₃; 3 - CF₃C₆H₄; alk C₆H₄

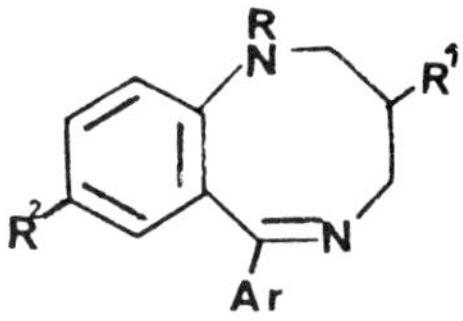

46

R=H; Me; PhCH₂
R¹ =OH; OAc; OCOPh; Cl
Ar = Ph; 2 halC₆H₄; 2, 4Cl₂C₆H₃; 3, 4, Cl₂C₆H₃; 3CF₃C₆H₄;
2CF₃C₆H₄; 3, 4, (OMe)₃C₆H₂; 2 - furyl; 2, 6 F₂C₆H₃
R² = H; Cl; Br; NO₂

sulfate to yield bis quaternary imminium compounds that are derivatives of **204** (see Section IV,A), as well as related monoquaternary salts. These salts were then split with glycine ethyl ester to afford the pharmacologically useful benzodiazepines **216** (67USP3297685).

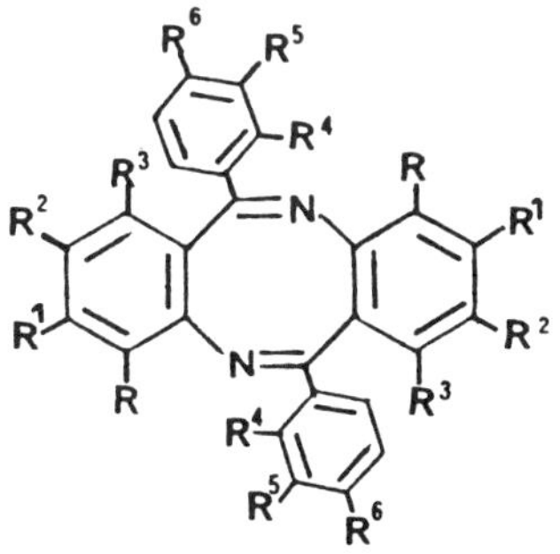

138

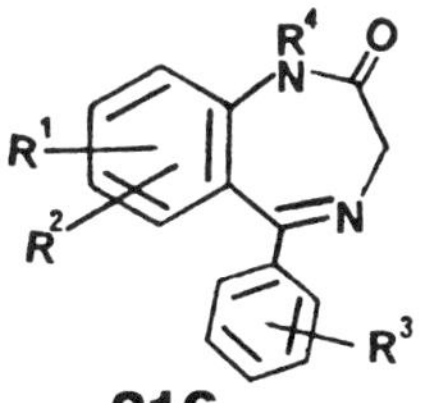

216

R¹ - R³ = H, hal, lower R, OR,
NO₂, CF₃, CN, SR, SO₂R,
R⁴ = H, lower alk, PhCH₂

Two groups found that dibenzodiazocine **35** (R = R^1 = H) undergoes ring contraction to benzoxazine **118** by heating in the presence of (67M12), or absence of (68CI(L)1809), polyphosphoric acid. Compound **35** (R = R^1 = H) at higher temperatures in polyphosphoric acid yielded a high melting compound whose elemental composition was consistent with the formula $C_{28}H_{10}N_2O_2$. It was suspected of containing a quinazolone moiety, but no definite structure was assigned (67MI2). Treatment of *cis*-**102** (R = R^1 = Ph, R^2 = H) with phosphorus tribromide yielded a mixture of isomeric piperazines (**217** and **218**) (71JOC3361). Reaction of

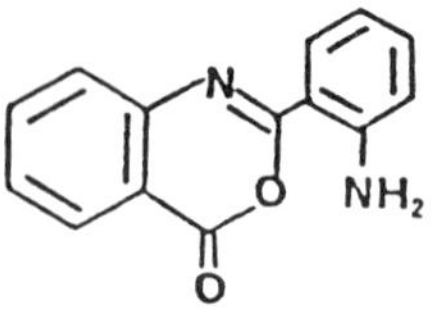

118 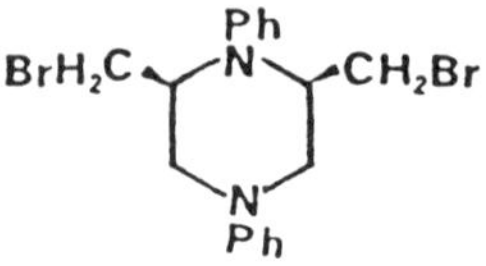   **217** 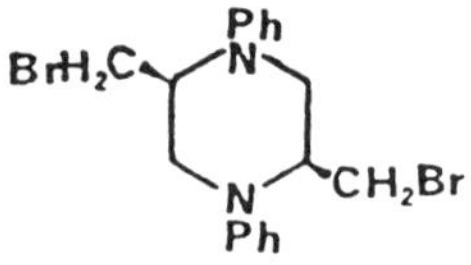   **218**

benzodiazocines **34** [R^1 = R^2 = R^4 = R^6 = R^7 = H; R^5 = Cl; R^3 = H (79CPB2589), CH_2CO_2Et (79CPB2608)] with methanolic hydrogen chloride afforded the quinazolines **219**[R = H, R^1 = Cl(79CPB2589); R = CH_2CO_2Et, R^1 = Cl(79CPB2608)], whereas the homologous benzodiazepine did not undergo ring contraction (79CPB2589).

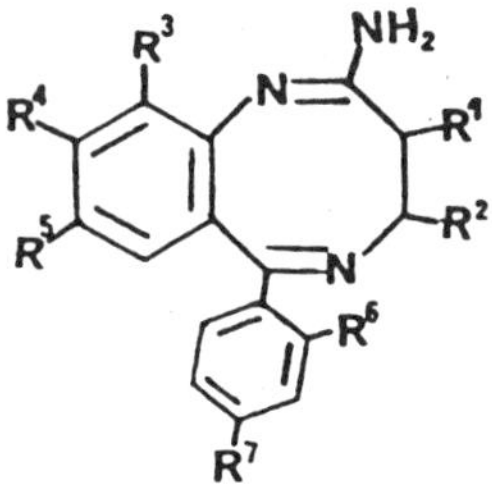

34 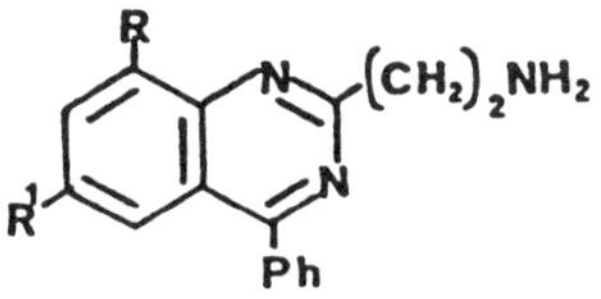   **219**

A common transformation of dibenzo-1,5-diazocines has been reductive ring contraction into isoindoloindoles. Sternbach and co-workers reported that diazocine **138** (R, R^1, R^3–R^6 = H; R^2 = Cl) reacted with zinc and acetic acid to give **220** (R = R^1 = H, R^2 = Cl), whereas the dihydro compounds **221** (R = H, Me), in the presence of sodium borohydride, afforded **220** (R = H, R^1 = Me, R^2 = Cl) and **220** (R = R^1 = H, R^2 = Cl), respectively (66JA1077, 66JOC3356). Catalytic hydrogenation of the N-methyl monoquaternary analogue of the bis salt **204** afforded **220** (R = R^1 = Me, R^2 = Cl) (67JHC435). Koch and Dessy found that electrochemical reduction of **138** (R, R^1, R^3–R^6 = H, R^2 = H, Cl) yielded **220** (R = R^1 = H, R^2 = H, Cl) (82JOC4452).

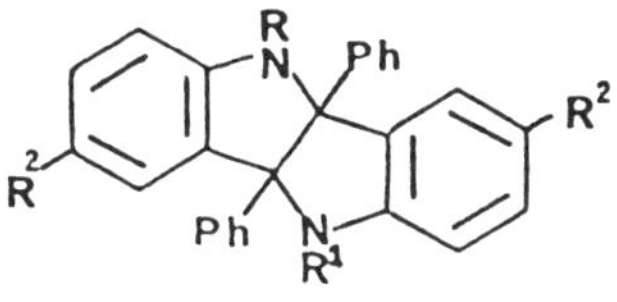

220

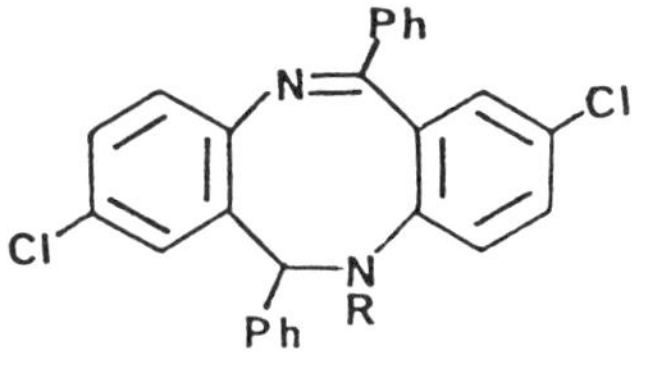

221

Fuchigami *et al.* reported the electrooxidative N—N coupling of diazocine **222**. Use of platinum, carbon, and nickel electrodes afforded the diazabicyclooctane **223** and dihydropyrazole **224**, whereas use of a silver electrode yielded only the latter compound (78CL1473; 80BCJ2040). (For the reverse of the **222** → **223** closure, see Section II,B,2).

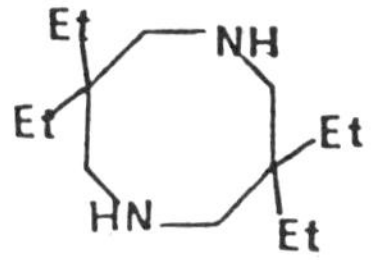

222

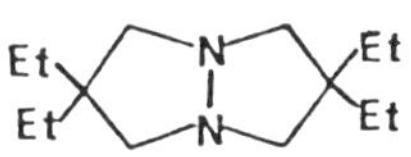

223

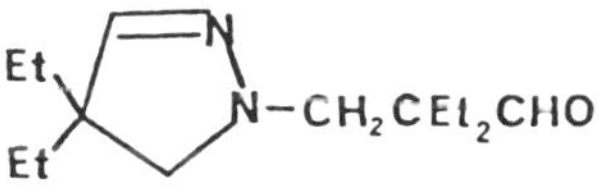

224

In the only reported 1,5-diazocine ring expansion, Crombie *et al.* utilized the *Homalium* alkaloid precursor **180** (R = R^1 = H) (see Section II,D) as a synthon for an intramolecular transamidative ring expansion in a synthesis of the 13-membered (±)-dihydroperiphylline [**225**, R = *(E)*-cinnamoyl], a natural product from *Peripterygia marginata* (87MI1). In this sequence [Eq. (8)], the diazocine **180** (R = R^1 = H) was alkylated with 1-bromo-4-chlorobutane to afford **180** [R = H, R^1 = $(CH_2)_4Cl$]. Liquid ammonia converted this compound into **180** [R = H, R^1 = $(CH_2)_4NH_2$]. Transamidation of the latter material to **225** (R = H) was accomplished using $KN(SiMe_3)_2$-THF. Selective acylation of this triazacyclotridecane with *(E)*-cinnamoyl chloride yielded (±)-dihydroperiphylline [**225**, R = *(E)*-cinnamoyl] (86TL5151). [For an alternate synthesis of dihydroperiphylline via transamidative ring expansion of a diazonine (as well as other lactam ring expansions to biologically interesting macrocyclic lactams and lactones), sec 87MI1.]

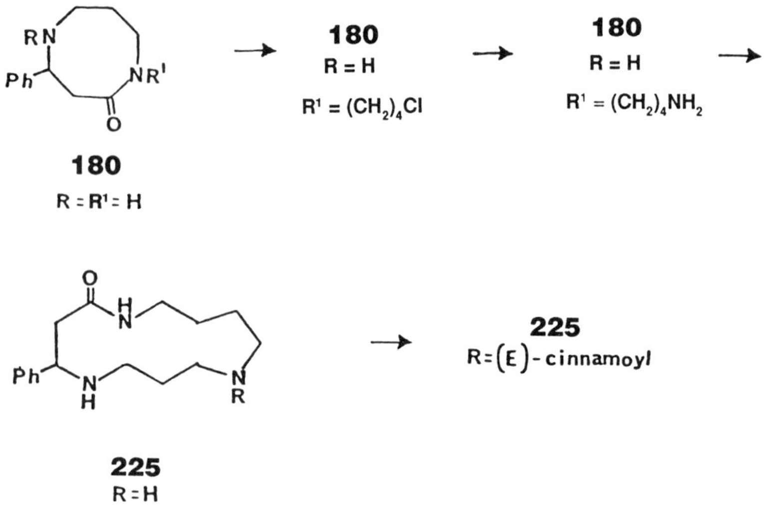

$$
\begin{array}{ccc}
\underset{\substack{\text{Ph} \\ 180 \\ R = R^1 = H}}{\overset{RN}{\bigcirc}} \underset{\text{NR}^1}{\longrightarrow} & \begin{array}{c} 180 \\ R = H \\ R^1 = (CH_2)_4 Cl \end{array} \longrightarrow & \begin{array}{c} 180 \\ R = H \\ R^1 = (CH_2)_4 NH_2 \end{array} \longrightarrow
\end{array}
\qquad (8)
$$

$$
\underset{\substack{\text{225} \\ R = H}}{\text{Ph}} \longrightarrow \begin{array}{c} 225 \\ R = (E)\text{-cinnamoyl} \end{array}
$$

E. RING OPENINGS

Early 1,5-diazocine ring openings utilized bis quaternary salts of perhydro-1,5-diazocines. Knorr and Roth treated the salt 97 (R = Me) with alkali and obtained N,N-dimethylallylamine and 2-propenyl ether (06CB1420). Two groups demonstrated the relative instability of an eight-membered heterocyclic ring relative to lower homologues such as piperidine (see Section III). Hörlein and Kneisel found that the bis salt 97 [RR = —$(CH_2)_4$—] cleaved in alkali to give N,N'-trimethylenedipiperidine (06CB1429), while von Braun and Goll reported that the bis(tetrahydroisoquinolinium) salt 98, when heated in tetrahydroisoquinoline, afforded N,N'-trimethylenebis(tetrahydroisoquinoline). This latter compound was also obtained when 98 was heated in piperidine, along with co-products N,N'-trimethylenedipiperidine and 1-(1-piperidinyl)-3-(2-tetrahydroisoquinolinyl)propane (27CB339). Müller *et al.* treated N,N'-ditosyl-1,5-diazacyclooctane (7, R = R^1 = Ts) with methyl iodide. Further reaction with sodium β-napthoxide yielded 1,3-bis(β-naphthoxy)propane (53M1206).

Sekiguchi hydrolyzed 1,5-diazocinedione 50 (R = R^1 = H) and obtained 1,3-diaminopropane and malonic acid, while the triazolo derivative 51 afforded 1,3-diaminopropane. The γ-aminobutyric acid also formed in

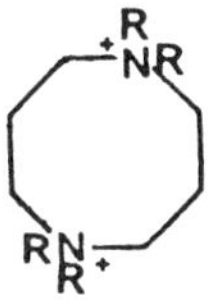

97

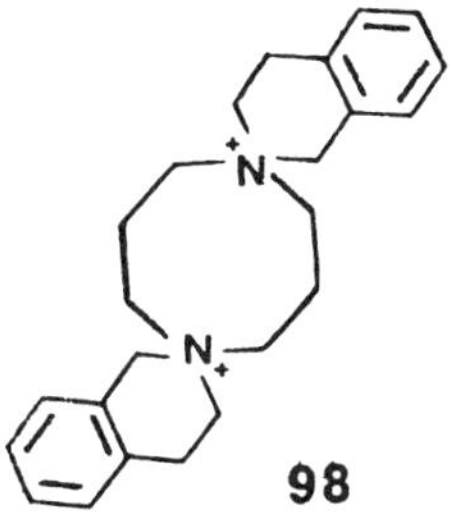

98

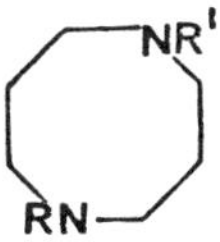

7

these hydrolyses was due to the presence of 1,4-diazocine isomers of **50** (R = R^1 = H) and **51** (65BSF691; 89AHC185). Hydrolysis of **49** afforded aspartic acid. In an unsuccessful attempt to separate the diasteriomers of **49,** the compound was heated with acetic anhydride. The resulting product, the result of a decarboxylative incorporation of 2 mol acetic acid and loss of 3 mol water, was said to be the oxazolyl aspartic anhydride **226** (54CB482).

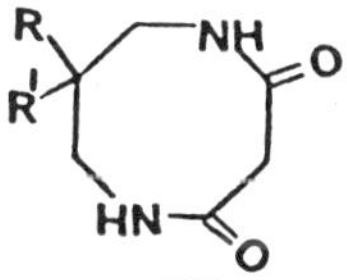

50

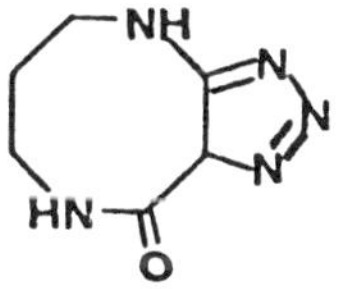

51

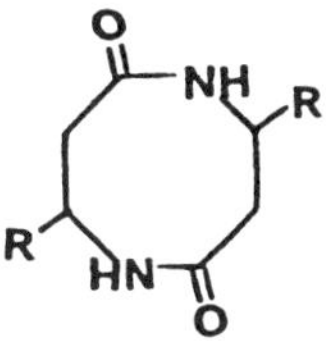

49

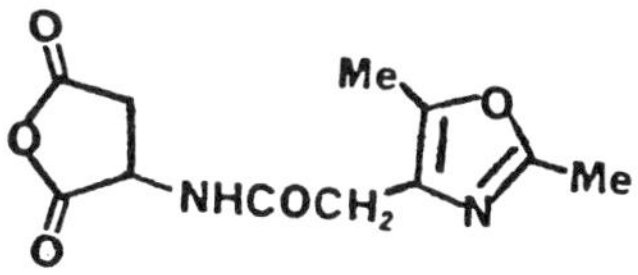

226

$$\left[NHCOCH_2CONHCH_2CR_2CH_2NHCOCH_2CONHCH_2CR_2CH_2 \right]_x$$

227

Iwakura *et al.* polymerized **50** (R = R^1 = H) in both solid and molten state to give poly(trimethylenemalonamide) (**227**, R = H). The dimethyl derivative **50** (R = R^1 = Me) could be polymerized to **227** (R = Me) in the molten state but not in the solid state (73MI1)

Ring opening of the diazocine ring in **192** to give indole **194** accompanied closure of **192** to the indole (see Section IV,A) (79CPB2618). The dinitrosamine **67** (R^1 = R^3 = H, R^4 = R^5 = NO) gave, on reduction with tin and hydrochloric acid, greater than one equivalent of 4-amino-*m*-xylene (35JA583); bislactam **35** (R = R^1 = SO$_2$Ph, Ts) reacted with Grignard reagents R^{2}MgBr to produce hydroxysulfonamides **228** (R^1 = SO$_2$Ph, Ts; R^2 = Me*p*MeC$_6$H$_4$, Ph) (49JCS384), while reduction of **35** (R = R^1 = SO$_2$Ph, Ts, SO$_2$-α-naphth) with lithium aluminum hydride afforded **228** (R^1 = SO$_2$Ph, Ts, SO$_2$-α-naphth; R^2 = H) (52JCS2435). Similar reduction of **35** (R = R^1 = Me) produced **228** (R^1 = Me, R^2 = H) (72BSF2868). Diazocine **35** (R = R^1 = 2,3Me$_2$C$_6$H$_3$) afforded, on alkaline hydrolysis, the analgesic *N*-(2',3'-dimethylphenyl)anthranilic acid (65GEP1190951). Hydrolysis of **35** (R = H, R^1 = Mes) in aqueous alcoholic alkali gave **229** (66YZ107). Solvolysis of the bis salt **204** (R = Me, Cl) in methanol gave the unstable dimethyl ketal of 2-(*N*-methylamino)-5-chlorobenzophenone, which underwent facile hydrolysis to the ketone (67JHC435), whereas aqueous hydrolysis of other benzene-substituted derivatives of these salts (**204**, R = alk, PhCH$_2$) (see Section IV,D) gave various *o*-aminobenzophenones (67USP3297685). Similar hydrolysis of **138** (R–R^6 = H) yielded 2,5-diaminobenzophenone (79M12).

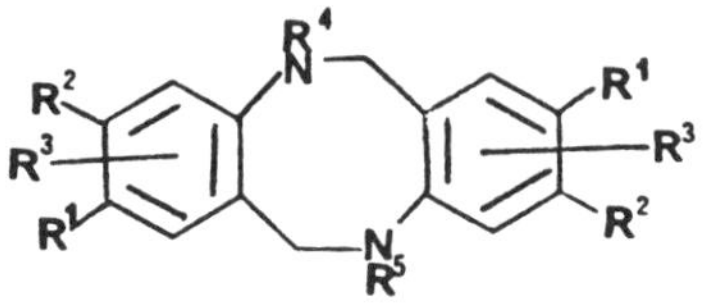

67

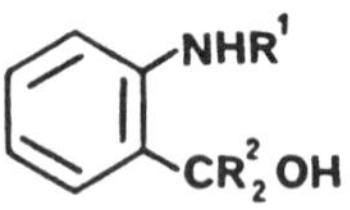

228

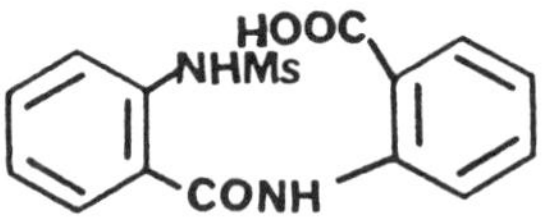

229

V. Applications

1,5-Diazacyclooctanes **7** have found application as diuretic intermediates (62BEP613566); as antitumor agents and for treatment of chorea in dogs or muscular spasms in other animals (64USP3117963); as antihistamines and/or anthelmintics (62BEP610375; 66USP3247206); as homologues of antihypertensive agents (63BRP926623); as neuromuscular blocking agents (**7**, R = alkyl, R^1 = alkylguanidine) (67SZP416654); and as general pharmaceutical intermediates (65SZP383392, 65SZP390925; 66USP3247206). Nonmedicinal uses include polymerization or vulcanization accelerators, antiknock agents, and scavengers in motor fuels and oils (62BEP610375). Also, compounds **7** (R = R^1 = chloromethyl, sulfamido) have found use as analgesics and fungicides (65USP3184447, 65USP3203952). Derivatives of 1,5-diazacyclooctane have also found use as tranquilizers (81EUP42354) and as coronary dilators [i.e., **7**, R = R^1 = 3,4,5(OMe)$_3$C$_6$H$_2$CO(CH$_2$)$_3$—] (68BRP1107470).

Diazocine **153** was an explosive analogue (83S830) (see Section 11,C,4). Lactams **146** found application as bacteriocides and as metal corrosion inhibitors (80MIP1; 81JAP81/154468, 81MI2). The isomeric 1,4-diazocines found similar uses (89AHC185). Antiinflammatory activity was found for bislactam **50** (but less than that for the homologous diazepine) (81FES135), whereas isomeric bislactam **49** polymerized to poly(β-alanine) (69MI1). The unsaturated diol **230** had applications as an insecticide and fungicide (77USP4001215). The acetate salt of **73** (R = H, R^1 = R^2 = Me) (see Section II,B,3) was polymerized with terephthaloyl chloride to give **231**. Diazocine **73** (R = H, R^1 = R^2 = Me) was also copolymerized with ethylene glycol bis(chloroformate) and similar diacids and bis(chloroformates) to prepare other polyamides and poly(esteramides).

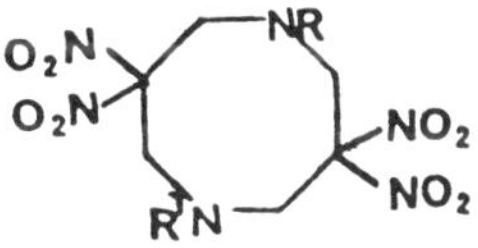

153

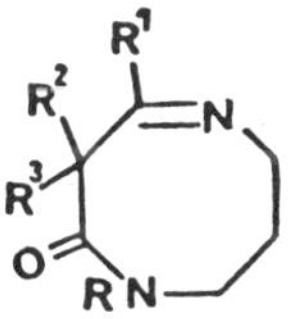

146
R = long chain
R^1, R^2, R^3, = H, Me

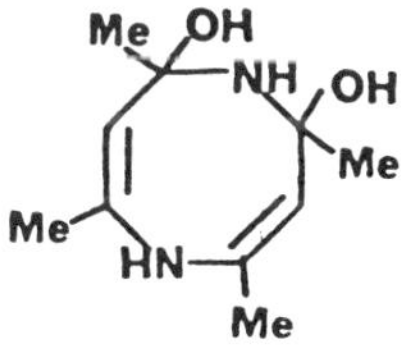

230

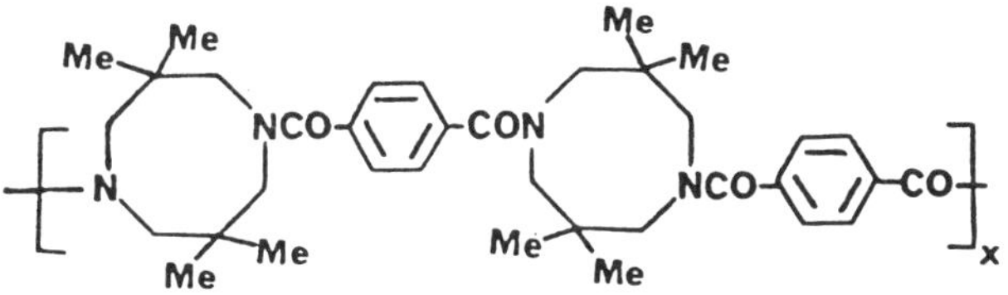

231

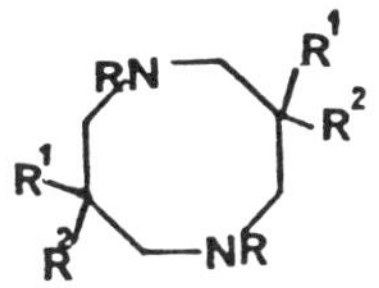

73

Polymer **231** and the other polymers have excellent thermal and photolytic stabilities, and are useful as fibers, films, coatings, moldings, adhesives, etc. (70USP3503939). The photoconductivities of the novel poly(1,5-benzodiazocines) **88** (see Section II,B,4) have been studied (78MI1). Ring opening of eight-membered bislactam **50** resulted in formation of polymer **227** (see Section IV,E). Benzo-1,5-diazocines such as **10** were found to lower blood pressure (70USP3488345). Bislactams **8** displayed unspecified pharmacological properties (70FES991). Compounds of type **12** have found use as central nervous system (CNS) depressants and/or have antiaggressive and/or other CNS effects (66USP3294782; 69JPS830; 71USP3577557; 80KFZ45), are sedatives and/or muscle relaxants and/or hypnotics and/or tranquilizers (69GEP1814332; 71JAP71/04176, 71JAP71/04177), and have anticonvulsant activity (69GEP1920908).

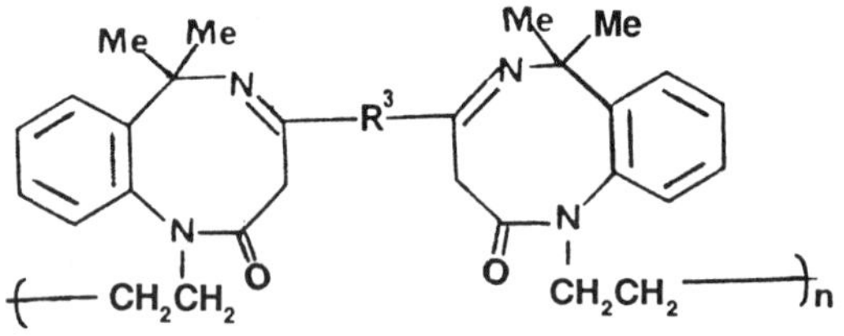

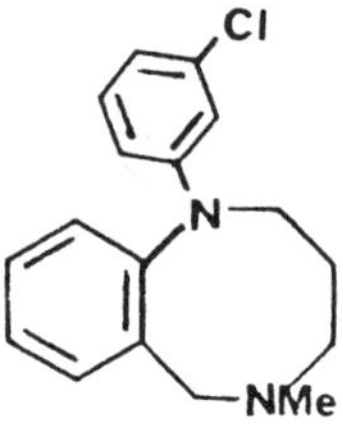

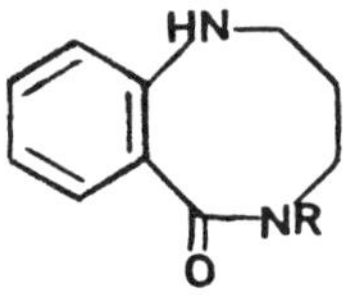

The imine N-oxides of pyrazolodiazocines **170** were antidepressants (74GEP2423642). Piperidinodiazocines **110** were found to be antihistamines (72GEP2141464). Completely unsaturated conjugated diazocines **75** are tranquilizers (76GEP2525094) and sedatives (76GEP262761). Kuwada *et al.* have prepared a series of benzo-1,5-diazocines (see Sections II,A,2, IV,A) that were useful as analgesics and/or antiinflammatory agents and/ or anticonvulsants and/or sleep-prolonging agents and/or diuretics and/or muscle relaxants. The compounds were the amidine **232** (73JAP73/99191),

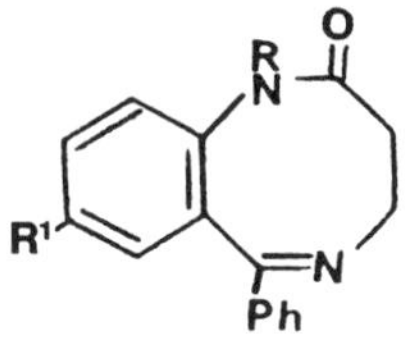

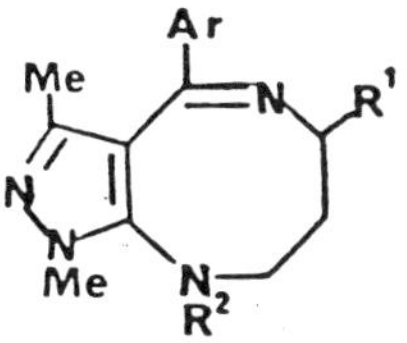

Sec. V] 1,5-DIAZOCINES 61

170
R^1 = various; R^2 = H
Ar = m - ClC_6H_4, m - FC_6H_4,
p - ClC_6H_4, m - $CF_3C_6H_4$, m - BrC_6H_4

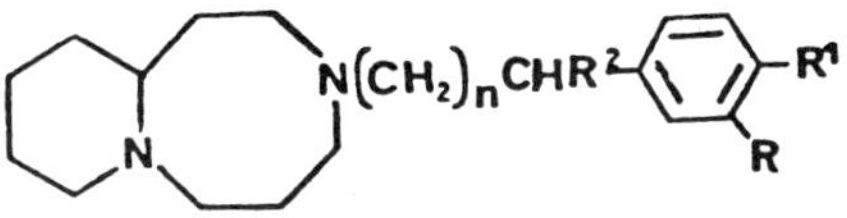

110
n = 0, 1, 2
R = H, OMe
R^1 = H, Cl, F, OMe
RR^1 = $-OCH_2O-$
R^2 = H, Ph, CH_2Ph

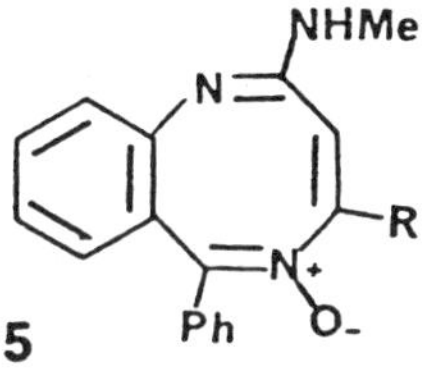

75
R = H, Me, Et, n-Pr, Ph, 2-furyl
R^1 = H, CH_2OH

the pyrimidinodiazocine **199** (73GEP2261777), the oxadiazolodiazocine **233** (74GEP2356308), and the triazolodiazocine **24** (74JAP74/85095).

The dibenzodiazocines **138** (see Section II,C,3) have been extensively studied as antigonadotropic (65MI1; 66FRP1463527, 66JMC633, 66USP3243430; 67MI1; 76ZOB1893), blood cholesterol-lowering (66USP3243430), hypotensive and psychotropic (76ZOB1893), and especially as estrogenic (contraceptive) agents (65MI1; 66USP324340; 67MI1; 70JAP70/06271; 71MI2). The lactams **23** were suitable as mild tranquilizers (67JMC642; 68USP3409608).

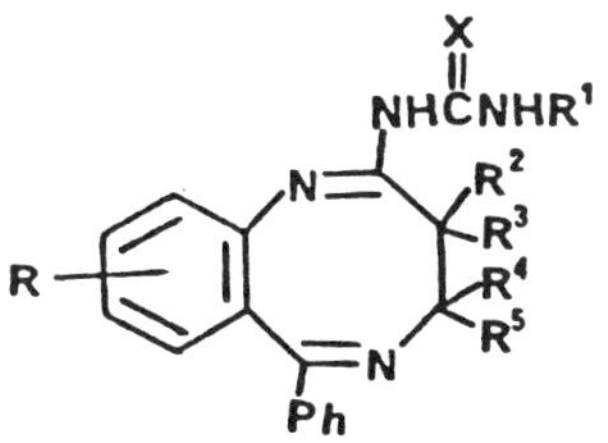

232
X = O,S
R - R^5 = various

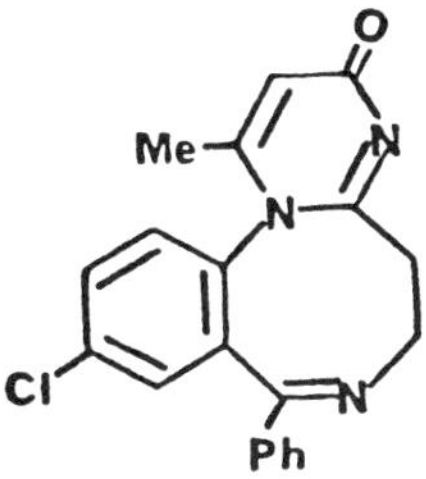

199

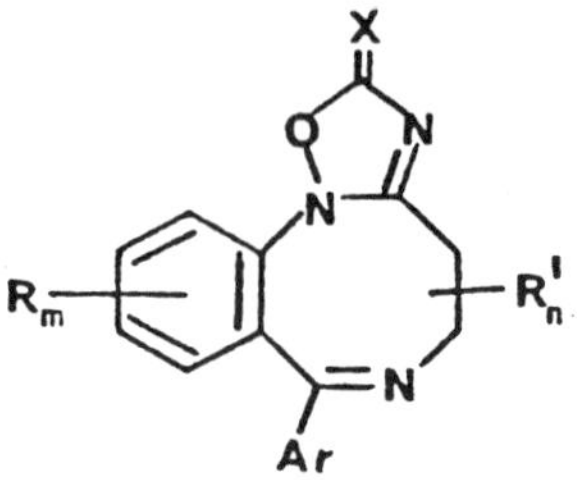

233

R, R¹Ar, m, n = various
X = O,S

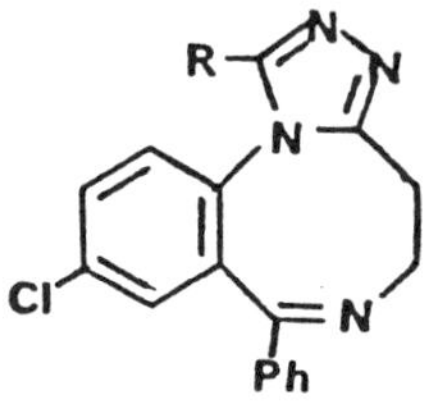

24

R = Me, CH$_2$NME$_2$

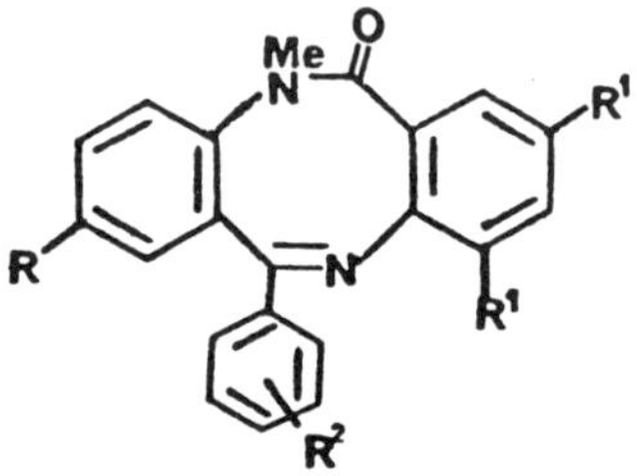

23

Acknowledgment

The author wishes to thank Mr. Daniel Tartaglia, who helped obtain and assemble the many articles used to write this review.

References

1888CB2669	S. Gabriel and J. Weiner, *Chem. Ber.* **21**, 2669 (1888).
1896CB1272	A. Sondheimer, *Chem. Ber.* **29**, 1272 (1896).
1896CB2381	S. Gabriel and R. Stelzner, *Chem. Ber.* **29**, 2381 (1896).
1898CB658	T. Posner, *Chem. Ber.* **31**, 658 (1898).
1898CB3264	W. Marckwald and A. F. von Droste-Huelshoff, *Chem. Ber.* **31**, 3264 (1898).
1899CB2031	C. C. Howard and W. Marckwald, *Chem. Ber.* **32**, 2031 (1899).
1899CB2038	C. C. Howard and W. Marckwald, *Chem. Ber.* **32**, 2038 (1899).
06CB1420	L. Knorr and P. Roth, *Chem. Ber.* **39**, 1420 (1906).
06CB1429	H. Hörlein and R. Kneisel, *Chem. Ber.* **39**, 1429 (1906).
09LA101	G. Schroeter and O. Eisleb, *Justus Liebigs Ann. Chem.* **367**, 101 (1909).
19CB2224	G. Schroeter, *Chem. Ber.* **52**, 2224 (1919).
23RC251	L. Klisiecki and E. Sucharda, *Rocz. Chem.* **3**, 251 (1923) [*CA* **19**, 72 (1925)].

27CB339	J. von Braun and O. Goll, *Chem. Ber.* **60B**, 339 (1927).
27MI1	F. P. Mazza and M. Ferrajolo, *Rend. Accad. Sci. Fis. Mat. Naples* [3] **33**, 229 (1927) [*CA* **23**, 2957 (1929)].
33G20	A. Giacalone, *Gazz. Chim. Ital.* **63**, 20 (1933).
33G764	A. Giacalone, *Gazz. Chim. Ital.* **63**, 764 (1933).
35G120	A. Giacalone, *Gazz, Chim. Ital.* **65**, 120 (1935).
35JA583	M. A. Spielman, *J. Am. Chem. Soc.* **57**, 583 (1935), and reference cited therein.
43JA29	E. L. Buhle, A. M. Moore, and F. Y. Wiselogle, *J. Am. Chem. Soc.* **65**, 29 (1943)
49JCS384	A. Mustafa and A. M. Gad, *J. Chem. Soc.*, 384 (1949).
51DOK1073	N. S. Dokunikhin, L. A. Gaeva, and I. D. Kraft, *Dokl. Akad. Nauk SSSR* **81**, 1073 (1951) [*CA* **46**, 8053 (1952)].
51ZOB268	A. P. Terent'ev, A. N. Kost, and K. I. Chursina, *Zh. Obshch. Khim.* **21**, 268 (1951) [*CA* **45**, 7008 (1951)].
52JCS2435	A. Mustafa, *J. Chem. Soc.*, 2435 (1952).
52RC182	T. Urbanski and E. Lipska, *Rocz. Chem.* **26**, 182 (1952) [*CA* **50**, 7820 (1956)].
53M1206	A. Müller, E. Funder-Fritzsche, W. Koner, and E. Rintersbacher-Wlasak, *Monatsh. Chem.* **84**, 1206 (1953).
54CB482	F. Weygand and H.-J. Dietrich, *Chem. Ber.* **87**, 482 (1954).
54HCA701	H. de Diesbach and M. Frossard, *Helv. Chim. Acta* **37**, 701 (1954).
54JCS3429	F. C. Cooper and M. W. Partridge, *J. Chem. Soc.*, 3429 (1954).
54MI1	T. Urbanski, *Chem. Technol.* **8**, 442 (1954) [*CA* **49**, 1536c (1955)].
54ZOB163	I. N. Nazarov and G. A. Shvekhgeimer, *Zh. Obshch. Khim.* **24**, 163 (1954) [*CA* **49**, 3034 (1955)].
55JCS991	F. C. Cooper and M. W. Partridge, *J. Chem. Soc.*, 991 (1955).
55RC379	T. Urbanski and H. Piotrowska, *Rocz. Chem.* **29**, 379 (1955) [*CA* **50**, 4967 (1956)].
56MI1	J. K. N. Jones, R. Kolinski, H. Piotrowska, and T. Urbanski, *Bull. Acad. Pol. Sci., Cl. 3* **4**, 521 (1956) [*CA* **51**, 8765f (1957)].
56MI2	J. Iwao and K. Tomio, *J. Pharm. Soc. Jpn.* **76**, 808 (1956) [*CA* **51**, 1177 (1957)].
57JA1983	L. Bauer and S. V. Miarka, *J. Am. Chem. Soc.* **79**, 1983 (1957).
57JCS1900	H. N. Rydon, N. H. P. Smith, and D. Williams, *J. Chem. Soc.*, 1900 (1957).
57RC101	J. K. N. Jones, R. Kolinski, H. Piotrowska, and T. Urbanski, *Rocz. Chem.* **31**, 101 (1957) [*CA* **51**, 14718a (1957)].
58CB1982	H. Stetter and H. Spangenberger, *Chem. Ber.* **91**, 1982 (1958).
58JCS205	V. Oakes and H. N. Rydon, *J. Chem. Soc.*, 205 (1958).
58JCS2319	R. Kolinski, H. Piotrowska, and T. Urbanski, *J. Chem. Soc.*, 2319 (1958).
59JCS2396	K. Butler and M. W. Partridge, *J. Chem. Soc.*, 2396 (1959).
59YZ1219	T. Yamazaki and M. Nagata, *Yakugaku Zasshi* **79**, 1219 (1959) [*CA* **54**, 4588b (1960)].
61CPB834	H. Watanabe, S. Kuwata, and S. Koyama, *Chem. Pharm. Bull.* **9**, 834 (1961).
61GEP1104516	C. A. Grob and O. Schier, Ger. Pat. 1,104,516 (1961) [*CA* **57**, 7288(1962)].
62BEP610039	Sandoz, Ltd., Belg. Pat. 610,039 (1962) [*CA* **58, 1469 (1963)**].

62BEP610375	Sandoz, Ltd., Belg. Pat. 610,375 (1962) [*CA* **58**, 2462 (1963)].
62BEP613566	Sandoz, Ltd., Belg. Pat. 613,566 (1962) [*CA* **58**, 1481 (1963)].
62CB783	M. Rothe and R. Timler, *Chem. Ber.* **95**, 783 (1962).
62JCS2545	M. W. Partridge, H. J. Vipond, and J. A. Waite, *J. Chem. Soc.*, 2545 (1962).
63BRP926623	CIBA Ltd., Br. Pat. 926,623 (1963) [*CA* **59**, 11526 (1963)].
63BRP926624	CIBA Ltd., Br. Pat. 926,624 (1963) [*CA* **59**, **11541 (1963)**].
63CB1868	H. Brederick, F. Effenberger, E. Henseleit, and E. H. Schweizer, *Chem. Ber.* **96**, 1868 (1963).
63HCA2970	M. Haering, *Helv. Chim. Acta* **46**, 2970 (1963).
64FRP1378964	W. L. Yost and R. B. Margerison, Fr. Pat 1,378,964 (1964) [*CA* **62**, 13159h (1965)].
64T207	T. Urbanski, D. Gurne, R. Kolinski, H. Piotrowska, A. Janczyk, B. Serafin, M. Szretter-Szmid, and M. Witanosaki, *Tetrahedron* **20**, Suppl. 1, 207 (1964).
64TL33	R. Barchet and K. W. Merz, *Tetrahedron Lett.*, 33 (1964).
64USP3117963	F. F. Blicke, U. S. Pat. 3,117,963 (1964) [*CA* **60**, 9298e (1964)].
64YZ656	S. Shiotani and K. Mitsuhashi, *Yakugaku Zasshi* **84**, 656 (1964) [*CA* **61**, 10685 (1964)].
65BSF691	H. Sekiguchi, *Bull. Soc. Chim. Fr.*, 691 (1965).
65CCC445	J. O. Jilek, J. Pomykarek, E. Svatek, V. Seidlova, M. Rajsner, K. Pelz, B. Hoch, and M. Protiva, *Collect. Czech. Chem. Commun.* **30**, 445 (1965) [*CA* **63**, 4257 (1965)].
65GEP1190951	R. A. Scherrer, Ger. Pat. 1,190,951 (1965) [*CA* **63**, 4209 (1965)].
65MI1	G. W. Duncan, S. C. Lyster, and J. B. Wright, *Proc. Soc. Exp. Biol. Med.* **120**, 725 (1965) [*CA* **64**, 8603 (1966)].
65SZP383392	E. Jucker, A. J. Lindenmann, and J. Gmuender, Swiss Pat. 383,392 (1965) [*CA* **62**, 16283 (1965)].
65SZP390925	R. P. Mull, Swiss Pat. 390,925 (1965) [*CA* **63**, 18130 (1960)].
65SZP393342	E. Jucker, A. J. Lindenann, and J. Gmuender, Swiss Pat. 393,342 (1965) [*CA* **64**, 5122 (1966)].
65USP3184447	L. A. Paquette, U. S. Pat. 3,184,447 (1965) [*CA* **63**, 8386 (1965)].
65USP3203952	L. A. Paquette, U. S. Pat. 3,203,952 (1965) [*CA* **63**, 18129 (1965)].
66CC427	R. B. Roy and G. A. Swan, *Chem. Commun.*, 427 (1966).
66DIS406B	G. R. Gapski, *Diss. Abstr.* **27**, 406B (1966).
66FRP1463527	Hoffman-La Roche and Co., A.-G., Fr. Pat. 1,463,527 (1966) [*CA* **68**, 13013 (1968)].
66IC1416	W. K. Musker and M. S. Hussain, *Inorg. Chem.* **5**, 1416 (1966).
66JA1077	W. Metlesics and L. H. Sternbach, *J. Am. Chem. Soc.* **88**, 1077 (1966).
66JHC527	H. Gurien, D. H. Malarek, and A. I. Rachlin, *J. Heterocycl. Chem.* **3**, 527 (1966).
66JMC633	W. Metlesics, T. Resnick, G. Silverman, R. Tavares, and L. H. Sternbach, *J. Med. Chem.* **9**, 633 (1966).
66JOC277	W. W. Paudler, G. R. Gapski, and J. M. Barton, *J. Org. Chem.* **31**, 277 (1966).
66JOC3356	W. Metlesics, R. Tavares, and L. H. Sternbach, *J. Org. Chem.* **31**, 3356 (1966).
66MI1	H. Mikolakewska and A. Kotelko, *Acta Pol. Pharm.* **23**, 425 (1966) [*CA* **66**, 85776 (1967)].
66NEP6602819	Upjohn Co., Neth. Pat. 6,602,819 (1966) [*CA* **66**, 37972 (1967)].

66USP3236837	V. R. Gaertner, U. S. Pat. 3,236,837 (1966) [*CA* **64**, 12706 (1966)].
66USP3243430	W. Metlesics and L. H. Sternbach, U. S. Pat. 3,243,430 (1966) [*CA* **64**, 17622 (1966)].
66USP3247206	W. L. Yost and R. B. Margerison, U. S. Pat. 3,247,206 (1966) [*CA* **64**, 19641 (1966)].
66USP3294782	T. S. Sulkowski, U. S. Pat. 3, 294,782 (1966) [*CA* **66**, 46443 (1967)].
66YZ107	T. Kametani, S. Asagi, S. Nakamura, M. Sato, N. Wagatsuma, and S. Takano, *Yakugaku Zasshi* **86**, 107 (1966) [*CA* **64**, 19522 (1966)].
67CC1077	W. W. Paudler and A. G. Zeiler, *J. C. S. Chem. Commun.*, 1077 (1967).
67CI(L)1644	W. V. Farrar, *Chem. Ind. (London)*, 1644 (1967).
67FRP1497272	K. Thomae, Fr. Pat. 1,497,272 (1967) [*CA* **70**, 20099 (1969)].
67JHC435	W. Metlesics, T. Anton, and L. H. Sternbach, *J. Heterocycl. Chem.* **4**, 435 (1967).
67JMC642	J. G. Topliss, E. P. Shapiro, and R. I. Taber, *J. Med. Chem.* **10**, 642 (1967).
67JOC2425	W. W. Paudler and A. G. Zeiler, *J. Org. Chem.* **32**, 2425 (1967).
67MI1	A. Boris, *Arch. Int. Pharmacodyn. Ther.* **166**, 374 (1967) [*CA* **67**, 18154 (1967)].
67M12	M. Kurihara and N. Yoda, *Makromol. Chem.* **107**, 112 (1967).
67NKZ102	M. Kawai, *Nippon Kagaku Zasshi* **88**, 102 (1967) [*CA* **67**, 73599 (1967)].
67SZP416654	E. Jucker, A. J. Lindenmann, and J. Gmuender, Swiss Pat. 416,654 (1967) [*CA* **67**, 3107 (1967)].
67SZP436288	A. J. Lindenmann and C. Brueschweiler, Swiss Pat. 436,288 (1967) [*CA* **69**, 10473 (1968)].
67T2123	V. Gaertner, *Tetrahedron* **23**, 2123 (1967).
67USP3297685	J. Hellerbach, W. Metlesics, and L. H. Sternbach, U. S. Pat. 3,297,685 (1967) [*CA* **66**, 55531 (1967)].
68BRP1107470	Asta-Werke A.-G. Chemische Fabrik, Br. Pat. 1,107,470 (1968) [*CA* **69**, 59297 (1968)].
68CC1376	G. A. Swan, *J. C. S. Chem. Commun.*, 1376 (1968).
68CI(L)1809	G. K. J. Gibson and A. S. Lindsey, *Chem. Ind. (London)*, 1809 (1968), and references cited therein.
68CR(C)37	M. Pais, G. Rattle, R. Sarfati, and F.-X. Jarreau, *C. R. Hebd. Seances Acad. Sci., Ser. C* **226**, 37 (1968).
68CR(C)82	M. Pais, G. Rattle, R. Sarfati, and F.-X. Jarreau, *C. R. Hebd. Seances Acad. Sci., Ser. C* **267**, 82 (1968).
68JA5030	J. I. Legg, D. O. Nielson, D. L. Smith, and M. L. Larson, *J. Am. Chem. Soc.* **90**, 5030 (1968).
68JCS(C)2862	B. J. Auret, M. F. Grundon, and I. T. McMaster, *J. Chem. Soc. C*, 2862 (1968).
68JOC3946	M. W. Williams, *J. Org. Chem.* **33**, 3946 (1968).
68TL513	D. S. Kemp, J. C. Charbala, and S. A. Mason, *Tetrahedron Lett.*, 513 (1968).
68TL547	D. S. Kemp, R. V. Punzar, and J. G. Chabala, *Tetrahedron Lett.*, 547 (1968).
68USP3409608	J. G. Topliss, U. S. Pat. 3,409,608 (1968) [*CA* **70**, 57921 (1969)].
69BCJ2066	K. Isagawa, T. Ishiwaka, M. Kawai, and Y. Yasaburo, *Bull. Chem. Soc. Jpn.* **42**, 2066 (1969).
69GEP1814332	H. Yamamoto, S. Inaba, T. Okamoto, K. Ishizumi, M. Yamamoto, I. Muruyama, T. Hirohashi, K. Mori, and T. Kobayashi, Ger. Pat. 1,814,332 (1969) [*CA* **72**, 21727 (1970)].

69GEP1920908	M. E. Derieg, R. I. Fryer, and L. H. Sternbach, Ger. Pat. 1,920,908 (1969) [*CA* **72,** 31862 (1970)].
69IC528	W. K. Musker and M. S. Hussain, *Inorg. Chem.* **8,** 528 (1969).
69JCS(B)544	A. Fischer, R. E. J. Hutchinson, R. D. Topsom, and G. J. Wright, *J. Chem. Soc. B,* 544 (1969).
69JCS(C)882	N. J. Harper and J. M. Sprake, *J. Chem. Soc. C,* 882 (1969).
69JOC179	M. E. Derieg, R. M. Schweininger, and R. I. Fryer, *J. Org. Chem.* **34,** 179 (1969).
69JOC183	M. Denzer and H. Ott, *J. Org. Chem.* **34,** 183 (1969).
69JPS830	M. Steinman and J. G. Topliss, *J. Pharm. Sci.* **58,** 830 (1969).
69MI1	Y. Iwakura, K. Uno, M. Akiyama, and K. Haga, *J. Polym. Sci., Part A-1* **7,** 657 (1969).
69TL4497	A. B. Evnin, A. Y. Lam, J. J. Maher, and J. J. Blyskal, *Tetrahedron Lett.,* 4497 (1969).
70FES830	F. Gatta and R. Landi-Vittory, *Farmaco, Ed. Sci.* **25,** 830 (1970) [*CA* **74,** 53753 (1971)].
70FES991	F. Gatta and R. Landi-Vittory, *Farmaco, Ed. Sci.* **25,** 991 (1970) [*CA* **75,** 35982 (1971)].
70GEP1802207	P. Wegener, Ger. Pat. 1,802,207 (1970) [*CA* **73,** 25549 (1970)].
70GEP1920908	M. E. Derieg, R. I. Fryer, and L. H. Sternbach, Ger. Pat. 1,920,908 (1969) [*CA* **72,** 31862 (1970)].
70JAP70/06270	H. Moriyama, H. Yamamoto, and H. Nagata, Jpn. Pat. 70/06270 (1970) [*CA* **72,** 132819 (1970)].
70JAP70/06271	H. Moriyama, H. Yamamoto, and H. Nagata, Jpn. Pat. 70/06271 (1970) [*CA* **73,** 3948 (1970)].
70JAP70/06272	H. Moriyama, H. Yamamoto, and H. Nagata, Jpn. Pat. 70/06272 (1970) [*CA* **72,** 132817 (1970)].
70MI1	M. P. Gupta and S. M. Prasad, *Indian J. Phys.* **44,** 141 (1970).
70MI2	H. Kutlu, *Istanbul Univ. Eczacilik Fak. Mecm.* **6,** 45 (1970) [*CA* **75,** 35980 (1971)].
7OUSP3488345	F. J. Petracek, U. S. Pat. 3,488,345 (1970) [*CA* **72,** 55535 (1970)].
70USP3503939	M. W. Williams, U. S. Pat. 3,503,939 (1970) [*CA* **72,** 122798 (1970)].
71BSF255	M. Pais, R. Sarfati, and F.-X. Jarreau, *Bull. Soc. Chim. Fr.,* 255 (1971).
71CR(C)1728	M. Pais, R. Sarfati, F.-X. Jarreau, and R. Goutarel, *C. R. Hebd. Seances Acad. Sci., Ser. C* **272,** 1728 (1971).
71GEP2024472	H. G. Greve, O. Graewinger, H. Kindler, P. E. Nitz, K. Resog, H. von Brachel, and H. Bender, Ger. Pat. 2,024,472 (1971) [*CA* **76,** 127034 (1972)].
71GEP2050344	M. K. Eberle and W. J. Houlihan, Ger. Pat. 2,050,344 (1971) [*CA* **75,** 49157 (1971)].
71JA1543	R. A. Olofson, R. K. Vander Meer, and S. Stournas, *J. Am. Chem. Soc.* **93,** 1543 (1971).
71JA5079	J. I. Legg, D. O. Nielson, D. L. Smith, and M. L. Larson, *J. Am. Chem. Soc.* **93,** 5079 (1971).
71JAP71/04176	H. Yamamoto, S. Inaba, T. Okamoto, T. Hirohashi, K. Ishizuki, M. Yamamoto, I. Maruyama, K. Mori, and T. Kobayashi, Jpn. Pat. 71/04176 (1971) [*CA* **74,** 125751 (1971)].
71JAP71/04177	H. Yamamoto, S. Inaba, T. Okamoto, T. Hirohashi, K. Ishizumi, M. Yamamoto, I. Maruyama, K. Mori, and T. Kobayashi, Jpn. Pat. 71/04177 (1971) [*CA* **74,** 141912 (1971)].

71JAP71/10861	M. Murakami, N. Inukai, and K. Nakano, Jpn. Pat. 71/10861 (1971) [*CA* **75**, 36172 (1971)].
71JCS(C)2880	G. A. Swan, *J. Chem. Soc.* C, 2880 (1971).
71JCS(D)372	D. Lloyd and K. Hideg, *J. Chem. Soc.* D, 372 (1971).
71JOC642	S. C. Pakrashi, *J. Org. Chem.* **36**, 642 (1971).
71JOC3361	D. A. Nelson, J. J. Worman, and B. Keen, *J. Org. Chem.* **36**, 3361 (1971).
71KGS1097	B. L. Moldaver and M. P. Papirnik, *Khim. Geterotsikl. Soedin.* **7**, 1097 (1971) [*CA* **76**, 59519 (1972).
71MI1	H. Mikolajewska, *Acta Pol. Pharm.* **28**, 253 (1971) [*CA* **76**, 34230 (1972)].
71MI2	L. Terenius, *Steroids* **17**, 653 (1971).
71MI3	B. M. Bolotin, M. V. Loseva, and I. P. Shepilov, *Tr. Vses. Nauchno-Issled Inst. Khim. Reakt. Osobo Chist. Khim. Veshchestv*, 104 (1971) [*CA* **77**, 164616 (1972)].
71USP3577557	T. S. Sulkowski, U. S. Pat. 3,577,557 (1971) [*CA* **75**, 49158 (1971)].
72BSF2868	J.-L. Aubagnac, J. Elguero, and R. Robert, *Bull. Soc. Chim. Fr.*, 2868 (1972).
72CB2898	G. Ege, E. Beisiegel, and P. Arnold, *Chem. Ber.* **105**, 2898 (1972.
72GEP2141464	H. Kato and T. Mori, Ger. Pat. 2,141,464 (1972) [*CA* **76**, 140927 (1972)].
72GEP2207291	H. Ono and H. Katsuyama, Ger. Pat. 2,207,291 (1972) [*CA* **77**, 152140 (1972)].
72IC2344	D. F. Averill, J. I. Legg, and D. L. Smith, *Inorg. Chem.* **11**, 2344 (1972).
72JCS(P1)1209	J. J. Worman, M. E. Kub, and M. Pearson, *J. C. S. Perkin 1*, 1209 (1972).
72JHC947	P. M. Schwartz and A. J. Saggiomo, *J. Heterocycl. Chem.* **9**, 947 (1972).
72JOC2208	E. H. Gold and E. Babad, *J. Org. Chem.* **37**, 2208 (1972).
72KGS1341	M. V. Loseva and B. M. Bolotin, *Khim. Geterotskil. Soedin.*, 1341 (1972) [*CA* **78**, 43392 (1973)].
72KGS1705	A. V. Bogatskii, O. P. Rudenko, S. A. Andronati, and T. K. Chumachenko, *Khim. Geterotsikl. Soedin.*, 1705 (1972) [*CA* **78**, 72087 (1973)].
72MI1	F. Gatta, R. Landi-Vittory, M. Tomassetti, and G. Nunez-Barrios, *Chim. Ther.* **7**, 480 (1972) [*CA* **77**, 61967 (1972).
72MI2	M. F. Sorokin, L. G. Shode, L. A. Dobrovinskii, and L. G. Stakhovskaya, *Tr.—Mosk. Khim.-Tekhnol. Inst. im. D. I. Mendeleeva* **70**, 77 (1972) [*CA* **79**, 5319 (1973).
73BSF331	M. Pais, R. Sarfati, and F.-X. Jarreau, *Bull. Soc. Chim. Fr.*, 331 (1973).
73CC571	W. Ollis and J. F. Stoddart, *J. C. S. Chem. Commun.*, 571 (1973).
73GEP2221558	W. Milkowski, R. Budden, S. Funke, R. Hueschens, H. G. Liepmann, W. Stuehmer, and H. Zuegner, Ger. Pat. 2,221,558 (1973) [*CA* **80**, 27306 (1974)].
73GEP2261777	Y. Kuwada, H. Natsugari, and K. Meguro, Ger. Pat. 2,261,777 (1973) [*CA* **79**, 78860 (1973)].
73JAP73/99191	Y. Kuwada, M. Hideaki, and K. Meguro, Jpn. Pat. 73/99191 (1973) [*CA* **80**, 96049 (1974)].
73JCS(P1)205	R. Crossley, A. P. Downing, M. Nógrádi, A. Braga de Oliveira, W. D. Ollis, and I. O. Sutherland, *J. C. S. Perkin I*, 205 (1973).
73JHC689	D. Misiti, V. Rimatori, and F. Gatta, *J. Heterocycl. Chem.* **10**, 689 (1973).

73JOC176	E. M. Burgess and J. P. Sanchez, *J. Org. Chem.* **38**, 176 (1973).
73MI1	Y. Iwakura, K. Uno, K. Haga, and K. Nakamura, *J. Polym. Sci., Polym. Chem. Ed.* **11**, 367 (1973).
73T1001	M. Pais, R. Sarfati, F.-X. Jarreau, and R. Goutarel, *Tetrahedron* **29**, 1001 (1973).
73TL2279	R. Heckendorn and A. R. Gagneux, *Tetrahedron Lett.*, 2279 (1973).
73USP3741969	H. Ott, U. S. Pat. 3,741,969 (1973) [*CA* **79**, 78861 (1973)].
74ACS(B)313	A. Osbirk and E. B. Pedersen, *Acta Chem. Scand., Ser. B* **33**, 313 (1974).
74CB1784	H. Wamhoff and C. Materne, *Chem. Ber.* **107**, 1784 (1974).
74GEP2314993	W. Milkowski, S. Funke, W. Stuehmer, R. Hueschens, H. G. Liepmann, and H. Zeugner, Ger. Pat. 2,314,993 (1974) [*CA* **82**, 4325 (1975)].
74GEP2353165	W. Milkowski, R. Budden, S. Funke, R. Hueschens, H. G. Liepmann, W. Stuehmer, and H. Zeugner, Ger. Pat. 2,353,165 (1974) [*CA* **82**, 73051 (1975)].
74GEP2356308	Y. Kuwada, H. Natsugari, and K. Meguro, Ger. Pat. 2,356,308 (1974) [*CA* **82**, 31364 (1975)].
74GEP2423642	H. A. DeWald, and Y. J. L'Italien, Ger. Pat. 2,423,642 (1974) [*CA* **83**, 206345 (1975)].
74H349	For a review, see T. Kametani, K. Kigasawa, M. Hiiragi, and K. Wakisaka, *Heterocycles* **2**, 349 (1974), and reference cited therein.
74JAP74/85095	Y. Kuwada, K. Meguro, and H. Tawada, Jpn. Pat. 74/85095 (1974) [*CA* **82**, 31364 (1975)].
74MI1	R. Ahmad, *J. Math. Sci.* **1**, 21 (1974) [*CA* **82**, 73460 (1975)].
74T1903	W. Ollis, J. F. Stoddart, and I. O. Sutherland, *Tetrahedron* **30**, 1903 (1974).
75AJC143	H. Singh and S. Singh, *Aust. J. Chem.* **28**, 143 (1975).
75AX(B)2571	A. Clearfield, R. D. G. Jones, A. C. Kellum, and C. H. Saldarriaga-Molina, *Acta Crystallogr., Sect. B* **B31**, 2571 (1975).
75BRP1413599	U. Shenoy, Br. Pat. 1,413,599 (1975) [*CA* **84**, 90191 (1976)].
75JMC803	J. R. Piper, L. M. Rose, T. P. Johnson, and M. M. Grenan, *J. Med. Chem.* **18**, 803 (1975).
75MI1	A. V. Bogatskii, S. A. Andronati, Y. I. Vikhlaev, Z. I. Zhilina, T. A. Klygul, O. P. Rudenko, N. Y. Golovenko, S. G. Soboleva, I. A. Starovoit, and A. I. Galatina, *Dokl. Soobshch–Mendeleevsk S'ezd. Obshch. Prikl. Khim., 11th* **2**, 27 (1975) [*CA* **88**, 152590 (1978)].
75MI2	M. Majchrzak, A. Koletko, and R. Guryn. *Acta Pol. Pharm.* **32**, 145 (1975) [*CA* **87**, 102295 (1977)].
75MI3	A. V. Bogatskii, I. A. Starovoit, S. A. Andronati, O. P. Rudenko, Z. I. Zhilina, V. V. Danilin, and N. I. Danilina, *Nov. Polyarogr. Tezisy Dokl. Vses. Soveshch. Polyarogr. 6th, 1975*, 142 (1975) [*CA* **86**, 43682 (1977)].
75M14	E. J. Billo, *Inorg. Nucl. Chem. Lett.* **11**, 491 (1975)].
75T295	B. A. Hess, Jr., L. J. Schaad, and C. W. Holyoke, Jr., *Tetrahedron* **31**, 295 (1975).
76AX(B)1305	O. Lefebvre-Soubeyran, *Acta Crystallogr., Sect. B* **B32**, 1305 (1976).
76CC48	J. C. Cass, A. R. Katritzky, R. L. Harlow, and S. H. Simonsen, *J. C. S. Chem. Commun.*, 48 (1976).

76CC1005	R. D. Chambers and J. R. Maslakiewicz, *J. C. S. Chem. Commun.*, 1005 (1976).
76CI(L)112	T. Demir and R. A. Shaw, *Chem. Ind. (London)*, 112 (1976).
76CSC905	C. Peters, W. P. Jensen, J. J. Worman, R. A. Jacobson, and D. A. Nelson, *Cryst. Struct. Commun.* **5**, 905 (1976).
76GEP2520937	W. Milkowski, R. Budden, S. Funke, R. Hueschens, H. G. Liepmann, W. Stuehmer, and H. Zeugner, Ger. Pat. 2,520,937 (1976) [*CA* **86**, 89913 (1977)].
76GEP2525094	U. Shenoy, Ger. Pat 2,525,094 (1976) [*CA* **86**, 106083 (1977)].
76GEP2627461	U. Shenoy, Ger. Pat. 2,627,461 (1976) (*CA* **86**, 106084 (1977)].
76JA6945	M. H. West and J. I. Legg, *J. Am. Chem. Soc.* **96**, 6945 (1976).
76JCS(P2)1172	J. R. Lindsay Smith and L. A. V. Mead, *J. C. S. Perkin 2*, 1172 (1976).
76JHC761	G. Roma, A. Ermili, and M. Mazzei, *J. Heterocycl. Chem.* **13**, 761 (1976).
76MI1	H. Liepmann, W. Milkowski, and H. Zeugner, *Eur. J. Med. Chem.—Chim. Ther.* **11**, 501 (1976) [*CA* **87**, 39446 (1977)].
76MI2	E. Finner, F. Russkopf, and W. Milkowski, *Eur. J. Med. Chem—Chim. Ther.* **11**, 508 (1976) [*CA* **87**, 22077 (1977)].
76ZOB1893	A. V. Bogatskii, S. A. Andronati, L. N. Vostrova, L. A. Litvinova, L. N. Yasinenko, E. I. Ivanov, and P. A. Sharbatyen, *Zh. Obshch. Khim.* **46**, 1893 (1976) [*CA* **85**, 177388 (1976)].
77BCJ957	M. Takahashi, N. Sugawara, and K. Yoshimura, *Bull. Chem. Soc. Jpn.* **50**, 957 (1977) [*CA* **87**, 68312 (1977)].
77CJC630	T. P. Ahern, T. Navratil, and K. Vaughan, *Can. J. Chem.* **55**, 630 (1977).
77FES33	F. Gatta and S. Chiavarelli, *Farmaco, Ed. Sci.* **32**, 33 (1977) [*CA* **86**, 140004 (1977)].
77IJC(B)36	V. D. Gaitonde and B. D. Hosangadi, *Indian J. Chem., Sect. B* **15B**, 36 (1977).
77IJC(B)786	H. Singh and R. K. Mehta, *Indian J. Chem., Sect. B* **15B**, 786 (1977).
77JCS(P1)1162	A. Benerji, J. C. Cass, and A. R. Katritzky, *J. C. S. Perkin 1*, 1162 (1977).
77JCS(P1)2347	T. Kametani, C. V. Loc, T. Higa, M. Ihaga, and K. Fukumoto, *J. C. S. Perkin 1*, 2347 (1977).
77JCS(P2)1732	J. R. Lindsay Smith and D. Masheder, *J. C. S. Perkin 2*, 1732 (1977).
77T2089	E. B. Pedersen and D. Carlsen, *Tetrahedron* **33**, 2089 (1977).
77USP4001215	T. J. Dietsche, U. S. Pat. 4,001,215 (1977) [*CA* **87**, 39313 (1977)].
77ZC174	K. Schulze, A. Vetter, W. Dietrich, and M. Muehlstaedt, *Z. Chem.* **17**, 174 (1977) [*CA* **87**, 102295 (1977)].
78CL1437	T. Fuchigami, T. Iwaoka, T. Nonaka, and T. Sekine, *Chem. Lett.*, 1437 (1978).
78GEP2729363	P. C. Bonsall, J. W. Morrison, E. S. Nicholson, and G. B. Smith, Ger. Pat. 2,729,363 (1978) {*CA* **88**, 190358 (1978)].
78IC3378	L. Campbell, J. J. McGarvey, and N. G. Samman, *Inorg. Chem.* **17**, 3378 (1978).
78IJC(B)393	D. M. Wakaner and B. D. Hosangadi, *Indian J. Chem., Sect. B* **16B**, 393 (1978) [*CA* **90**, 22998 (1979)].
78JOC622	F. J. Weigert, *J. Org. Chem.* **43**, 622 (1978).
78MI1	S. Watarai, H. Katsuyama, A. Umehara, and H. Sato, *J. Polym. Sci., Polym. Chem. Ed.* **16**, 2039 (1978).

78TL547	D. S. Kemp, R. V. Punzar, and J. C. Chabala, *Tetrahedron Lett.*, 547 (1978).
79ACS(B)313	A. Osbirk and E. B. Pedersen, *Acta Chem. Scand., Ser. B* **B33**, 313 (1979).
79CPB2589	H. Natsugari, K. Meguro, and Y. Kuwada, *Chem. Pharm. Bull.* **27**, 2589 (1979).
79CPB2608	H. Natsugari, K. Meguro, and Y. Kuwada, *Chem. Pharm. Bull.* **27**, 2608 (1979).
79CPB2618	H. Natsugari and Y. Kuwada, *Chem. Pharm. Bull.* **27**, 2618 (1979).
79CPB2927	H. Natsugari, K. Meguro, and Y. Kuwada, *Chem. Pharm. Bull.* **27**, 2927 (1979).
79DOK1140	S. A. Andronati, Yu. A. Dvorkin, Yu. A. Simonov, and V. Danilin, *Dokl. Akad. Nauk SSSR* **248**, 1140 (1979) [*CA* **92**, 110983 (1980)].
79JHC935	S. Plescia, G. Daidone, and V. Spiro, *J. Heterocycl. Chem.* **16**, 935 (1979).
79JOC3559	T. Latowski and B. Zelent, *J. Org. Chem.* **44**, 3559 (1979), and reference cited therein.
79JOC4473	D. S. Kemp, M. D. Sidell, and T. J. Shortridge, *J. Org. Chem.* **44**, 4473 (1979).
79MI1	A. Costa, A. Garcia Raso, J. V. Sinisterra, and J. M. Marinas, *An. Quim.* **75**, 381 (1979) [*CA* **91**, 175055 (1979)].
79MI2	A. V. Bogatskii, E. I. Ivanov, S. A. Andronati, and L. N. Vostrova, *Khim. Prom-st. Ser.: Reakt. Osobo Chist. Veshchestva*, 14 (1979) [*CA*, **92**, 22199 (1980)].
80AX(B)452	P. Murray-Rust, J. Murray-Rust and R. Clay, *Acta Crystallogr., Sect. B* **B36**, 452 (1980).
80BCJ2040	T. Fuchigami, T. Iwaoka, T. Nonaka, and T. Sekine, *Bull. Chem. Soc. Jpn.* **53**, 2040 (1980).
80CL1331	T. Sakakibara and H. Matsuyama, *Chem. Lett.*, 1331 (1980).
80IJC(B)703	D. M. Wakanar and B. N. Hosangadi, *Indian J. Chem., Sect. B* **19B**, 703 (1980).
80JA4983	E. M. Kosower and B. Pazhenchevsky, *J. Am. Chem. Soc.* **102**, 4983 (1980).
80JMC392	J. B. Hester, Jr., A. D. Rudzik, and P. F. Voigtlander, *J. Med. Chem.* **23**, 392 (1980).
80KFZ45	A. V. Bogatskii, S. A. Andronati, T. A. Voronina, V. V. Danilin, and L. N. Nerobkova, *Khim.-Farm. Zh.* **14**, 45 (1980) [*CA* **94**, 84081 (1981)].
80MI2	T. Latowska, E. Latowska, B. Poplawska, M. Przytarska, M. Walczak, and B. Zelent, *Pol. J. Chem.* **54**, 1073 (1980), and references cited therein.
80MI3	A. V. Bogatskii, E. I. Ivanov, L. N. Vostrova, S. A. Andronati, and A. M. Arlinskaya, *Khim. Prom-st. Ser.: Reakt. Osobo Chist. Veshchestva*, 35 (1980) [*CA* **94**, 4001 (1981)].
80MIPI	M. Shinano and H. Hamanaka, PCT Int. Pat. 80/02155 (1980) [*CA* **94**, 175183 (1981)].
81ACH171	I. Bitter, L. Szocs, and L. Toke, *Acta Chim. Acad. Sci. Hung.* **107**, 171 (1981) [*CA* **95**, 220056 (1981)].
81EUP42354	I. Vlattas, Eur. Pat. EP42354 (1981) [*CA* **96**, 162759 (1982)].
81FES135	B. Bobranski, J. Przytocka-Balik, M. Wilimowski, and J. Barczynska, *Farmaco, Ed. Sci.* **36**, 135 (1981) [*CA* **94**, 167625 (1981)].

81FES425	S. Masso, F. Corelli, G. De Martino, G. C. Pantaleoni, D. Fanini, and G. Palumbo, *Farmaco, Ed. Sci.* **36**, 425 (1981) [*CA* **95**, 132842 (1981)].
81IZV1676	R. G. Gafurov, E. Sugumonyan, G. V. Lagodzinskaya, and L. T. Eremenko, *Izv. Akad. Nauk SSSR, Ser. Khim.*, 1676 (1981) [*CA* **95**, 203910)1981)].
81JAP81/151468	Toho Chemical Industry Co., Ltd., Jpn. Pat. 81/151468 (1981) [*CA* **96**, 217891 (1982)].
81JHC1153	E. Aiello, G. Dattalo, G. Cirrincione, A. M. Almerico, and I. D'Asdia, *J. Heterocycl. Chem.* **18**, 1153 (1981).
81MI1	F. Gatta and F. Ponti, *Boll. Chim. Farm.* **120**, 102 (1981) [*CA* **95**, 132843 (1981)].
81MI2	Toho Chemical Industry Co., Ltd., *CEER, Chem. Econ. Eng. Rev.* **13**, 32 (1981) [*CA* **95**, 115496 (1981)].
82AHC115	H. D. Perlmutter and R. B. Trattner, *Adv. Heterocycl. Chem.* **31**, 115 (1982).
82AX(B)638	A. A. Dvorkin, Yu. A. Simonov, T. I. Malinowski, S. S. Andronati, A. V. Bogatskii, and V. V. Danilin, *Acta Crystallogr., Sect. B* **B38**, 638 (1982).
82CL1823	T. Sakakibara and H. Hamakawa, *Chem. Lett.*, 1823 (1982).
82IJC(B)899	D. Konwar, R. C. Boruah, J. S. Sandhu, and J. N. Baruah, *Indian J. Chem., Sect. B* **21B**, 899 (1982).
82JCR(S)80	A. F. Cuthbertson, C. Glidewell, and D. Lloyd, *J. Chem. Res. Synop.*, 80 (1982).
82JCS(D)2545	M. S. Hussain, *J. C. S. Dalton*, 2545 (1982).
82JCS(P2)477	R. W. Alder, M. R. Bryce, and N. C. Goode, *J. C. S. Perkin 2*, 477 (1982).
82JOC4452	R. W. Koch and R. E. Dessy, *J. Org. Chem.* **47**, 4452 (1982).
82MI1	S. Plescia, G. Daidone, and V. Spiro, *Boll. Chim. Farm.* **122**, 190 (1982) [*CA* **100**, 22650 (1984).
82MI2	M. S. Hussain and S. Rehman, *Inorg. Chim. Acta* **60**, 231 (1982).
82TL465	H. H. Wasserman, G. D. Berger, and K. R. Cho, *Tetrahedron Lett.* **23**, 465 (1982).
82TL4181	R. W. Alder, P. Eastment, R. E. Moss, R. B. Sessions, and M. A. Stringfellow, *Tetrahedron Lett.* **23**, 4181 (1982).
82UKZ1077	A. V. Bogatskii, L. N. Vostrova, E. I. Ivanov, M. V. Grenaderova, P. A. Sharbatyan, and I. A. Starovoit, *Ukr. Khim. Zh. (Russ. Ed.)* **48**, 1077 (1982) [*CA* **98**, 126051 (1983)].
83CC959	L. Crombie, R. C. F. Jones, S. Osborne, and A. R. Mat-Zin, *J. C. S. Chem. Commun.*, 959 (1983).
83CC960	L. Crombie, R. C. F. Jones, A. R. Mat-Zin, and S. Osborne, *J. C. S. Chem. Commun.*, 960 (1983).
83MI1	M. N. Glukhovtsev and B. Ya. Simkin, *Zh. Struct. Khim.* **24**, 31 (1983); *J. Struct. Chem. USSR* **24**, 356 (1883) [*CA* **99**, 157447 (1983)].
83S830	D. A. Cichra and H. G. Adolph, *Synthesis*, 830 (1983).
83T2459	H. H. Wasserman and G. D. Berger, *Tetrahedron* **39**, 2459 (1983).
84JOC1127	A. Greenberg, N. Molinaro, and M. Lang, *J. Org. Chem.* **49**, 1127 (1984).
85AG579	C. Schneiders, W. Huber, J. Lex, and K. Muellen, *Angew. Chem.* **97**, 579 (1985).

85HCA750	E. Askitoglu, A. Guggisberg, and M. Hesse, *Helv. Chim. Acta* **68**, 750 (1985).
85MI1	W. Milkowski, H. Liepmann, H. Zeugner, M. Ruhland, and M. Tulp, *Eur. J. Med. Chem.—Chim. Ther.* **20**, 345 (1985) [*CA* **105**, 60588 (1986)].
86AG190	J. Barluenga, M. Tomas, A. Ballestreros, V. Gotor, and C. Krueger, *Angew. Chem.* **98**, 190 (1986).
86TL5147	L. Crombie, R. C. F. Jones, and D. Haigh, *Tetrahedron Lett.* **42**, 5147 (1986).
86TL5151	L. Crombie, R. C. Jones, and D. Haigh, *Tetrahedron Lett.* **42**, 5151 (1986), and references cited therein.
86USP4595531	W. Milkowski, R. Budden, S. Funke, R. Hueschens, H. Liepmann, W. Stuehmer, and H. Zuegner, U. S. Pat. 4,595,531 (1986) [*CA* **105**, 133919 (1986)].
87MI1	H. H. Wasserman, *Aldrichim. Acta* **20**, 63 (1987).
89AHC185	H. D. Perlmutter, *Adv. Heterocycl. Chem.* **45**, 185 (1989).

Behavior of Monocyclic 1,2,4-Triazines in Reactions with C-, N-, O-, and S-Nucleophiles

V. N. CHARUSHIN, S. G. ALEXEEV, AND O. N. CHUPAHKIN

Laboratory of Organic Chemistry, The Urals Polytechnical Institute, Sverdlovsk, 620002, U. S. S. R.

H. C. VAN DER PLAS

Laboratory of Organic Chemistry, Agricultural University, Dreijenplein 8, 6703 BC, Wageningen, The Netherlands

I. Introduction 74
II. Some Physical and Spectroscopic Properties of 1,2,4-Triazines 75
   A. Uncharged 1,2,4-Triazines 75
   B. *NH*-1,2,4-Triazinium salts 77
   C. 1,2,4-Triazine *N*-Oxides 78
   D. *N*-Alkyl-1,2,4-triazinium Cations 82
III. Nucleophilic Addition Reactions 84
   A. Site Selectivity in Addition Reactions with Nucleophiles 85
     1. Adduct Formation with C-Nucleophiles 85
     2. Adduct Formation with N-Nucleophiles 89
     3. Adduct Formation with O-Nucleophiles 90
     4. Adduct Formation with S-Nucleophiles 93
   B. Diadduct Formation in Reactions with Mononucleophiles 94
   C. Cyclizations with Bifunctional Nucleophiles 94
IV. Nucleophilic Substitution Reactions 97
   A. Displacement of One Nucleofugic Group 98
     1. Substitution at C-3 98
     2. Substitution at C-5 103
     3. Substitution at C-6 106
   B. Displacement of Two Nucleofugic Groups 106
     1. Substitution Reactions at C-3 and C-5 106
     2. Substitution Reactions at C-5 and C-6 110
     3. Substitution Reactions at C-3 and C-6 111
   C. Displacement of Three Nucleofugic Groups 113
   D. Intramolecular Nucleophilic Substitution Reactions 117
V. Nucleophilic Substitution of Hydrogen ($S_N H$ Reactions) 119
   A. $S_N H$ Reactions in the Presence of Oxidant 119
   B. Vicarious Nucleophilic Substitution 122
   C. Other $S_N H$ Reactions 124

VI. Transformations of the 1,2,4-Triazine Ring 125
   A. Ring-Degenerate Transformations 125
   B. Transformations of 1,2,4-Triazines into Other Azines 127
   C. Transformations of 1,2,4-Triazines into Azoles 130
      1. Transformations Leading to 1,2,4-Triazoles 131
      2. Formation of 1,2,3-Triazoles and Imidazoles 132
VII. Conclusion 134
   References 135

I. Introduction

Among six-membered nitrogen-containing heterocycles the 1,2,4-triazines have attracted the attention of chemists for a long time. This is due to the fact that many 1,2,4-triazine derivatives are biologically active and are used in medicine and agriculture (78HC(33)189; 84MI3; 84OPP199). Moreover, 1,2,4-triazines are of chemical interest. They proved to be very susceptible to attack by all kinds of nucleophiles, leading to addition and subsequently either substitution or cyclization, and ring transformation (78HC(33)189; 83AHC305, 83T2869; 84MI3; 85T237; 86CRV781; 87MI1; 88AHC301, 88T1). During the last two decades, 1,2,4-triazines have also been shown to be useful starting materials for the synthesis of a great variety of functionalized pyridines, pyrimidines, and other heterocyclic systems by means of Diels–Alder cycloaddition reactions (83T2869; 86CRV781; 87MI1). Discovery of intramolecular transformations of 1,2,4-triazines has given a strong impetus to new investigations on intramolecular Diels–Alder reactions of 1,2,4-triazines (84AP379, 84CZ331; 85AP1048, 85AP1051, 85TL2419, 85TL4355; 86TL431, 86TL1967, 86TL2107, 86TL2747; 87TL2747; 87JOC4280, 87JOC4287, 87T5145, 87T5159). For this reason the synthesis of numerous 1,2,4-triazines containing alkenyl or alkynyl side-chain substituents has been developed.

This continuing and lively interest in this area of heterocyclic chemistry has produced a great number of publications and patent applications. Since 1978, several excellent monographs and reviews covering different aspects of preparation, chemical, and physical properties of 1,2,4-triazines have been published (78HC(33)189; 80S165; 83T2869; 84MI3, 84OPP199; 86CRV781; 87MI1). No review dealing with reactivity of simple 1,2,4-triazines toward nucleophilic reagents is so far available, although a huge amount of literature has appeared dealing with the dependency of the reactivity of the 1,2,4-triazine ring on type of activation, nature of substituents and leaving groups, character of nucleophile employed, stability of σ-adducts, site selectivity, and other factors. This article does not pretend to be an exhaustive compilation of all data on the reactivity of 1,2,4-triazines. Rather, it concentrates on the behavior of monocyclic 1,2,4-triazines in reactions with C-, N-, O-, and S-nucleophiles, because important features of general character established for simple 1,2,4-triazines can evidently be applied to other derivatives, including condensed 1,2,4-triazines. We cover, with some exceptions, the literature from 1968 through 1987, with some 1988 references. References appearing before 1968 can be found in the literature mentioned in this articles.

II. Some Physical and Spectroscopic Properties of 1,2,4-Triazines

A. UNCHARGED 1,2,4-TRIAZINES

1,2,4-Triazine is a very π-deficient heterocyclic system. Quantitative data on π-deficiency of azaaromatic compounds obtained by ^{13}C-NMR spectroscopy indicate that 1,2,4-triazines are more electron-deficient compounds than pyridines, pyrimidines, pyrazines, or pyridazines (Scheme 1) (82OMR192).

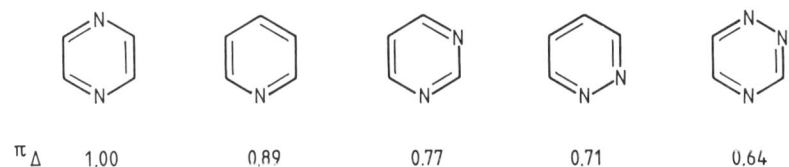

π_Δ 1.00   0.89   0.77   0.71   0.64

SCHEME 1. π-Deficiency values (π_Δ) for some azines (a lower value refers to a more π-deficient system).

Indeed, 1,2,4-triazines have a profound tendency to react with nucleophilic reagents. All carbon atoms in 1,2,4-triazines are vulnerable to attack by nucleophiles (78HC(33)189; 84MI3, 84OPP199); however, reactivities of C-3, C-5, and C-6 positions differ greatly, depending on the substituents present in the ring.

Theoretical calculations of ($\sigma + \pi$)-charge distribution for the parent compound using semiempirical methods show that in the ground state C-3 is the most electrophilic of all positions (Table I).

An insight into the electronic structure of 1,2,4-triazines can also be provided by ^{1}H-, ^{13}C-, and ^{15}N-NMR spectroscopy data (for review, see 88MI2). There are many examples in which ^{13}C chemical shifts correlate satisfactorily with total ($\sigma + \pi$)-electron charge densities on carbon

TABLE I

($\sigma + \pi$) AND π-CHARGE DISTRIBUTION[a] IN MOLECULES OF 1,2,4-TRIAZINE AND SOME 3-SUBSTITUTED 1,2,4-TRIAZINES (CNDO/2)

1,2,4-Triazine	N-1	N-2	C-3	N-4	C-5	C-6	References
Unsubstituted	-0.043	-0.101	0.155	-0.134	0.099	0.040	88KGS525
	(-0.009)	(-0.072)	(0.055)	(-0.047)	(0.064)	(0.012)	88MI1
			0.150		0.093	0.017)	75OMR194
	(-0.002)	(-0.069)		(-0.049)			78BAP285
			0.135		0.091	0.035	70JA7154
3-Methoxy-	-0.024	-0.140	0.309	-0.184	0.114	0.017	88KGS525
	(0.033)	(-0.140)	(0.099)	(-0.100)	(0.089)	(-0.040)	88MI1
3-Methylthio-	-0.034	-0.119	0.201	-0.159	0.105	0.026	88KGS525
	(0.024)	(-0.113)	(0.078)	(-0.078)	(0.078)	(-0.078)	88MI1
3-(N,N-Dimethyl-	-0.018	-0.167	0.287	-0.207	0.118	0.004	88KGS525
amino	(0.042)	(-0.157)	(0.130)	(-0.110)	(0.092)	(-0.055)	88MI1
3-Phenyl-	-0.048	-0.116	0.183	-0.163	0.111	0.026	88MI1
	(-0.021)	(-0.094)	(0.069)	(-0.062)	(0.072)	(-0.001)	88MI1

[a] Values of π-charge distribution are given in parentheses

atoms in 1,4-diazines and 1,2,4-triazines calculated by the CNDO/2 method (75OMR194; 80OMR172; 88KGS525). From Table II, showing the ^{13}C chemical shifts of some 1,2,4-triazines, it is evident that the C-3 carbon is mostly deshielded and, as a rule, it resonates at the lowest field, in full agreement with calculation. Thus, according to both theoretical

TABLE II

^{13}C-CHEMICAL SHIFTS FOR SELECTED 1,2,4-TRIAZINES IN CDCl$_3$

Compound	C-3	C-5	C-6	References
1,2,4-Triazine	158.1	149.6	150.8	75OMR194
3-Methyl-1,2,4-triazine	167.7	148.9	147.7	75OMR194
5-Methyl-1,2,4-triazine	157.0	160.5	150.9	75OMR194
6-Methyl-1,2,4-triazine	155.8	149.6	159.3	75OMR194
3-Phenyl-1,2,4-triazine	164.0	148.7	147.7	75OMR194
5-Phenyl-1,2,4-triazine	157.5	155.5	146.8	75OMR194
6-Phenyl-1,2,4-triazine	156.1	146.6	157.8	75OMR194
3-Methylthio-1,2,4-triazine	174.3	148.0	145.2	88KGS525
3-Amino-1,2,4-triazine[a]	163.3	149.8	140.6	86H951
3-(N,N-Dimethylamino)-1,2,4-triazine	160	148	138	77JOC546
3-Morpholino-1,2,4-triazine	161.1	148.6	140.1	86KGS1535
3-Amino-5,6-diphenyl-1,2,4-triazine[a]	162.1	158.9	147.1	86H951

[a] In DMSO-d_6.

calculations and ^{13}C-NMR data, the C-3 position of the 1,2,4-triazine ring is the most electron-deficient and seems to be the preferential site for a nucleophilic attack, provided, of course, that reactions of 1,2,4-triazines with nucleophiles belong to a charge-controlled interaction. On the other hand, π-charge distributions in molecules of 1,2,4-triazines show that C-3 and C-5 have rather similar π-deficiency indices (Table I), suggesting that both of these positions are likely to be attacked with nucleophiles.

B. NH-1,2,4-Triazinium Salts

The basicity of nitrogen atoms in 1,2,4-triazine and its 3-substituted derivatives increased as follows: N-4 > N-2 > N-1, as can be seen from both calculated ($\sigma + \pi$)-charge distribution indices (Table I) (78BAP285; 88KGS525) and experimental ^{14}N- and ^{15}N-NMR data for some 1,2,4-triazines (Table III) (78BAP285; 79JMR227; 84SA(A)637; 88KGS525, 88MI2). The distribution of π-charge over the ring nitrogens is somewhat different: N-2 > N-4 > N-1 (Table I). Therefore, protonation of 1,2,4-triazines can be expected to occur at either N-4 or N-2. On the other hand, N^1H- and N^2-protonated salts are thermodynamically more favored, as protonation of one of the two neighboring nitrogen atoms eliminates repulsion of electron pairs, which is a quite important feature of the basicity in 1,2-diazine derivatives (85MI1). A theoretical study on protonation of 1,2,4-triazines (87MI3) predicts that the N^2H-1,2,4- triazinium cation is the most stable because of the N^2-H $\cdots$ N^1 hydrogen bond formation.

Experimental data concerning protonation sites for monocyclic 1,2,4-triazines are still not sufficient to make any generalization of this phenomenon. On the basis of ultraviolent (UV) spectroscopic studies it has been

TABLE III

^{15}N- and ^{14}N-Chemical Shifts for Selected 1,2,4-Triazines

1,2,4 Triazine	Solvent	Nuclei	N-1	N-2	N-4	References
Unsubstituted	DMSO-d_6	^{15}N	420.0	382.0	318.0	84SA(A)637
	Ether	^{14}N	422	378	299	78BAP285
3-Methoxy-	DMSO-d_6	^{15}N	416.0	322.0	253.6	84SA(A)637
	CDCl$_3$	^{14}N	435	335	260	88KGS525
3-Methylthio-	DMSO-d_6	^{15}N	412.0	351.0	282.0	84SA(A)637
	CDCl$_3$	^{14}N	430	366	288	88KGS525
3-Amino-	DMSO-d_6	^{15}N	415.7	319.0	250.0	84SA(A)637
3-Morpholino-	CDCl$_3$	^{14}N	432	338	265	88KGS525

concluded that protonation of 3-methyl-5-methoxy-1,2,4-triazine occurs at either N-2 or N-4 (72LA111). However, experimental ^{1}H-, ^{13}C-, and ^{14}N-NMR studies performed on some 3-substituted 1,2,4-triazines have revealed that a mixture of interconverting prototropic forms is present in solution with preferential contribution of the N^1H-isomeric salt (Scheme 2) (88KGS525).

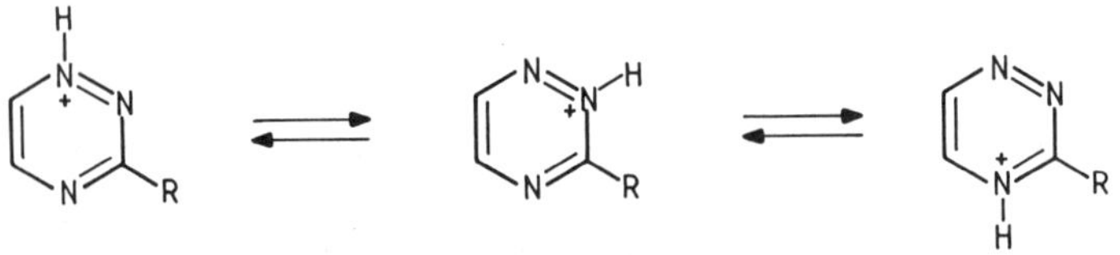

SCHEME 2

As far as reactivity of isomeric NH-triazinium salts toward nucleophilic reagents is concerned it appears that N^2H- and N^4H-triazinium cations are more appropriate substrates than N^1H-triazinium cations. The addition of nucleophiles in these N^2H- and N^4H-salts seems to be favored at C-5 (ortho or para positions relative to the charged nitrogen atom) for both kinetic and thermodynamic reasons and gives rise to rather stable σ-adducts (67YZ1501; 70JHC767; 78HC(33)189; 88KGS525). Unfortunately, no reliable data on the kinetically controlled addition of nucleophiles to NH-triazinium cations have so far appeared in the literature.

C. 1,2,4,-TRIAZINE N-OXIDES

The three isomeric 1,2,4-triazine N-oxides can be employed as substrates in reactions with nucleophilic reagents. However, they are specific with respect to the fact that two opposing factors affect charge distribution in their molecules. The electron-withdrawing ability of the N-oxide function seems to activate the ring carbon atoms for a nucleophilic attack (**1a**), however, this activation effect is partially compensated by the back-donation effect, which is especially significant for 1,2,4-triazine N-oxides (**1b**) (Scheme 3) (82H93; 84SA(A)637; 85SA(A)1135).

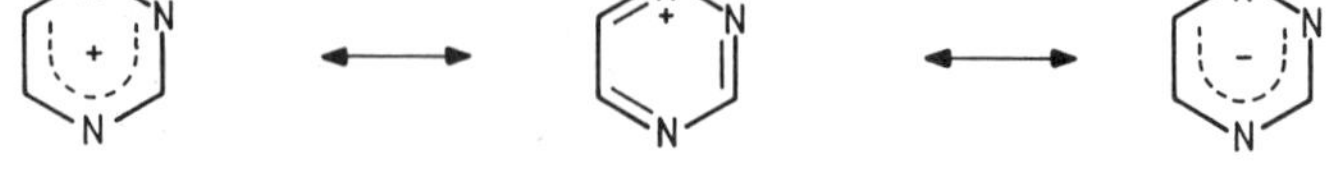

1a   1b

SCHEME 3

When comparing the ^{1}H- and ^{13}C-NMR data of 1,2,4-triazines with their corresponding N-oxides one reaches the conclusion that resonance structure **1b** contributes considerably to the overall picture of charge distribution in these molecules (Tables IV and V) (76LA153; 77JOC546; 82H93; 84SA(A)637; 85SA(A)1135). Indeed, the ^{1}H-NMR spectra of 1,2,4-triazine N-oxides exhibit the H-α and H-γ signals at significantly higher field relative to those for the corresponding 1,2,4-triazines (Scheme 4, Table IV) (71JOC787, 71LA12; 77JHC1221, 77JOC546; 78JOC2514, 78RTC273; 80JOC5421).

TABLE IV

^{1}H-NMR Spectral Data for Selected 1,2,4-Triazines and Their N-Oxides in CDCl$_3$[a]

Compound	H-3	H-5	H-6	References
1,2,4-Triazine	9.63	8.53	9.24	77JOC546
1,2,4-Triazine 1-oxide	9.00	8.57	8.04	71JOC787; 77JOC546
	(−0.63)	(0.04)	(−1.20)	
1,2,4-Triazine 2-oxide	8.82	8.00	8.42	77JOC546
	(−0.81)	(−0.53)	(−0.82)	
3-Methoxy-1,2,4-triazine	—	8.56	9.16	70JHC767
1-Oxide	—	8.37	7.83	71JOC787
		(−0.19)	(−1.33)	
2-Oxide	—	7.70	8.12	77JOC546
		(−0.86)	(−1.04)	
3-Amino-1,2,4-triazine[b]	—	8.53	8.88	77JOC546
2-Oxide[b]	—	8.19	8.23	77JOC546
		(−0.34)	(−0.65)	
3-(N,N-Dimethylamino)-1,2,4-triazine	—	8.14	8.15	77JOC546
2-Oxide	—	7.76	7.86	77JOC546
		(−0.38)	(−0.65)	
3-Morpholino-1,2,4-triazine	—	8.14	8.54	86KGS1535
2-Oxide	—	7.81	8.02	77JOC546
		(−0.33)	(−0.52)	
6-Methyl-1,2,4-triazine	9.55	8.55	—	71LA12
4-Oxide	9.28	8.22	—	71LA12
	(−0.27)	(−0.33)		
6-Phenyl-1,2,4-triazine[c]	9.52	8.91	—	71LA12
4-Oxide	9.31	8.60	—	71LA12
	(−0.21)	(−0.31)		

[a] The numbers in parentheses refer to the shielding (−) or deshielding effect of the N-oxide function.
[b] In DMSO-d6.
[c] In CCl$_4$.

TABLE V

^{13}C-NMR SPECTRAL DATA FOR SELECTED 1,2,4-TRIAZINE N-OXIDES[a]

Compound		C-3	C-5	C-6	References
1,2,4-Triazine	1-oxide	158.5	152.7	129.7	86H951
		(0.4)	(3.1)	(−21.1)	
	2-oxide	143.5	132.5	146.0	86H951
		(−14.6)	(−17.1)	(−4.8)	
3-Methoxy-1,2,4-triazine					
	1-oxide	166.5	154.0	124.5	77JOC546
		(2.5)	(4.0)	(−18.5)	
	2-oxide	152.5	130.0	135.5	77JOC546
		(−11.5)	(−20.0)	(−7.5)	
3-Amino-1,2,4-triazine					
	1-oxide[b]	165.0	155.9	120.7	86H951
		(1.7)	(6.1)	(−19.9)	
	2-oxide[b]	151.6	132.4	134.9	86H951
		(−11.7)	(−17.4)	(−5.7)	
3-(N,N-Dimethylamino)-					
1,2,4-triazine	1-oxide	161	152	120	77JOC546
		(1)	(4)	(−18)	
	2-oxide	151	132	133	77JOC546
		(−9)	(−16)	(−5)	
3-Amino-5,6-dimethyl-					
1,2,4-triazine	1-oxide[b]	166.0	164.3	119.4	86H951
		(3.9)	(5.4)	(−27.7)	
	2-oxide[b]	148.9	140.2	144.2	86H951
		(−13.2)	(−18.7)	(−2.9)	
	4-oxide[b]	154.4	143.0	148.7	86H951
		(−7.7)	(−15.9)	(1.6)	
5-Methyl-1,2,4-triazine					
	1-oxide	158.7	166.1	129.1	86H951
		(1.7)	(5.6)	(−21.8)	

[a] In CDCl$_3$, unless indicated otherwise. Shielding or deshielding effect of the N-oxide function is given in parentheses. For the ^{13}C chemical shifts of the parent 1,2,4-triazines, see Table II.
[b] In DMSO-d_6.

Comparison of the ^{13}C-NMR spectra of 1,2,4-triazines with those of their isomeric N-oxides also reveals a great influence of the N-oxide function on the resonance signals of neighboring α-carbon atoms. Oxidation at N-1 results in a large upfield shift of the C-6 resonances (18–27 ppm) (77JHC1221, 77JOC546; 86H951), while oxidation at N-2 gives a large influence on both C-3 (9–15 ppm upfield) and C-5 (16–20 ppm upfield) (77JOC546; 86H951). In the ^{13}C-NMR spectra of 1,2,4-triazine 4-oxides, the C-3 and C-5 resonance signals are also shifted upfield (Scheme 4, Table V) (79JHC1389; 86H951).

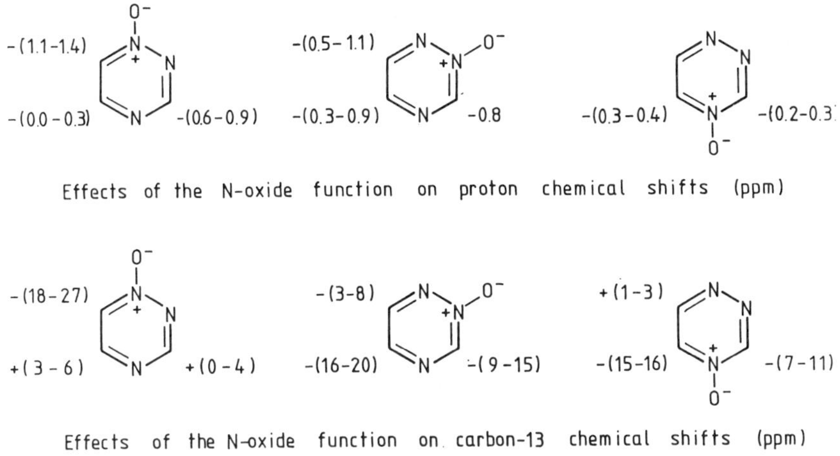

Effects of the N-oxide function on proton chemical shifts (ppm)

Effects of the N-oxide function on carbon-13 chemical shifts (ppm)

SCHEME 4

Strong upfield shifts for the resonance signals of the carbon atoms, located in the α- and γ-positions toward the N-oxide group in 1,2,4-triazine N-oxides, are attributed to the back-donation effect of the N-oxide group. Contributions of structure **1b** to the resonance stabilization (Scheme 3) in 1,2,4-triazine 1-oxide and 1,2,4-triazine 2-oxide were estimated to amount to 85 and 75% respectively, as shown by the ^{15}N-NMR measurements (82H93). Because of this back-donation effect, 1,2,4-triazine N-oxides may be expected to be less reactive toward nucleophilic reagents than are the corresponding triazines. However, the reactivity of N-oxides is enhanced considerably when protonating, alkylating, or acylating agents are present in the reaction mixture. The process is supposed to begin with an electrophilic attack of E$^+$ (proton, alkyl, acyl, or any other electrophilic group) at the N-oxide oxygen to generate *in situ* the corresponding azinium cation **2**, followed by interaction of the latter with a nucleophilic reagent Nu$^-$ (Scheme 5).

SCHEME 5

This method of activation of N-oxides for nucleophilic attack is also well known in the chemistry of pyridine N-oxides (86H161) and other azine N-oxides (71MI1; 86MI2). It can be exemplified by the reaction of 1,2,4-triazine 2-oxides with hydro halides, which proceeds via the intermediacy of 2-hydroxy-1,2,4-triazinium cations **3** (Scheme 6) (78JOC2514).

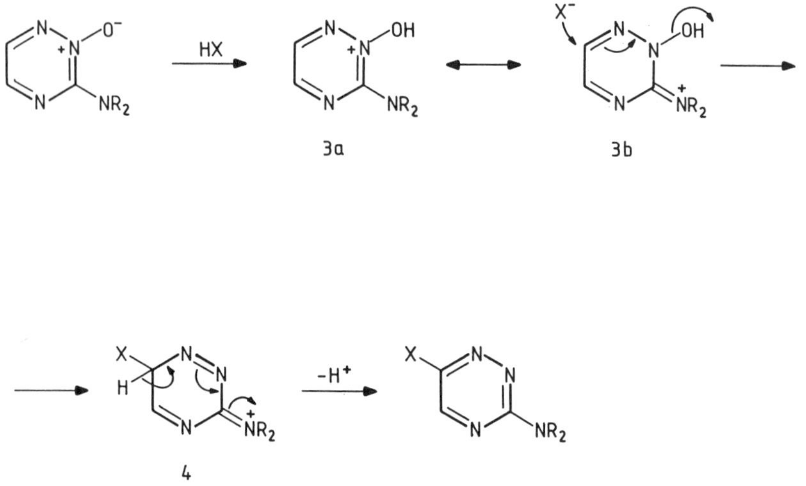

SCHEME 6

This example shows that azinium cations generated from N-oxides are not always attacked by a nucleophile at C-α or C-γ positions. In the above-mentioned reaction, it is the C-β position that is the preferential site for nucleophilic attack. Reactions of this type are usually accompanied by loss of the N-oxide function. The presence of a substituent able to stabilize the positively charged species (e.g., **4**) seems to be an important feature in determining the site selectivity for these deoxidative nucleophilic substitutions (Scheme 6) (78JOC2514). Similar substitution reactions at C-3 (C-β) of the pyridine ring have also been observed in pyridine N-oxides (86H161).

D. N-ALKYL-1,2,4-TRIAZINIUM CATIONS

Quaternary N-alkyl-1,2,4-triazinium salts are undoubtedly more reactive toward nucleophilic reagents than uncharged 1,2,4-triazines and are, therefore, attractive substrates. Only a few reports on the synthesis and reactions of N-alkyl-1,2,4-triazinium cations have been published (71CC1636; 78HC(33)189; 84LA283; 86KGS1535; 88KGS525).

As already mentioned, nitrogen atoms in 1,2,4-triazines have different basicities: N-4 > N-2 > N-1 (Tables I and III). Therefore, in principle three isomeric N-alkyl-1,2,4-triazinium salts can possibly be formed. It has been established, however, that the formation of 1,2,4-triazinium salts is governed predominantly by steric effects of the ring substituents. Triazines containing substituents at C-5 and C-6 are usually quaternized at N-2 (78HC(33)189), while alkylation of 3-substituted or 3,5-disubstituted 1,2,4-triazines occurs exclusively at N-1 (Scheme 7) (71CC1636; 78HC(33)189; 86KGS1535; 88KGS525, 88MI2). In particular, the synthesis of 1-ethyl-1,2,4-triazinium salts containing an alkynylthio group at C-3 has been developed (Scheme 7) (88H(ip)). The latter proved to be more active in intramolecular Diels–Alder cycloaddition reactions than the corresponding neutral triazines (see Section VI,B).

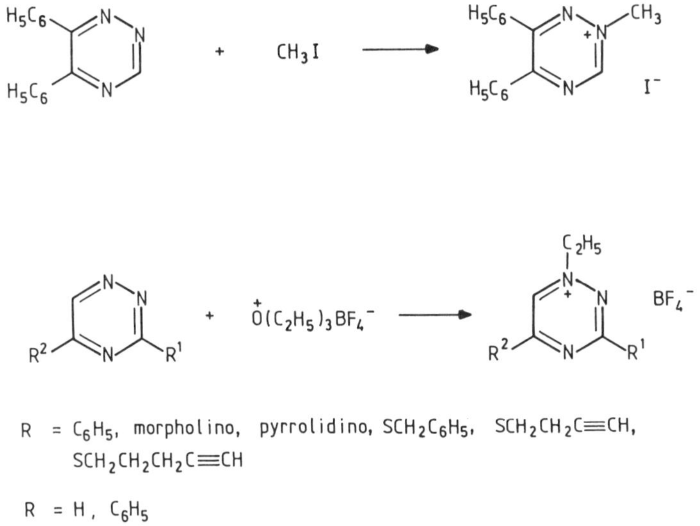

R = C_6H_5, morpholino, pyrrolidino, $SCH_2C_6H_5$, $SCH_2CH_2C{\equiv}CH$, $SCH_2CH_2CH_2C{\equiv}CH$

R = H, C_6H_5

SCHEME 7

Molecular orbital (MO) calculations using the CNDO/2 method revealed that for the three isomeric N-methyl-1,2,4-triazinium cations, regardless of the position of the N-methyl group, the charge distribution is C-3 > C-5 > C-6 (Scheme 8) (88KGS525). Thus, methylation did not change the order of charge density compared to 1,2,4-triazine. However, alkylation of N-1 activates the C-6 position more than the C-5 position thus making the difference in charge density between C-5 and C-6 less than for the parent compound (Scheme 8).

This also has been verified experimentally: 1-alkyl-1,2,4-Triazinium salts are capable of reacting with nucleophiles both at C-5 and C-6 giving

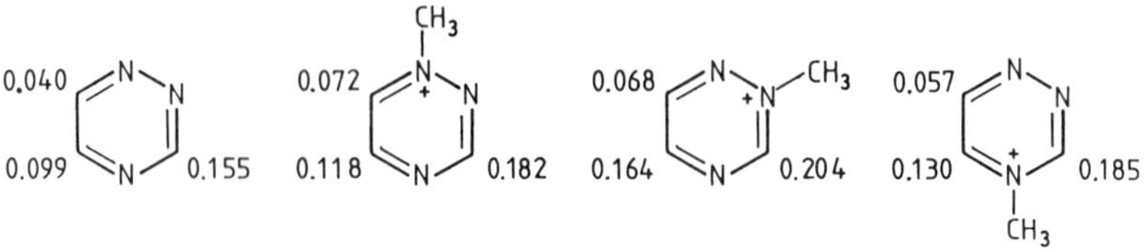

SCHEME 8

rise to diadducts **5** (86KGS1535) or cyclization products **6** if bifunctional nucleophiles are used (88AHC301, 88TL1431) (Scheme 9).

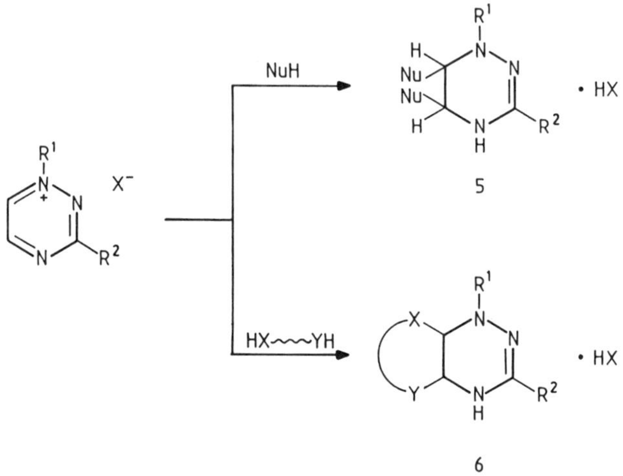

SCHEME 9

III. Nucleophilic Addition Reactions

It is well recognized that many reactions between azaaromatic substrates and nucleophilic reagents are initiated by an addition step, leading to the formation of σ-adducts (83AHC305). As far as 1,2,4-triazines are concerned, the formation of σ-adducts with nucleophiles is especially favored in this series because of the very low aromatic character of the 1,2,4-triazine ring (74AHC255). Since these σ-adducts play an important role as key intermediates in $S_N(AE)^{ipso}$ (65AHC145; 78HC(33)189; 84MI3), S_N(ANRORC) (78ACR462), S_NH (88T1), and other nucleophilic reactions, several aspects of σ-adduct formation in the series of 1,2,4-triazines will be discussed in detail in the following sections; first the adduct formation with mononucleophilic species (Sections III,A and B), then cyclizations with bifunctional nucleophiles (Section III,C).

A. SITE SELECTIVITY IN ADDITION REACTIONS WITH NUCLEOPHILES

In principle, three anionic σ-adducts (**7–9**), commonly featuring a tetrahedral center, can be formed from reactions of 1,2,4-triazines with anionic nucleophiles (Scheme 10). Their neutral analogues, the corresponding dihydro-1,2,4-triazines, may also be obtained on protonation or when triazines react with uncharged nucleophiles or on addition of anionic reagents to cationic 1,2,4-triazinium salts. A considerable amount of literature data is available (78HC(33)189; 83AHC305; 84MI3; 85T237; 88MI2) showing that these σ-adducts can indeed be formed but that they differ vastly in their stabilities. The vast majority of C-, N-, and O-, and S-nucleophiles have been reported to add easily at C-5 of the 1,2,4-triazine ring. The C-5 adducts, both anionic and neutral, are the most stable ones. They can usually be registered by NMR spectroscopy (83AHC305; 85T237; 88MI2) and can quite often be isolated from the reaction mixture as neutral species. The C-3 adducts are found to be less stable, while C-6 adduct formation is rarely observed. There are only a few examples of C-6 monoadduct formation in reactions with C-nucleophiles, although in some cases their existence can be suggested based on the structure of reaction products. Examples are given in the following subsections.

Thus, despite the fact that C-3 is the most electron-deficient position in the 1,2,4-triazine ring, it is the C-5 adduct which is the most stable. The greater stability of the C-5 adducts **8** is probably due to the *p*-quinoid contribution, leading to a negative charge at N-2, in contrast to the C-3 and C-6 adducts, **7** and **9** which feature the *o*- quinoid structure (Scheme 10) (65AHC145).

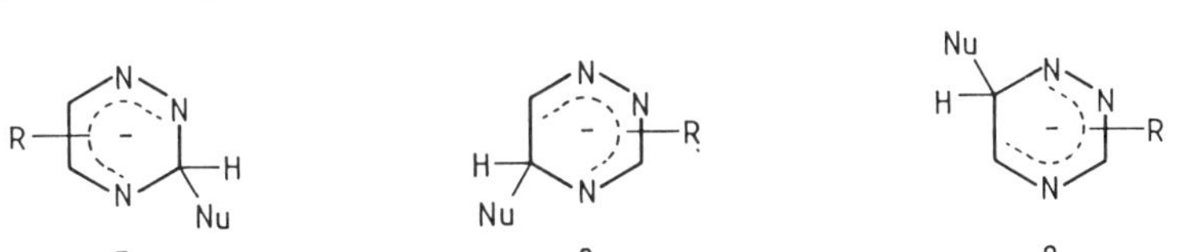

7   8   9

SCHEME 10

1. Adduct Formation with C-Nucleophiles

C-Nucleophilic reagents, both neutral and anionic, have been reacted with 1,2,4-triazine substrates. In particular, the interaction of uncharged, 1,2,4-triazines with Grignard reagents leading to the formation of rather stable σ-adducts at C-5 of the triazine ring has been described by many authors (Scheme 11) (71JPR699; 73BSF2493; 79JHC427; 83IJC(B)559;

85H2807; 86H239; 87CPB1378, 87H3111). The reaction mechanism can be depicted as given in Scheme 11, suggesting that the reaction is initiated by the formation of 1,2,4-triazinium cations **10** and **13**, generated *in situ* from 1,2,4-triazines and organomagnesium compounds, followed by addition of the carbanion at C-5 (73BSF2493).

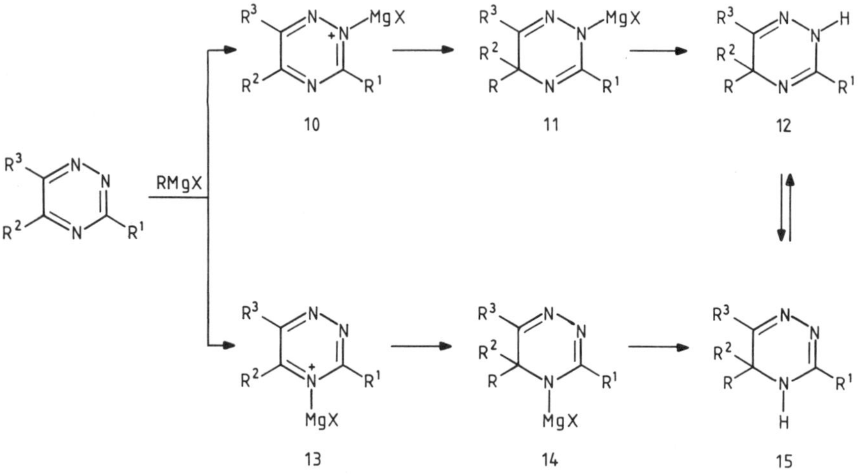

SCHEME 11

Dihydro-1,2,4-triazines resulting from the addition of Grignard reagents at C-5 are supposed to exist in solution in two tautomeric forms, **12** and **15**; however, the structure of 2,5-dihydro-1,2,4-triazines **12** is considered to be preferred over the 4,5-dihydro structure **15**. This is probably due to the fact that when isolated from the reaction mixture as crystalline products these C-5 adducts proved to have the 2,5-dihydro structure (84MI3; 85AHC1). Indeed, unequivocal evidence for the structure of 2,5-dihydro-3-methylthio-5-phenyl-1,2,4-triazine in a crystalline state has been obtained by X-ray analysis (80JOC4587; 81BCJ41).

Analogously, 2,3,4,5-tetrahydro-1,2,4-triazines **16** are formed on addition of C-nucleophiles at C-5 when 1,2,4-triazin-3-ones and 1,2,4-triazine-3-thiones react with alkyl- or arylmagnesium compounds

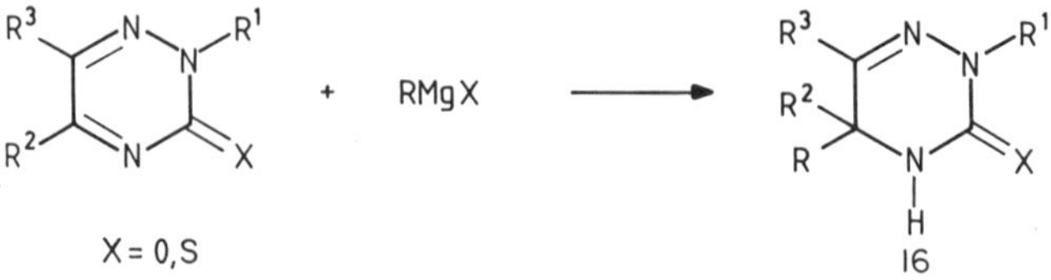

SCHEME 12

(Scheme 12) (70CR1201, 70LA177; 72BSF4637; 73BSF2818; 74ZN(B)792; 78HC(33)189; 83IJC(B)559; 84JHC905). Thus, the reaction of 1,2,4-triazines with organometallic compounds can be regarded as a common way to introduce an alkyl or aryl substituent at C-5 of the 1,2,4-triazine ring.

The addition of active methylene compounds of C-5 of 6-methyl-3-phenyl-1,2,4-triazine demonstrates a wider applicability of the reactions with carbanionic reagents for the synthesis of 5-substituted triazines (Scheme 13) (87CPB1378).

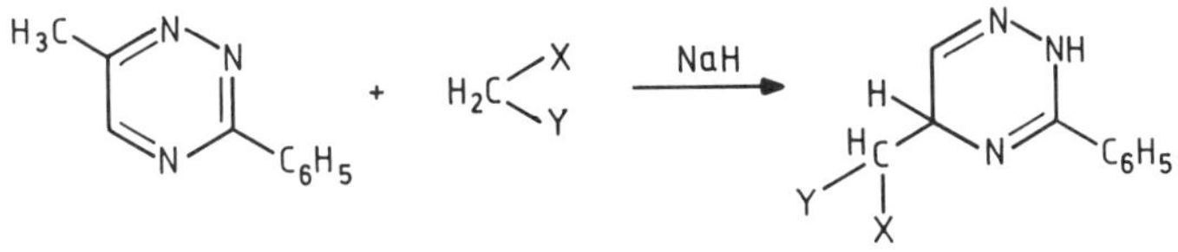

SCHEME 13

The C-3 and C-6 positions of the 1,2,4-triazine ring are also active in the addition of C-nucleophiles. The reaction of unsubstituted 1,2,4-triazine with phenylmagnesium bromide has been investigated to establish the relative reactivity of the ring carbons. The reactivity decreases in the order C-5 > C-6 > C-3 (87H3111). Evidence for both C-3 and C-6 adducts **18** and **19** has been obtained from the reaction of 4,6-dimethyl-1,2,4-triazin-5-one with phenylmagnesium bromide (Scheme 14) (73BSF559). Quaternization of triazin-5-one **17** with dimethyl sulfate yields the N^2-triazinium salt **20**, which, as might be expected, easily adds Grignard reagents at C-3 (Scheme 14) (74BSF999).

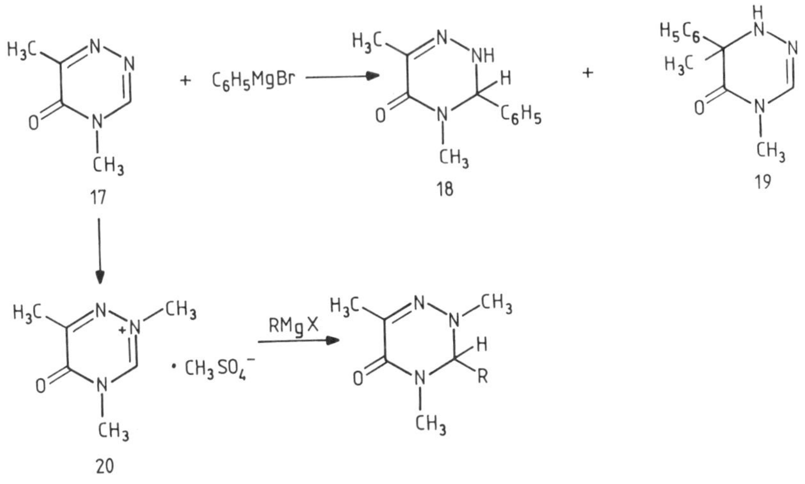

SCHEME 14

Enamines, ketene-*N,N*-, -*O,O*-, and *N,S*-acetals and other electron-rich alkenes usually react with 1,2,4-triazines as dienophiles to cause transformation of the ring. These inverse electron-demand Diels—Alder cycloaddition reactions have been extensively reviewed (80S165; 83T2869; 86CRV781; 87MI1). Therefore, in this article we do not pay too much attention to these transformations. There are, however, several known example in which enamines, enols, and thiols exhibit their carbanionic nature and, when reacted with appropriately substituted 1,2,4-triazines, give (as well as cycloaddition) addition of these reagents at C-3 (75TL2897) or C-5 (85LA1263; 86AP798) of the 1,2,4-triazine ring (Scheme 15).

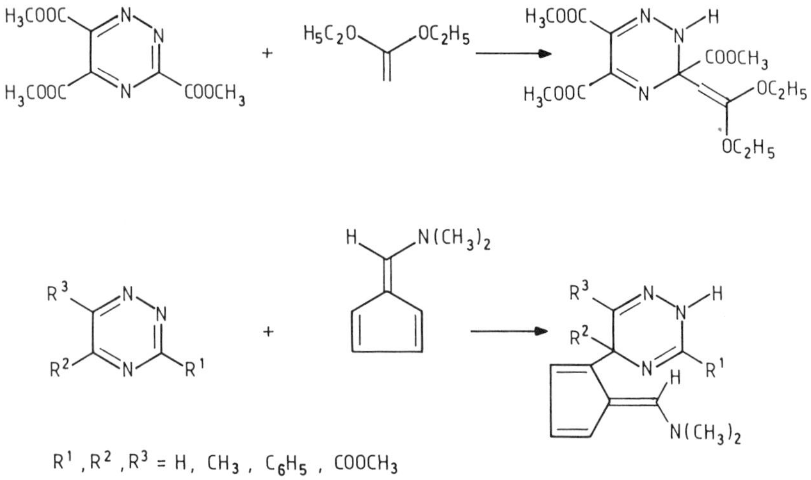

SCHEME 15

Protonation of the 1,2,4-triazine ring enhances its electrophilicity and, therefore, facilitates the addition of carbanionic nucleophiles. However, only a limited number of C-nucleophiles may be used in reactions with *NH*-triazinium substrates, because of possible proton transfer from substrate to reagent. Acetone seems to be an appropriate reagent for this kind of reaction, in spite of its very low C-nucleophilic character. Thus, 6-phenyl-1,2,4-triazin-3-one dissolved in acetone in the presence of hydrochloric acid gives the C-5 adduct **21** (Scheme 16) (85ACS(B)235). It is of interest that 5-phenyl-1,2,4-triazin-3-one remains unreactive under identical conditions, indicating that the addition reaction is sensitive to steric effects and can be completely blocked when the most reactive C-5 position carries a substituent.

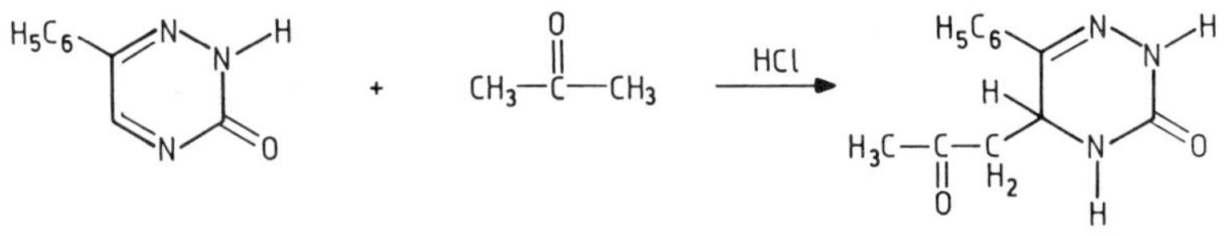

Sec. III.A] MONOCYCLIC 1,2,4-TRIAZINES 89

SCHEME 16

Another example of interaction between *NH*-triazinium salts and C-nucleophiles is the reaction of 3-substituted 1,2,4-triazines with indoles in the presence of trifluoroacetic acid (Scheme 17) (87MI2; 88MI1).

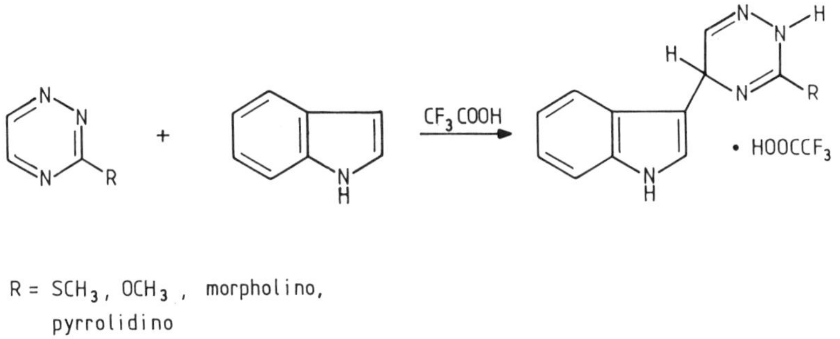

R = SCH$_3$, OCH$_3$, morpholino, pyrrolidino

SCHEME 17

2. Adduct Formation with N-Nucleophiles

1,2,4-Triazines containing substituents at C-3 and/or C-6 have been found to react easily with liquid ammonia yielding the C-5 amino adducts **22** (Scheme 18). This addition reaction is characterized by high site selectivity. Only C-5 adducts are registered in the course of low-temperature NMR studies (78H1490, 78RTC273; 85S884, 85T237; 87JOC71; 88MI2). If the 1,2,4-triazine ring contains a substituent at C-5, the addition of ammonia does not occur (85T237). Also 3-amino-1,2,4-triazine does not show the occurrence of a C-5 σ-adduct with liquid ammonia.

With potassium amide, in addition to the major reaction of addition at C-5, the C-3 and C-6 positions of the 1,2,4-triazine ring are also attacked by the amide ion, resulting in very unstable C-6(C-3)-aminodihydro-1,2,4-triazinides **23** and **24** (Scheme 18) (85T237).

Stability of amino adducts depends not only on the site to which the amino group is attached but also on the nature of substituents in the ring. For example, the C-5 adducts 5-azacycloalkyl-2,3,4,5-tetrahydro-1,2,4-

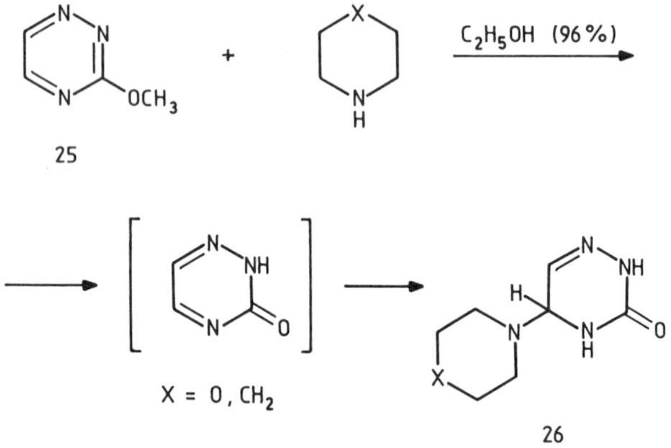

SCHEME 18

triazin-3-ones **26** are formed when 3-methoxy-1,2,4-triazine reacts with piperidine or morpholine in refluxing 96% ethanol (Scheme 19) (87MI2; 88MI1). The oxo group at C-3 stabilizes these amino adducts sufficiently to make their isolation as crystalline products possible. The adduct formation is supposed to be preceded by hydrolysis of the methoxy compound **25** into 1,2,4-triazin-3-one, which then undergoes a subsequent addition reaction with amines. To substantiate this result, no reaction was observed in pure tetrahydrofuran (THF) solution (87MI2; 88MI1).

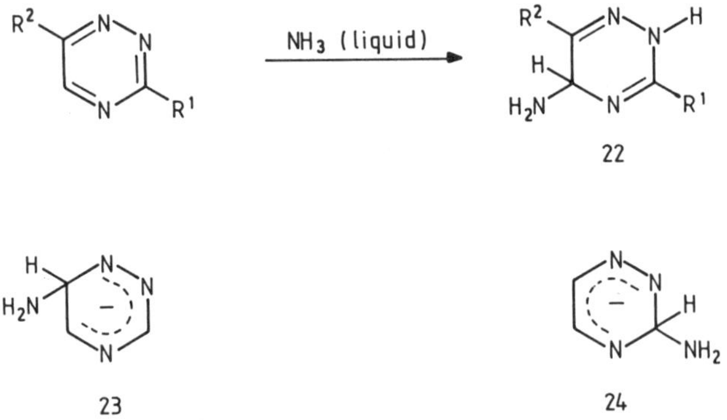

SCHEME 19

3. Adduct Formation with O-Nucleophiles

Most 1,2,4-triazines do not interact with water or alcohols unless activated by positive charge or by the presence of an electron-withdrawing substituent. *NH*-1,2,4-Triazinium cations can react with water or methanol to yield the C-5 hydroxy and methoxy adducts **27** (Scheme 20) (68YZ1501; 70JHC767; 73T2495; 78HC(33)189; 87MI2; 88KGS525, 88MI1, 88MI2). However, when electron-donating groups such as methylthio, methoxy, and morpholino and attached to C-3 of the 1,2,4-triazine ring, in acid solution equilibria between *NH*-triazinium salts and alkoxy adducts **27** usually exist. A large excess of trifluoroacetic acid is needed to shift the equilibrium completely to the adduct side (88KGS525, 88MI1).

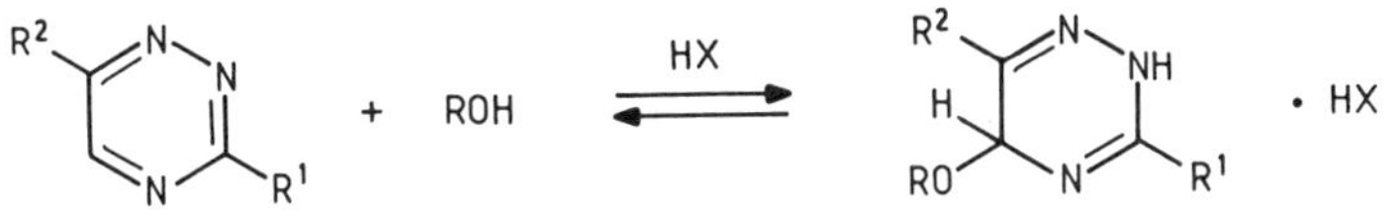

R = H, Alk
R^1 = H, OCH$_3$, SCH$_3$, NR$_2$
R^2 = H, CH$_3$, Ar, Het

SCHEME 20

Due to the electron-withdrawing character of the carbonyl and thiocarbonyl groups, substituted 1,2,4-triazin-3-ones and 1,2,4-triazin-3-thiones react smoothly with alcohols without acid to form the C-5 alkoxy adducts **28**. These adducts are quite stable and can be isolated (Scheme 21) (70CR1201; 71JOC3921; 72BSF4637; 85ACS(B)235; 86YZ54).

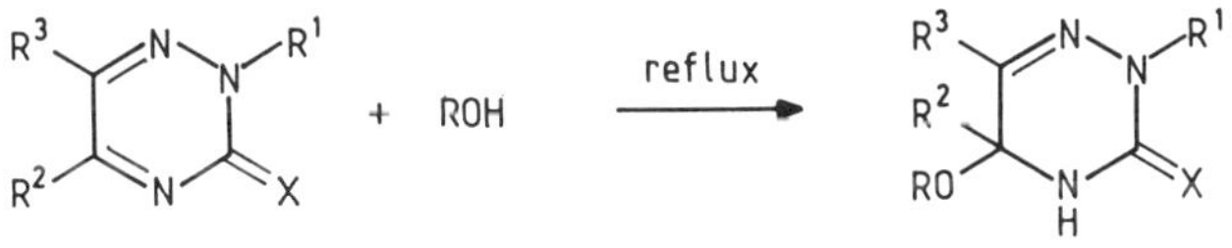

X = O, S

SCHEME 21

1,2,4-Triazin-3-one is so reactive toward water that it cannot be isolated except in the C-5 hydrated form **30**. Treatment of potassium salt **29** with an acid immediately results in hydrate **30** (Scheme 22) (71JOC3921; 79JHC1649). This result is also in agreement with what has been observed

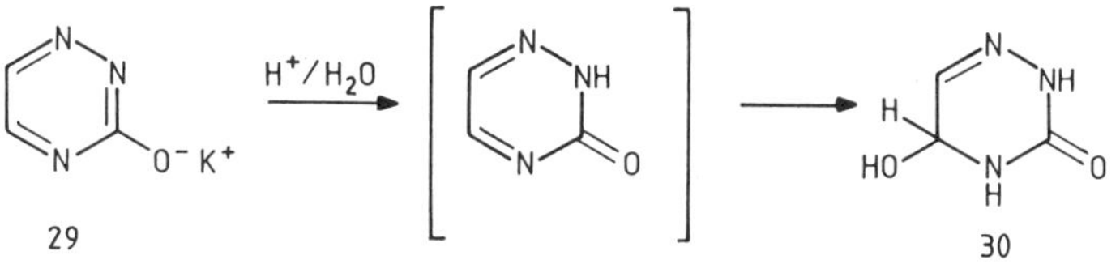

SCHEME 22

in the reaction between 3-methoxy-1,2,4-triazine and azacycloalkanes (Scheme 19).

A rare example of O-adduct formation at C-3 is the formation of **32**, observed by ^{1}H-NMR spectroscopy when pyrimido[5,4-*e*]1,2,4-triazine **31** is transformed into ethyl-5-amino-1,2,4-triazine-6-carboxylate by action of bromine in ethanolic solution (Scheme 23) (71JOC2974).

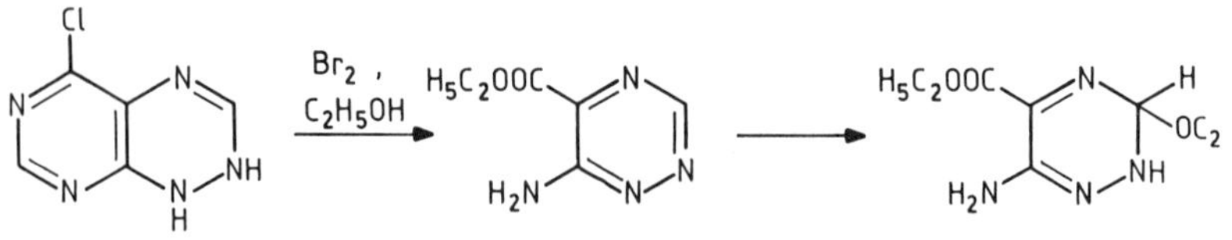

SCHEME 23

Another example of a nucleophilic addition at C-3 is provided by the reaction of the quaternary salt **20** with sodium methoxide yielding the relatively stable adduct **33**, (Scheme 24) (74BSF999). The C-3 adducts **35** and **36** have been suggested as intermediates in the conversion of 2-methyl-3-methylthio-1,2,4-triazin-5-ones **34** into the corresponding 3-methoxy derivatives; however, no spectroscopic evidence for adduct formation has been presented (Scheme 24) (72BSF1511).

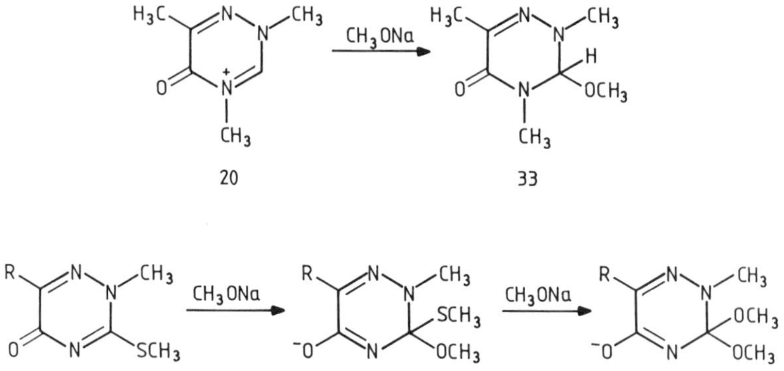

SCHEME 24

The C-3 adduct **37**, presumed earlier to be the structure of the product obtained on refluxing 3-chloro-5,6-diphenyl-1,2,4-triazine in a water–ethanolic solution, was not detected; the product was found later to be 5,6,-diphenyl-1,2,4-triazin-3-one (Scheme 25) (82CPB152).

No other reports on the formation of C-3 alkoxy adducts have appeared in the literature.

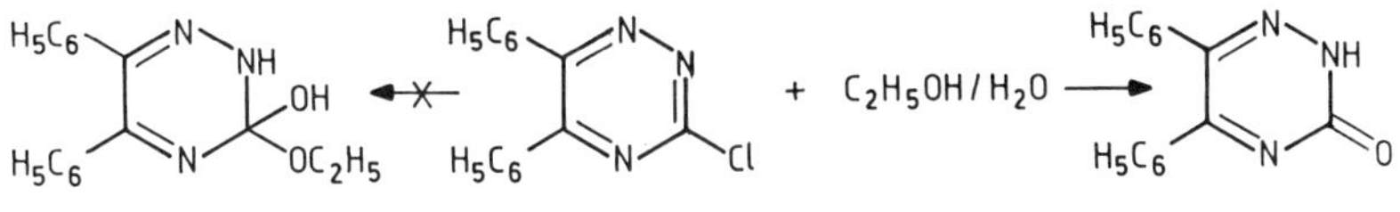

37

SCHEME 25

4. *Adduct Formation with S-Nucleophiles*

There are only a few examples known of S-adduct formation. They all feature addition at C-5 of the 1,2,4-triazine ring. Thus, 2,5-dihydro-1,2,4-triazines **38** are formed in good yield when sulfur dioxide is passed through a solution of 1,2,4-triazine in water (water–methanol) (Scheme 26) (78HC(33)189; 86H1243; 87CPB1378).

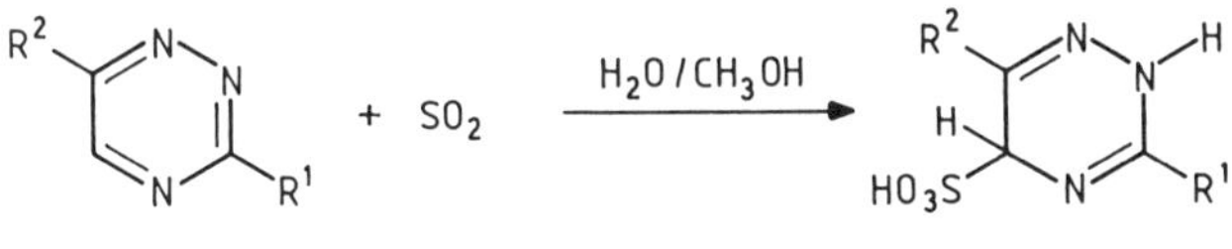

38

SCHEME 26

Cysteine has also been found to add at C-5 of the 1,2,4-triazine ring. Panfuran, the pharmacologically active 1,2,4-triazine derivative **39**, gives adduct **40** when mixed with cysteine (Scheme 27) (78HC(33)189).

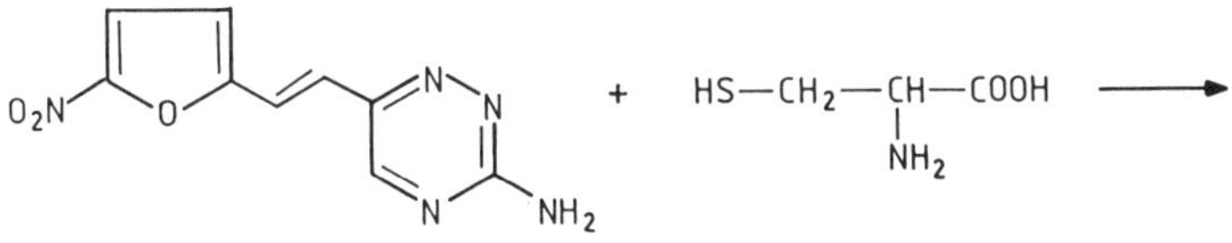

39

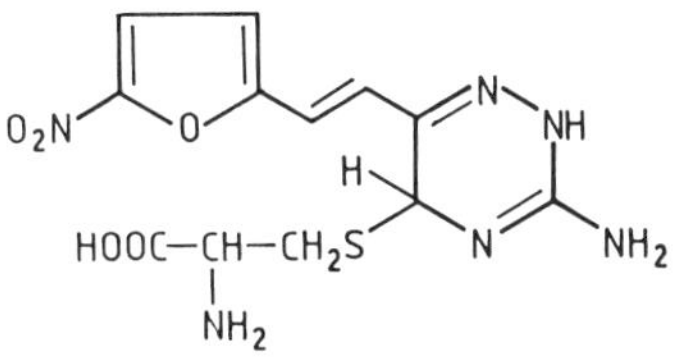

40

SCHEME 27

B. Diadduct Formation in Reactions with Mononucleophiles

When reacting with simple nucleophiles, 1,2,4-triazines generally participate in monoaddition reactions at C-5. However, sometimes the nucleophilic addition process is not complete at this stage and a subsequent addition reaction takes place resulting in diaddition products at C-5 and C-6. In this respect the behavior of 1,2,4-triazines is somewhat similar to that of 1,4-diazines, which are known to exhibit a marked tendency to add two molecules of nucleophilic reagents (78MI1; 84UK1648; 85KGS1011; 88AHC301, 88MI2). This mode of nucleophilic addition reaction usually requires as prerequisite a 1,2,4-triazine substrate strongly activated either by positive charge and/or an electron acceptor as substituent.

The first example of nucleophilic diaddition reaction in the series of monocyclic 1,2,4-triazines has been found when refluxing 5,6-diphenyl-1,2,4-triazine-3-thione in ethanol, yielding the 5,6-diethoxy adduct **41** (Scheme 28) (72BSF4637).

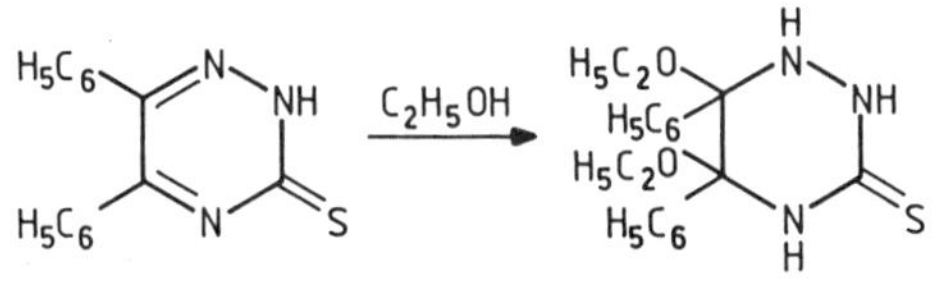

SCHEME 28

It has been shown that 1-alkyl-1,2,4-triazinium salts **42** add two molecules of indole at C-5 and C-6 to afford 5,6-diindolyl-substituted 1,4,5,6-tetrahydro-1,2,4-triazines **43** in good yield (Scheme 29) (86KGS1535; 87MI2; 88MI1). An attempt to extend this reaction to *NH*-triazinium salts failed. According to the ^{1}H-NMR spectra an equilibrium mixture of mono- and diadducts, **44** and **45**, is formed; even with an excess of indole, diadducts **45** are present only as minor products. With one equivalent of indole, only the C-5 monoadduct could be registered, as discussed above (Section III,A,1) (Scheme 29) (87MI2; 88MI1).

Similarly, in the reaction of 3-methylthio-1,2,4-triazine with methanol, the 5,6-dimethoxy adduct **46** is registered by ^{1}H NMR as only a minor adduct when a large excess of trifluoroacetic acid (TFA) is present (Scheme 30) (87MI2; 88MI1).

C. Cyclizations with Bifunctional Nucleophiles

The tendency of 1,2,4-triazines to give diadducts under appropriate conditions can be applied successfully to the synthesis of condensed

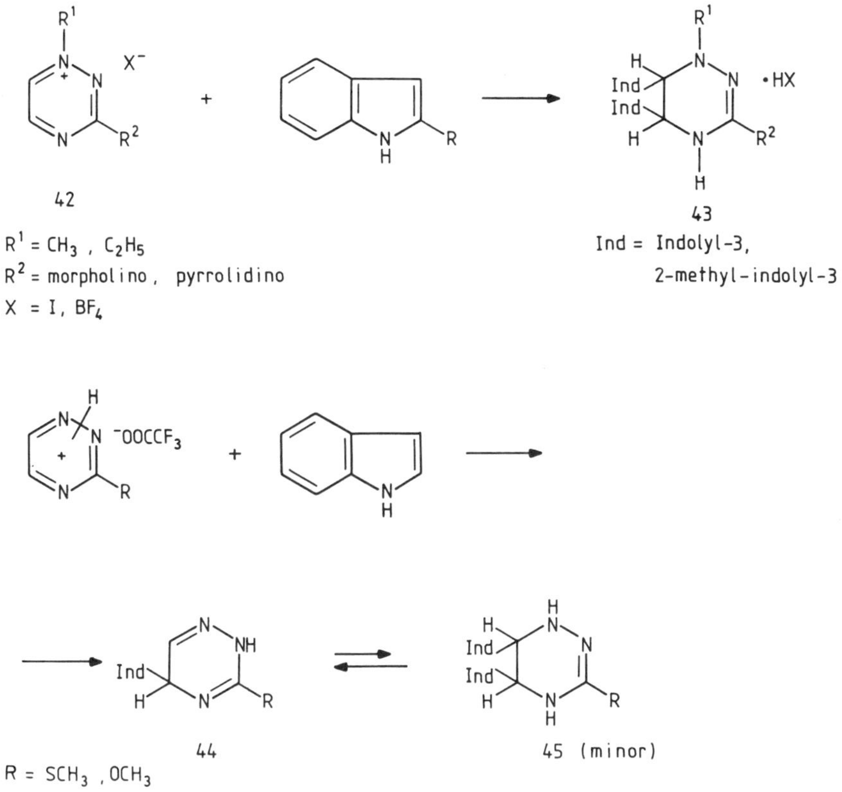

Sec. III.C]   MONOCYCLIC 1,2,4-TRIAZINES   95

42
$R^1 = CH_3$, C_2H_5
R^2 = morpholino , pyrrolidino
X = I, BF_4

43
Ind = Indolyl-3,
2-methyl-indolyl-3

44
R = SCH_3 , OCH_3

45 (minor)

SCHEME 29

46
minor

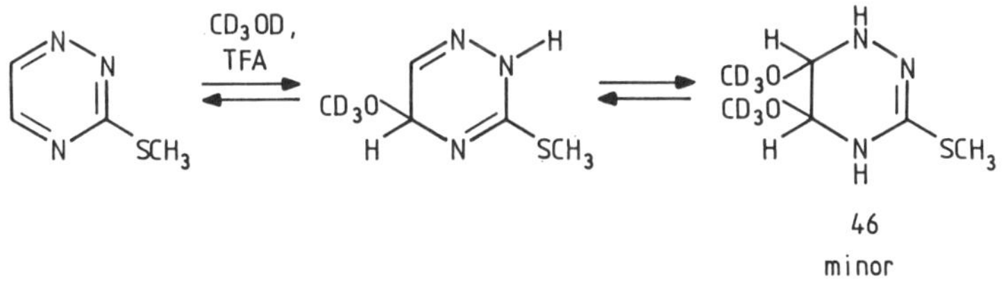

SCHEME 30

1,2,4-triazines. The use of bifunctional nucleophiles can lead to intramolecular adduct formation. Although this type of cyclization is well known in the chemistry of 1,4-diazines and has been exploited extensively for the synthesis of condensed pyrazines, quinoxalines, and pteridines (84UK1648; 85KGS1011; 88AHC301, 88MI2), only recently have examples of intramolecular diaddition reactions on the 1,2,4-triazine ring been reported in the literature (87MI2; 88AHC301, 88TL1431, 88TZ4(ip)). In particular, amides of acetoacetic acid proved to be effective C,N-bifunctional reagents for the synthesis of fused pyrrolo[3,2-e]-1,2,4-triazines 47 and **48** by direct annelation of the pyrrole ring to C(5)—C(6) of the triazine substrate in reactions with both *N*-protonated and *N*-alkyl-1,2,4-triazinium salts (Scheme 31) (87MI2; 88TL1431).

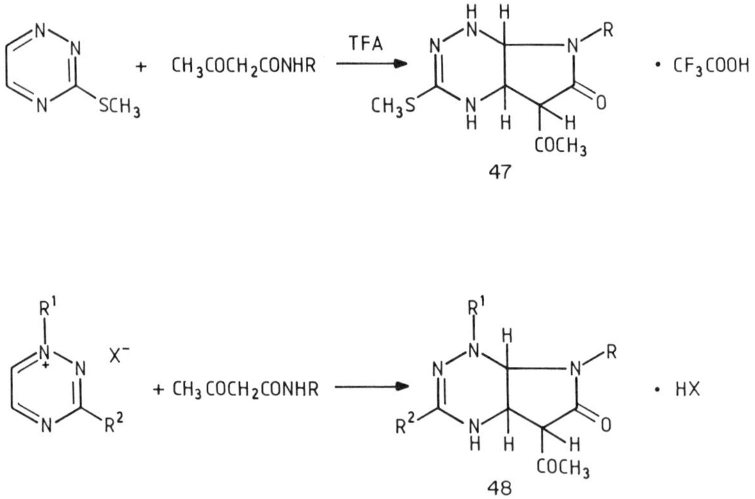

SCHEME 31

It has also been found that pyrrolo[3,2-*e*]triazines **50** are formed by cyclization of 1,2,4-triazinium salts with acylated ketenaminals **49** (Scheme 32) (88TZV(ip)). It is of interest that uncharged 1,2,4-triazines react with ketenaminals and enamines in an inverse electron demand Diels–Alder cycloaddition across C(3)—C(6) leading to ring transformation products (83T2869; 86CRV781; 87MI1) (see Section VI,B) rather than giving diaddition at C-5 and C-6.

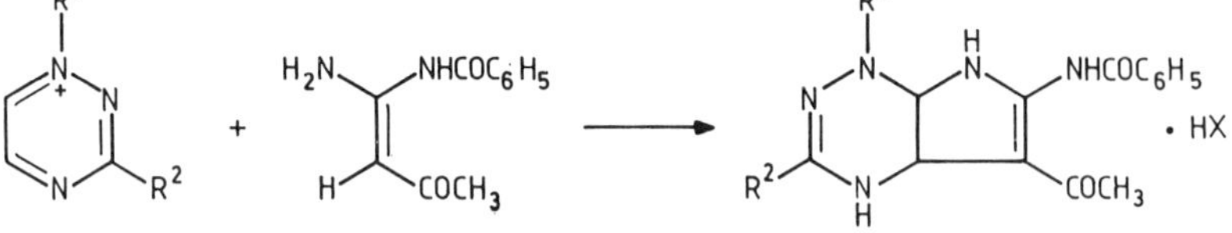

SCHEME 32

Cyclization products resulting from diaddition of bifunctional reagents at C-5 and C-6 of the 1,2,4-triazine ring are sometimes so unstable that they cannot be isolated, but only detected by NMR spectroscopy. For example, when reacting with *o*-phenylenediamine or other aromatic 1,2-

diamines, 1,2,4-triazinium salts give cycloadducts **51**, as measured in the ^{1}H-NMR spectra. It is of interest that the adduct **51** with the morpholino group at C-3 is found to be unstable and dissociates into quinoxaline (Scheme 33) (87KGS280). This is an interesting ring transformation reaction! However, cycloadduct **51** with the phenyl group at C-3 resulting from the reaction of 1-ethyl-3-phenyl-1,2,4-triazinium tetrafluoroborate with o-phenylenediamine is more stable and can be oxidized by potassium permanganate into 1,2,4-triazino[5,6-*b*]quinoxaline **52** (Scheme 33) (88UP1).

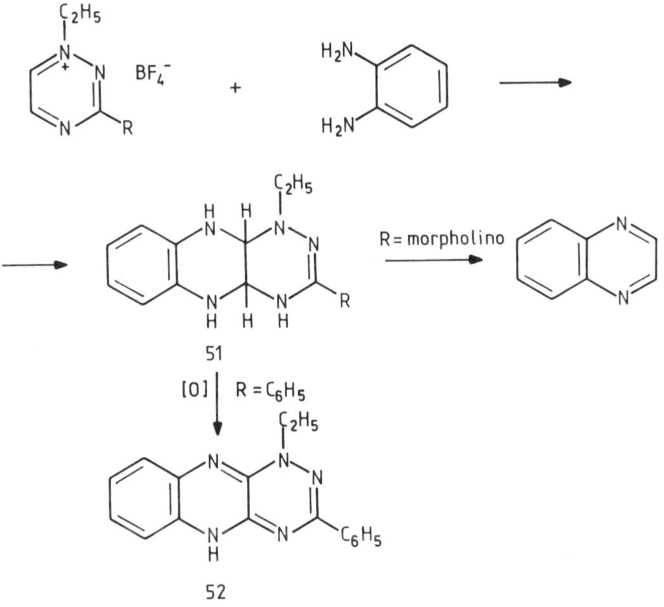

SCHEME 33

IV. Nucleophilic Substitution Reactions

When a substituent with a good leaving ability is present in the 1,2,4-triazine ring at C-3, C-5, and/or C-6, nucleophilic substitution reactions can occur. Many S_N reactions are described in the literature. In Sections IV,A–C, examples of displacement of one, two, and three nucleofugic groups are discussed. In this article, the displacement reactions are not exhaustively reviewed. We show only the fundamental reaction features, their value, and scope for application in the synthesis of functionalized 1,2,4-triazine derivatives. As far as the mechanisms of nucleophilic substitution reactions in the series of 1,2,4-triazines are concerned, the classical $S_N(AE)^{ipso}$ mechanism (68MI1) seems to be dominating in an overwhelming majority of cases although there are also many examples known where S_N(ANRORC) (78ACR462) and other mechanistic pathways are concurrently involved. The $S_N(AE)^{pso}$ mechanism suggests the formation of σ-adducts like **53**, which, however, have never been registered by spectroscopy due to the very fast departure of the leaving group from the sp^3 carbon (Scheme 34).

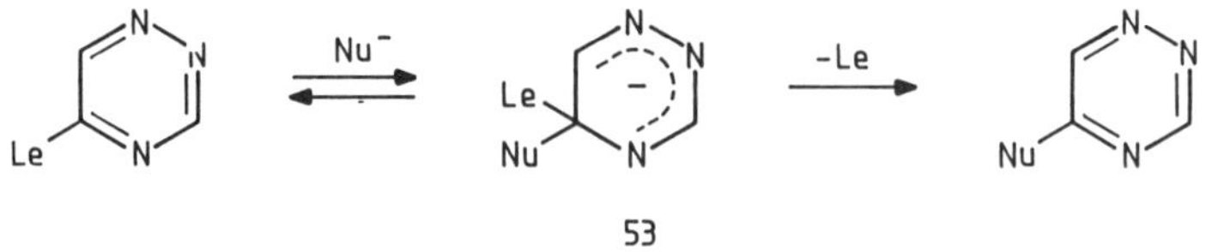

SCHEME 34

A. DISPLACEMENT OF ONE NUCLEOFUGIC GROUP

1. *Substitution at C-3*

The $S_N(AE)^{ipso}$ reactions at C-3 usually proceed very smoothly since the C-3 carbon is very susceptible to a nucleophilic attack and elimination of the nucleofugic substituent from the C-3 adduct allows the triazine ring to gain its 6π-electron system. In some cases, however, the concurrent addition at C-5 occurs, resulting in a lower yield of the $S_N(AE)^{ipso}$ product at C-3 or causing transformations of the ring (Scheme 35) (see also Section VI).

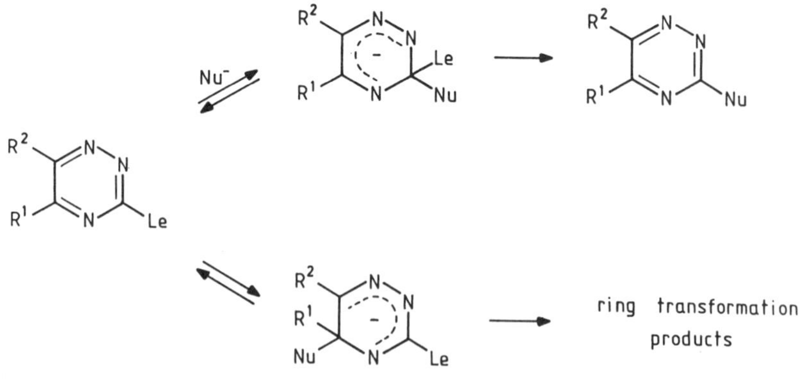

SCHEME 35

Methysulfonyl, arylsulfonyl, methoxy, and methylthio groups, as well as halogen atoms, are among the most frequently used substituents at C-3 and by analyzing literature data we can conclude that their leaving abilities decrease in the order methylsulfonyl > arylsulfony > halogen > methoxy > methylthio. To illustrate this difference in leaving group abilities we mention that both methylsulfonyl and p-tolylsulfonyl groups at C-3 are easily substituted by action of such nucleophiles as sodium ethoxide, hydrazine, amines, potassium cyanide, and CH-active methylene compounds. However, 3-(p-tolylsulfonyl)-1,2,4-triazines remain inert toward ketones, whereas the 3-methylsulfonyl group is displaced by even weak C-nucleophiles, such as acetophenone and cyclohexanone (Scheme 36) (82CPB152).

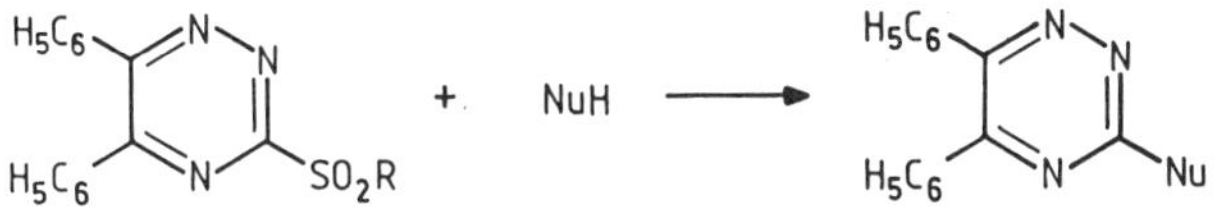

NuH = H_2N-NH_2, C_2H_5ONa, $n-C_4H_9-NH_2$, $C_6H_5NH_2$,
KCN, $NC-CH_2-COOC_2H_5$, $CH_3COCH_2COOC_2H_5$, $C_6H_5COCH_3$,
cyclohexanone

SCHEME 36

The ability of the 3-methylsulfonyl group to undergo the displacement reaction by action of C-, O-, and N-nucleophiles has been successfully employed for the synthesis of 3-alkynyl-substituted 1,2,4-triazines **54–56** (Scheme 37) (86TL431). The latter compounds proved to be very useful starting materials for providing a variety of fused pyridines via intramolecular Diels–Alder reactions (86TL431; 87JOC4280, 87T5145, 87T5159, 87TL379) (see Section VI,B).

Halogen atoms in 1,2,4-triazine substrates are also very widely used in nucleophilic substitutions. Thus, the C-3 chlorine atom is reported to be easily displaced by heteroatomic nucleophiles such as amines (72JCS(P1)1221; 75JAP27874; 77USP3979516, 77USP4008232; 78JPS737; 82CPB152; 83MI2), hydrazine (79JHC1393, 79USP4159375; 82CPB152; 86IJC(B)815; 87IJC(B)496), alkoxy (77USP4008232, 77USP4021553, 78HC(33)189; 82CPB152), azido (77USP3979516), and iodide (84H2245) anions. Some examples are given in Scheme 38. At the same time, 3-chloro-1,2,4-triazines remain unchanged when exposed to potassium cyanide solution or on treatment with CH-active compounds (82CPB152). It has been shown that substitution of the chlorine atom in 3-chloro-1,2,4-triazines by action of CH-active compounds can be reached under more severe conditions, for example, reflux in pyridine (86IJC(B)815). Substi-

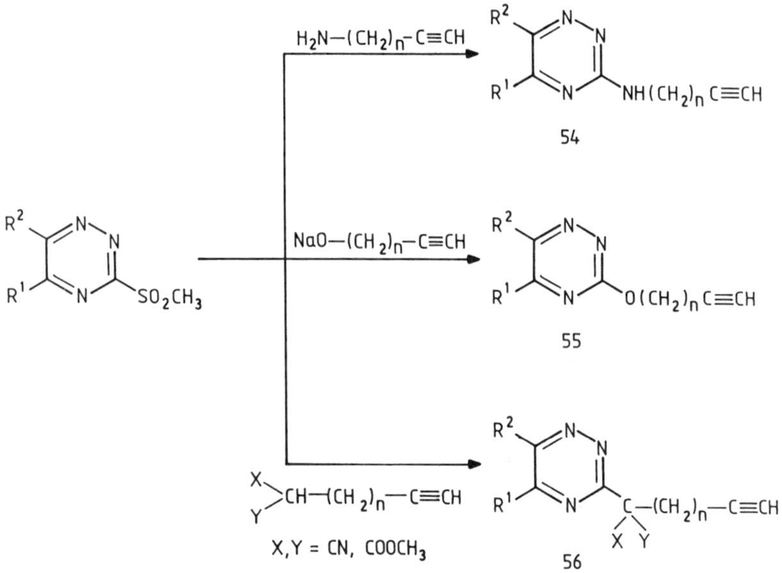

SCHEME 37

tution of the chlorine atom requires more carefully controlled conditions than displacement of the methylsulfonyl group. Sometimes it requires a certain substrate/reagent ratio, as exemplified by the reactions of 3-chloro-1,2,4-triazine with hydrazine and ethoxide. Hydrazine in excess is needed to obtain 3-hydrazino-1,2,4-triazine **57**, but only one equivalent of the ethoxide ion has to be used for the synthesis of the 3-ethoxy derivative **58**; otherwise by-products are formed (Scheme 38) (82CPB152).

Nevertheless, substitution of halogen in 1,2,4-triazines is a convenient way to introduce various substituents at C-3. The 3-hexenyloxy-1,2,4-triazine **59** has been obtained from the 3-chloro-1,2,4-triazine (Scheme 38) (86TL2747). Triazines with these type of unsaturated side chains are used extensively for the synthesis of condensed heterocycles by means of intramolecular Diels–Alder reactions (87MI1).

3-Bromo- and 3-iodo-1,2,4-triazines react with nucleophiles in a similar manner, but more easily than their chloro analogues. For example, 3-iodo-5,6-diphenyl-1,2,4-triazine is converted into 3-(2-trimethylsily-1-ethynyl)-5,6-diphenyl-1,2,4-triazine (**61**) by action of trimethylsilylacetylene (**60**) (Scheme 39), whereas no reaction between 3-chloro-5,6,-diphenyl-1,2,4-triazine and acetylene (**60**) has been observed under identical conditions (84H2245).

The reactivity of 3-halo-1,2,4-triazines can be enhanced by introduction of the N-oxide function. Nucleophilic substitutions on 3-bromo-1,2,4-tria-

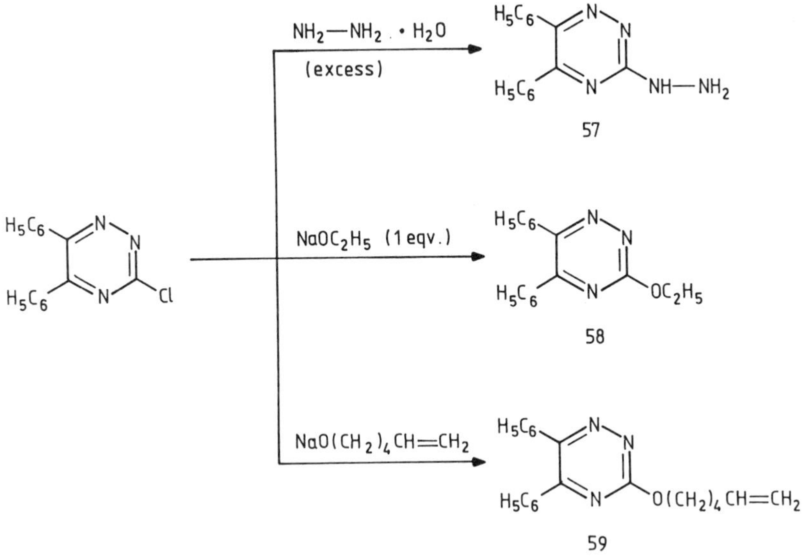

SCHEME 38

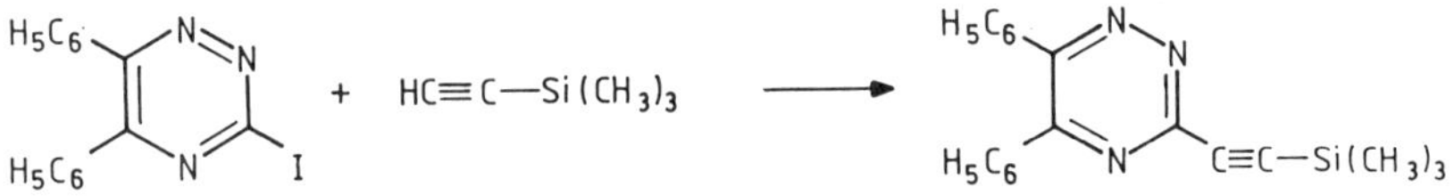

SCHEME 39

zin 2-oxide (62) with amines (77JOC546, 77JOC3498), hydrazines (77JOC546), thiols (77JOC3498), or the azide anion (77JHC1221) proceed under milder conditions relative to the corresponding 1,2,4-triazine substrate (Scheme 40).

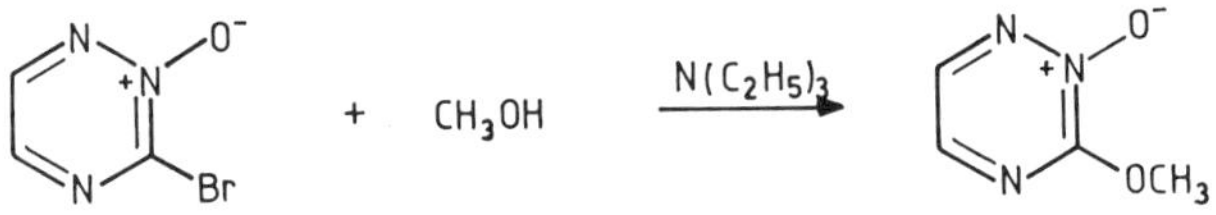

SCHEME 40

3-Methoxy- and 3-methylthio-1,2,4-triazines and their N-oxides are less active in the $S_N(AE)^{ipso}$ reactions than the 3-halo-1,2,4-triazines, but they are still suitable substrates to form 3-amino- (70JHC767; 71JHC689; 74ZN(B)792; 76CB1113, 76JHC807; 77JOC3498, 77USP3979516,

77USP4008232; 86KGS1535), 3-hydrazino- (70JHC767; 79JHC1393, 79JMC671, 79USP4159375; 80JOC5421; 87YZ301), and 3-ethoxy-1,2,4-triazines (70JHC767; 73BSF2493, 73JHC343; 77USP4013654, 77USP4021553; 78RTC273; 79USP4105434; 80JPS282).

Many authors report the conversion of 3-substituted 1,2,4-triazines and their N-oxides into the corresponding 1,2,4-triazin-3-ones in acidic or alkaline water solution (71JOC3921; 73JOC3277; 78HC(33)189; 79JHC817; 82CPB152; 83IJC(B)559). This hydrolysis reaction is not restricted to only typical nucleofugic substituents, such as chloro, alkylthio and alkoxy, but also 3-amino and 3-hydrazino groups can easily be replaced by the hydroxy function (Scheme 41).

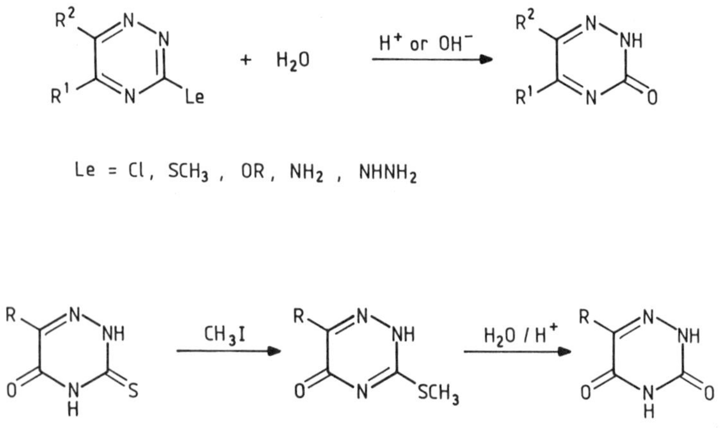

SCHEME 41

Similarly, 3-substituted 1,2,4-triazin-5-ones are converted into triazine-3,5-diones (65CCC3134; 66TL3115; 67CCC1295, 67CCC3572; 73MI2; 83JOC4585). In order to hydrolyze triazine-3-thiones the latter are usually methylated on the sulfur atom and then hydrolyzed. Since hydrolysis of the methylthio group is fast, it is a common practice to combine these two reactions in one step (Scheme 41) (65CCC3134; 67CCC1295, 67CCC3572).

Many aminations at C-3 have been found to occur via addition at C-5. Ring opening and ring closure constitute such reactions, which occur according to the S_N(ANRORC) mechanism. These reactions feature a so-called ring degenerate transformation (85T237), and will be dealt with in Section VI,A.

2. Substitution at C-5

Most publications on nucleophilic substitution reactions at C-5 deal with 5-chloro-1,2,4-triazines as substrates. The S_N reactions with amines (72LA173; 87JOC4287), hydrazine (78HC(33)189), sodium hydrosulfide (78HC(33)189), alkoxide anions (87JOC4287), sodium iodide (84H2245), and carbanionic reagents (84H2241, 84H2245; 86H239, 86H1243; 87JOC4287) have been reported in the literature. Among them is a novel and convenient route for the preparation of 5-alkyl-1,2,4-triazines **66** (84H2241; 86H239). Interaction of 3-phenyl-5-chloro-1,2,4-triazines **63** with ethylidenetriphenylphosphoranes **64** has been found to give, in excellent yield, 5-alkyl-1,2,4-triazines **66** via the intermediacy of **65** (Scheme 42).

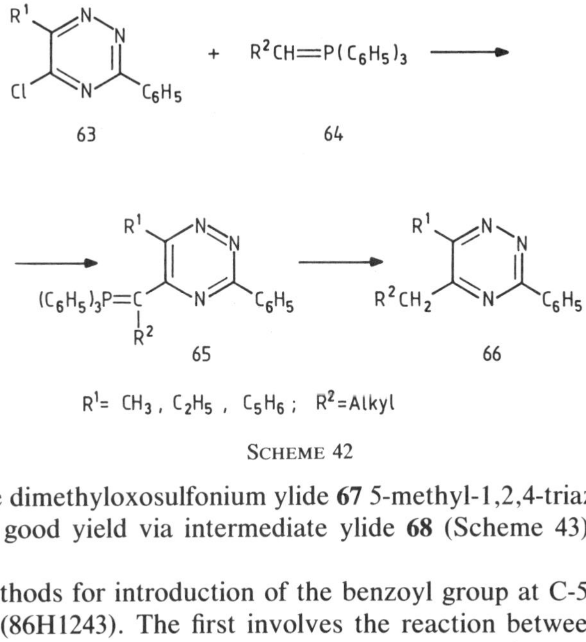

$R^1 = CH_3, C_2H_5, C_5H_6$; $R^2 = Alkyl$

SCHEME 42

With the dimethyloxosulfonium ylide **67** 5-methyl-1,2,4-triazines **69** are formed in good yield via intermediate ylide **68** (Scheme 43) (84H2241; 86H239).

Two methods for introduction of the benzoyl group at C-5 have been described (86H1243). The first involves the reaction between 5-chloro-1,2,4-triazine **70** and benzaldehyde in the presence of the catalyst 1,3-dimethylbenzimidazolium iodide (**71**) and sodium hydride. The C-nucleophilic ylide **72** formed under these conditions reacts with 5-chloro-3-phenyl-1,2,4-triazine **70** to yield the 5-benzoyl-substituted 1,2,4-triazine **73**, simultaneously regenerating the catalyst **71** (Scheme 44) (86H1243).

The second method for the synthesis of a 5-benzoyl-1,2,4-triazine (**75**) makes use of 1-benzoyl-1-phenylacetonitrile (**74**) as reagent. The reaction is described in Scheme 45 (86H1243).

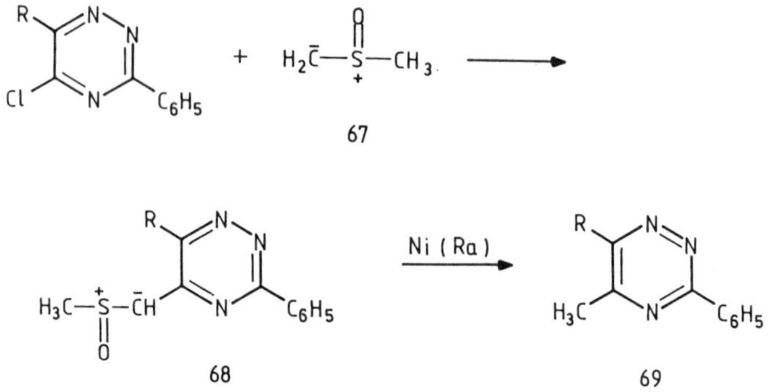

SCHEME 43

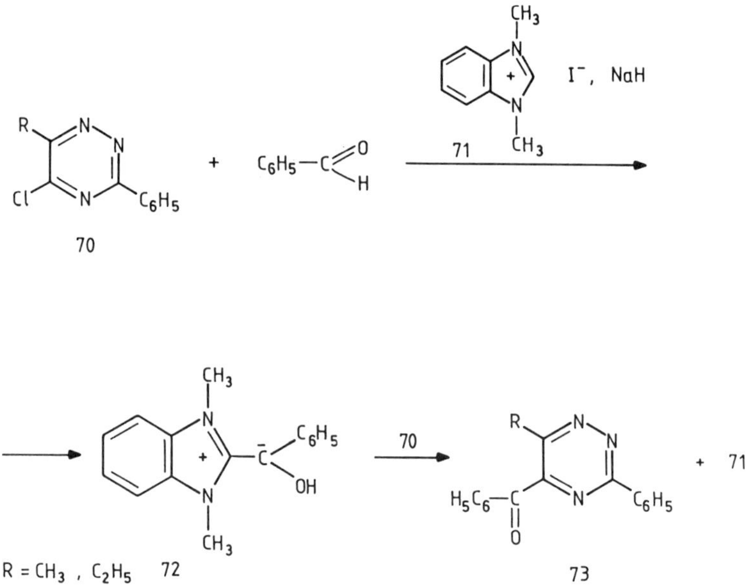

SCHEME 44

A synthesis of 5-(alkyn-1-yl)-1,2,4-triazines **77** has been developed, in which, by activation with the palladium-containing catalyst, acetylenes effectively react with 5-chloro- and 5-iodo-1,2,4-triazines **76** (Scheme 46) (84H2245). The synthesis of 5-(ω-alkynyl)-1,2,4-triazines **79** has been described using as nucleophile acetylenes **78** featuring a nucleophilic group (e.g., hydroxy or CH-activated group) (Scheme 46).

Besides displacement of the halogeno atom at C-5, many other substituents are easily replaced by nucleophiles. A large variety of N-, S-, O-,

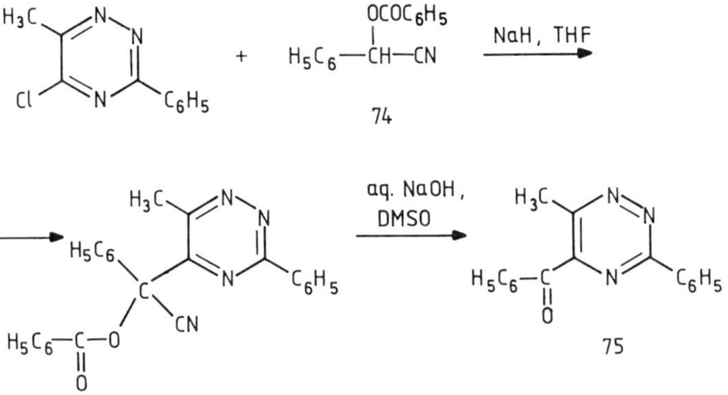

Sec. IV.A] MONOCYCLIC 1,2,4-TRIAZINES 105

SCHEME 45

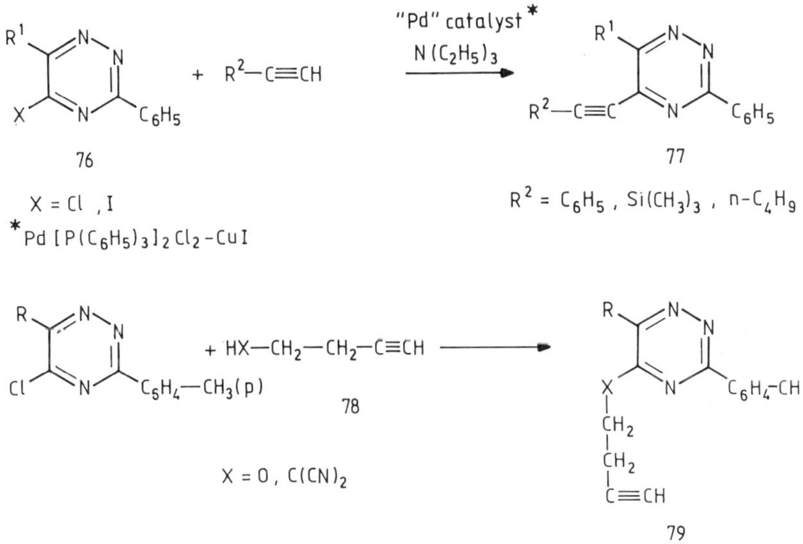

SCHEME 46

and S-substituents at C-5 in 1,2,4-triazines are obtained by the S_N reactions using 5-cyano-1,2,4-triazines as substrates (86MI3). There are also several reports on substitution of methylthio, mercapto, and alkoxy groups at C-5 by action of amines (70JAP25903; 76CB1113), hydrazines (85LA857), and other nucleophiles (78HC(33)189). Acid hydrolysis of 5-amino-, 5-hydrazino-, and 5-methylthio-1,2,4-triazines proceeds smoothly as well (72CCC2221; 79JHC817).

3. Substitution at C-6

Relatively little experimental work is known concerning the displacement of nucleofugic groups at C-6 of the 1,2,4-triazine ring. A few papers report the formation of 6-methoxy- (76CB1113, 76CCC465; 78JOC2514), 6-methylthio- (85JHC1329), and 6-amino-1,2,4-triazines (76CB1113) on treatment of 6-chloro- and 6-bromo-1,2,4-triazines with sodium salts of methanol, mercaptomethane, and amines, respectively.

B. DISPLACEMENT OF TWO NUCLEOFUGIC GROUPS

Of all the nucleophilic substitutions in 1,2,4-triazines those which contain the two leaving groups at C-3 and C-5 have been studied most extensively. Triazines in which both C-3 and C-5 contain the same nucleofugic group have proved to be of particular interest. This is probably due to the fact that these starting materials are easily available.

1. Substitution Reactions at C-3 and C-5

Halogen atoms at both C-3 and C-5 in 1,2,4-triazines can be replaced by N-, O-, and S-nucleophilic reagents under appropriate conditions. 3,5-Diamino- (76CB1113), 3,5-dimethylthio- (78HC(33)189), and 3,5-dimethoxy-1,2,4-triazines (75CCC2340) are examples of compounds which have been obtained on heating 3,5-dichloro-1,2,4-triazines with a large excess of the corresponding nucleophilic reagents. Under milder conditions and with a small excess of amines or thiols only 5-substituted 3-chloro-1,2,4-triazines are formed (78HC(33)189). It is evident from these results and other investigations (76CB1113; 78HC(33)189; 84MI3) that the chlorine atom at C-5 is more easily displaced than the chlorine atom at C-3 (Scheme 47).

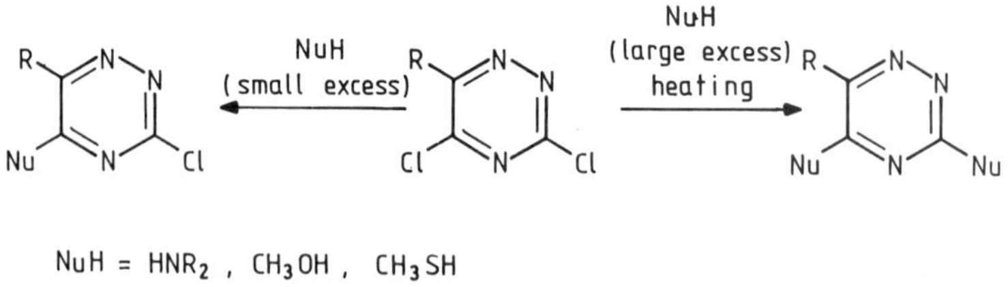

NuH = HNR$_2$, CH$_3$OH , CH$_3$SH

SCHEME 47

This reaction course is common for all the 1,2,4-triazine substrates containing the same leaving groups at C-3 and C-5. Mono- or disubstitution reactions with amines and hydrazines yield 5-amino- (hydrazino)

(72ZN(B)818; 78HC(33)189; 79JHC817; 84MI4) or 3,5-diamino-(dihydrazino)-1,2,4-triazines (Scheme 48) (70RRC1409; 72AG348; 74MIP1; 76CB1113, 76H1341; 78HC(33)189; 87MI4).

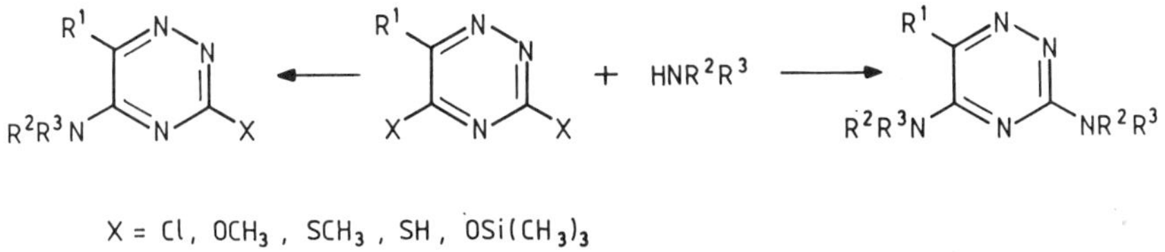

X = Cl, OCH_3, SCH_3, SH, $OSi(CH_3)_3$

SCHEME 48

In a similar way, by the action of sodium methoxide, both mono- and disubstitution reactions have been performed on 3,5-dimethylthio-1,2,4-triazines; substitution of the methylthio groups at C-5 and C-3 has been found to occur sequentially (Scheme 49) (73GEP2256604; 75CCC2340; 78HC(33)189).

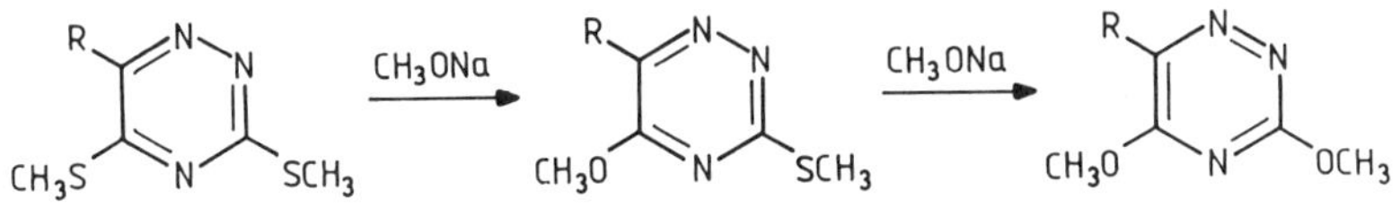

SCHEME 49

All experimental data on the behavior of 3,5-disubstituted 1,2,4-triazines in the S_N reactions indicate a greater reactivity of the C-5 position. An exception is the reaction between ethyl-3,5-dimethylthio-1,2,4-triazine-6-carboxylate (**80**) and hydrazine or amines, which results in a mixture of 3- and 5-hydrazino (amino) derivatives, **81** and **82** (Scheme 50) (76JCS(P2)2521, 76JMC845; 77JHC729; 87MI4).

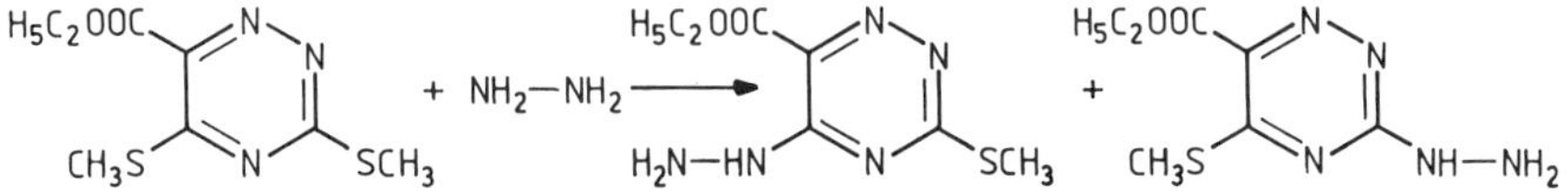

SCHEME 50

The greater reactivity of the C-5 position is not predicted by the order of ground-state charge distribution found in the 1,2,4-triazine ring (see Section II,A). The transition state leading to C-5 adduct formation seems to be promoted, since the C-5 adduct has a greater stability than the C-3 adduct. This is due to the important *p*-quinoid contribution, putting a considerable negative charge on N-2, being apparently a major factor determining the composition of the S_N reaction products. This is substantiated by the fact that when a nucleophilic addition at C-3 is facilitated by an electron acceptor at C-6, like in the case of compound **80,** the formation of both C-3 and C-5 substitution products may occur concurrently (Scheme 50).

A hydroxy group attached to the 1,2,4-triazine ring exists predominantly in the oxo form (78HC(33)189; 84MI3, 84OPP199). Therefore, it can be expected that 1,2,4-triazin-3-ones containing a nucleofugic group at C-5 react with nucleophiles exclusively at C-5 to afford 5-amino- (70JAP09546, 70KGS986; 72GEP2206395; 78HC(33)189), 5-hydrazino(72ZN(B)818; 78HC(33)189), and 5-sulfamido-1,2,4-triazin-3-ones (70JAP09546) in good yield (Scheme 51). In all these reactions no indication for the replacement of the C-3 oxygen was reported.

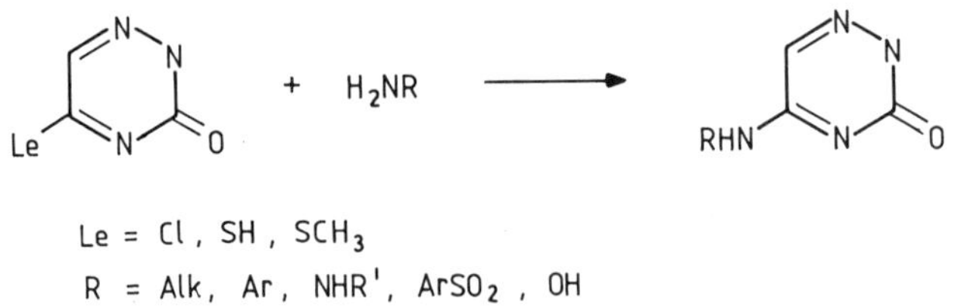
N
N
+ H2NR
Le
N
O
RHN
N
O
Le = Cl, SH, SCH3
R = Alk, Ar, NHR', ArSO2, OH

SCHEME 51

Similarly, 1,2,4-triazin-5-ones and 1,2,4-triazine-5-thiones **84** containing a good leaving group at C-3 react with nucleophiles very smoothly to produce the corresponding C-3 S_N products **85** (Scheme 52) (78HC(33)189; 84MI3, 84S983; 86JCS(P1)2037; 87MI5).

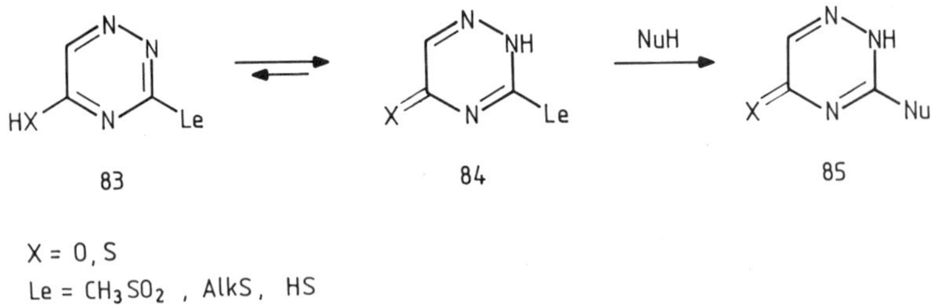
HX
Le
83
X
NH
Le
84
NuH
X
NH
Nu
85
X = O, S
Le = CH3SO2, AlkS, HS

SCHEME 52

Many publications deal with the substitution of methylsulfonyl (80MI2), alkylthio (68ACH319, 68CCC2513; 69BSF2492, 69MI1, 69MIP1, 69USP3412083; 70JAP09545, 70JMC288, 70RRC1409; 72JHC995; 73JPR221; 74JPR163, 74JPR667, 74ZN(B)792; 75CCC2340; 82JHC1583;

84CB1077, 84CB1083, 84S983), and mercapto (69ACH181, 69FRP1519180; 71CCC4000; 74GEP2236340; 77MI1) groups at C-3 in 1,2,4-triazin-5-ones by the action of N- and O-nucleophiles. In all of these reactions the oxygen group at C-5 remains unchanged (Scheme 52). However, substitution of both the methylthio group at C-3 and the hydroxy group at C-5 occurs in 3-methylthio-1,2,4-triazin-5-ones by the action of amino alcohols (Scheme 53) (69ACH181).

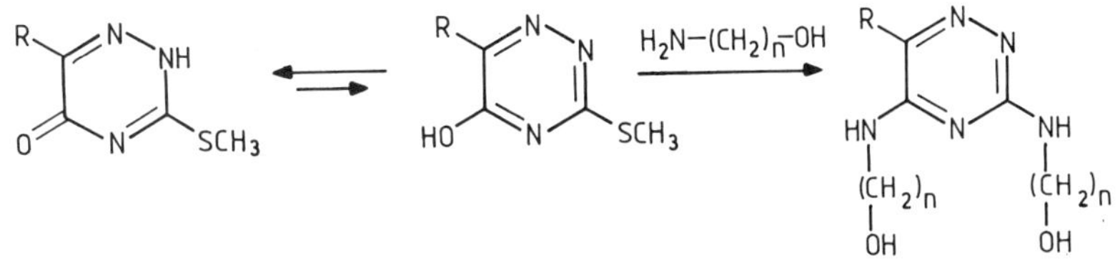

SCHEME 53

Many examples available in the literature show that most of the nucleophilic displacement reactions with 1,2,4-triazines containing different nucleofugic groups at C-3 and C-5 take place at C-5 (72JCS(P1)2316; 77JHC729; 80MI1; 84MI5; 85JOC2293; 87JOC4287). In particular, azinyl groups attached to C-5 in 3-methylthio-1,2,4-triazines **86** and **87** are found to have easily displaceable groups when reacted with aromatic amines and phenols (Scheme 54) (84MI5).

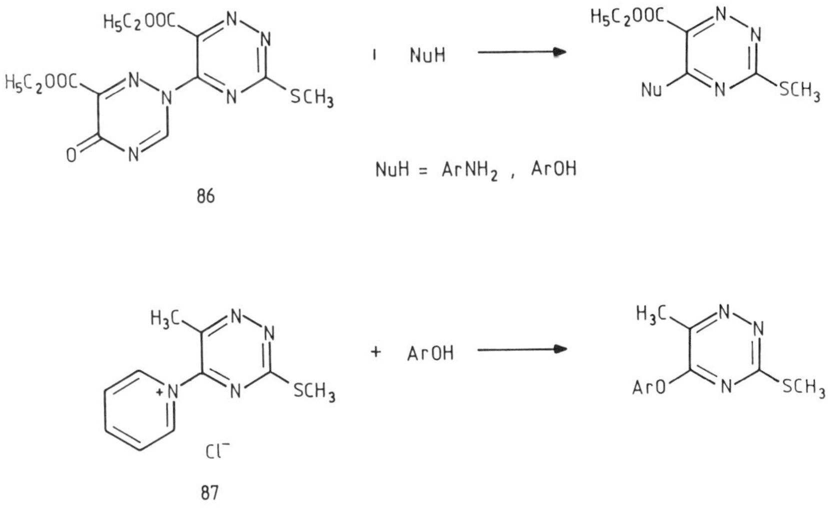

NuH = ArNH$_2$, ArOH

86

87

SCHEME 54

Disubstitution reactions on 1,2,4-triazine substrates with different leaving groups at C-3 and C-5 can also be performed under appropriate conditions, as is illustrated by amination of a number of 1,2,4-triazines **88** on heating with excess amine (Scheme 55) (72AG348; 73GEP2163873).

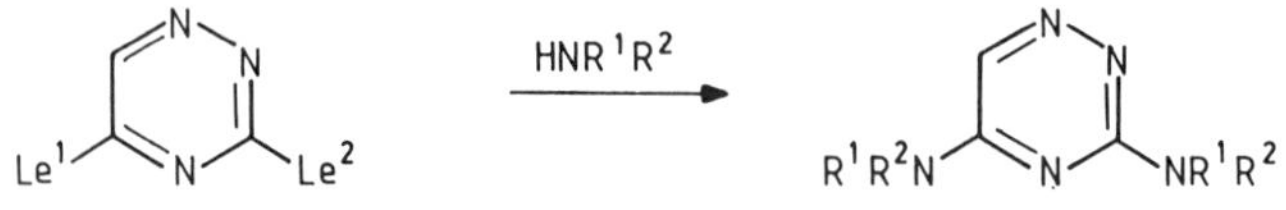

SCHEME 55

2. Substitution Reactions at C-5 and C-6

Only a few papers report on S_N reactions with 1,2,4-triazines containing nucleofugic groups at C-5 and C-6. One example is the reaction of 5,6-dimethoxy-1,2,4-triazine with dimethylamine at 20°C to give a mixture of 5-dimethylamino-6-methoxy-1,2,4-triazine (**89**) and 5-methoxy-6-dimethylamino-1,2,4-triazine (**90**) in the ratio 4 : 5 (Scheme 56) (76CB1113). Attempts to obtain the 5,6-bis(dimethylamino) compound by carrying out the reaction at higher temperature have been unsuccessful. Instead, 5-dimethylamino-1,2,4-triazin-6-one was obtained (Scheme 56) (76CB1113). Although the evidence is not overwhelming, this result seems to indicate that C-5 is more reactive than C-6, in accordance with the discussion presented before (see Section IV,B,1).

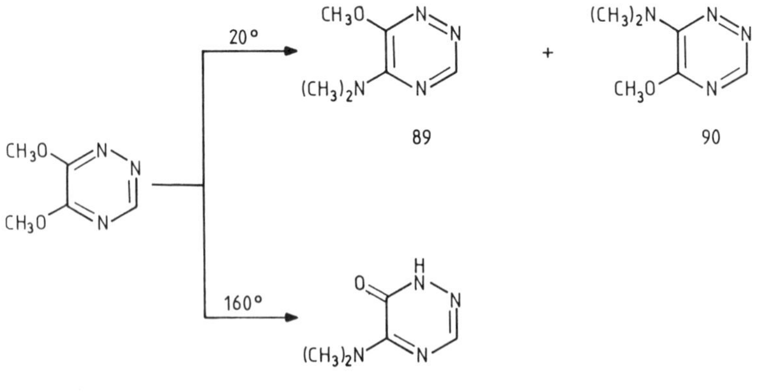

SCHEME 56

The reaction of 3-amino-6-bromo-1,2,4-triazin-5-one with hydrazines results in 6-hydrazino-substituted 1,2,4-triazin-5-ones (Scheme 57). No indication for substitution at C-5 has been observed (75CCC2680).

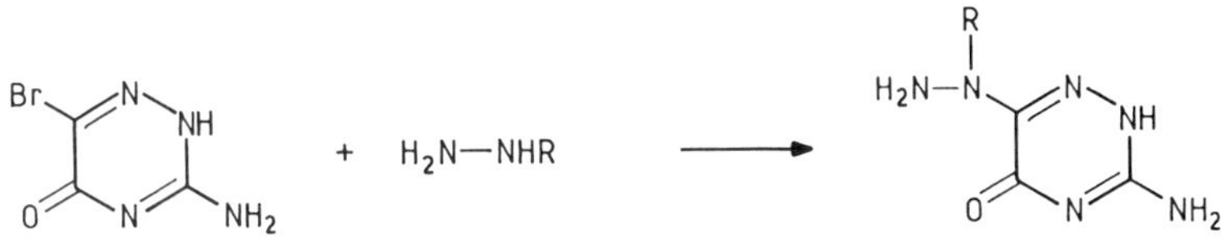

SCHEME 57

3. Substitution Reactions at C-3 and C-6

Several examples of S_N reactions with 1,2,4-triazine substrates containing at C-3 and C-6 the same leaving groups indicate that the displacement of a nucleofugic group at C-3 is somewhat more preferred. Thus, only the substituted 3-ethoxy-1,2,4-triazine **92** is formed from reaction of 3,6-dimethylthio-1,2,4-triazine-5-carboxamide (**91**) with sodium methoxide (Scheme 58) (85JHC1329). In the reaction between triazine **91** and ammonia the 3-amino derivative **93** is the major product, and 6-amino-3-methylthio-1,2,4-triazine-5-carboxamide **94** is formed only in poor yield (Scheme 58) (85JHC1329).

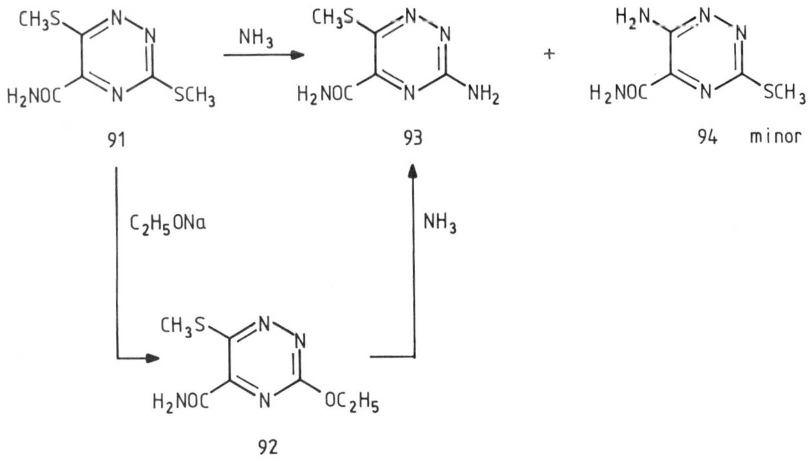

SCHEME 58

By contrast, when comparing the relative reactivity of the C-3 and C-6 positions in 3,6-dichloro-5-amino-1,2,4-triazine (**95**) with sodium methoxide, a different reaction pattern was observed. Substitution takes place at C-6 and not at C-3, yielding 5-amino-3-chloro-6-methoxy-1,2,4-triazine (**96**). When the reaction is carried out in methanol in the presence of hydrochloric acid, C-3 substitution is observed, yielding 5-amino-6-chloro-3-methoxy-1,2,4-triazine (**97**) (Scheme 59) (75CCC2680). In this case we deal with N^2H- and N^4H-triazinium salts stabilized by conjugation with the amino group. In these azinium salts the electrophilicity at C-3 is likely more enhanced than at C-6.

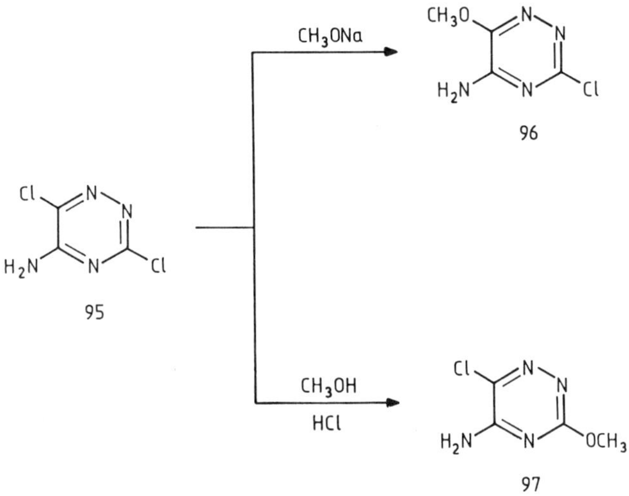

SCHEME 59

A rather complex mixture of products, i.e., **99**, **100**, **93**, and **94**, was obtained in the reaction of 3-methylthio-6-chloro-1,2,4-triazin-5-carboxamide (**98**) with ammonia (Scheme 60) (85JHC1329). It has been established that either of the nucleofugic groups in **98** are replaced by ammonia to form the 1,2,4-triazines **99** and **100** (Scheme 60). Substitution of the methylthio group at C-3 generates the nucleophilic methylthio anion, which causes a conversion of unreacted **98** into 3,6-dimethylthio-1,2,4-triazine **101**. The latter undergoes subsequent reaction with ammonia to produce both isomeric S_N products **93** and **94** (Scheme 60, see also Scheme 58).

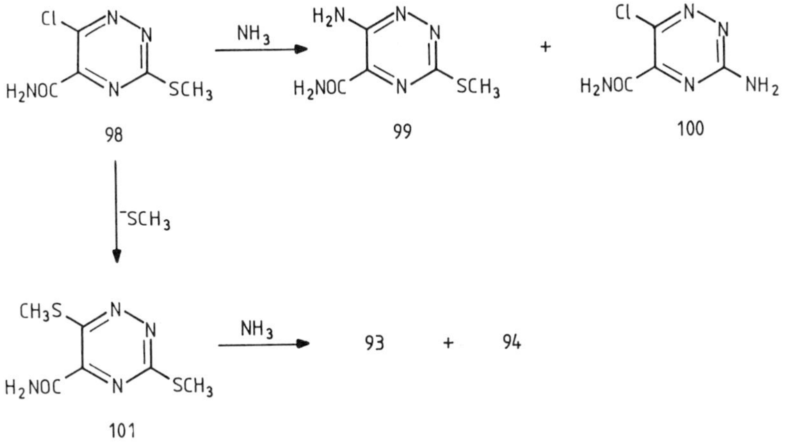

SCHEME 60

C. DISPLACEMENT OF THREE NUCLEOFUGIC GROUPS

The behavior of 3,5,6-trihalo-1,2,4-triazines in S_N reactions is the most studied nucleophilic substitution reaction in trisubstituted 1,2,4-triazines. It has been shown that in all these substitution reactions C-5 is, as expected, the most reactive position. However, the relative reactivity at C-3 and C-6 depends on the nature of the nucleophile employed (76CCC465). In the reaction of 3,5,6-trichloro-1,2,4-triazine (**102**) with some uncharged nucleophiles, reactivity decreases in the order C-5 > C-3 > C-6, but in other cases the reactivity order is observed to be C-5 > C-6 ≅ C-3 <76CCC465>. Thus, only 5-methoxy-(75CCC2680; 76CB1113) or 5-amino-3,6-dichloro-1,2,4-triazines (75CCC2680; 76CB1113, 76CCC465) are formed with 1 equivalent of sodium methoxide or amine, respectively (Scheme 61).

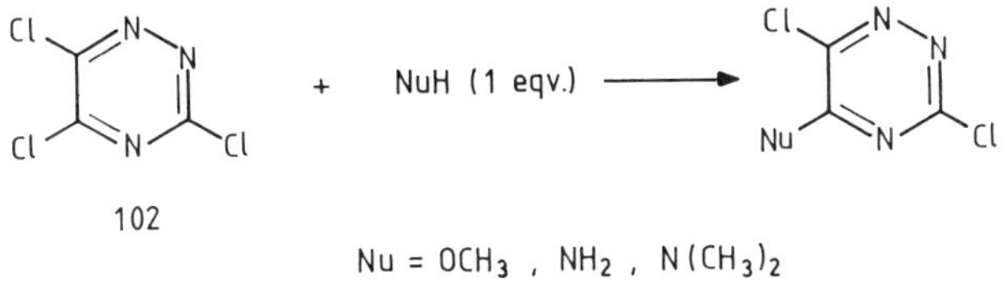

Nu = OCH_3 , NH_2 , $N(CH_3)_2$

SCHEME 61

With 2 equivalents of sodium methoxide a mixture of 3,5-dimethoxy-6-chloro-1,2,4-triazine (**103**) and 5,6-dimethoxy-3-chloro-1,2,4-triazine (**104**) is formed (75CCC2680; 76CB1113), while a larger excess of sodium methoxide leads to the introduction of three methoxy groups into the 1,2,4-triazine ring (Scheme 62) (75CCC2680). Full substitution of the halogen atoms in triazine **102** by fluorine atoms has been attained with potassium fluoride at high temperature (82JCS(P1)1251). Amination of **102** with 2–5 equivalents of aliphatic amines results in 3,5-diamino derivatives **105** (Scheme 62) (69CCC1104; 77JHC1221), showing the preference of the neutral nucleophile for attack at C-3 rather than at C-6.

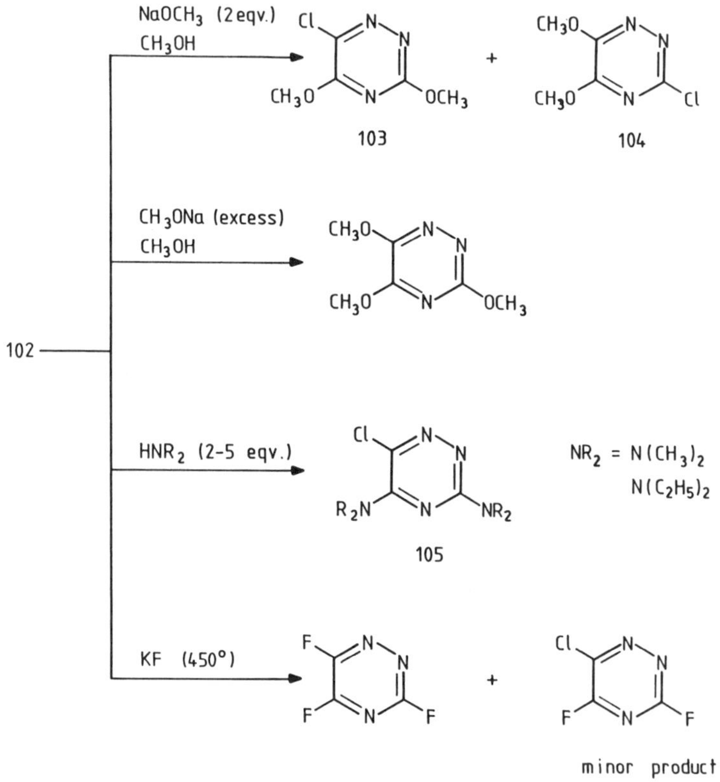

SCHEME 62

3,5,6-Trifluoro-1,2,4-triazine (**106**) behaves similarly to the trichloro compound toward nucleophilic reagents, but is more reactive. When treated with methanol at room temperature it gives a mixture of 5,6-dimethoxy-3-fluoro-1,2,4-triazine (**109**) and 3,5-dimethoxy-6-fluoro-1,2,4-triazine (**110**) in the ratio 2 : 1 (82JCS(P1)1251). Reactions of **106** with di-ethylamine and *p*-chloroaniline result in 3,5-diamino derivatives **107**. However, only 3,6-difluoro-5-amino-1,2,4-triazine (**108**) is formed on passing ammonia through the solution of triazine **106** in THF (Scheme 63) (82JCS(P1)1251).

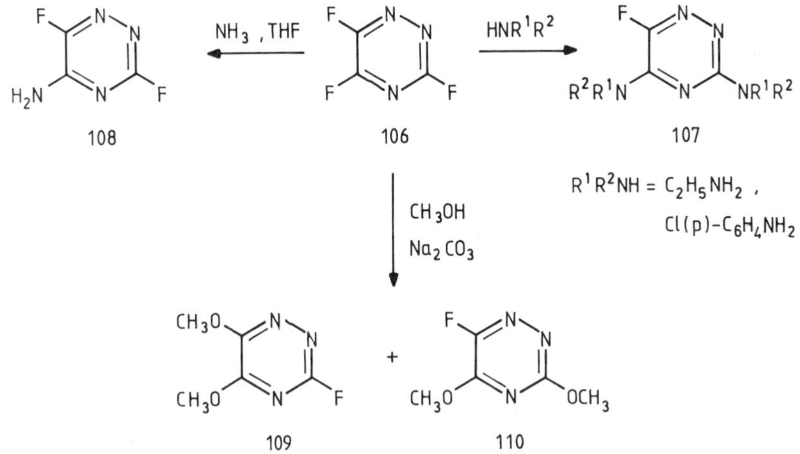

SCHEME 63

3,5,6-Trimethoxy-1,2,4-triazine reacts with dimethylamine in methanolic solution at room temperature exclusively at C-5 to give the corresponding amino compound **111** (76CCC465). In addition, the reaction of 3,5,6-trimethylthio-1,2,4-triazine with hydrazine yields exclusively the C-5 substitution product (Scheme 64) (71RRC135, 71RRC311).

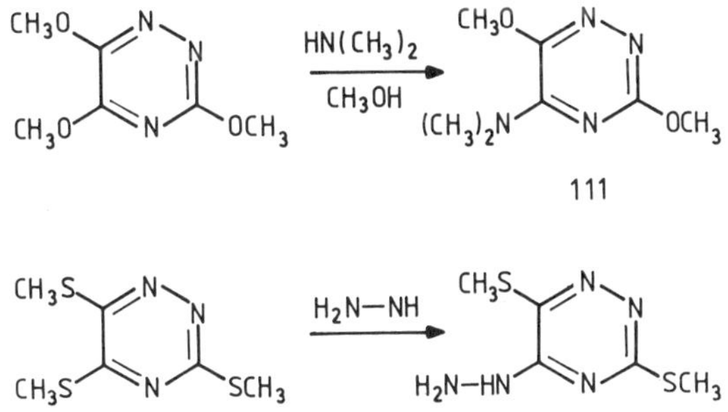

SCHEME 64

The preference of the C-5 position for nucleophilic substitution is further supported by the results of reactions of 1,2,4-triazines that posess groups with better leaving abilities at C-3 and/or C-6 than at C-5. For instance, substitution of the methoxy group at C-5, but not of halogen atoms at C-3 or C-6, has been found to occur in the reaction of 3,6-dichloro-5-methoxy-1,2,4-triazine with ammonia (Scheme 65) (75CCC2680). Analogously, 6-chloro-5-cyano-3-methylthio-1,2,4-triazine undergoes the S_N reaction at C-5 on treatment with ammonia, sodium alkoxides, and dimethylaniline (Scheme 65) (85JOC2293).

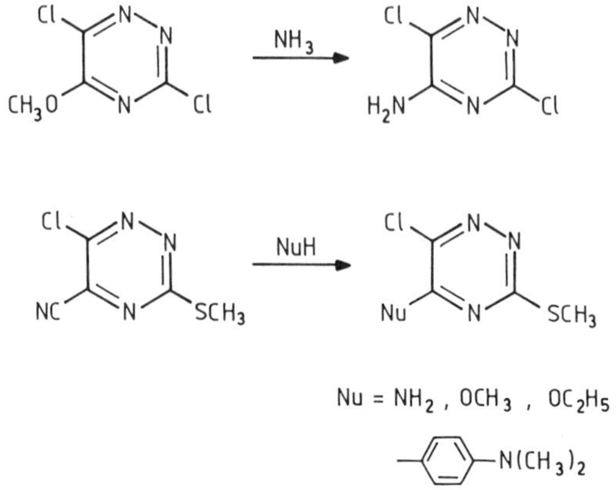

Nu = NH_2 , OCH_3 , OC_2H_5

—⟨⟩—$N(CH_3)_2$

SCHEME 65

The reactions of 5-chloro-6-cyano and 5,6-dialkylthio-substituted 1,2,4-triazin-3-ones with amines and hydrazines also confirms the preference of the displacement reaction at C-5 (69RRC135; 71CCC3507, 71RRC135, 71RRC311; 73MI1, 73MIP1).

In reactions of 5-chloro-6-cyano-3-methoxy- and 6-amino-5-cyano-3-ethoxy-1,2,4-triazines with water, displacement also takes place at C-5 yielding the corresponding 1,2,4-triazin-5-ones (74AJC1781; 85JOC2293). When hydrolysis of triazines containing three potential leaving groups was carried out in acid solution, formation of triazine-3,5-diones was observed (71RRC135, 71RRC311; 74MIP2; 75CCC2680; 84KGS557; 85RRC233). These examples show that nucleofugic groups at C-6, such as chloro, cyano, methoxy, alkylthio, amino, and dialkylamino, usually do not undergo displacement by water (71RRC135, 71RRC311; 74AJC1781, 74MIP2; 75CCC2680; 82JCS(P1)1251; 84KGS557; 85JOC2293, 85RRC233). Exceptional cases are the conversion of 6-amino-1,2,4-triazine-3,5-dione into 1,2,4-triazine-3,5,6-trione (61JOC1118) and the hydrolysis of 3-chloro-5-dimethylamino-6-methoxy-1,2,4-triazine **(112)** (Scheme 66). In the latter case, 3-chloro-5-dimethylamino-1,2,4-triazin-6-one **(113)** is formed as the minor product in 18% yield, together with 6-methoxy-1,2,4-triazine-3,5-dione **(114)** (Scheme 66) (76CCC465).

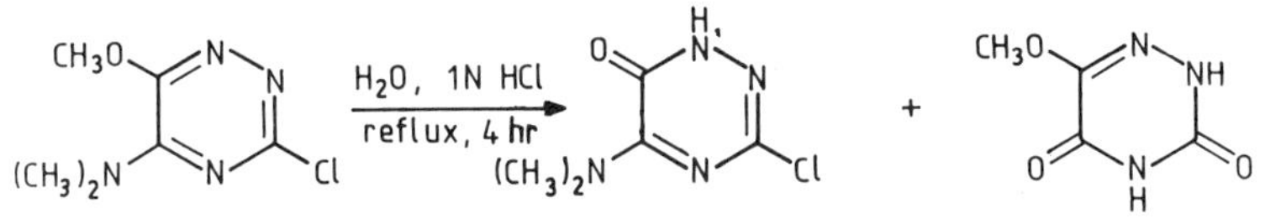

SCHEME 66

D. INTRAMOLECULAR NUCLEOPHILIC SUBSTITUTION REACTIONS

As discussed above, a number of triazines containing nucleofugic groups at C-5 and C-6 can undergo disubstitution reactions (see Schemes 62 and 63). If the appropriate conditions are chosen, these S_N reactions could be applied for synthesizing condensed 1,2,4-triazines, provided bifunctional reagents are used as nucleophiles. In fact, cyclizations with bifunctional nucleophiles have proved to be a very useful methodology in the 1,4-diazine series for the synthesis of many types of condensed pyrazines (85KGS1011; 88AHC301). Surprisingly, only a few examples of cyclizations with bifunctional nucleophiles based on the displacement of nucleofugic groups at C-5 and C-6 of the 1,2,4-triazine ring have been reported (85JOC2293). In particular, interaction of 6-chloro-5-cyano-3-methylthio-1,2,4-triazine with benzamidine results in imidazo[4,5-e]-1,2,4-triazine 115 via a two-step nucleophilic disubstitution reaction (Scheme 67) (85JOC2293). In addition, several publications report intramolecular cyclization of 1,2,4-triazines, in which a nucleophilic sidechain substituent attached to C-5 or C-6 of the triazine ring reacts with a nucleofugic group at C-6 or C-5. Using this methodology the synthesis of pyrazolo[4,3-e]-1,2,4-triazines (84MI4), pyrimido[5,4-e]-1,2,4-triazines 116 (85JOC2293), and a number of tricyclic system derivatives 117 (75JAP48697, 75JAP48698; 82JHC313; 84CCC2628) has been developed (Scheme 67).

Formation of imidazo[4,5-e]-1,2,4-triazines 119 has been observed to occur when 6-ethylamino-5-methylamino-1,2,4-triazines 118 react with benzamidine. However, in this case only the methylamino group at C-5 is substituted by the amidine, while the ethylamino group at C-6 participates itself as a nucleophile in an intramolecular cyclization reaction (Scheme 68) (78CPB3154).

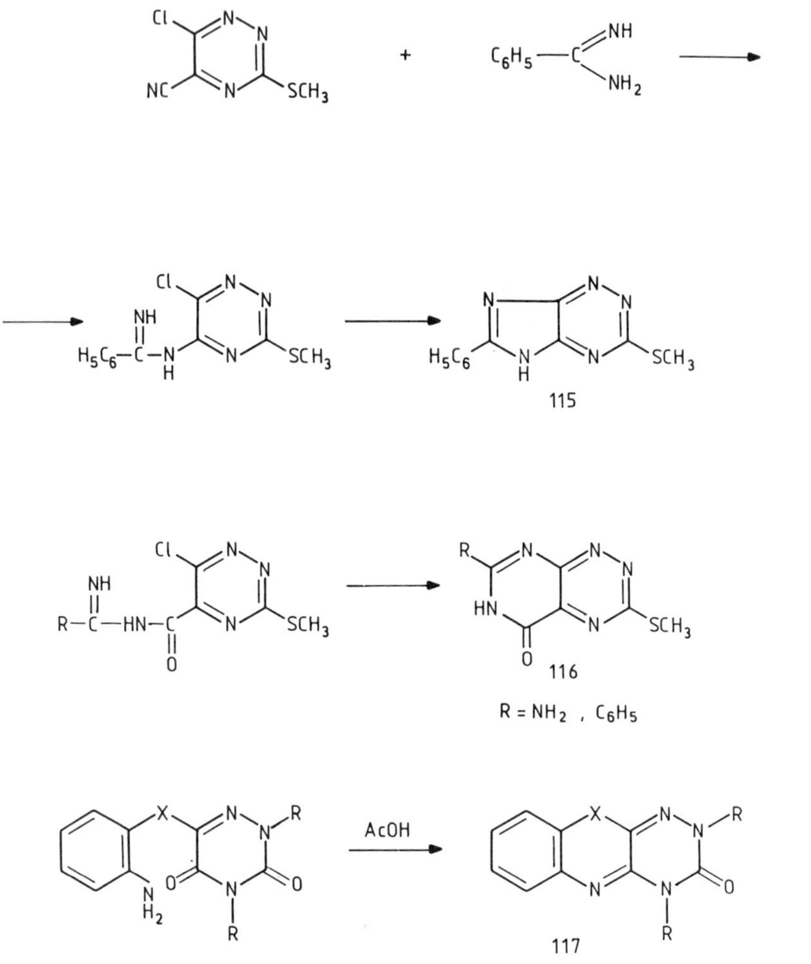

SCHEME 67

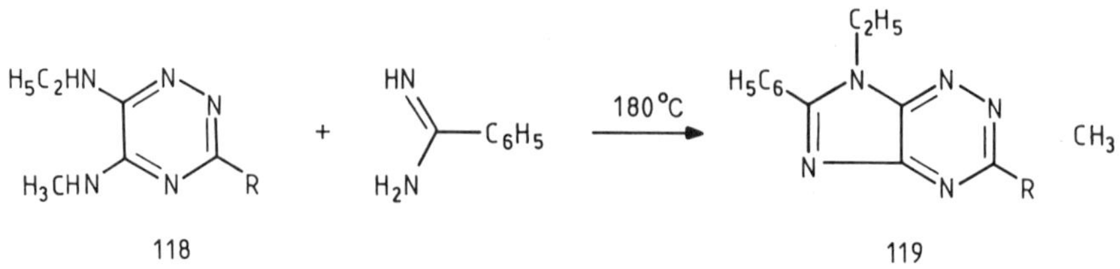

SCHEME 68

V. Nucleophilic Substitution of Hydrogen ($S_N H$ Reactions)

As we have seen in previous Sections, reactions of 1,2,4-triazines with C-, N-, O-, and S-nucleophiles often involve σ-adduct formation. When the addition of nucleophiles takes place at a ring carbon carrying a nucleofugic group, usually a fast elimination of the leaving group from the tetrahedral center completes the S_N process. In such cases the σ-adducts proposed as intermediates could not be registered by NMR spectroscopy. However, σ-adducts carrying hydrogen at the sp^3 carbon are quite stable because of the very poor leaving ability of the hydride ion. In many cases, they can be registered spectroscopically or even isolated from the reaction mixture (see Section III). These σ-adducts can be converted into aromatic compounds by an oxidant or by inter- or intramolecular hydrogen shifts. Thus, the formation of σ-adducts like **120** and removal of hydrogen from the σ-adducts **120** may be regarded as two steps of specific S_N reactions for which the symbol $S_N H$ has recently been suggested (Scheme 69) (for review on the S_N reactions, see 88T1).

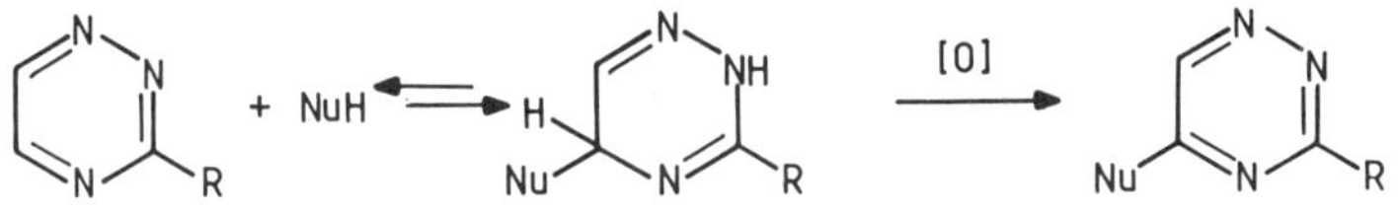

120

SCHEME 69

A. $S_N H$ Reactions in the Presence of Oxidant

Most reports on the $S_N H$ reactions in the series of 1,2,4-triazines concern substitution of hydrogen at C-5. For instance, a synthesis of 5-alkyl-substituted 1,2,4-triazines has been developed involving oxidation of the C-5 adducts **121** and **122**, resulting from the reaction of 3-methyl-6-phenyl-1,2,4-triazine with carbanionic reagents. Potassium permanganate, potassium ferric cyanide, or air oxygen were used as oxidizing agents (Scheme 70) (85H2807; 86H239; 87CPB1378).

Similarly, the C-5 adduct **123**, obtained from reaction of 3-methylthio-1,2,4-triazine with phenylmagnesium bromide, was oxidized by potassium ferricyanide under alkaline conditions into 3-methylthio-5-phenyl-1,2,4-triazine (Scheme 71) (87H3111).

3-Methyl-6-phenyl-1,2,4-triazine reacts with hydrogen cyanide in the

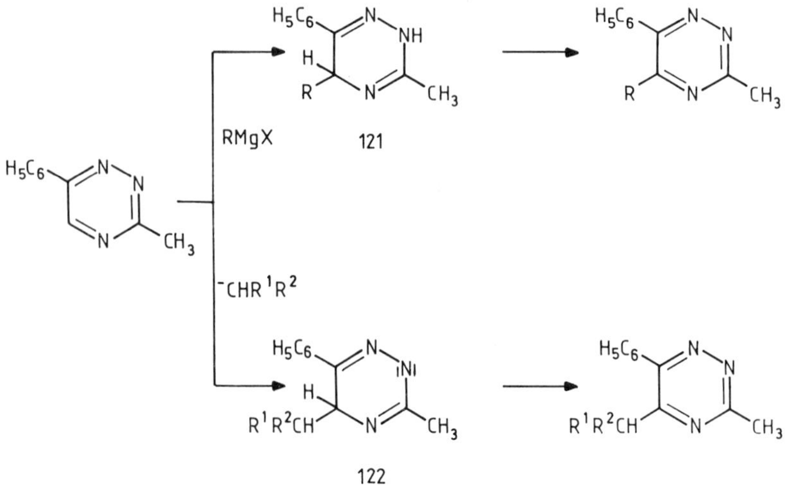

SCHEME 70

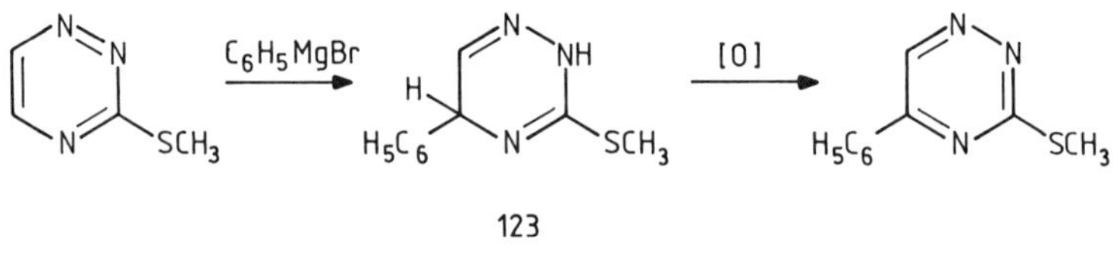

SCHEME 71

presence of air oxygen to give the S_nH product **125**, via the short-lived intermediate **124** (Scheme 72) (86H1243).

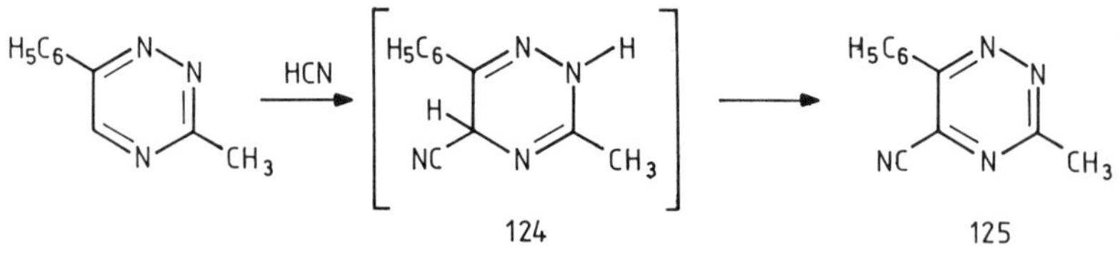

SCHEME 72

Amination of azaaromatic compounds using liquid ammonia–potassium permanganate (85S884; 86MI1) has been found to be a successful method for preparing a number of 5-amino-1,2,4-triazines (Scheme 73). This S_NH reaction is based on the ability of 3- and 6-substituted triazines to form amino adducts **22** at C-5 in liquid ammonia (see Section III,A,2 and Scheme 18). These adducts can be oxidized into 5-amino-1,2,4-triazines **126**. A number of 1,2,4-triazines containing substituents at C-3 and C-6, including 6-phenyl-1,2,4-triazine 4-oxide, were aminated according to this procedure (Scheme 73) (85S884).

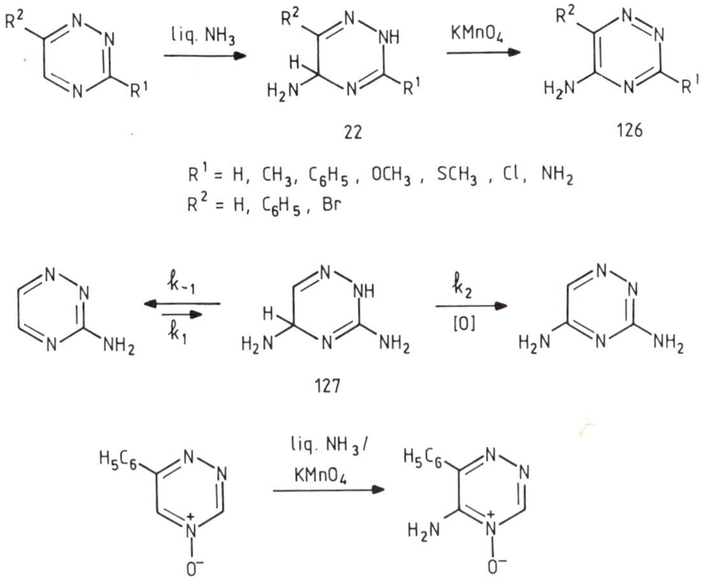

SCHEME 73

Although 3-amino-1,2,4-triazine does not give a C-5 adduct detectable by ^{1}H-NMR spectroscopy (see Section III,A,2), it gives an excellent yield of 3,5-diamino-1,2,4-triazine when treated with potassium permanganate in liquid ammonia. This reaction is an interesting case in which the equilibrium between substrate and C-5 adduct is far to the left, but in which oxidation still gives the product in good yield. The C-5 adduct **127** is apparently present as intermediate in a small steady-state concentration (Scheme 73) (86MI1; 88T1). Despite the presence of good leaving groups at C-3 and C-6, this amination reaction still occurs at C-5. Introduction of a bulky substituent at C-5 makes the process impossible. No amination reaction at C-3 or C-6 of the 1,2,4-triazine ring has been observed (85S884).

A rare example of an S_NH reaction at C-6 has been found when triazinium salts **128** were reacted in alkaline solution in the presence of air oxygen. It yielded the 1,2,4-triazin-6-ones **130**, the formation of which must be initiated by the addition of the hydroxide ion at C-6, i.e., via the intermediacy of the C-6 adducts **129** (Scheme 74) (71CC1636).

Similarly, when treated with bromine in aqueous solution, 4-amino-3-methylthio-1,2,4-triazin-5-one is converted into the corresponding 1,2,4-

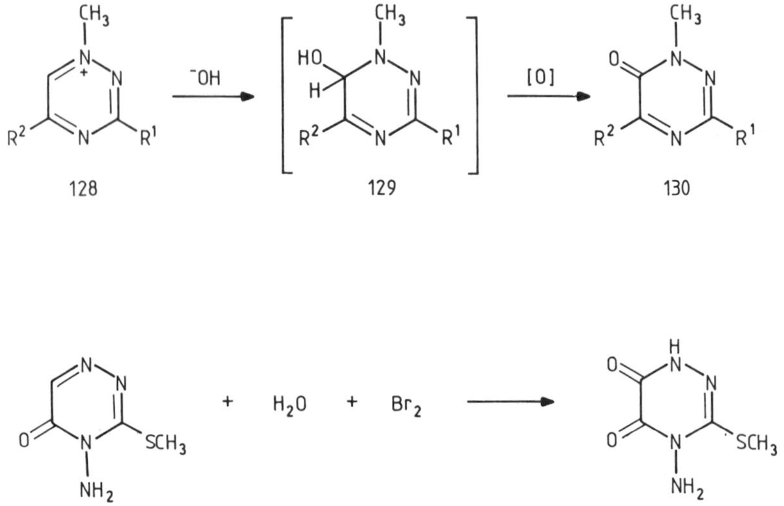

SCHEME 74

triazine-5,6-dione (Scheme 74) (86JCR(S)320). In some cases, nucleophilic substitution of hydrogen in 1,2,4-triazines is accompanied by a dimerization reaction giving rise to bis-1,2,4-triazin-5-yl derivatives (73BSF2493; 73JHC343; 74JHC43; 75RTC204; 79USP4105434; 82JHC653; 86H1243; 87CPB1378).

B. VICARIOUS NUCLEOPHILIC SUBSTITUTION

An interesting process for nucleophilic substitution of hydrogen in π-deficient aromatic systems is the so-called vicarious nucleophilic substitution of hydrogen (84MI2). The methodology is based on the properties of carbon nucleophiles containing a good leaving group X attached to the carbanionic center, to facilitate β-elimination of HX in σ-adducts **131**. Thus, it makes the S_NH process possible without using any oxidant (Scheme 75). Discovered to be useful in the series of nitroarene substrates, this methodology has been extended to the synthesis of functionalized 1,2,4-triazine derivatives (83TL3277; 84MI1, 84MI2, 84TL4795). Thus, 3-substituted 1,2,4-triazines were found to react with chloromethylsulfones under super basic (KOH–DMSO) conditions to afford 5-substituted 1,2,4-triazines **132** (Scheme 75). If the C-5 position is occupied by a phenyl group, the reaction takes place at C-3. When both C-5 and C-3 bear substituents the vicarious nucleophilic substitution occurs at C-6 of the 1,2,4-triazine ring (83TL3277; 84MI1, 84MI2). This example again

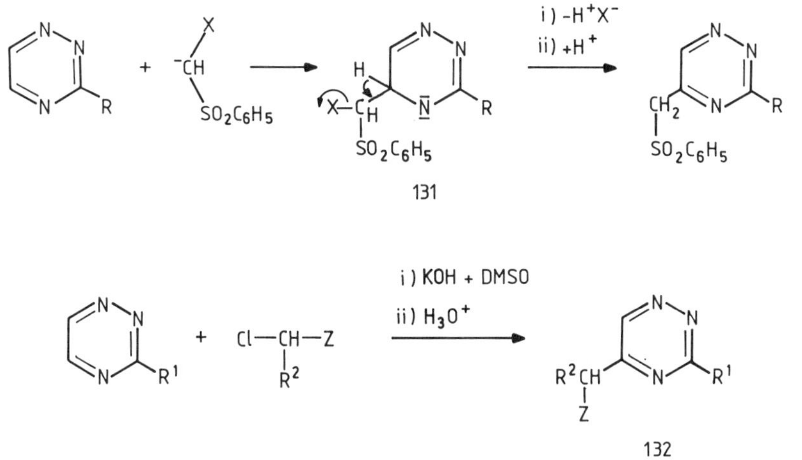

Sec. V.B] MONOCYCLIC 1,2,4-TRIAZINES 123

131

R^1 = H, CH_3, C_6H_5, SCH_3, OCH_3
R^2 = H, CH_3, Cl; Z = $SO_2C_6H_5$, $SO_2-N(CH_3)C_6H_5$, $SO_2OCH_2C(CH_3)_3$

SCHEME 75

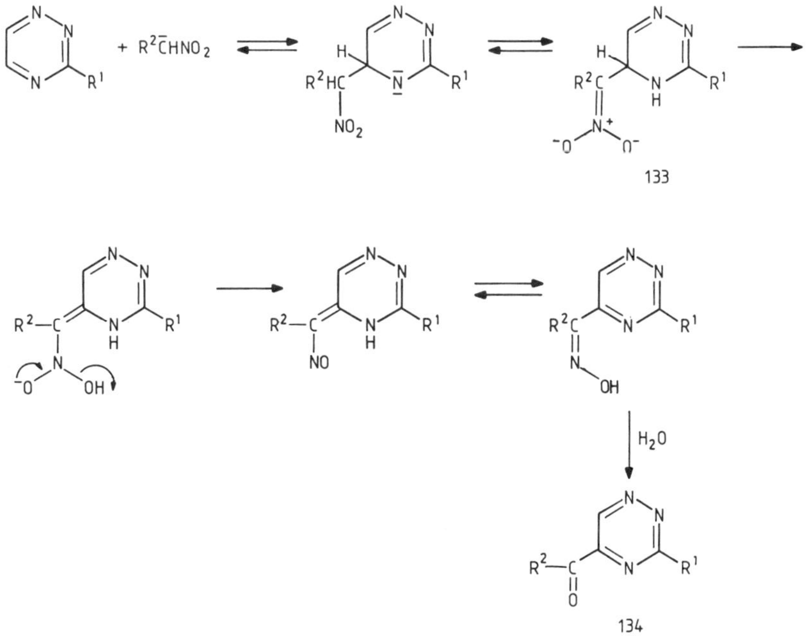

SCHEME 76

shows that the relative reactivity of the ring carbons in the S_N reactions decreases in the order C-5 > C-3 > C-6.

Using nitroalkanes as nucleophilic reagents, an elegant synthesis of 5-acyl-1,2,4-triazines **134** has been developed (Scheme 76) (84TL4795). In this reaction, nitroalkanes are supposed to act as hidden vicarious reagents causing the intramolecular shift of hydrogen in the adduct **133** from C-5 to the nitronate moiety (Scheme 76) (84TL4795).

C. OTHER $S_N H$ REACTIONS

The specific nature of 1,2,4-triazine N-oxides discussed in Section II,C is also manifested in the $S_N H$ reactions. Treatment of 1,2,4-triazine 2-oxides containing electron-donating groups at C-3 with methanol, ethanol, or isopropanol in the presence of hydrochloric acid results in 6-alkoxy-substituted 1,2,4-triazines **137** (Scheme 77) (77JOC3498). The reaction is proposed to involve the addition of alcohol to C-6 of the hydroxy-1,2,4-triazinium cation **135**, followed by elimination of water. The presence of an electron-donating group at C-3 seems also to be important because it contributes considerably to the stabilization of the intermediate cationic species **136** (Scheme 77).

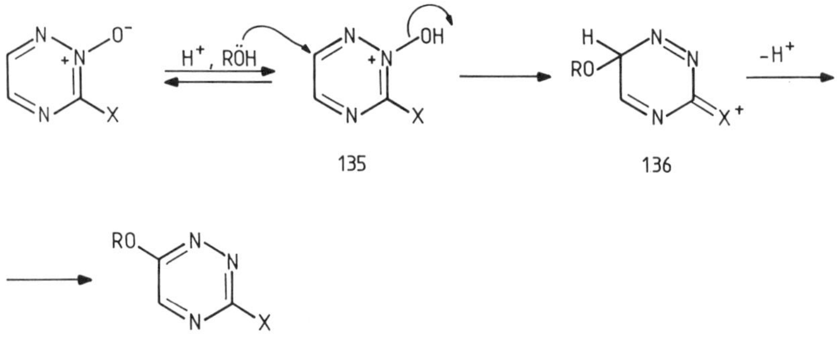

R = CH_3, C_2H_5, $CH(CH_3)_2$
X = NH_2, $NHCH_3$, $N(CH_3)_2$, SCH_3

SCHEME 77

The reaction of 3-amino-1,2,4-triazine 2-oxide with alcohols in the presence of acylating agents proceeds in a similar way (Scheme 78) (77JOC3498).

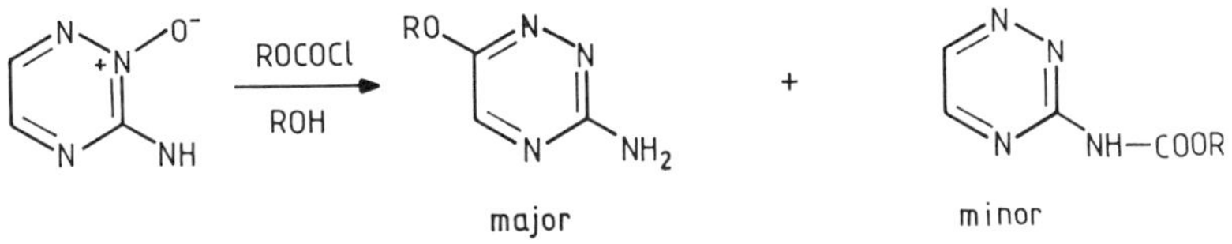

SCHEME 78

VI. Transformations of the 1,2,4-Triazine Ring

A. Ring-Degenerate Transformations

In Section IV,A,1 it was shown that triazines containing a nucleofugic group at C-3 readily undergo $S_N(AE)^{ipso}$ reactions. On the other hand, all the data on the reactivity of 1,2,4-triazines toward nucleophilic reagents discussed above indicate that the C-5 position is the preferential site for σ-adduct formation. Taking this into consideration, it is not astonishing to observe that amination of 1,2,4-triazines containing a nucleofugic group at C-3 with potassium amide proceeds via two alternative pathways, one involving nucleophilic attack on C-3 and the other one at C-5 (85T237). ^{15}N-Labeling experiments performed on a number of 1,2,4-triazines (Table VI) (75RTC204; 80JOC881; 82JHC653; 83MI1, 83MI3) unequivocally prove that not only the classical $S_N(AE)^{ipso}$ but also the $S_N(ANRORC)$ mechanism involving addition of ammonia at C-5, ring opening, and ring closure is operating in this amination reaction (Scheme 79) (85T237). It is clear that, in the latter case, the amino group at C-3 does not originate from ammonia but from the triazine ring (Table VI, Scheme 79). This amination is an example of the so-called ring-degenerate transformation reactions (85T237).

The percentage of the participation of the $S_N(ANRORC)$ pathway in the amination with potassium amide in liquid ammonia depends strongly, as can be expected, on the nature of the leaving group at C-3, although there is no linear dependency found between the leaving group ability and the contribution of the $S_N(ANRORC)$ mechanism: methylthio > chloro > bromo > iodo > methylsulfonyl > trimethylammonium > fluoro > hydroxy (Table VI) (85T237).

Ring-degenerate transformations initiated by the nucleophilic addition at C-5 of the 1,2,4-triazine ring have also been reported to occur in reactions of 4-aryl-substituted 3-methylthio-1,2,4-triazine-3,5-diones with hydrazine hydrate (Scheme 80). This ANRORC mechanism involves the open-chain compound **138** as intermediate (80JHC1733; 81JHC953).

TABLE VI
AMINATION OF 3-SUBSTITUTED 1,2,4-TRIAZINES[a]

Substituents			Yield of 3-amino compound (%)	S_N(ANRORC) (%)	References
Le group	R^1	R^2			
Methylthio[b]	H	H	11	93	75RTC204
Methylthio	Phenyl	H	71	100	82JHC653
Methylthio	H	Phenyl	51	83	82JHC653
Methylthio	Phenyl	Phenyl	82	94	83MI1
Methylthio	T-Butyl	H	62	95	82JHC653
Methylsulfonyl	Phenyl	H	65	33	82JHC653
Methylsulfonyl	Phenyl	Phenyl	48	53	83MI1
Methylsulfonyl	T-Butyl	H	59	54	82JHC653
Fluoro	Phenyl	H	54	18	80JOC881
Chloro	Phenyl	H	40	96	80JOC881
Chloro[b]	H	Phenyl	52	83	82JHC653
Chloro	Phenyl	Phenyl	58	56	83MI1
Chloro	T-Butyl	H	49	95	82JHC653
Bromo	Phenyl	H	29	93	80JOC881
Iodo	Phenyl	H	31	63	80JOC881
Trimethylammonium	Phenyl	H	42	34	82JHC653
Trimethylammonium	Phenyl	Phenyl	29	25	83MI1
Hydroxy[c]	Phenyl	H	61	10	82JHC653
Hydroxy[c]	T-Butyl	H	39	0	82JHC653

[a]Reaction with potassium amide in liquid ammonia at $-33°C$ (Scheme 79).
[b]At $-75°C$.
[c]With $H_5C_6OP(=O)(NH_2)_2$ at 235–240°C.

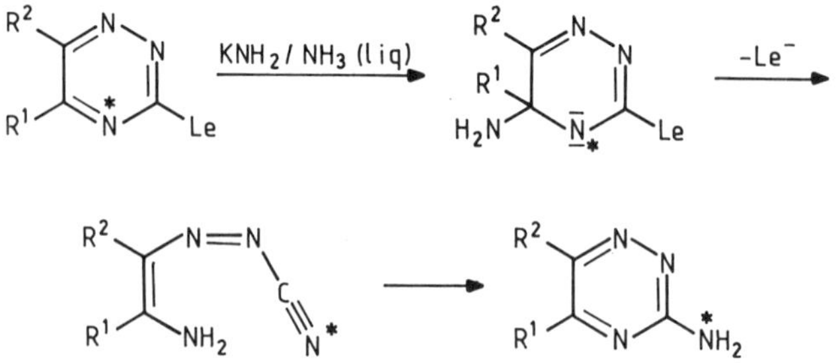

SCHEME 79

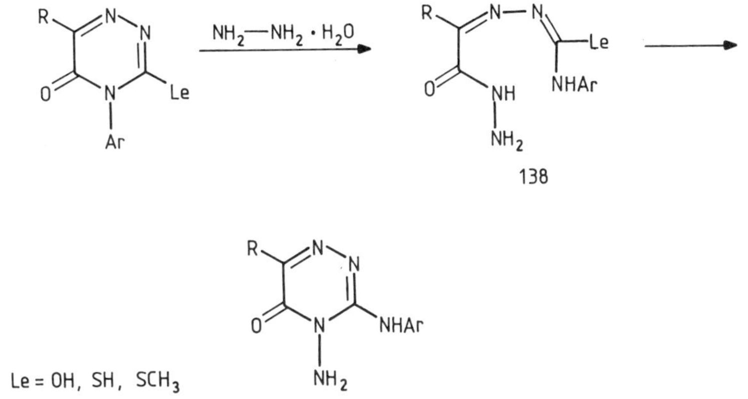

SCHEME 80

Le = OH, SH, SCH₃

B. TRANSFORMATIONS OF 1,2,4-TRIAZINES INTO OTHER AZINES

There is overwhelming experimental evidence of the ability of 1,2,4-triazines to undergo inverse electron-demand Diels–Alder reactions by action of electron-rich ethylenes, such as ketene-*O,O*-, -*O,S*-, -*S,S*-, -*S,N*-, -*N,N*-, and -*N,O*-acetals. In these cycloaddition reactions, the initial attack of the dienophile is across C-3 and C-6 of the triazine ring, whereupon, with evolution of nitrogen, from the pyridine derivatives **140** and **142** are formed from the cycloadducts **139** and **141**: (Scheme 81) (69TL3151; 83T2869; 86CRV781; 87MI1).

Much attention has been paid to the intramolecular version of this Diels–Alder reaction. The synthesis of a large number of condensed pyridines and pyrazines has been developed based on intramolecular transformations of 1,2,4-triazines containing alkenyl or alkynyl side-chain substituents (84AP379, 84CZ331; 85AP1048, 85AP1051, 85TL2419, 85TL4355; 86TL431, 86TL1967, 86TL2107, 86TL2747; 87JOC4280, 87JOC4287, 87T5145, 87T5159, 87TL379). In particular, triazines **54–56** and **59**, the preparation of which has been discussed in Section IV,A,1 (Schemes 37 and 38), are transformed on heating into the corresponding pyrrolo, furo, pyrano, or cycloalkeno annelated pyridines. An example is given in Scheme 82. Also, the intramolecular triazine to pyrazine transformation has been observed to occur in the reaction of 3-methylsulfonyl-substituted 1,2,4-triazines with 2-cyanophenol (Scheme 82) (87T5159). Thus, using 1,2,4-triazines with a nucleofugic group at C-3 as substrate and nu-

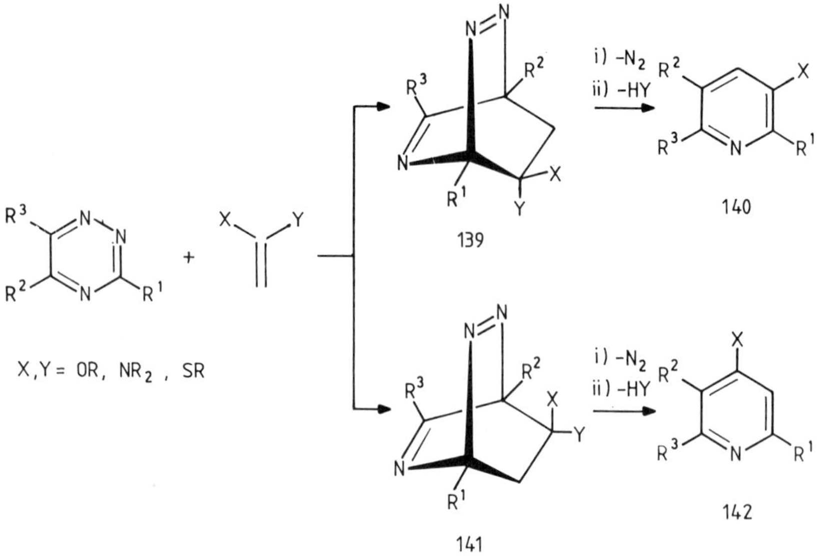

SCHEME 81

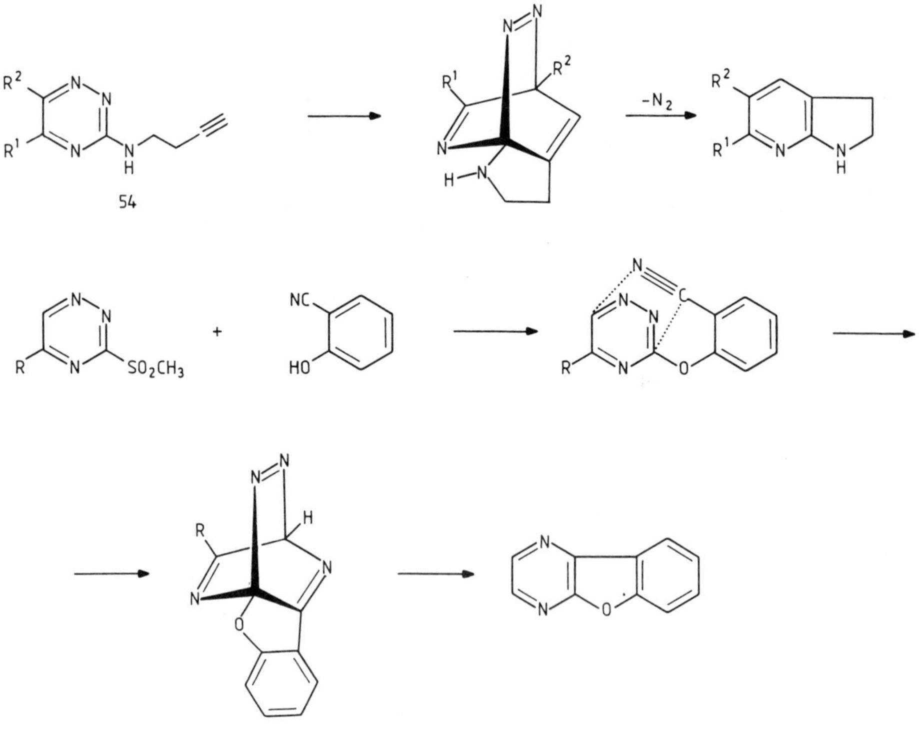

SCHEME 82

cleophilic reagents containing the cyano group provides an elegant synthesis of condensed pyrazines!

These inter- and intramolecular ring transformations of triazines have been reviewed extensively (83T2869; 86CRV781; 87MI1); thus, there is no need to dwell on this subject in more detail. We wish to mention here one particular example which will likely stimulate new investigations into the area of cycloaddition reactions.

It has been reported that the intramolecular transformation of 5-phenyl-3-(3-butynylthio)-1,2,4-triazine (**143**) into 6-phenyl-2,3-dihydrothieno[2,3-*b*] pyridine (**145**) requires heating in refluxing dioxane (100°C) for 24 hr (Scheme 83) (87JOC4280). Our own experiments show that the same result can also be reached under considerably milder conditions if the triazinium salt **146** is used as the starting material (88H(ip)). The salt **146**, easily obtained by treatment of triazine **143** with triethyloxonium tetrafluoroborate at room temperature in dichloromethane solution, gives on heating in acetone (60°C) for 7 hr the borate salt **147** in nearly quantitative yield (Scheme 83) (88H(ip)).

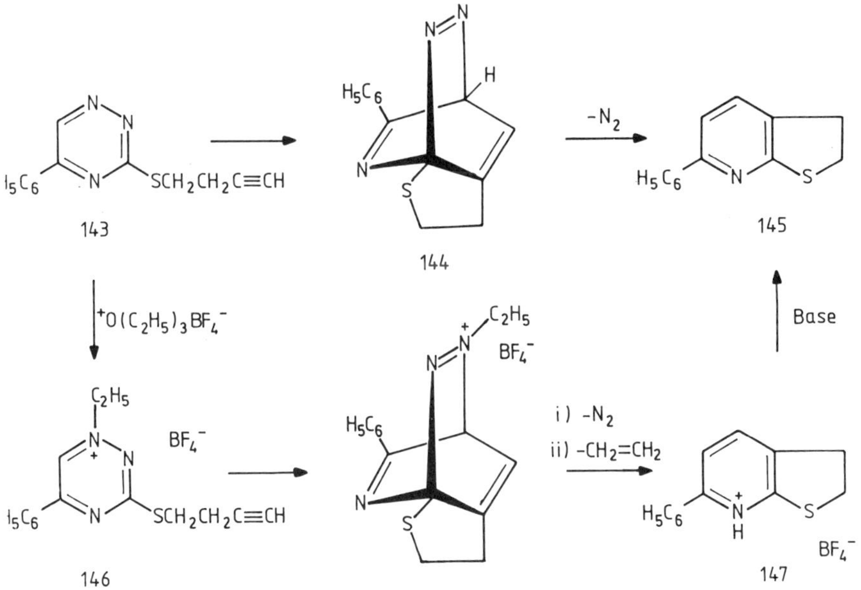

SCHEME 83

It is well known that introduction of electron-withdrawing substituents increases electron deficiency in the 1,2,4-triazine substrate and enhances its reactivity in Diels–Alder reactions with electron-rich dienophiles (83T2869; 86CRV781; 87MI1). The reaction under consideration is the first example in which electron deficiency of the 1,2,4-triazine ring is enhanced by quaternization of the N-1 nitrogen. The cycloaddition reaction on the quaternary salt **146** occurs in the same manner as in the case of neutral 1,2,4-triazines, i.e., initial addition takes place across the C-3/C-6 positions of the 1,2,4-triazine ring (Scheme 83) (88H(ip)).

Several papers report on transformations of 1,2,4-triazines into symmetrical 1,3,5-triazine derivatives (80JOC881; 81TL1393; 83IJC(B)559; 87JOC71). For instance, 5-phenyl-1,2,4-triazine is transformed into 2-amino-4-phenyl-1,3,5-triazine on treatment with potassium amide in liquid ammonia. The scheme advanced on the basis of ^{1}H-NMR and ^{15}N-labeling experiments suggests that the reaction is initiated by the addition of the amide ion at C-5, which provokes ring opening, followed by intramolecular rearrangements resulting in the formation of 1,3,5-triazine **148** (Scheme 84) (87JOC71).

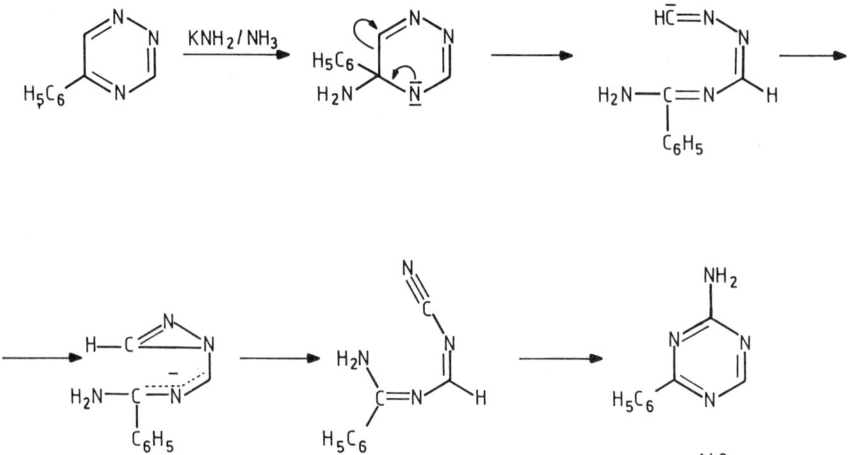

SCHEME 84

Formation of 1,3,5-triazine derivatives has also been observed in the reaction of 3-halo-5-phenyl-1,2,4-triazines with potassium amide in liquid ammonia (80JOC881) and on treatment of 3-chloro-5,6-diphenyl-1,2,4-triazine with phenylmagnesium bromide (83IJC(B)559). Both transformations were assumed to be initiated by the addition of the nucleophile employed at C-5 of the 1,2,4-triazine ring.

C. Transformations of 1,2,4-Triazines into Azoles

There are several reports in which the addition of nucleophilic reagents at C-5 or C-6 does not lead to ring transformations into azines but causes contraction of the 1,2,4-triazine ring into five-membered azoles.

1. Transformations Leading to 1,2,4-Triazoles

Reaction of alkali hydroxide with 1-methyl-1,2,4-triazinium salts **149** provokes transformation into 1,2,4-triazole derivatives **153** (Scheme 85) (71CC1636). The reaction is presumed to be initiated by addition of the nucleophile on the carbon adjacent to the quaternary nitrogen (i.e., at C-6). Attempts to obtain spectroscopic evidence for the intermediary σ-adduct **150** failed; however, some support for their formation is provided by the fact that 1-methyl-1,2,4-triazin-6-ones **152** were isolated as by-products when 6-unsubstituted quaternary salts **149** (R^3 = H) were treated with alkali under oxidative conditions (Scheme 85) (71CC1636). Transformation into triazoles **153** occurs in the presence of nucleofugic groups at C-3 (R^1 = OCH_3, SCH_3) in salts **149** (Scheme 85).

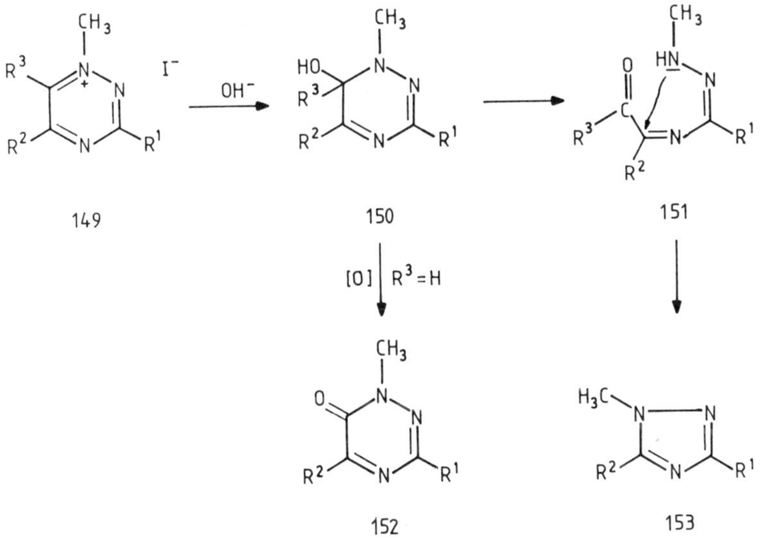

SCHEME 85

5-Phenyl- and 3,5-diphenyl-substituted triazines **154** undergo to a small extent ring contraction by action of potassium amide in liquid ammonia to yield 1,2,4-triazoles **155**. Again, nucleophilic addition at C-6 is presumed to facilitate ring opening (Scheme 86) (87JOC71).

There are also a number of reports on 1,2,4-triazine-to-1,2,4-triazole ring transformations initiated by the addition of nucleophiles at C-5 (68CR9041, 68CR1726; 70BSF1590, 70BSF1599; 72BSF1511; 79JHC199). For instance, when treated with aqueous sodium hydroxide, 6-amino-1,2,4-triazine-3,5-dione (**156**) is converted into 3-hydroxy-1,2,4-triazole-5-carboxylic acid (**157**), as shown in Scheme 87. In a similar way, the

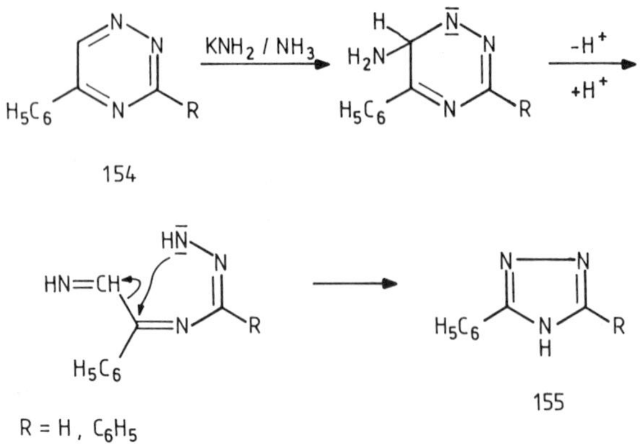

SCHEME 86

methoxide ion causes the transformation of 1,2,4-triazine-3-thiones **158** into 3-substituted 1,2,4-triazole-5-thiones **159** (Scheme 87) (72BSF1511).

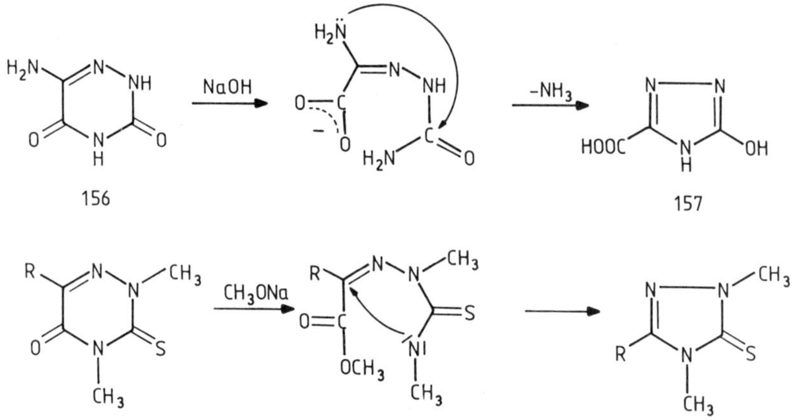

SCHEME 87

2. Formation of 1,2,3-Triazoles and Imidazoles

There are several reports on conversion of 1,2,4-triazines into 1,2,3-triazole derivatives (66JOC3914; 76LA153; 78HC(33)189). Acidic hydrolysis of 3-methyl-6-phenyl-1,2,4-triazine 4-oxide (**160**) affords 4-phenyl-1,2,3-triazole (**162**). The reaction occurs via an ANRORC route initiated by the addition of water at C-3 of the protonated triazine *N*-oxide **160,** followed by the formation of the acyclic intermediate **161** and ring closure (Scheme 88) (76LA153). The proposed route is confirmed by the fact that the hydrolysis of other triazine 4-oxides results only in acyclic ring-opened products (76LA153).

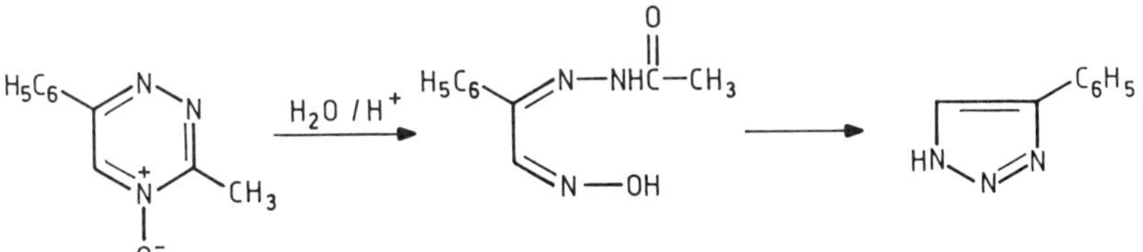

160 161 162

SCHEME 88

Interesting conversions of 1,2,4-triazine 2-oxides **163** and **165** into 1,2,3-triazole or imidazole derivatives **164** and **166,** by action of nucleophiles such as aniline and phenylhydrazine, occur under rather drastic degradative conditions, but the mechanism of these transformations remains unsolved (Scheme 89) (66JOC3914; 78HC(33)189).

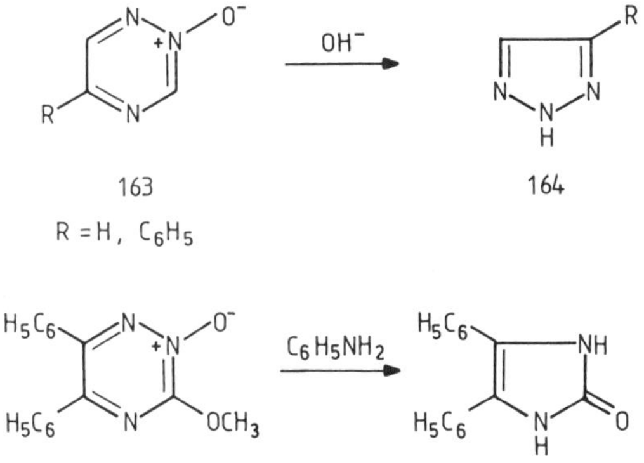

163 164

R = H, C_6H_5

165 166

SCHEME 89

Reaction of 3-chloro-5,6-diphenyl-1,2,4-triazine with phenylmagnesium bromide has been found to result in a rather complex mixture of 12 products(!), including tetraphenylimidazole **(169)** and triphenylimidazolone **172.** 2,5-Dihydro-1,2,4-triazines **168** and **171,** resulting from the addition of the Grignard reagent at C-5 of 3,5,6-triphenyl-1,2,4-triazine **167** and 5,6-diphenyl-1,2,4-triazin-3-one **170,** are postulated as the key intermediates in these ring-contraction reactions (Scheme 90) (83IJC(B)559). Formation of tetraphenylimidazole (**169**) from 5,6-diphenyl-1,2,4-triazin-3-one (**170**) and 3,5,6-triphenyl-1,2,4-triazine (**167**) by phenylmagnesium bromide has also been observed by other authors (71JPR699; 73BSF2818).

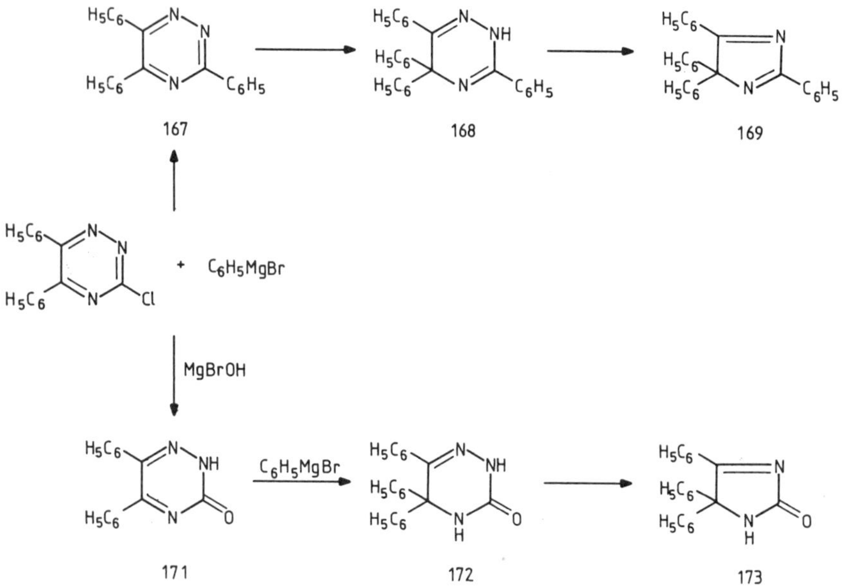

SCHEME 90

VII. Conclusion

In this article we discussed the behavior of simple 1,2,4-triazines in reactions with nucleophilic reagents, paying special attention to the features of common character and using the data published between 1968 and 1987. Unfortunately, many aspects of the reactivity of triazines have to be discussed qualitatively because of lack of quantitative data. Although the chemistry of 1,2,4-triazines has been studied extensively for a long period of time and many papers have appeared on this heterocyclic system, many details still remain to be investigated. In particular, very little is known about the reactivity of 1,2,4-triazinium cations toward both mono- and bifunctional nucleophilic reagents and their participation in both inter- and intramolecular cyclization reactions. Recent investigations of intramolecular cycloaddition reactions with triazinium cations indicate that it might be a very promising area for the development of new syntheses of heterocyclic compounds.

ACKNOWLEDGMENTS

Mr. J. Menkman (University of Wageningen) is gratefully acknowledged for his valuable help in preparing the drawings.

References

65AHC145	R. G. Shepherd and J. L. Fedric, *Adv. Heterocycl. Chem.* **4**, 145 (1965).
65CCC3134	M. Bobec, J. Farkas, and F. Sorm, *Collect. Czech. Chem. Commun.* **30**, 3134 (1965).
66JOC3914	T. Sasaki and K. Minamoto, *J. Org. Chem.* **31**, 3914 (1966).
66TL3115	M. Bobec, J. Farkas, and F. Sorm, *Tetrahedron Lett.*, 3115 (1966).
67CCC1295	M. Bobec, J. Farkas, and J. Gut, *Collect. Czech. Chem. Commun.* **32**, 1295 (1967).
67CCC3572	M. Bobec, J. Farkas, and F. Sorm, *Collect. Czech. Chem. Commun.* **32**, 3572 (1967).
67YZ1501	I. Saikawa and T. Maeda, *Yakugaku Zasshi* **87**, 1501 (1967).
68ACH319	G. Doleschall, M. Hornyak-Hamori, and K. Lempert, *Acta Chim. Acad. Sci. Hung.* **55**, 319 (1968).
68CCC2513	K. Kalfus, *Collect. Czech. Chem. Commun.* **33**, 2513 (1968).
68CR904	M. Pesson and M. Antoine, *C. R. Hebd. Seances Acad. Sci.* **267**, 904 (1968).
68CR1726	M. Pesson and M. Antoine, *C. R. Hebd. Seances Acad. Sci.* **267**, 1726 (1968).
68MI1	J. Miller, "Aromatic Nucleophilic Substitution." Elsevier, Amsterdam, 1968.
68YZ1501	I. Saikawa and T. Maeda, *Yakugaku Zassi* **87**, 1501 (1968).
69ACH181	G. Hornyak, K. Lempert, and K. Zauer, *Acta Chim. Acad. Sci. Hung.* **61**, 181 (1969).
69BSF2492	J. Daunis, R. Jacquier, and P. Viallenfont, *Bull. Soc. Chim. Fr.*, 2492 (1969)
69CCC1104	M. Prystas and F. Sorm, *Collect. Czech. Chem. Commun.* **34**, 1104 (1969).
69FRP1519180	Farbenfabriken Bayer A.-G., Fr. Pat. 1,519,180 (1969) [*CA* **70**, 106570 (1969)].
69MI1	J. Jenik, J. Kralovsky, F. Renger, and A. Miketa, *Sb. Ved. Pr. Vys. Sk. Chemickox technol. Pardubice*, 67 (1969) [*CA* **74**, 150833 (1971)].
69MIP1	H. Timmler, R. Wegler, L. Eue, and H. Hack, Afr. Pat. 04,409 (1969) [*CA* **71**, 39014 (1969)].
69RRC135	C. Cristescu and V. Badea, *Rev. Roum. Chim.* **14**, 135 (1969).
69TL3151	H. Neunhoeffer and H. W. Fruhauf, *Tetrahedron Lett.*, 3151 (1969).
69USP3412083	A. R. Restivo, U.S. Pat. 3,412,083 (1969) [*CA* **70**, 47807 (1969)].
70BSF1590	M. Pesson and M. Antoine, *Bull. Soc. Chim. Fr.*, 1590 (1970).
70BSF1599	M. Pesson and M. Antoine, *Bull. Soc. Chim. Fr.*, 1599 (1970).

70CR1201	J.-P. M, Packo and N. Vinot, *C. R. Hebd. Seances Acad. Sci.* **271**, 1201 (1970).
70JA7154	K. B. Wiberg and T. P. Lewis, *J. Am. Chem. Soc.* **92**, 7154 (1970).
70JAP09545	I. Saikawa and T. Maeda, Jpn. Pat. 09,545 (1970) [*CA* **72**, 132791 (1970)].
70JAP09546	I. Saikawa and T. Maeda, Jpn. Pat. 09,546 (1970) [*CA* **72**, 132798 (1970)].
70JAP25903	I. Saikawa and T Maeda, Jpn. Pat. 25,903 (1970) [*CA* **73**, 131041 (1970)].
70JHC767	W. W. Paudler and T.-K. Chen, *J. Heterocycl. Chem.* **7**, 767 (1970).
70JMC288	H. A. Burch, *J. Med. Chem.* **13**, 288 (1970).
70KGS986	V. P. Cherneckij, D. V. Semenuyk, and I. K. Vatutina, *Khim. Geterotsikl. Soedin.*, 986 (1970).
70LA177	A. Mustafa, A. K. Mansour, and H. A. Zaher *Justus Liebigs. Ann. Chem.* **733**, 177 (1970).
70RRC1409	C. Cristescu, *Rev. Roum. Chim.* **15**, 1409 (1970).
71CC1636	J. Lee and W. W. Paudler, *J. C. S. Chem. Commun.*, 1636 (1971).
71CCC3507	A. Novacek and P. Fiedler, *Collect. Czech. Chem. Commun.* **36**, 3507 (1971).
71CCC4000	M. Beran, M. Semonsky, and E. Svatek, *Collect. Czech. Chem. Commun.* **36**, 4000 (1971).
71JHC689	I. Lalezari, A. Shafiee, and M. Yalpani, *J. Heterocycl. Chem.* **8**, 689 (1971).
71JOC787	W. W. Paudler and T.-K. Chen, *J. Org. Chem.* **36**, 787 (1971).
71JOC2974	C. Jr. Temple, C. L. Kussner, and J. A. Montgomery, *J. Org. Chem.* **36**, 2974 (1971).
71JOC3921	W. W. Paudler and J. Lee, *J. Org. Chem.* **36**, 3921 (1971).
71JPR699	A. Mustafa, A. K. Mansour, and H. A. Zaher, *J. Prakt. Chem.* **313**, 699 (1971).
71LA12	H. Neunhoeffer, F. Weischedel, and V. Bohnisch, *Justus Liebigs Ann. Chem.* **750**, 12 (1971).
71MI1	A. R. Katritzky and J. M. Lagowski (Eds.), "Chemistry of the Heterocyclic N-Oxides." Academic Press, London, 1971.
71RRC135	C. Cristescu and S. Sitaru, *Rev. Roum. Chim.* **16**, 135 (1971).
71RRC311	C. Cristescu, *Rev. Roum. Chim.* **16**, 311 (1971).
72AG348	H. Vorbruggen, *Angew. Chem.* **84**, 348 (1972).
72BSF1511	J. Daunis, Y. Guindo, R. Jacquier, and P. Viallefont, *Bull. Soc. Chim. Fr.*, 1511 (1972).
72BSF4637	N. Vinot and J.-P. M, Packo, *Bull. Soc. Chim. Fr.*, 4637 (1972).
72CCC2221	V. Uchytilova, P. Fiedler, and J. Gut, *Collect. Czech. Chem. Commun.* **37**, 2221 (1972).
72GEP2206395	M. W. Miller, Ger. Pat. 2,206,395 (1972) [*CA* **77**, 152237 (1972)].
72JCS(P1)1221	M. F. G. Stevens, *J. C. S Perkin 1*, 1221 (1972).
72JCS(P1)2316	D. J. Brown and J. R. Kershaw, *J. C. S. Perkin 1*, 2316 (1972).
72JHC995	J. Lee and W. W. Paudler, *J. Heterocycl. Chem.* **9**, 995 (1972).
72LA111	H. Neunhoeffer and H. W. Frühauf, *Justus Liebigs Ann. Chem.* **758**, 111 (1972).
72LA173	M. Brugger, H. Wamhoff, and F. Korte, *Justus Liebigs Ann. Chem.* **758**, 173 (1972).
72ZN(B)818	A. Spassov, E. Golovinski, N. Spassovska, and L. Maneva, *Z. Naturforsch., B: Anorg. Chem., Org. Chem., Biochem., Biophys., Biol.* **27B**, 818 (1972).
73ACH345	G. Doleschall and K. Lempert, *Acta Chim. Acad. Sci. Hung.* **77**, 345 (1973).

73BSF559	J. Daunis and R. Jacquier, *Bull. Soc. Chim. Fr.*, 559 (1973).
73BSF2493	J. Daunis and C. Pigiere, *Bull. Soc. Chim. Fr.*, 2493 (1973).
73BSF2818	J. Daunis and C. Pigiere, *Bull. Soc. Chim. Fr.*, 2818 (1973).
73GEP2163873	H. Vorbruggen, Ger. Pat. 2,163,873 (1973) [*CA* **79**, 66405 (1973)].
73GEP2256604	A. Piskala and F. Sorm, Ger. Pat. 2,256,604 (1973) [*CA* **79**, 32264 (1973)].
73JHC343	D. K. Krass, T.-K. Chen, and W. W. Paudler, *J. Heterocycl. Chem.* **10**, 343 (1973).
73JOC3277	G. L. Szekeres, P. K. Robins, P. Dea, M. P. Schweizer, and R. A. Long, *J. Org. Chem.* **38**, 3277 (1973).
73JPR221	A. K. Mansour and Y. A. Ibrahim, *J. Prakt. Chem.* **315**, 221 (1973).
73MI1	J. Slouka and V. Svecova, *Acta Univ. Palacki. Olomuc., Fak. Rerum Nat.* **41**, 143 (1973).
73MI2	J. Slouka and Z. Stransky, *Pharmazie* **28**, 309 (1973).
73MIP1	C. Cristescu and V. Badea, Roum. Pat. 54,172 (1973) [*CA* **79**, 53379 (1973)].
73T2495	W. W. Paudler, J. Lee, and T.-K. Chen, *Tetrahedron* **29**, 2495 (1973).
74AHC225	M. J. Cook, A. R. Katritzky, and P. Linda, *Adv. Heterocycl. Chem.* **17**, 255 (1974).
74AJC1781	D. J. Brown and R. K. Lynn, *Aust. J. Chem.* **27**, 1781 (1974).
74BSF999	J. Daunis, *Bull. Soc. Chim. Fr.*, 999 (1974).
74GEP2236340	M. Jautelat, K. Ley, and L. Eue, Ger. Pat. 2,236,340 (1974) [*CA* **80**, 108583 (1974)].
74JHC43	D. K. Krass and W. W. Paudler, *J. Heterocycl. Chem.* **11**, 43 (1974).
74JPR163	M. I. Ali, A. A. El-Sayed, and H. A. Hammouda, *J. Prakt. Chem.* **316**, 163 (1974).
74JPR667	L. Heinisch, *J. Prakt. Chem.* **316**, 667 (1974).
74MIP1	C. Cristescu, Roum. Pat. 53,275 (1974) [*CA* **80**, 37179 (1974)].
74MIP2	C. Cristescu, Roum. Pat. 55,849 (1974) [*CA* **80**, 96030 (1974)].
74ZN(B)792	A. K. Mansour, S. B. Awad, and S. Antoon, *Z. Naturforsch., B: Anorg. Chem., Org. Chem.* **29B**, 792 (1974).
75CCC2340	A. Piskala, *Collect. Czech. Chem. Commun.* **40**, 2340 (1975).
75CCC2680	A. Piskala, J. Gut, and F. Sorm, *Collect. Czech. Chem. Commun.* **40**, 2680 (1975).
75JAP27874	H. Tahi, M. Ohtani, and M. Nara, Jpn. Pat. 27,874 (1975) [*CA* **82**, 171081 (1975)].
75JAP48697	K. Kaji, H. Nagashima, J. Masaki, M. Joshida, and K. Kamiya, *Jpn. Pat.* 48,697 (1975) [*CA* **82**, 43470 (1975).
75JAP48698	K. Kaji, H. Nagashima, J. Masaki, M. Joshida, and K. Kamiya, *Jpn. Pat.* 48,698 (1975) [*CA* **82**, 43471 (1975).
75OMR194	S. Braun and G. Frey, *Org. Magn. Reson.* **7**, 194 (1975).
75RTC204	A. Rykowski and H. C. van der Plas, *Recl. Trav. Chim. Pays-Bas* **94**, 204 (1975).
75TL2897	B. Burg, W. Dittmar, H. Reim, A. Steigel, and J. Sauer, *Tetrahedron Lett.*, 2897 (1975).
76CB1113	H. Neunhoeffer and B. Lehmann, *Chem. Ber.* **109**, 1113 (1976).
76CCC465	A. Piskala and F. Sorm, *Collect. Czech. Chem. Commun.* **41**, 465 (1976).
76H1341	K. Wasti and M. M. Joullie, *Heterocycles* **4**, 1341 (1976).
76JCS(P2)2521	K. Wasti and M. M. Joullie, *J. C. S. Perkin 2*, 2521 (1976).
76JHC807	B. T. Keen, D. K. Krass, and W. W. Paudler, *J. Heterocycl. Chem.* **13**, 807 (1976).

76JMC845	K. Wasti and M. M. Joullie, *J. Med. Chem.* **19**, 845 (1976).
76LA153	H. Neunhoeffer and V. Bohnisch, *Liebigs Ann. Chem.*, 153 (1976).
77JHC729	J. Daunis and M. Follet, *J. Heterocycl. Chem.* **14**, 729 (1977).
77JHC1221	M. M. Goodman and W. W. Paudler, *J. Heterocycl. Chem.* **14**, 1221 (1977).
77JOC546	R. J. Radel, B. T. Keen, C. Wong, and W. W. Paudler, *J. Org. Chem.* **42**, 546 (1977).
77JOC3498	B. T. Keen, R. J. Radel, and W. W. Paudler, *J. Org. Chem.* **42**, 3498 (1977).
77MI1	N. A. Spassovska, A. V. Spassov, and E. V. Golovinski, *Dokl. Bolg. Akad. Nauk* **30**, 65 (1977).
77USP3979516	W. B. Lacefield, U.S. Pat. 3,979,516 (1977) [*CA* **86**, 72707 (1977)].
77USP4008232	W. B. Lacefield, U.S. Pat. 4,008,232 (1977) [*CA* **86**, 190021 (1977)].
77USP4013654	W. B. Lacefield, U.S. Pat. 4,013,654 (1977) [*CA* **87**, 39541 (1977)].
77USP4021553	W. B. Lacefield and P. P. K. Ho, U.S. Pat. 4,021,553 (1977) [*CA* **87**, 29030 (1977)].
78ACR462	H. C. van der Plas, *Acc. Chem. Res.* **11**, 462 (1978).
78BAP285	M. Witanowski, L. Stefaniak, S. Szymanski, and G. A. Webb, *Bull. Acad. Pol. Sci., Ser. Sci, Chim.* **26**, 285 (1978).
78CPB3154	F. Yoneda, M. Noguchi, M. Noda, and Y. Nitta, *Chem. Pharm. Bull.* **26**, 3154 (1978).
78H1490	A. Rykowski, P. Nantka-Namirski, and H. C. van der Plas, *Heterocycles* **9**, 1490 (1978).
78HC(33)189	H. Neunhoeffer, *Chem. Heterocycl. Compd.* **33**, 189 (1978).
78JOC2514	R. J. Radel, J. L. Atwood, and W. W. Paudler, *J. Org. Chem.* **43**, 2514 (1978).
78JPS737	M. W. Dawidson and D. W. Boykin, Jr., *J. Pharm. Sci.* **67**, 737 (1978).
78MI1	M. Wozniak, *Wiad. Chem.* **32**, 283 (1978).
78RTC273	A. Rykowski, H. C. van der Plas, and A. van Veldhuizen, *Recl. Trav. Chim. Pays-Bas* **97**, 273 (1978).
79JHC199	J. Schwan, T. J. Sanford, R. L. Jr. White, and N. J. Miles, *J. Heterocycl. Chem.* **16**, 199 (1979).
79JHC427	J. Daunis, L. Djouai-Hifdi, and H. Lopex, *J. Heterocycl. Chem.* **16**, 427 (1979).
79JHC817	M. Daneshtabab, A. Khalaj, and I. Lalezari, *J. Heterocycl. Chem.* **16**, 817 (1979).
79JHC1389	G. B. Bennett, A. D. Kahle, H. Minor, and M. J. Shapiro, *J. Heterocycl. Chem.* **16**, 1389 (1979).
79JHC1393	R. I. Trust, J. D. Albright, F. M. Lovell, and N. A. Perkinson, *J. Heterocycl. Chem.* **16**, 1393 (1979).
79JHC1649	C. A. Lovelette, *J. Heterocycl. Chem.* **16**, 1649 (1979).
79JMC671	W. P. Heilman, R. D. Heilman, J. A. Scozzie, R. J. Wayner, and J. M. Gullo, *J. Med. Chem.* **22**, 671 (1979).
79JMR227	M. Witanowski, L. Stefaniak, and G. A. Webb, *J. Magn. Reson.* **36**, 227 (1979).
79USP4105434	W. W. Paudler, R. E. Moser, and N. M. Pollack, U.S. Pat. 4,105,434 (1979) [*CA* **90**, 54991 (1979)].
79USP4159375	R. I. Trust and J. D. Albright, U.S. Pat. 4,159,375 (1979) [*CA* **91**, 123758 (1979)].
80JHC1733	Y. A. Ibrahim, M. M. Eid, and E. A. L. Abdel-Hady, *J. Heterocycl. Chem.* **17**, 1733 (1980).

80JOC881	A. Rykowski and H. C. van der Plas, *J. Org. Chem.* **45**, 881 (1980).
80JOC4587	T. Sasaki, M. Minamoto, and K. Harada, *J. Org. Chem.* **45**, 4587 (1980).
80JOC5421	W. W. Paudler and R. M. Sheets, *J. Org. Chem.* **45**, 5421 (1980).
80JPS282	W. P. Heilman, R. D. Heilman, J. A. Scozzie, R. J. Wayner, and J. M. Gullo, *J. Pharm. Sci.* **69**, 282 (1080).
80MI1	M. Pesson, M. Antoine, J.-L. Benichon, P. de Lajudie, E. Horvath, B. Leriche, and S. Patte, *Eur. J. Med. Chem.—Chim. Ther.* **15**, 269 (1980).
80OMR172	M. Matsuo, S. Matsumoto, T. Kurihara, Y. Akita, T. Watanabe, and A. Ohta, *Org. Magn. Reson.* **13**, 172 (1980).
80S165	T. Wagner-Jauredy, *Synthesis,* 165 (1980).
81BCJ41	H. Ayato, I. Tanaka, T. Yamane, T. Ashida, T. Sasaki, K. Minamoto, and K. Harada, *Bull. Chem. Soc. Jpn.* **54**, 41 (1981).
81JHC953	Y. A. Ibrahim, M. M. Eid, M. A. Badawy, and S. A. L. Abdel-Hady, *J. Heterocycl. Chem.* **18**, 953 (1981).
81TL1393	H. P. Figeys and A. Mathy, *Tetrahedron Lett.* **22**, 1393 (1981).
82CPB152	S. Konno, M. Yokoyama, A. Kaite, I. Yamatsuta, S. Ogawa, M. Mizngaku, and H. Yamanaka, *Chem. Pharm. Bull.* **30**, 152 (1982).
82H93	W. W. Paudler and M. V. Jovanovic, *Heterocycles* **19**, 93 (1982).
82JCS(P1)1251	M. G. Barlow, R. N. Naszeldine, C. Simon, D. J. Simpkin, and G. Ziervogel, *J. C. S. Perkin 1,* 1251 (1982).
82JHC313	I. Hajpal and E. Berenyi, *J. Heterocycl. Chem.* **19**, 313 (1982).
82JHC653	A. Rykowski and H. C. van der Plas, *J. Heterocycl. Chem.* **19**, 653 (1982).
82JHC1583	Y. Sanemitsu, Y. Nakayama, and M. Shirishita, *J. Heterocycl. Chem.* **19**, 1583 (1982).
82OMR192	W. W. Paudler and M. V. Jovanovic, *Org. Magn. Reson.* **19**, 192 (1982).
83AHC305	G. Illuminati and F. Stegel, *Adv. Heterocycl. Chem.* **34**, 305 (1983).
83IJC(B)559	H. A. Zaher, Y. A. Ibrahim, O. Sherif, and R. Mohammady, *Indian J. Chem., Sect. B* **22B**, 559 (1983).
83JOC4585	M. Mizutani, Y. Sanemitsu, Y. Tamaru, and Z. Yoshida, *J. Org. Chem.* **48**, 4585 (1983).
83MI1	A. Rykowski, *Pol. J. Chem.* **57**, 467 (1983).
83MI2	A. Rykowski, *Pol. J. Chem.* **57**, 631 (1983).
83MI3	A. Rykowski, *Pol. J. Chem.* **57**, 1033 (1983).
83T2869	D. L. Boger, *Tetrahedron* **39**, 2869 (1983).
83TL3277	M. Makosza, J. Golinski, and A. Rykowski, *Tetrahedron Lett.* **24**, 3277 (1983).
84AP379	G. Seitz and S. Dietrich, *Arch. Pharm. (Weinheim, Ger.)* **317**, 379 (1984).
84CB1077	S. A. L. Abdel-Hady, M. A. Badawy, Y. A. Ibrahim, and W. Pfleiderer, *Chem. Ber.* **117**, 1077 (1984).
84CB1083	M. A. Badawy, S. A. L. Abdel-Hady, and M. M. Eid, *Chem. Ber.* **117**, 1083 (1984).
84CCC2628	J. Slouka, V. Bekarek, K. Nalepa, and A. Lycka, *Collect. Czech. Chem. Commun.* **49**, 2628 (1984).
84CZ331	G. Seitz and L. Gorge, *Chem. Ztg.* **108**, 331 (1984).
84H2241	S. Konno, M. Sagi, M. Agata, Y. Aizawa, and H. Yamanaka, *Heterocycles* **22**, 2241 (1984).
84H2245	S. Konno, S. Fujimura, and H. Yamanaka, *Heterocycles* **22**, 2245 (1984).

84JHC905	H. A. Zaher, R. Mohammady, and Y. A. Ibrahim, *J. Heterocycl. Chem.* **21**, 905 (1984).
84KGS557	V. L. Rusinov, T. V. Dragunova, V. A. Zyraynov, G. G. Alexandrov, N. A. Kluev, and O. N. Chupakhin, *Khim. Geterotsikl. Soedin.*, 557 (1984).
84LA283	H. Neunhoeffer and H. Hammann, *Liebigs Ann. Chem.*, 283 (1984).
84MI1	A. Rykowski and M. Makosza, *Proc. Symp. Chem. Chem. Heterocycl. Compd.*, 8th; *Nucl. Acid Components*, 6th, 173 (1984).
84MI2	M. Makosza, in "Current Trends in Organic Synthesis" (H. Nozaki, ed.), p. 401. Pergamon, Oxford, 1984.
84MI3	H. Neunhoeffer, in "Comprehensive Heterocyclic Chemistry" (A. J. Boulton and A. McKillop eds.), Vol 3, p. 385 Pergamon, New York, 1984.
84MI4	K. Nalepa and J. Slouka, *Pharmazie* **39**, 504 (1984).
84MI5	M. Cai, T. Cheng, W. Zhu, W. Chen, and S. Wang, *Kexue Tongbao* **29**, 1641 (1984) [*CA* **103**, 87843 (1985)].
84OPP199	M. J. Hearn and F. Levy, *Org. Prep. Proced. Int.* **16**, 199 (1984).
84S983	P. Molina, M. Alajarin, and J. R. Saez, *Synthesis*, 983 (1984).
84SA(A)637	M. V. Jovanovic, *Spectrochim. Acta, Part A* **40A**, 637 (1984).
84TL4795	A. Rykowski and M. Makosza, *Tetrahedron Lett.* **25**, 4795 (1984).
84UK1648	V. N. Charushin and O. N. Chupakhin, *Usp. Khim.* **53**, 1648 (1984) [*CA* **102**, 45795 (1985)].
85ACS(B)235	A. Rasmussen, F. Rise, and K. Undhein, *Acta Chem. Scand., Ser. B* **B39**, 235 (1985).
85AHC1	A. L. Weis, *Adv. Heterocycl. Chem.* **38**, 1 (1985).
85AP1048	G. Seitz and S. Dietrich, *Arch. Pharm. (Weinheim, Ger.)* **318**, 1048 (1985).
85AP1051	G. Seitz and S. Dietrich, *Arch. Pharm. (Weinheim, Ger.)* **318**, 1051 (1985).
85H2807	S. Konno. M. Sagi, Y. Yuki, and H. Yamanaka, *Heterocycles* **23**, 2807 (1985).
85JHC1329	J. J. Huang, *J. Heterocycl. Chem.* **22**, 1329 (1985).
85JOC2293	J. J. Huang, *J. Org. Chem.* **50**, 2293 (1985).
85KGS1011	V. N. Charushin, M. G. Ponizovski, and O. N. Chupakhin, *Khim. Geterotsikl. Soedin.*, 1011 (1985).
85LA857	H. Neunhoeffer and M. Hamana, *Liebigs Ann. Chem.*, 857 (1985).
85LA1263	H. Neunhoeffer and M. Bachmann, *Liebigs Ann. Chem.*, 1263 (1985).
85MI1	A. F. Pozharskii, "Teoreticheskije Osnovy Khimii Geterotsiklov." Khimiya, Moskva, 1985.
85RRC233	C. Cristescu and S. Sitaru, *Rev. Roum. Chim.* **30**, 233 (1985).
85S884	A. Rykowski and H. C. van der Plas, *Synthesis*, 884 (1985).
85SA(A)1135	M. V. Jovanovic, *Spectrochim. Acta, Part A* **41A**, 1135 (1985).
85T237	H. C. van der Plas, *Tetrahedron* **41**, 237 (1985).
85TL2419	E. C. Taylor and J. E. Macor, *Tetrahedron Lett.* **26**, 2419 (1985).
85TL4355	G. Seitz, L. Gorge, and S. Dietrich, *Tetrahedron Lett.* **26**, 4355 (1985).
86AP798	G. Seitz, S. Dietrich, R. Dhar, and W. Massa, *Arch. Pharm. (Weinheim, Ger.)* **319**, 798 (1986).
86CRV781	D. L. Boger, *Chem. Rev.* **86**, 781 (1986).
86H161	L. Bauer and S. Prachayasittikul, *Heterocycles* **24**, 161 (1986).

86H239	S. Konno, M. Sagi, M. Agata, Y. Yuki, and H. Yamanaka, *Heterocycles* **24**, 239 (1986).
86H951	M. V. Jovanovic, *Heterocycles* **24**, 951 (1986).
86H1243	S. Konno, S. Ohba, M. Sagi, and H. Yamanaka, *Heterocycles* **24**, 1243 (1986).
86IJC(B)815	R. M. Abdel-Rahman, *Indian J. Chem., Sect. B* **25B**, 815 (1986).
86JCR(S)320	F. Sauter, U. Jordis, and S. M. Siddigi, *J. Chem. Res., Synop.*, 320 (1986).
86JCS(P1)2037	P. Molina, M. Alajarin, J. R. Saez, M. de la C. Foces-Foces, F. H. Cano, R. M. Claramunt, and J. Elguero, *J. C. S. Perkin 1*, 2037 (1986).
86KGS1535	S. G. Alexeev, V. N. Charushin, O. N. Chupakhin, S. V. Shorshnev, A. I. Cherbyshev, and A. I. Kluev, *Khim. Geterotsikl. Soedin.*, 1535 (1986).
86MI1	H. C. van der Plas and M. Wozniak, *Croat. Chim. Acta* **59**, 33 (1986).
86MI2	M. Hamana, *Croat. Chim. Acta* **59**, 89 (1986).
86MI3	A. Rykowski, P. V. Ly, and M. Makosza, *Programme Abstr. Pap. Int. Conf. Org. Synth., 6th*, 103 (1986).
86TL431	E. C. Taylor and J. E. Macor, *Tetrahedron Lett.* **27**, 431 (1986).
86TL1967	E. C. Taylor and L. G. French, *Tetrahedron Lett.* **27**, 1967 (1986).
86TL2107	E. C. Taylor and J. E. Macor, *Tetrahedron Lett.* **27**, 2107 (1986).
86TL2747	G. Seits, S. Dietrich, L. Gorge, and J. Richter, *Tetrahedron Lett.* **27**, 2747 (1986).
86YZ54	S. Shibamoto, T. Nishimura, and H. Fukuyasu, *Yakugaku Zasshi* **106**, 54 (1986).
87CPB1378	S. Konno, S. Ohba, M. Sagi, and H. Yamanaka, *Chem. Pharm. Bull.* **35**, 1378 (1987).
87H3111	S. Konno, M. Sagi, N. Yoshida, and H. Yamanaka, *Heterocycles* **25**, 3111 (1987).
87IJC(B)496	R. M. Abdel-Rahman and M. Ghareib, *Indian J. Chem., Sect. B* **26B**, 496 (1987).
87JOC71	A. Rykowski and H. C. van der Plas, *J. Org. Chem.* **52**, 71 (1987).
87JOC4280	E. C. Taylor and J. E. Macor, *J. Org. Chem.* **52**, 4280 (1987).
87JOC4287	E. C. Taylor and J. L. Pont, *J. Org. Chem.* **52**, 4287 (1987).
87KGS280	S. G. Alexeev, V. N. Charushin, and O. N. Chupakhin, *Khim. Geterotsikl. Soedin.*, 280 (1986).
87MI1	D. L. Boger and S. N. Weinreb, *Org. Chem. (N.Y.)* **47** (1987).
87MI2	O. N. Chupakhin, S. G. Alexeev, and V. N. Charushin, *Proc. Int. Congr. Heterocycl. Chem., 11th, 1987*, 443 (1987).
87MI3	O. Mo, J. L. de Paz, and M. Yanez, *J. Mol. Struct. (Theochem.)* **150**, 135 (1987).
87MI4	M. Cai, W. Chen, and T. Cheng, *Chem. J. Chin. Univ.* **8**, 523 (1987).
87MI5	S.-M. Cai, T.-M. Cheng, and J. Qi, *Acta Chim. Sin.* **45**, 185 (1987) [*CA* **107**, 175999 (1987)].
87T5145	E. C. Taylor, J. E. Macor, and J. L. Pont, *Tetrahedron* **43**, 5145 (1987).
87T5159	E. C. Taylor, J. E. Macor, and J. L. Pond, *Tetrahedron* **43**, 5159 (1987).
87TL379	E. C. Taylor and J. L. Pont, *Tetrahedron Lett.* **28**, 379 (1987).
87YZ301	S. Shibamoto and T. Nishimura, *Yakugaku Zasshi* **107**, 301 (1987).
88AHC301	V. N. Charushin, O. N. Chupakhin, and H. C. van der Plas, *Adv. Heterocycl. Chem.* **43**, 301 (1988).

88H(ip)	V. N. Charushin and H. C. van der Plas, *Heterocycles* (to be published) (1988).
88IZV(ip)	S. G. Alexeev, V. N. Charushin, O. N. Chupakhin, M. F. Gordeev, and V. A. Dorokhov, *Izv. Akad. Nauk SSSR, Ser. Khim. Nauk* (to be published) (1988).
88KGS525	S. G. Alexeev, P. A. Torgashev, M. A. Phedotov, A. I. Rezvukhin, S. V. Shorshnev, A. V. Belik, V. N. Charushin, and O. N. Chupakhin, *Khim. Geterotsikl. Soedin.*, 525 (1988).
88MI1	S. G. Alexeev, Dissertation, Sverdlovsk (1988).
88MI2	O. N. Chupakhin, V. N. Charushin, and A. I. Chernyshev, *Prog. NMR Spectrosc.* **20**, 95 (1988).
88T1	O. N. Chupakhin, V. N. Charushin, and H. C. van der Plas, *Tetrahedron* **44**, 1 (1988).
88TL1431	S. G. Alexeev, V. N. Charushin, O. N. Chupakhin, and G. G. Alexandrov, *Tetrahedron Lett.* **29**, 1431 (1988).
88UP1	V. N. Charushin and H. C. van der Plas, unpublished data.

Boron-Substituted Heteroaromatic Compounds

MASANAO TERASHIMA AND MINORU ISHIKURA

*Faculty of Pharmaceutical Sciences,
Higashi-Nippon-Gakuen University,
Ishikari-Tobetsu, Hokkaido, Japan*

I. Introduction 143
II. Syntheses 144
III. Reactions 147
    A. Borylheteroaromatic Compounds 148
        1. Palladium-Catalyzed Cross-Coupling Reaction with Organic Halides . . 148
        2. Oxidation 151
        3. Miscellaneous Reactions 152
            a. Protodeboronation 152
            b. Bromination 152
            c. Protonolysis 153
    B. Heteroaryl Borate Complexes 153
        1. Intramolecular 1,2-Anionotropic Shift from Boron to Carbon 153
        2. Reactions Promoted by Copper(I) Ion 158
        3. Miscellaneous Reactions 160
IV. Nuclear Magnetic Resonance Spectra 161
    A. Carbon-13 NMR Spectra 161
    B. Proton NMR Spectra 162
    C. Boron-11 NMR Spectra 163
    D. Nitrogen-14 NMR Spectra 165
    References 165

I. Introduction

As is well-recognized, the enormous developments in the chemistry of organoboranes and related compounds during the last two decades have contributed greatly to the recent progress of synthetic organic chemistry. Several comprehensive and excellent reviews and books have been published (72MI1; 73MI1; 75MI1; 76MI2; 79MI3; 80MI1, 80MI2; 82ACR178, 82CSR191).

The chemistry of boron-containing heterocycles is diverse. A variety of heteroaromatic compounds having a boron atom in the ring, borazaromatics for example, has been investigated, mostly due to their interesting physical and chemical properties (76MI1; 77HC381; 82MI1). A number of derivatives of boron have been synthesized and evaluated for potential use in cancer therapy (64MI1; 83MI1; 85JOC841). In some cases, the reaction of heterocycles with various boron reagents involves their derivatives as crucial reaction intermediates.

This article, therefore, will deal only with the chemistry of heteroaromatic compounds directly substituted on a ring carbon with a boron atom. The primary chemical literature published between 1965 and the middle of 1987 is surveyed, but patents are not included. Earlier relevant literatures are also referenced.

II. Syntheses

The application of general reactions available for the synthesis of arylboranes to synthesis of boron-substituted heteroaromatic compounds is somewhat restricted by the presence of a sensitive heteroaromatic ring. The substitution reaction of heteroaryl–metallic compounds (1) with haloboranes (2) has been used to synthesize five-membered π-excessive heteroarylboranes (3) (76CB1075) (Scheme 1).

$$R-\underset{Z}{\underset{|}{\boxed{}}}-M \;+\; R'_2BX \;\longrightarrow\; R-\underset{Z}{\underset{|}{\boxed{}}}-BR'_2$$

(1)　　　　　(2)　　　　　　　　　(3)

Z = O, S, NMe ; R = alkyl ; R' = alkyl, halogen

M = Li , SiMe$_3$, SnMe$_3$

SCHEME 1

The redox reaction between aryl iodides and triiodoborane to give aryldiiodoboranes (70JOM305, 70JOM315) can also be applied to the synthesis of thienylborane derivatives (5) from 5-methyl-2-iodothiophene (4) (70CB2308) (Scheme 2).

As special care must be taken in handling haloboranes, these methods are limited and yields are usually unsatisfactory. The reaction of metalloaromatics with trialkoxyboranes has been commonly used for the preparation of arylboronic acids (56CRV959; 59MI1; 65RTC439; 75MI2; 76BSF(2)1999). The method is successfully applied to the synthesis of six-membered π-deficient heteroarylborane derivatives from the corresponding lithioheteroaromatics (6) and methoxyboranes (7). The procedure provides a facile and versatile synthetic route to dialkyl(pyridyl or benzo-fused pyridyl)boranes (8) (84H2471) (Scheme 3).

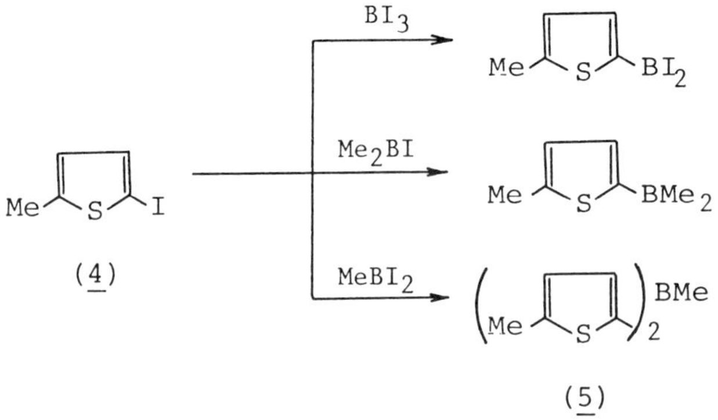

SCHEME 2

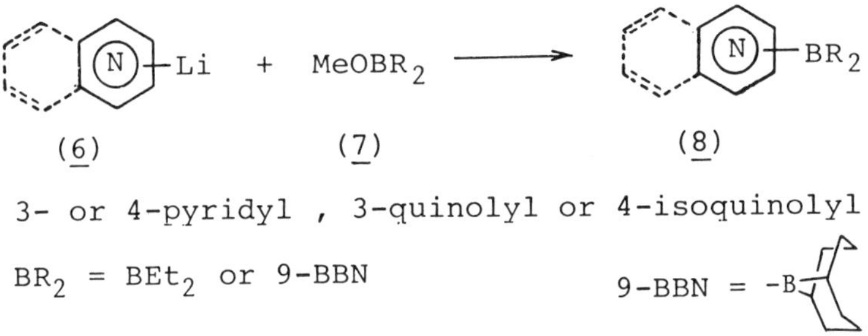

SCHEME 3

3- or 4-pyridyl, 3-quinolyl or 4-isoquinolyl

BR_2 = BEt_2 or 9-BBN   9-BBN =

Diethyl-(3-pyridyl)borane can also be prepared from 3-lithiopyridine by the sequence shown (83CPB4573) in Scheme 4.

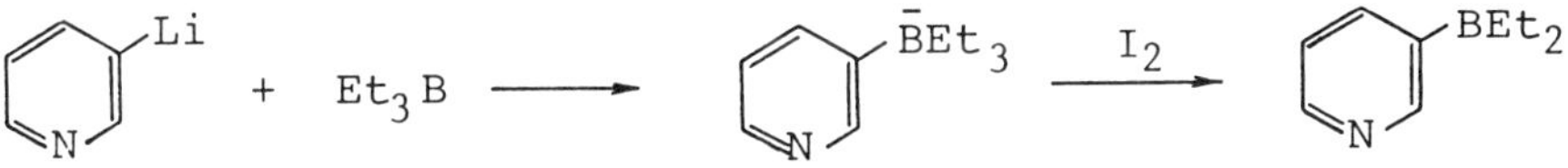

SCHEME 4

The reaction of 2-lithiopyridines (**9**) with **10a,b** affords borates **11**, whereas reaction of **9b** with **10b** gives **13**, suggesting some steric effect of the 6-methyl group on the pyridine ring in the reaction. Upon heating, **11a** is converted to dimeric borane **12**, which is obtainable alternatively by iodinolysis of the corresponding borate derived from **9** and triethylborane (84H2471) (Scheme 5).

In the quinoline series, remarkable differences between the 2-, 3-, and 4-isomers are observed. Thus, contrary to the smooth formation of diethyl-(3-quinolyl)borane mentioned previously, the products from reaction of **14** with **10a** are the corresponding ethylquinolines (**16**), presumably derived from intermediate borate **15** via 1,2-migration of an ethyl group

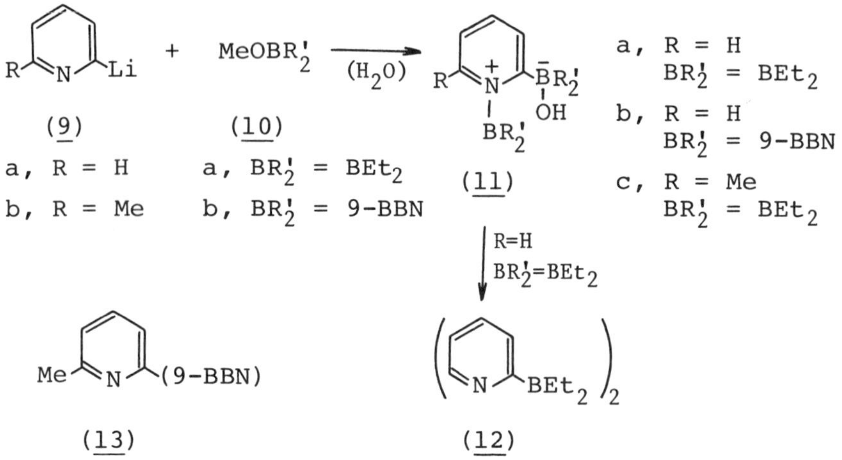

SCHEME 5

followed by oxidation (Section III,B,1). Similar treatment of **14a** with **10b** gives *B*-(2-quinolyl)-9-borabicyclo[3.3.1]nonane (-9-BBN) in moderate yield (84H2471) (Scheme 6).

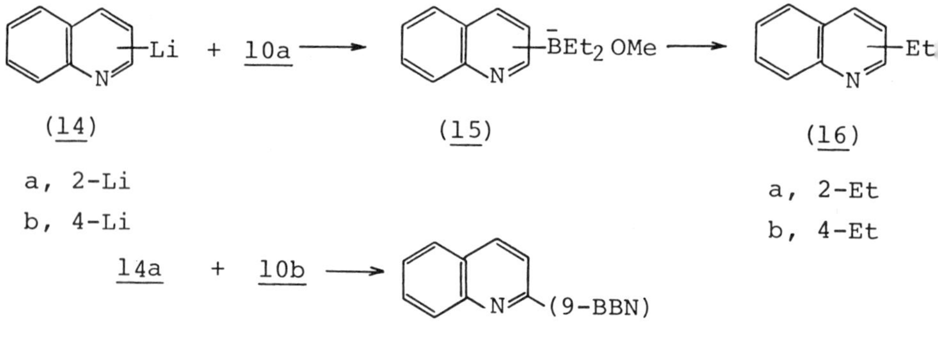

SCHEME 6

Several pyrimidine derivatives having the dihydroxyboryl group, such as 5- (64JA1869) or 6-uracil- (**17**) (78TL4981; 85JOC841), 2,4-dialkoxy-5-pyrimidinyl- (**18**) (64JA1869), and 2′-deoxy-5-uridinylboronic acid (**19**) (78TL4981; 85JOC841), are similarly prepared by a halogen–metal exchange reaction of the corresponding bromo derivatives, followed by boronation, giving possible boron-substituted antimetabolites.

Tetracoordinated borate complexes are usually formed as intermediates, which are not always isolated. Convenient methods for the synthesis of aryl borates consist of the reaction of boron halides with an appropriate organometallic compound or the complexation of boranes with a nucleophile (76JOM281; 78JCS(P2)1225; 79MI2). Application of the former

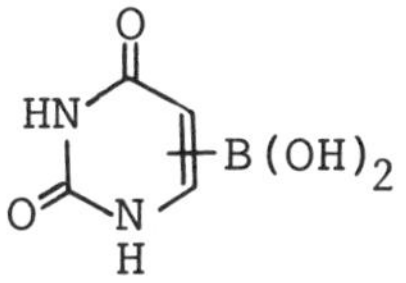

(17) 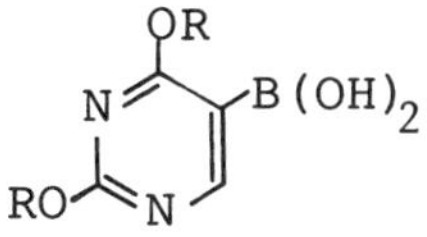   (18) 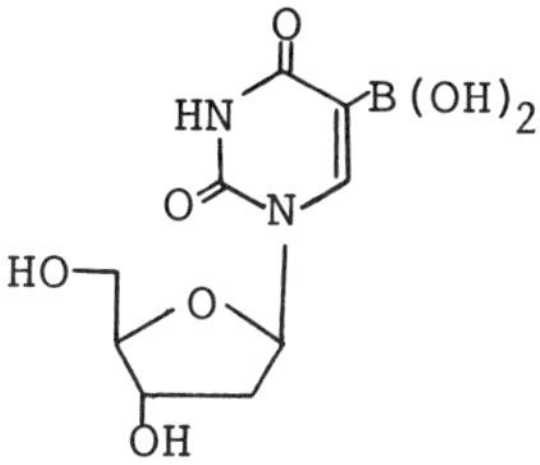   (19)

a, 5-substd.   b, 6-substd.

method, however, might be unsuitable for the synthesis of heteroaryl borates because of the use of reactive boron halides.

π-Excessive heteroaryl borate salts are usually prepared *in situ* at low temperature by the reaction of metalloheteroaromatics with the desired boranes. In the π-deficient N-heteroaromatic series, the same type of method is used for the formation of borates (74JA5601; 76T769). The synthetic versatility of these borates is discussed in Section III,B.

Some examples of attempted isolation of resulting borates such as tributyl 3-pyridyl-, 3-quinolyl-, or 4-isoquinolyl borates by quaternization of the ring nitrogen with various allylic bromides have been reported (87JHC377). Alternatively, diethyl(3-pyridyl)borane obtained by the method mentioned previously reacts with potassium cyanide, and then, with alkyl halides, to afford cyanoborate betaines (20) in high yields (83CPB4573) (Scheme 7).

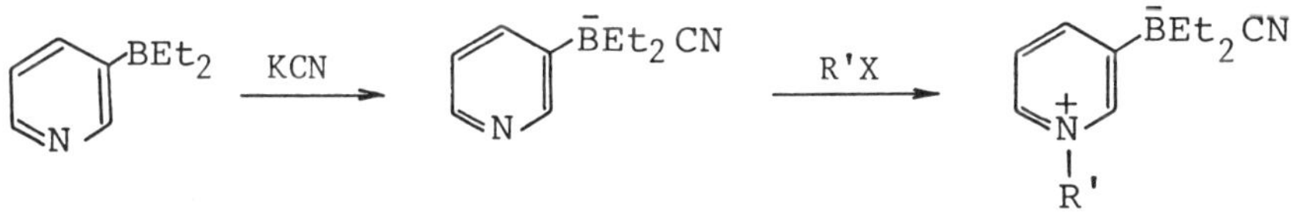

SCHEME 7

III. Reactions

Compared with other organometallics, organoborane compounds are relatively inert to chemical transformations, particularly to ionic reactions, because of the less polar nature of the carbon–boron bond.

The reactions of boron-substituted heteroaromatic compounds may be formally classified into two types: (1) reactions relating to the carbon–boron bond, and (2) all other reactions. Very few investigations have been reported concerning the latter.

Section III,A emphasizes versatile and convenient reactions of boron-substituted heteroaromatic compounds for synthetic heterocyclic chemistry.

A. BORYLHETEROAROMATIC COMPOUNDS

1. *Palladium-Catalyzed Cross-Coupling Reaction with Organic Halides*

Cross-coupling reactions between 1-alkenylboranes and organic halides such as an alkenyl, alkynyl (79TL3437), allyl, benzyl (80TL2865), and aryl (79CC866) halides in the presence of a catalytic amount of tetrakis-(triphenylphosphine)palladium and an appropriate base have been extensively studied (82ACR178). The reaction has further been successfully extended to cross-coupling of phenylboronic acid with haloarenes to provide an efficient synthesis of biaryls (81SC513).

Arylboronic acids are stable and unreactive toward various functional groups in general, so this method has greater advantages than preceding ones that will poorly tolerate reactive groups. The procedure has been widely used as a versatile method for the synthesis of a large number of arylated heteroaromatics from aryl- or heteroarylboronic acids.

The synthesis of unsymmetrical bithienyls (**23**) from **21** and **22** was real-

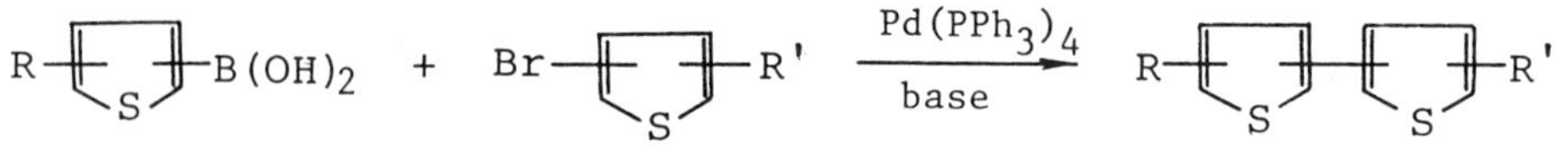

$$R\text{-}[\text{thienyl}]\text{-}B(OH)_2 + Br\text{-}[\text{thienyl}]\text{-}R' \xrightarrow[\text{base}]{Pd(PPh_3)_4} R\text{-}[\text{thienyl}]\text{-}[\text{thienyl}]\text{-}R'$$

(**21**)   (**22**)   (**23**)

2- or 3-boronic acid   2- or 3-bromo

R = H, 2- or 3-CHO ; R' = 2-CHO, NO_2, SMe, CO_2H, 3-Me, 2,4-Me_2

SCHEME 8

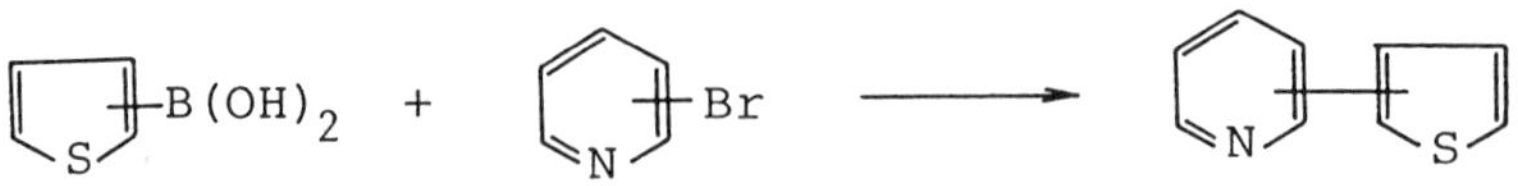

$$[\text{thienyl}]\text{-}B(OH)_2 + [\text{pyridyl}]\text{-}Br \longrightarrow [\text{pyridyl}]\text{-}[\text{thienyl}]$$

(**24**)   (**25**)   (**26**)

2- or 3-substd.   2-, 3- or 4-bromo   six isomers

SCHEME 9

ized by utilizing this procedure (83CS265; 84CS(23)120; 86MI2) (Scheme 8). Likewise, all six isomeric thienylpyridines (**26**), examples of a combination of a π-excessive thiophene ring with a π-deficient pyridine ring previously obtained in low yields (69JOC3175), were prepared from thienylboronic acids (**24**) and bromopyridines (**25**) in fair yield (84CS(24)5) (Scheme 9).

Despite numerous procedures available for the synthesis of 2- and 4-arylpyridines, versatile methodology is scarce for 3-arylpyridines. 5-Thienyl nicotinates (**28**) of biological interest are synthesized by the reaction of **24** and 5-bromo nicotinates (**27**) in good yield (84JOC5237) (Scheme 10).

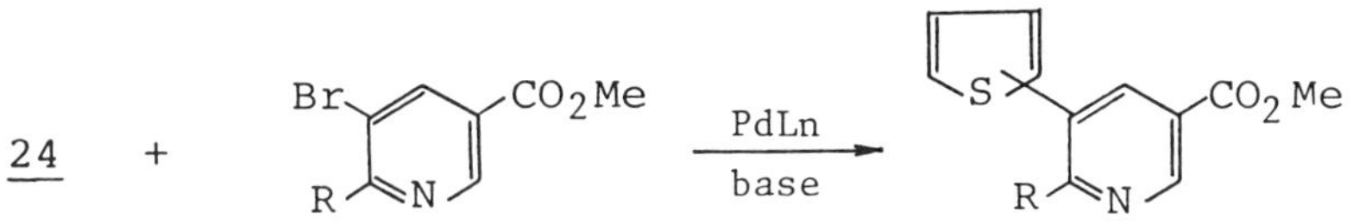

$$\mathbf{24} + \underset{(27)}{\text{Br-pyridine-CO}_2\text{Me}} \xrightarrow[\text{base}]{\text{PdLn}} \underset{(28)}{\text{S-pyridine-CO}_2\text{Me}}$$

R = Me, OMe, SPh

SCHEME 10

The coupling reaction of arylboronic acids with aryl halides offers a direct and excellent route for the synthesis of a variety of biaryls. The same type of reaction proceeds with pyridylborane derivatives to provide an efficient alternative for the arylated pyridines as well. For example, diethyl(3- or 4-pyridyl)boranes (**29a,b**) were coupled with aryl or vinylic bromides (**30**) under similar conditions to give 3- or 4-substituted pyridines (**31**) in moderate to good yield (84H265, 84H2475, 84S936) (Scheme 11). The reaction will tolerate a wide variety of functional groups.

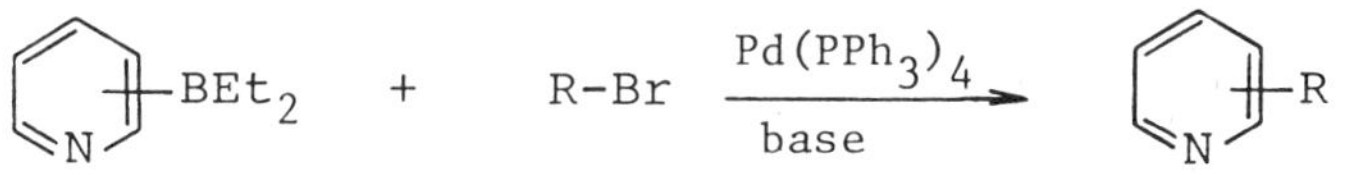

$$\underset{(29)}{\text{pyridine-BEt}_2} + \underset{(30)}{\text{R-Br}} \xrightarrow[\text{base}]{\text{Pd(PPh}_3)_4} \underset{(31)}{\text{pyridine-R}}$$

a, 3-substd.   R = aryl,   3- or 4-
b, 4-substd.   alkenyl,   substd.
                     heteroaryl,

SCHEME 11

Further applications of this methodology to an efficient preparation of terpyridyls (**32, 33**) are realized by the reaction of **29a** with 2,6-dibromopyridine in 50–60% yield. Selective preparation of desired terpyridyls can be performed by altering the pyridylborane: dibromopyridine molar ratios employed in the reaction (85CPB4755) (Scheme 12).

Although much attention has been given to the synthesis of substituted quinolines due to their valuable chemotherapeutic, tumor-inhibiting, and fungicidal properties, general methods for the regioselective introduction of an alkenyl, aryl, or heteroaryl group onto the quinoline nucleus are rare (82HC1). The following reaction may overcome these difficulties.

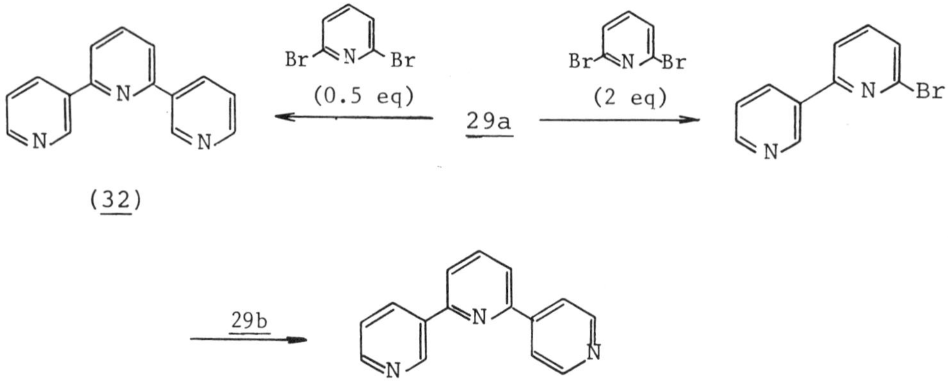

SCHEME 12

The reaction of dialkylquinolylboranes (**34**) with organic halides (**35**) furnishes substituted quinolines (**36**) in moderate to fair yield, as in the case of pyridines (85H2375) (Scheme 13). Similarly, 4-substituted isoquinolines (**38**), one of the common structural units of isoquinoline alkaloids, have been prepared from diethyl-(4-isoquinolyl)borane (**37**) (87H1603) (Scheme 14).

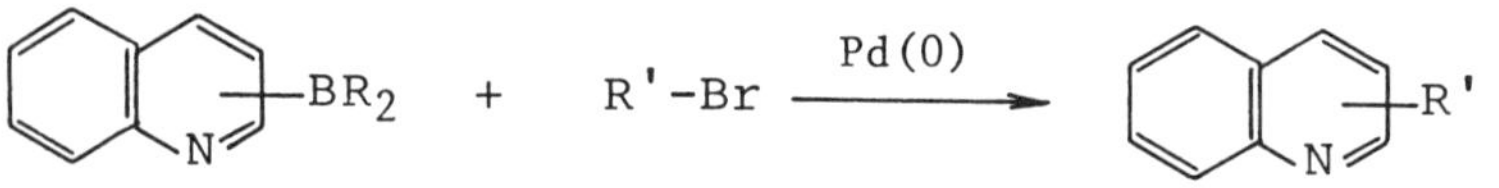

(34) (35) (36)

a, BR_2 = 2-(9-BBN)   R' = aryl,   a, 2-substd.
b, BR_2 = 3-BEt$_2$ heteroaryl,   b, 3-substd.
                         alkenyl

SCHEME 13

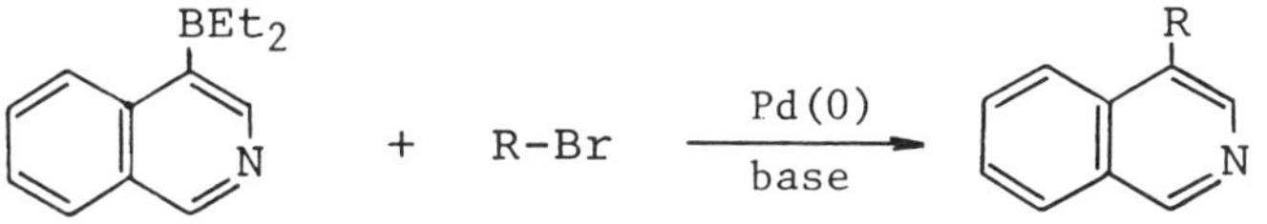

Sec. III.A]   BORON-SUBSTITUTED HETEROAROMATICS   151

SCHEME 14

2. Oxidation

The oxidative cleavage of a boron–carbon bond is one of the important and well-documented processes in organoboron chemistry. With thiopheneboronic acids (**39**), oxidation is conveniently performed by aqueous hydrogen peroxide, with or without alkali, to afford the corresponding hydroxyl (**40**) or the predominant tautomeric oxo derivatives (**41, 42**), depending upon the ring substituent (63AK239; 64AK211; 66ACS261; 74JHC291) (Scheme 15).

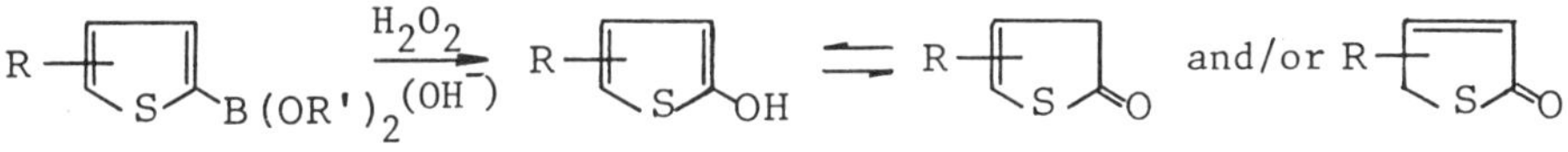

R = H, 3- or 4-Br, 3- or 5-alkyl, 3-MeO, 5-Ph, or thienyl

R' = H or Me

SCHEME 15

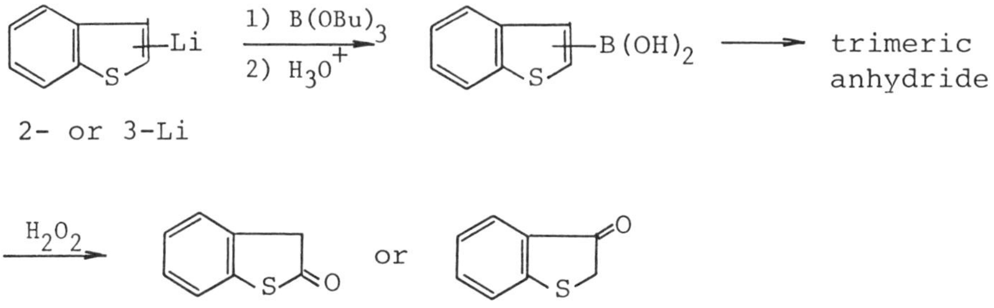

SCHEME 16

A simple two-step synthesis of benzo[*b*]thiophene-2(3*H*)- (**43**) and -3(2*H*)-ones (**44**) has been reported (70JCS(C)1926) (Scheme 16). Similar results are also obtained by starting with *N*-methylindole (**45**). Compound **46** is obtained in a one-pot reaction in moderate yield (87MI1) (Scheme 17).

More recently, an efficient direct conversion of 5-substituted 2-lithiofurans (**47**) to corresponding butenolides (**48**) has been attained in a related sequence (87TL1203) (Scheme 18).

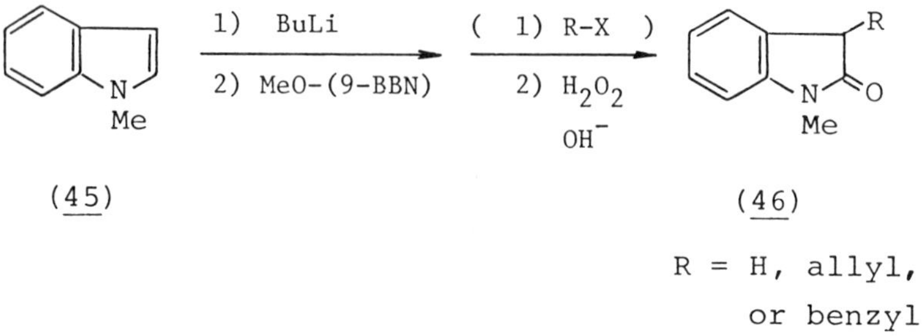

$$\text{(45)} \xrightarrow[\text{2) MeO-(9-BBN)}]{\text{1) BuLi}} \xrightarrow[\text{2) } H_2O_2 / OH^-]{\text{(1) R-X)}} \text{(46)}$$

R = H, allyl, or benzyl

SCHEME 17

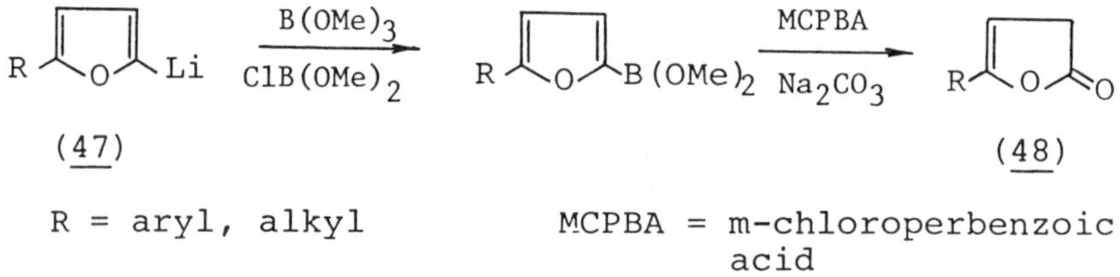

$$\text{(47)} \xrightarrow[\text{ClB(OMe)}_2]{B(OMe)_3} \xrightarrow[\text{Na}_2CO_3]{\text{MCPBA}} \text{(48)}$$

R = aryl, alkyl   MCPBA = m-chloroperbenzoic acid

SCHEME 18

3. Miscellaneous Reactions

a. *Protodeboronation.* Studies on protodeboronation of furan- and thiopheneboronic acids to their parent heterocycles and boric acid are carried out using a conventional aqueous perchloric acid method to estimate the relative reactivities of heteroaromatic rings for electrophilic substitution. The quantitative assignment of the reactivities of the ring to substitution is α-furan > α-thiophene > β-furan > β-thiophene (57AK387; 65AJC1521; 75JHC195).

b. *Bromination.* The directing effect of the boronic acid group in the bromination of 2- or 3-thiopheneboronic acid has been explored. Exclusive formation of 5-bromo derivatives from the 2-isomer, despite the meta-directing effect of the boronic acid group, reveals that the reaction is completely controlled by the sulfur atom in the ring (57AK387).

c. *Protonolysis.* Although the protonolysis of a boron–carbon bond by a carboxylic acid is a well-established process, predominant cleavage of the ethyl group of diethyl-(3-pyridyl)borane is observed upon refluxing in propionic acid to give 3-pyridylboronic acid (83CPB4573), originally synthesized by the reaction of 3-pyridylmagnesium bromide and butyl borate followed by hydrolysis (65RTC439).

B. HETEROARYL BORATE COMPLEXES

1. *Intramolecular 1,2-Anionotropic Shift from Boron to Carbon*

New C—C bond formation through a spontaneous 1,2-anionotropic rearrangement is the most fundamental and important process in tetravalent, negatively charged boron compounds ("ate" complexes) (76JOM281; 80MI2; 82ACR178; 82CSR191). The process may be examined using heteroaryl borates, allowing the regiospecific functionalization of the aromatic ring. This concerted 1,2-shift works well with π-electron excessive five-membered heteroaryl borates through a rapid interaction with added electrophiles.

Treatment of 2-lithiothiophene or 2-lithio-1-methylpyrrole (49) with R_3B (R = alkyl, phenyl), then iodine or *N*-chlorosuccinimide (NCS), gives 2-substituted derivatives of thiophene or 1-methylpyrrole (50), respectively (Scheme 19). The reaction can be rationalized by considering the initial electrophilic attack on the heteroaryl ring with spontaneous 1,2-migration of the R group, followed by dehaloboration (79BCJ1865, 79TL2313).

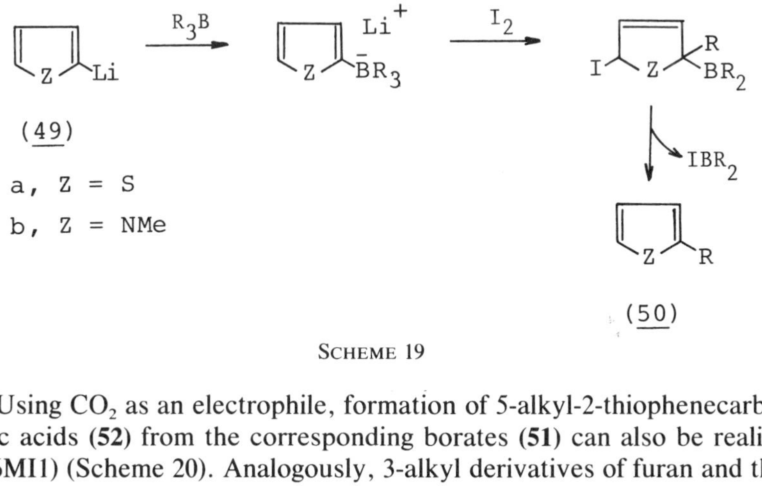

(49)
a, Z = S
b, Z = NMe

SCHEME 19

Using CO_2 as an electrophile, formation of 5-alkyl-2-thiophenecarboxylic acids (52) from the corresponding borates (51) can also be realized (86MI1) (Scheme 20). Analogously, 3-alkyl derivatives of furan and thiophene (54) are formed from the corresponding borates (53) and iodine or bromine (81BCJ1587) (Scheme 21).

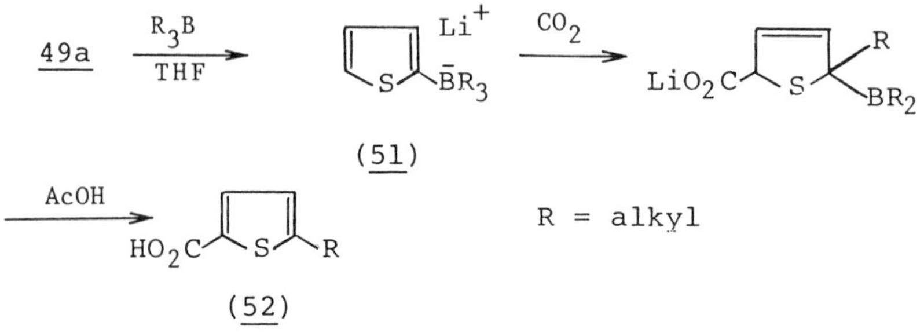

SCHEME 20

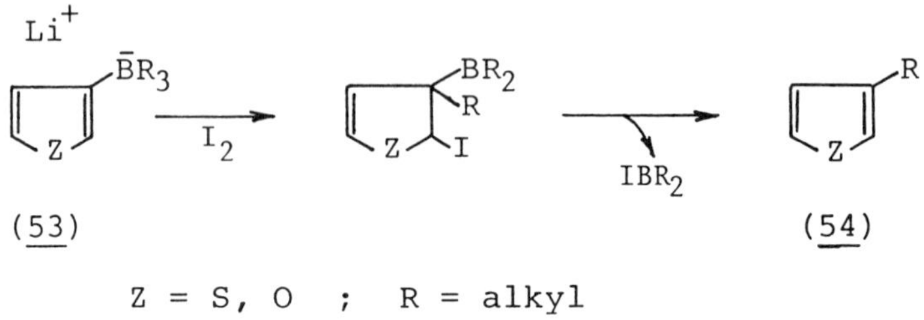

SCHEME 21

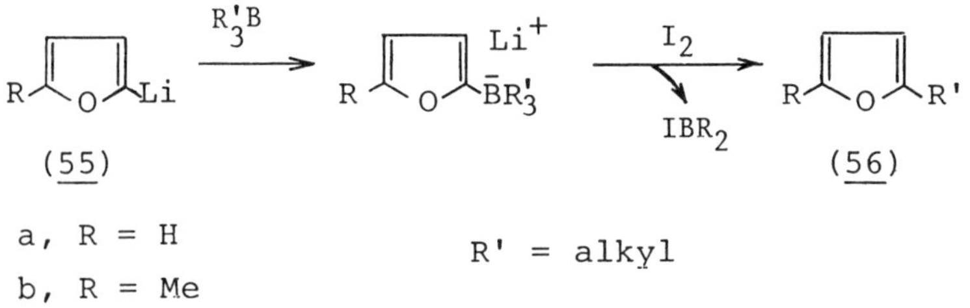

SCHEME 22

When equimolar amounts of 2-lithiofurans (**55a,b**) and R_3B are treated with iodine, 2-alkylfurans (**56**) are formed. Bulky alkyl groups give higher yields of **56** (79S146) (Scheme 22).

With a 2:1 molar ratio of 2-lithiofuran (**55a**) to R_3B, a cyclic borate (**57**) is isolated after treatment with AcOH at room temperature; **57** is convertible to 1,4-diols (**58**) by oxidation. Severe steric hindrance by a bulky R group in R_3B lowers the yield of **58** (71T2775) (Scheme 23). This principle has been developed for the selective formation of substituted indole derivatives (**60, 61**) through 2-indolyl borates (**59**) under the action of various electrophiles (78JOC4684; 79TL4021; 87JHC377) (Scheme 24). Biaryl synthesis is also possible on the basis of this principle. The action of bromine or N-bromosuccinimide (NBS) on heteroarylboric acid esters (**62**) produces 2-aryl derivatives of thiophene and furan (**63**) (76TL795; 84TL453) (Scheme 25).

Oligomers of thiophene (**67**) are obtained by the sequential reaction of

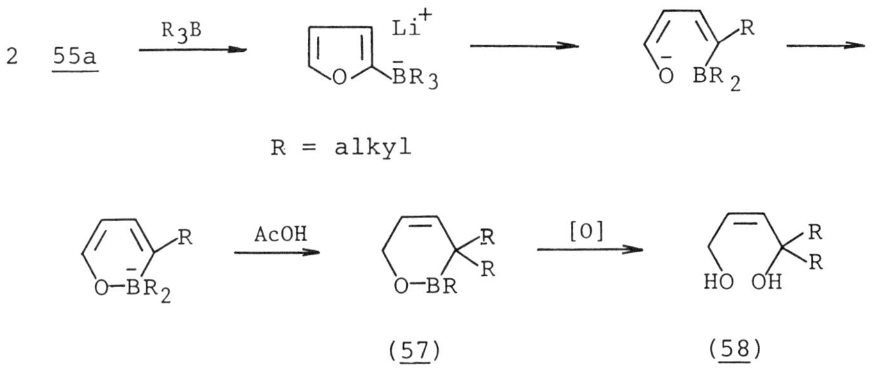

Sec. III.B] BORON-SUBSTITUTED HETEROAROMATICS 155

SCHEME 23

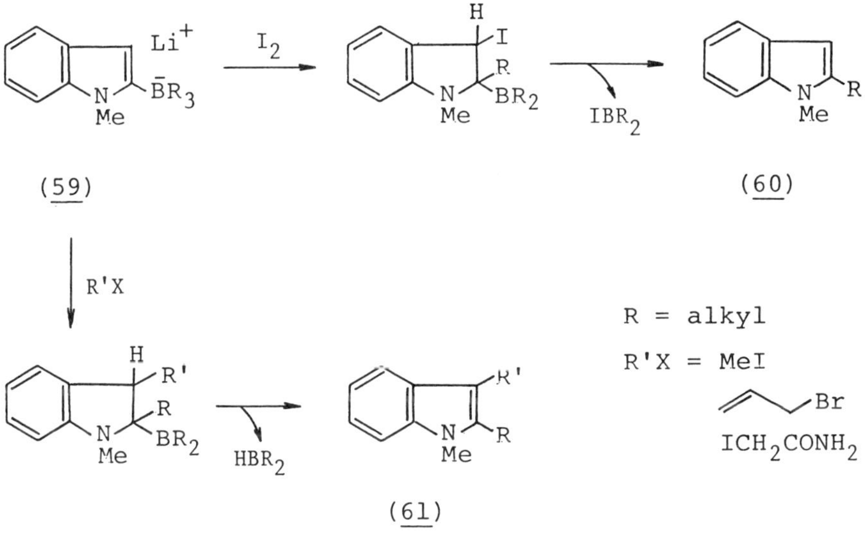

SCHEME 24

borates **64–66** (83TL4043) (Scheme 26). Intramolecular 1,2-migration of a heteroaryl group from boron to carbon containing a leaving group is also possible.

Treatment of B-(2-thienyl)-9-BBN (**68**) and a Schiff base in the presence of a base, accompanied by 1,2-thienyl migration leading to **69**, gave amino acid **70** following hydrolysis (85CC1168) (Scheme 27).

Borate **72**, derived from **49a** and 3-chloro-1,2-oxaborinane (**71**), readily undergoes migration to yield **73**. Oxidation of **73** gives the 1,4-alkanediol (**74**) (87TL2599) (Scheme 28).

Furyl borinate **75** (TMS = tetramethylsilyl) is converted smoothly to

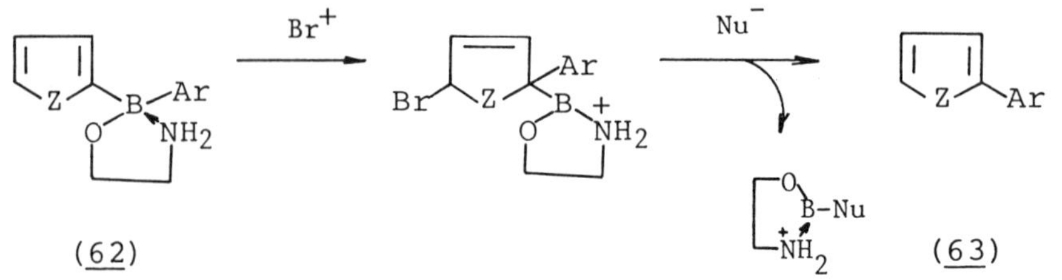

Z = O,S ; Ar = phenyl,2-furyl,2-thienyl

SCHEME 25

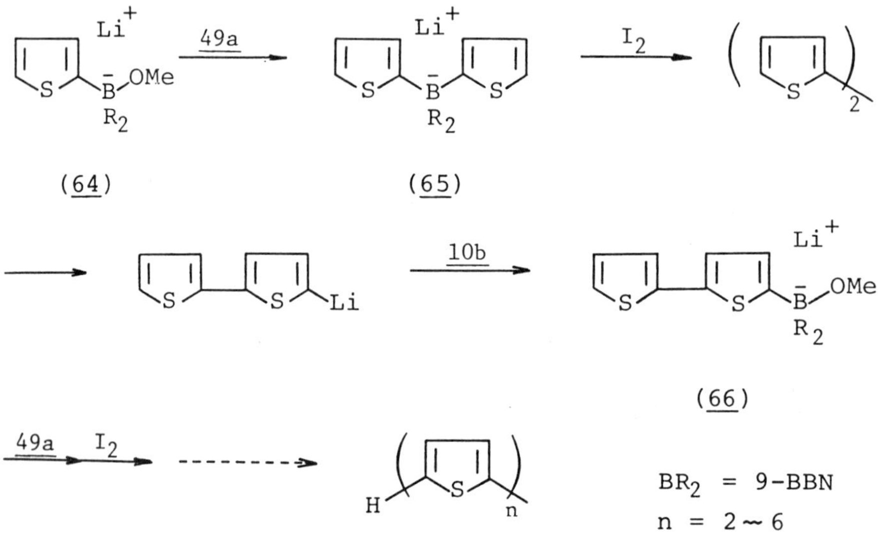

BR_2 = 9-BBN
n = 2 ~ 6

SCHEME 26

furyl ketone **76** by the action of dichloromethyl methyl ether in the presence of a base. The natural product *Perilla* ketone (**77**) has also been synthesized (87JA5420) (Scheme 29).

On reacting π-electron-deficient six-membered heteroaryl borates, a rapid 1,2-alkyl migration can be realized. Exposure of 6-bromo-2-lithiopyridine (**78**) to R_3B causes a smooth concerted alkyl migration with ring opening to produce unsaturated nitriles (**79**) (74JA5601; 76T769) (Scheme 30). Similarly, a very rapid 1,2-alkyl migration is found in the reaction between **14a** and R_3B, leading to 2-alkylquinolines (**80**) (87SC959) (Scheme 31). The reaction of Et_2BOMe (**10a**) with 2- or 4-lithioquinoline (**14**) produces 2- or 4-ethylquinoline (**16**), respectively (84H2471) (Section

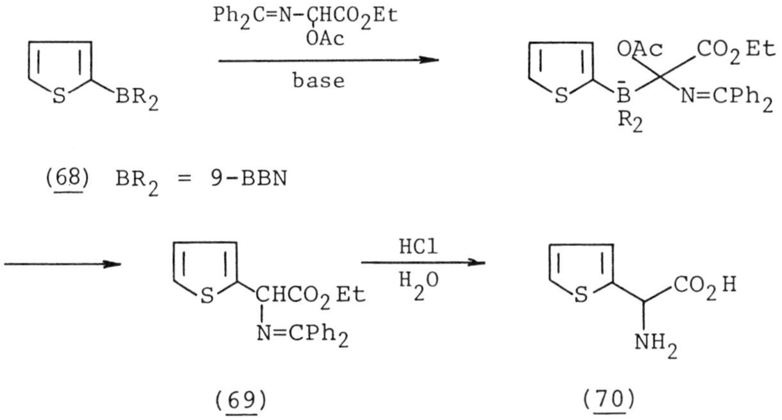

SCHEME 27

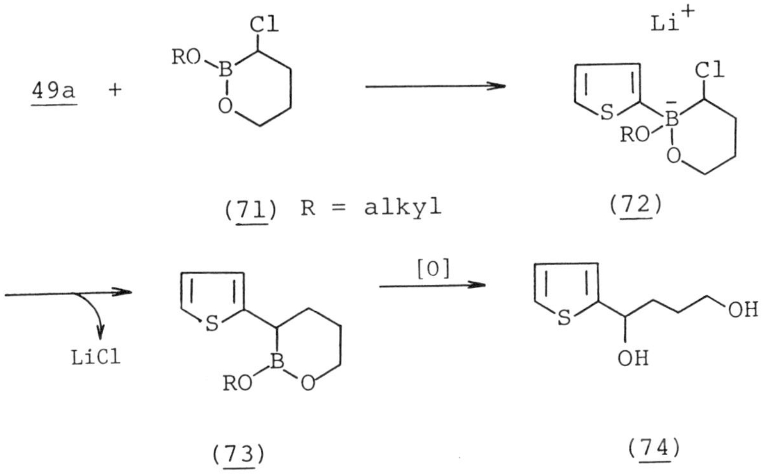

SCHEME 28

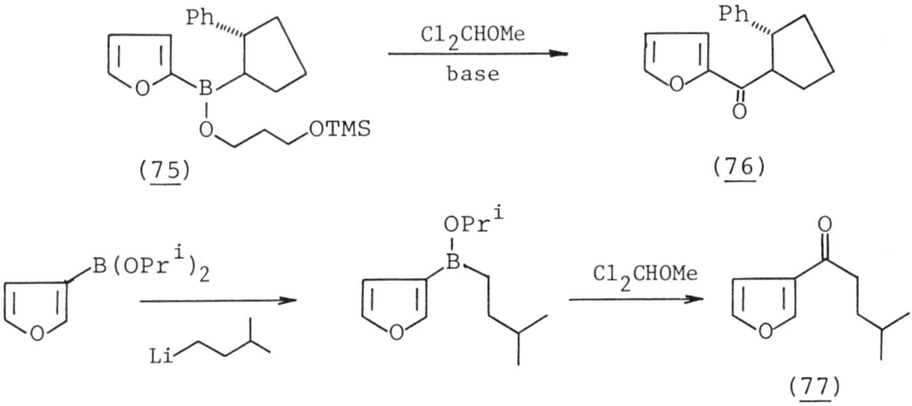

SCHEME 29

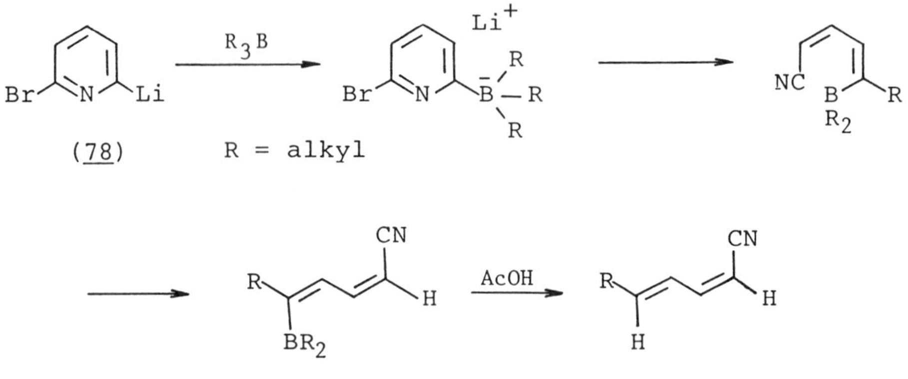

SCHEME 30

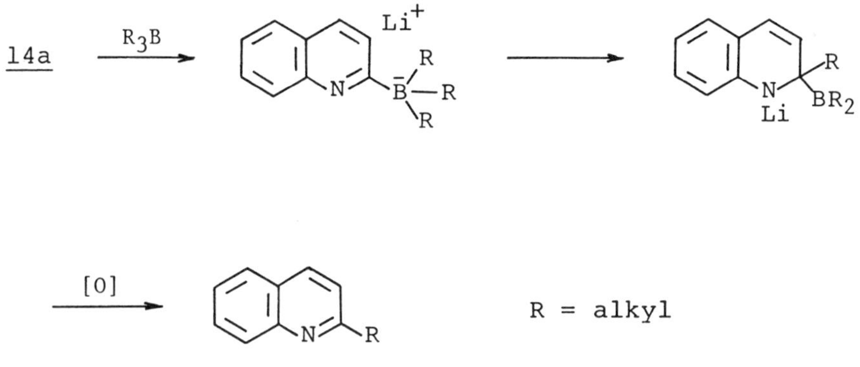

SCHEME 31

II, Scheme 6). In the case of 4-pyridyl borates (**81**), a similar alkyl shift is feasible following attack of a Lewis acid or an acyl halide on the nitrogen to afford 4-alkylated products **82** and **83** (86H2793) (Scheme 32).

2. *Reactions Promoted by Copper(I) Ion*

Combination of copper(I) ion with organoborates can bring about useful synthetic reactions (82ACR178).

A noticeable effect of Cu^+ on the reaction of 3-pyridyl borates **84** with allylic halides can be found; regioselective allylation at the 3-position of pyridines **85** results (85H117). This is a common feature with benzo-fused pyridyl borates (87JHC377) (Scheme 33).

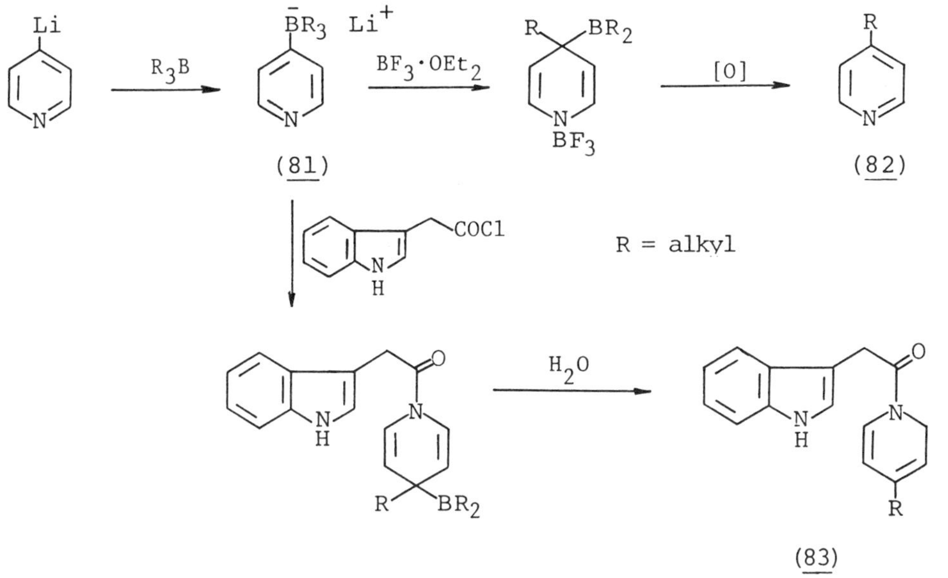

SCHEME 32

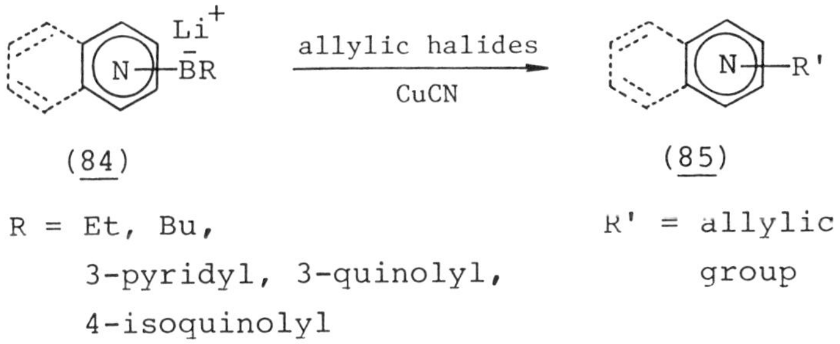

SCHEME 33

R = Et, Bu,
3-pyridyl, 3-quinolyl,
4-isoquinolyl

R' = allylic group

By contrast, 4-quinolyl borate **86** gives 3-allyl-4-alkylquinolines **(87)** on similar treatment (87SC959), probably through a pathway analogous to the case of tributyl 1-naphthyl borate (74CC860) (Scheme 34).

The presence of Cu^+ markedly enhances the formation of allylated compounds **(89, 91)** from five-membered heteroaryl borates **(88, 90)** and allyl bromide, as mentioned above (87JHC377) (Scheme 35). Coupling of **59** with allyl bromide under the action of Cu^+ occurs selectively at position -2 of the indole ring to give **93**, in marked contrast to the formation of 2,3-disubstituted indole **(92)** described previously (Section III,B,1) (87JHC377) (Scheme 36).

14b

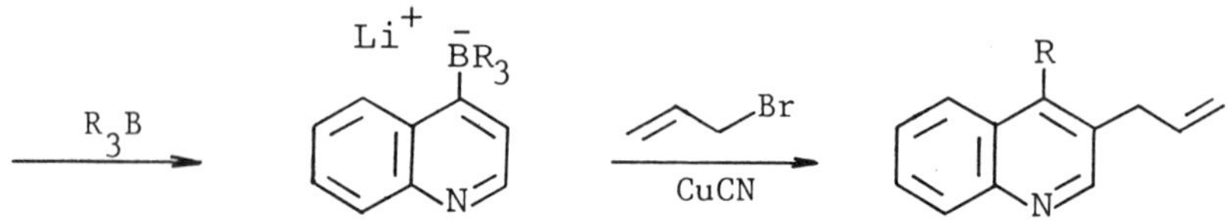

R = alkyl (86) (87)

SCHEME 34

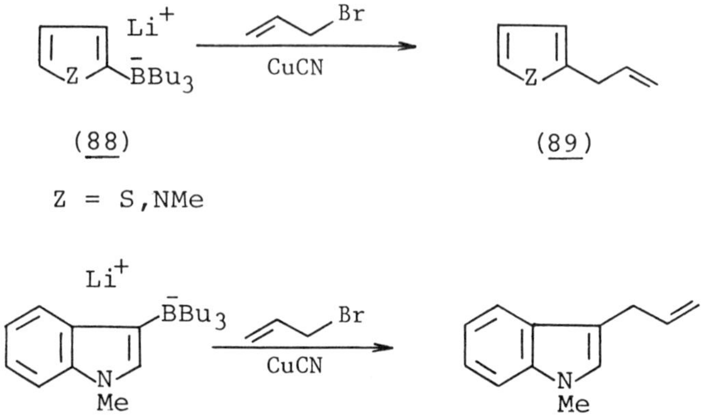

(88) (89)

Z = S,NMe

(90) (91)

SCHEME 35

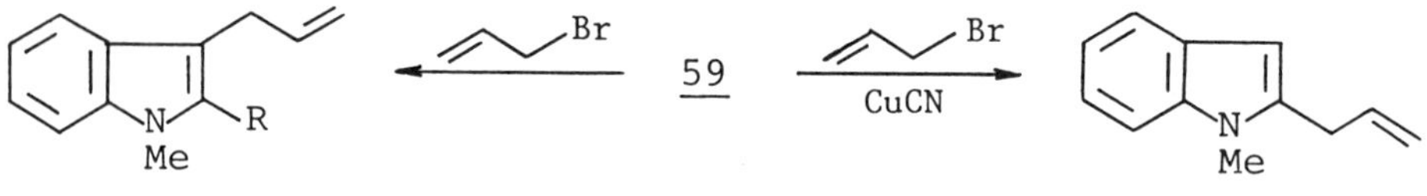

(92) (93)

R = alkyl

SCHEME 36

3. *Miscellaneous Reactions*

The action of iodine on triethyl 2- or 3-pyridyl borate resulted in dealkylation, leading to the corresponding diethylpyridylboranes (Section II), in a process somewhat similar to that of phenyl borate (78JCS(P2)1225;

83CPB4573; 84H2471). In contrast to 3-pyridylborane, borates **94** are susceptible to quarternization with various allyl and alkyl halides to form stable betaines (**95**). Likewise, betaines **96** and **97** are obtained from 3-quinolyl and 4-isoquinolyl borates, respectively (83CPB4573; 87JHC377) (Scheme 37).

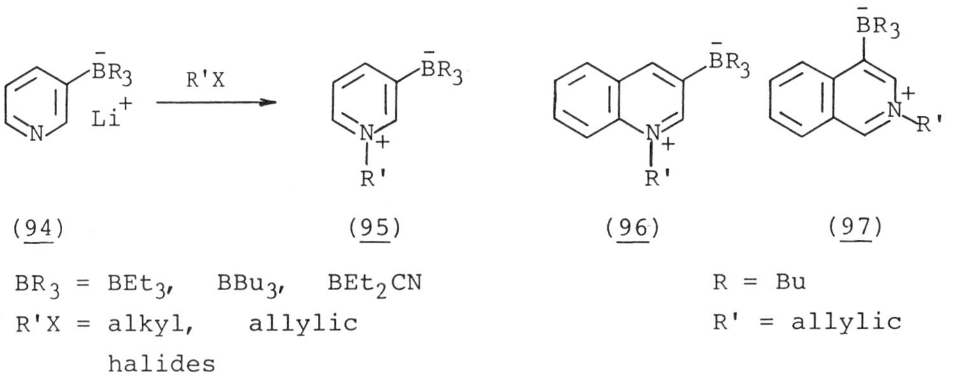

(**94**)   (**95**)   (**96**)   (**97**)

BR_3 = BEt_3, BBu_3, BEt_2CN
R'X = alkyl, allylic halides

R = Bu
R' = allylic

SCHEME 37

IV. Nuclear Magnetic Resonance Spectra

A. CARBON-13 NMR SPECTRA

Carbon-13 NMR (^{13}C-NMR) signals of aryl carbon atoms bonded directly to boron can be broad and/or low-intensity peaks. In some cases, they may not be observed, possibly depending on the quadrupolar relaxation rate of boron nuclei (79MI1) (Fig. 1). Some special techniques are known to increase the intensity of ^{13}C signals of these carbon atoms (77JOM(132)29; 78JOMC34).

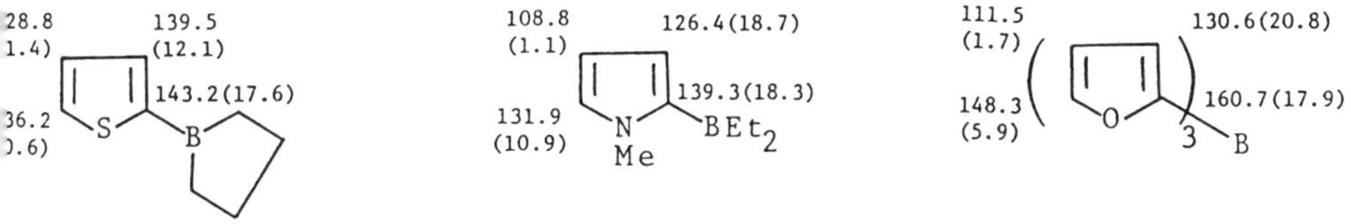

FIG. 1. ^{13}C-NMR signals for some heteroaryl boranes. Numbers in parentheses refer to ^{13}C chemical shifts relative to the parent heteroaryl compound.

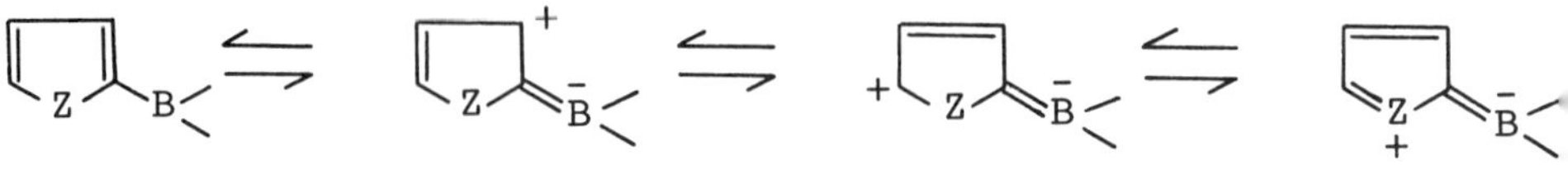

FIG. 2. Resonance interactions between boron atom and heteroaromatic ring in heteroaryl boranes.

Resonance interaction between the boron atom and the five-membered heteroaromatic ring accounts for the downfield shift of the ring carbons, particulaly the strong deshielding of the C-3 and C-5 carbon atoms (79JOM15) (Fig. 2).

Tabulated ^{13}C chemical shift data for pyridylboranes show that deshielding of ring carbons is less signficant in comparison with those of five-membered heteroarylboranes, but the C-3 carbon in 3-pyridylborane is conspicuously deshielded (31 ppm) (83CPB4573; 87UP1) (Fig. 3).

B. Proton NMR Spectra

Lower field shifts of ^{1}H signals at the 3- and 5-positions in 2-borylated five-membered heteroaromatics have been observed (76CB1075), suggesting that the π-interaction to boron is also significant (Table I).

Proton NMR data for pyridylboranes are given (83CPB4573; 87UP1) in Fig. 4; those of heteroarylboronic acids are available in the literature (70CR(C)1608; 71MI1; 76BSF(2)1999; 77CS76).

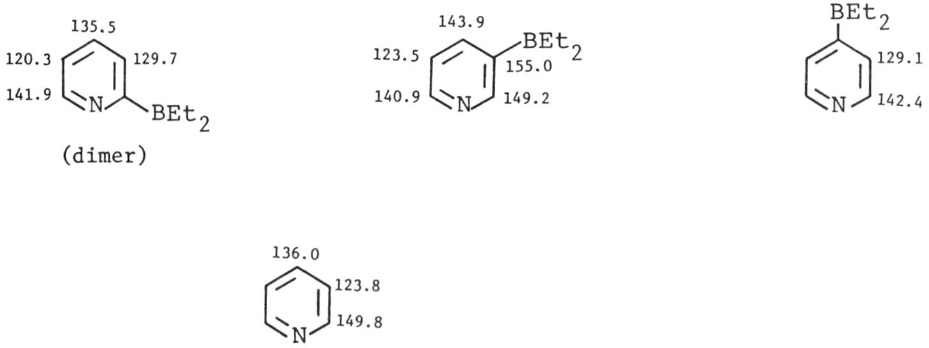

FIG. 3. ^{13}C chemical shift data for diethyl(2-, 3-, and 4-pyridyl)boranes and for pyridine. δ_{ppm} given in CDCl$_3$.

Sec. IV.C] BORON-SUBSTITUTED HETEROAROMATICS 163

TABLE I
PROTON CHEMICAL SHIFTS OF BORYLATED
FIVE-MEMBERED HETEROAROMATICS

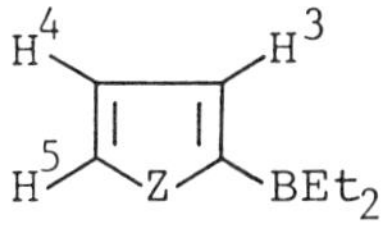

	^{1}H-chemical shifts (δ ppm)[a]		
Z	H^3	H^4	H^5
S	7.80(0.84)	7.25(0.29)	7.80(0.60)
O	7.30(1.00)	6.50(0.20)	7.75(0.37)
N Me	7.05(0.94)	6.12(0.01)	6.84(0.28)

[a] Numbers in parentheses refer to the differences in ^{1}H chemical shifts relative to the parent heteroaromatic compounds.

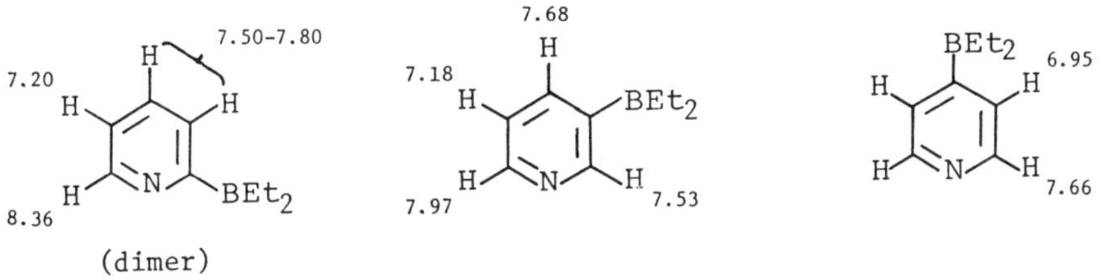

FIG. 4. Proton NMR data for diethyl(2-, 3-, and 4-pyridyl)boranes. δ$_{ppm}$ with respect to TMS in CDCl$_3$.

C. BORON-11 NMR SPECTRA

As is obvious from the ^{11}B-NMR spectral data of borylated thiophene, N-methylpyrrole, and furan in Fig. 5, ^{11}B signals appear at higher field in comparison with phenylborane. This suggests that B–C (p_π–p_π) interaction is much more significant in five-membered heteroaromatic rings than in a phenyl ring (76CB1075).

In a pentacarbonylchromium complex of σ-s-bonded 2-diethylboryl-5-methylthiophene, a donor–acceptor interaction between boron and a carbonyl group is suggested in order to rationalize the higher field shift of its ^{11}B signal (77JOM(125)155) (Fig. 6).

A brief review of ^{11}B-NMR spectroscopic studies is available (64MI2), along with an extensive listing of ^{11}B-NMR chemical shift data (78MI1; 83JOM269).

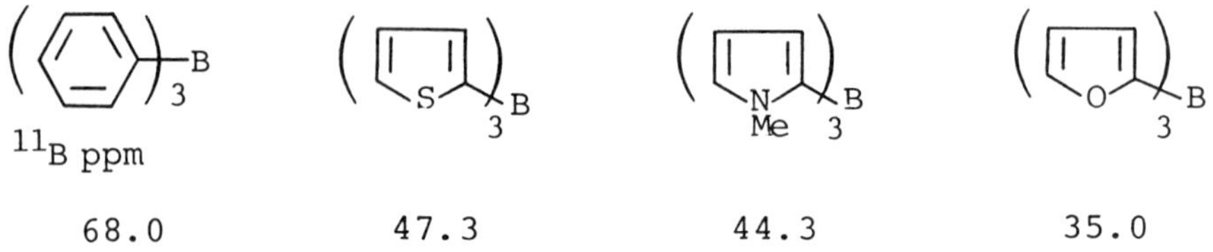

FIG. 5. ^{11}B NMR data of phenylborane, as well as borylated thiophene, N-methylpyrrole, and furan.

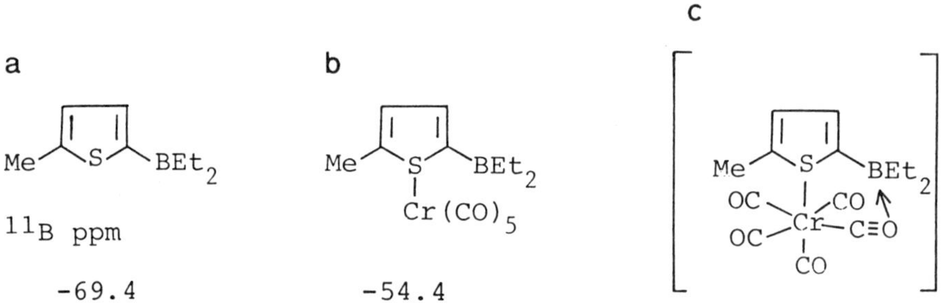

FIG. 6. ^{11}B-NMR data for 2-diethylboryl-5-methyl thiophene (a) and its pentacarbonylchromium complex (b). In the latter, donor–acceptor relationship between boron and the carbonyl group is postulated to account for its higher field ^{11}B shift (c).

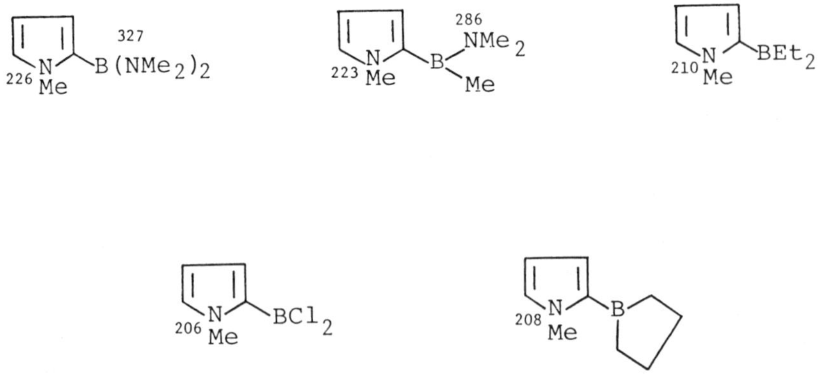

FIG. 7. ^{14}N-NMR data (in ppm) 2-borylated N-methylpyrroles.

D. Nitrogen-14 NMR Spectra

Few ^{14}N-NMR data for 2-borylated N-methylpyrroles show a ^{14}N chemical shift dependency on the π-accepting ability of a boryl group (76CB1075) (Fig. 7).

References

56CRV959	M. F. Lappert, *Chem. Rev.* **56,** 959 (1956).
57AK387	S. O. Lawesson, *Ark. Kemi* **11,** 387 (1957).
59MI1	B. M. Mikhailov and T. K. Kozminskaya, *Izv. Akad. Nauk SSSR, Otd. Khim. Nauk,* 80 (1959).
63AK239	A. B. Hörnfeldt and S. Gronowitz, *Ark. Kemi* **21,** 239 (1963).
64AK211	A. B. Hörnfeldt, *Ark. Kemi* **22,** 211 (1964).
64JA1869	T. K. Liao, E. G. Podrebarac, and C. C. Cheng, *J. Am. Chem. Soc.* **86,** 1869 (1964).
64MI1	A. H. Soloway, *Prog. Boron Chem.* **1,** 203 (1964).
64MI2	R. Schaeffer, in "Progress in Boron Compounds" (H. Steinberg and A. L. McCloskey, ed.), Vol. 1, p. 417. Pergamon, Oxford, 1964.
65AJC1521	R. D. Brown, A. S. Buchanan, and A. A. Humffray, *Aust. J. Chem.* **18,** 1521 (1965).
65RTC439	F. C. Fischer and E. Havinga, *Recl. Trav. Chim. Pays-Bas* **84,** 439 (1965).
66ACS261	S. Gronowitz and A. Bugge, *Acta Chem. Scand.* **20,** 261 (1966).
69JOC3175	H. Wynberg, T. J. van Bergen, and R. M. Kellogg, *J. Org. Chem.* **34,** 3175 (1969).
70CB2308	W. Siebert, *Chem. Ber.* **103,** 2308 (1970).
70CR(C)1608	D. Florentin and B. Roques, *C. R. Hebd. Seances Acad. Sci., Ser. C* **270,** 1608 (1970).
70JCS(C)1926	R. P. Dickinson and B. Iddon, *J. Chem. Soc. C,* 1926 (1970).
70JOM305	W. Siebert, K. J. Schaper, and M. Schmidt, *J. Organomet. Chem.* **25,** 305 (1970).
70JOM315	W. Siebert, K. J. Schaper, and M. Schmidt, *J. Organomet. Chem.* **25,** 315 (1970).
71MI1	S. Gronowitz, T. Dahlgren, J. Namtvedt, C. Roos, G. Rosen, B. Sjöberg, and U. Forsgren, *Acta Chem. Suec.* **8,** 623 (1971).
71T2775	A. Suzuki, N. Miyaura, and M. Itoh, *Tetrahedron* **27,** 2775 (1971).
72MI1	H. C. Brown, "Boranes in Organic Chemistry." Cornell Univ. Press, Ithaca, New York, 1972.
73MI1	G. M. Cragg, "Organoboranes in Organic Synthesis." Dekker, New York, 1973.
74CC860	E. Negishi and R. E. Merrill, *J. C. S. Chem. Commun.,* 860 (1974).
74JA5601	K. Utimoto, N. Sakai, and H. Nozaki, *J. Am. Chem. Soc.* **96,** 5601 (1974).
74JHC291	R. T. Hawkins, *J. Heterocycl. Chem.* **11,** 291 (1974).
75JHC195	B. P. Roques, D. Florentin, and M. Callanquin, *J. Heterocycl. Chem.* **12,** 195 (1975).

75MI1	H. C. Brown, "Organic Syntheses via Boranes." Wiley, New York, 1975.
75MI2	T. Onak, "Organoborane Chemistry." Academic Press, New York, 1975.
76BSF(2)1999	D. Florentin, B. P. Roques, and M. C. Fournie-Zaluski, *Bull. Soc. Chim. Fr.* **11–12**, Pt, 2, 1999 (1976).
76CB1075	B. Wrackmeyer and H. Nöth, *Chem. Ber.* **109**, 1075 (1976).
76JOM281	E. Negishi, *J. Organomet. Chem.* **108**, 281 (1976).
76MI1	S. Gronowitz, *Lect. Heterocycl. Chem.* **3**, 17 (1976).
76MI2	E. Negishi, in "New Applications of Organometallic Reagents in Organic Synthesis" (D. Seyferth, ed.), p. 93. Elsevier, Amsterdam, 1976.
76T769	K. Utimoto, N. Sakai, M. Obayashi, and H. Nozaki, *Tetrahedron* **32**, 769 (1976).
76TL795	G. M. Davies, P. S. Davies, W. E. Paget, and J. M. Wardleworth, *Tetrahedron Lett.*, 795 (1976).
77CS76	S. Gronowitz, and C. Glennow, *Chem. Scr.* **11**, 76 (1977).
77HC381	A. J. Fritsch, *Chem. Heterocycl. Compd.* **30**, 381 (1977).
77JOM(125)155	H. Nöth and U. Schuchardt, *J. Organomet. Chem.* **125**, 155 (1977).
77JOM(132)29	B. R. Gragg, W. J. Layton, and K. Niedenzu, *J. Organomet. Chem.* **132**, 29 (1977).
78JCS(P2)1225	E. Negishi, M. J. Idacavage, K. Chiu, T. Yoshida, A. Abramovitch, M. E. Goettel, A. Silveira, and H. D. Bretherick, *J. C. S. Perkin 2*, 1225 (1978).
78JOC4684	A. B. Levy, *J. Org. Chem.* **43**, 4684 (1978).
78JOMC34	C. Brown, R. H. Cragg, T. Miller, D. O'N. Smith, and A. Steltner, *J. Organomet. Chem.* **149**, C34 (1978).
78MI1	H. Nöth and B. Wrackmeyer, "Nuclear Magnetic Resonance Spectroscopy of Boron Compounds." Springer-Verlag, Berlin, 1978.
78TL4981	R. F. Schinazi and W. H. Prusoff, *Tetrahedron Lett.*, 4981 (1978).
79BCJ1865	T. Sotoyama, S. Hara, and A. Suzuki, *Bull. Chem. Soc. Jpn.* **52**, 1865 (1979).
79CC866	N. Miyaura and A. Suzuki, *J. C. S. Chem. Commun.*, 866 (1979).
79JOM15	J. D. Odom and T. F. Moore, *J. Organomet. Chem.* **173**, 15 (1979).
79MI1	B. Wrackmeyer, *Prog. Nucl. Magn. Spectrosc.* **12**, 227 (1979).
79MI2	A. Pelter and K. Smith, in "Comprehensive Organic Chemistry" (D. N. Jones, ed.) Vol. 3, p. 687. Pergamon, Oxford, 1979.
79MI3	A. Pelter and K. Smith, in "Comprehensive Organic Chemistry" (D. N. Jones, ed.), Vol. 3, p. 883. Pergamon, Oxford, 1979.
79S146	I. Akimoto and A. Suzuki, *Synthesis*, 146 (1979).
79TL2313	E. R. Marinelli and A. B. Levy, *Tetrahedron Lett.*, 2313 (1979).
79TL3437	N. Miyaura, K. Yamada, and A. Suzuki, *Tetrahedron Lett.*, 3437 (1979).
79TL4021	A. B. Levy, *Tetrahedron Lett.*, 4021 (1979).
80MI1	H. C. Brown, "Hydroboration." Benjamin, Reading, Massachusetts, 1980.
80MI2	A. Pelter, in "Boron Chemistry 4" (R. W. Parry and G. Kodama, eds.), p. 49. Pergamon, Oxford, 1980.
80TL2865	N. Miyaura, T. Yano, and A. Suzuki, *Tetrahedron Lett.* **21**, 2865 (1980).
81BCJ1587	I. Akimoto, M. Sano, and A. Suzuki, *Bull. Chem. Soc. Jpn.* **54**, 1587 (1981).
81SC513	T. Yanagi, N. Miyaura, and A. Suzuki, *Synth. Commun.* **11**, 513 (1981).

82ACR178	A. Suzuki, *Acc. Chem. Res.* **15**, 178 (1982).
82CSR191	A. Pelter, *Chem. Soc. Rev.*, 191 (1982).
82HC1	P. A. Claret and A. G. Osborne, *Chem. Heterocycl. Compd.* **32**, Pt, 2 (1982).
82MI1	J. H. Morris, in "Comprehensive Organometallic Chemistry" (Sir G. Wilkinson, ed.), Vol. 1, p. 311. Pergamon, Oxford, 1982.
83CPB4573	M. Terashima, H. Kakimi, M. Ishikura, and K. Kamata, *Chem. Pharm. Bull.* **31**, 4573 (1983).
83CS265	S. Gronowitz and K. Lawitz, *Chem. Scr.* **22**, 265 (1983).
83JOM269	G. W. Kabalka, U. Sastry, and K. A. R. Sastry, *J. Organomet. Chem.* **259**, 269 (1983).
83MI1	R. F. Schinazi, B. H. Laster, R. G. Fairchild, and W. H. Prusoff, *Proc. Int. Symp. Neutron Capture Ther.*, First, 260 (1983).
83TL4043	J. Kagan and S. K. Arora, *Tetrahedron Lett.* **24**, 4043 (1983).
84CS(23)120	S. Gronowitz, V. Bobos, and K. Lawitz, *Chem. Scr.* **23**, 120 (1984).
84CS(24)5	S. Gronowitz and K. Lawitz, *Chem. Scr.* **24**, 5 (1984).
84H265	M. Ishikura, M. Kamada, and M. Terashima, *Heterocycles* **22**, 265 (1984).
84H2471	M. Ishikura, T. Mano, I. Oda, and M. Terashima, *Heterocycles* **22**, 2471 (1984).
84H2475	M. Ishikura, M. Kamada, T. Ohta, and M. Terashima, *Heterocycles* **22**, 2475 (1984).
84JOC5237	W. J. Thompson and J. Gaudino, *J. Org. Chem.* **49**, 5237 (1984).
84S936	M. Ishikura, M. Kamada, and M. Terashima, *Synthesis*, 936 (1984).
84TL453	A. Pelter, H. Williamson, and G. M. Davies, *Tetrahedron Lett.* **25**, 453 (1984).
85CC1168	M. J. O'Donnell and J. B. Falmagne, *J. C. S. Chem. Commun.*, 1168 (1985).
85CPB4755	M. Ishikura, T. Ohta, and M. Terashima, *Chem. Pharm. Bull.* **33**, 4755 (1985).
85H117	M. Ishikura,, M. Kamada, I. Oda, and M. Terashima, *Heterocycles* **23**, 117 (1985).
85H2375	M. Ishikura, I. Oda, and M. Terashima, *Heterocycles* **23**, 2375 (1985).
85JOC841	R. F. Schinazi and W. H. Prusoff, *J. Org. Chem.* **50**, 841 (1985).
86H2793	M. Ishikura, T. Ohta, and M. Terashima, *Heterocycles* **24**, 2793 (1986).
86MI1	Y. Tang, M. Deng, and W. Xu, *Acta Chim. Sin.* **44**, 276 (1986).
86MI2	S. Gronowitz, A. B. Hörnfeldt, and Y. Yang, *Croat. Chem. Acta* **59**, 313 (1986).
87H1603	M. Ishikura, I. Oda, and M. Terashima, *Heterocycles* **26**, 1603 (1987).
87JA5420	H. C. Brown, M. Srebnik, R. K. Bakshi, and T. E. Cole, *J. Am. Chem. Soc.* **109**, 5420 (1987).
87JHC377	M. Ishikura, M. Kamada, I. Oda, T. Ohta, and M. Terashima, *J. Heterocycl. Chem.* **24**, 377 (1987).
87MI1	M. Terashima, I. Oda, K. Kanazawa, and M. Ishikura, *Abstr. Pap., 107th Annu. Meet. Pharm. Soc. Jpn.*, 199 (1987).
87SC959	M. Ishikura, I. Oda, M. Kamada, and M. Terashima, *Synth. Commun.* **17**, 959 (1987).
87TL1203	A. Pelter and M. Rowland, *Tetrahedron Lett.* **28**, 1203 (1987).
87TL2599	D. Hongxun, Z. Weike, and B. Junchai, *Tetrahedron Lett.* **28**, 2599 (1987).
87UP1	M. Ishikura and M. Terashima, unpublished results (1987).

1,2,4-Triazolines

PANKAJA K. KADABA

*Division of Medicinal Chemistry and Pharmacognosy, College of Pharmacy,
A. B. Chandler Medical Center,
University of Kentucky, Lexington, Kentucky 40536*

I. Introduction 170
   Nomenclature 171
II. Synthesis of the Triazoline Ring 171
   A. Cyclization Reactions of Hydrazones 172
     1. Cyclocondensation of Hydrazones with Monocarbonyl Compounds . . 173
       a. Amidrazones. 173
       b. N,N'-Diamidines 178
       c. Dihydroformazans 178
       d. α-Aminoazo Compounds 178
       e. Imidate–Hydrazine Interactions 181
       f. Hydrazino Heterocycles 182
       g. Azo Heterocycles 185
     2. Tautomeric Transformation of Amidrazonium Salts 187
       a. N^1-Alkylidene- and Arylideneamidrazones 187
       b. Thiosemicarbazone S,S,S-Trioxides 190
     3. Cyclization of Hydrazone Derivatives by
       Reagents Susceptible to Nucleophilic Attack 191
       a. Cyclizations Induced by Ethoxymethylenemalononitrile
         and Ethoxymethylene Cyanoacetate 192
       b. Isocyanate- and Isothiocyanate-Induced Cyclizations 194
       c. Rearrangement of O-Acetyl Derivatives of
         1,2-Hydroxylaminohydrazones and Thiosemicarbazones 196
   B. 1,3-Dipolar Cycloaddition Reactions 197
     1. Nitrilimine Addition to Carbon–Nitrogen Double Bonds 198
       a. Acyclic C=N Bonds 199
       b. Cyclic C=N Bonds 204
     2. Nitrile Ylide Addition to Azo Compounds 211
     3. Azomethinimine–Nitrile Interactions 214
       a. C-Aryl-N,N'-dialkylazomethinimines 214
       b. 3,4-Dihydroisoquinolinium-N-imides 215
       c. Heteroaromatic N-Imides 216
       d. N-β-Cyanoazomethinimines 219
       e. Pyrazolidineazomethinimines 219
     4. Reactions of Azomethine Ylides 220
       a. 3,4-Dihydroisoquinolinium N-Ylides 221
       b. 1,2,4-Triazolium N-Ylides 222

5. Thiocarbonyl Ylides 223
  6. Diazoalkane Addition to Imines 223
 C. From Triazolium and Mesoionic Triazoles 225
  1. Sodium Borohydride Reduction of Triazolium Salts 226
  2. Reduction of Mesoionic Triazolium Compounds 230
  3. Reaction of Triazolium Salts with Nucleophiles 231
  4. Reaction of Triazolium Salts with Electron-Rich Multiple Bonds . . . 232
 D. Other Routes to Triazoline Synthesis. 233
 E. Synthesis of Δ^1-1,2,4-Triazolines 234
III. Structure and Physical Properties 235
 A. Structure 235
 B. Spectroscopic Properties 237
  1. Nuclear Magnetic Resonance Spectra 237
  2. Ultraviolet and Infrared Spectra 247
  3. Mass Spectral Fragmentations 248
 C. X-Ray Crystallography 249
 D. Other Physical Properties 254
IV. Reactivity of 1,2,4-Triazolines 254
 A. Aromatization to Triazoles 255
  1. Oxidative Dehydrogenation Reactions 255
  2. Elimination of Stable Fragments 257
  3. Isomerization Reactions 260
  4. Retro Diels–Alder Reactions 260
 B. Ring Cleavage Reactions 261
  1. Hydrolytic Cleavage by Acids:
    A Synthetic Route to Aldehydes 261
  2. Oxidative Ring Cleavage 264
    a. Oxidation by Oxygen in Organic Solvents:
      C—N Bond Cleavage via 5-Hydroxytriazolines 265
    b. Oxidation by Acidic Oxidizing Agents Followed
      by Nucleophilic Attack 265
 C. Other Reactions 269
  1. Oxidation by Acidic Oxidizing Agents Followed by Electrophilic Attack 269
  2. Alkali-Induced Reactions 270
    a. Dealkoxycarbonylations with Double Bond Isomerization 270
    b. Intramolecular Cyclizations to Fused-Ring Systems 271
  3. 1,1′-Carbonyldiimidazole-Induced Cyclization Reactions 271
  4. Acid Hydrolysis 272
  5. Photolytic Reaction with Trifluoro Acetate 272
  6. Ring Alkylation Reactions 273
 References 274

I. Introduction

Formation of a 1,2,4-triazoline heterocycle was first observed by Pinner (1892MI1; 1895CB465; 1897LA(297)221, 1897LA(298)1) in the course of his classic work on the synthesis of amidrazones (50JA2783; 65MI1; 70CRV151) by the reaction of an imidate with hydrazine; 15 years later an example was synthesized by Busch and Ruppenthal (10MI1) by the cyclocondensation of an amidrazone with an aldehyde. Since then, the birth of the concept of 1,3-dipolar cycloadditions (84MI1, 84MI2), along with the discovery of several new classes of 1,3-dipoles from the laboratories of Huisgen and associates (63AG(E)565, 63AG(E)633; 68AG(E)321, 68JOC2291), has contributed greatly to the development of 1,2,4-triazoline chemistry.

Unlike 1,2,3-triazolines (84AHC217), there are no previous reviews on the chemistry of 1,2,4-triazolines, although the major synthetic routes are outlined as parts of related reviews on nitrogen heterocycles (75MI1; 81MI1; 83MI1; 84MI3). This article surveys the synthesis, physical properties, and chemical reactions of 1,2,4-triazolines and covers the literature through 1986 to the middle of 1987. Triazolines bearing ring-linked functions that are potentially aromatic (84MI3), such as $=O$, $=S$, $=N-R$, and $=CRR^1$, are not considered.

NOMENCLATURE

Three classes of 1,2,4-triazolines can be recognized (Scheme 1). These are identified in *Chemical Abstracts* since 1972 as dihydro derivatives of 1,2,4-triazoles. Names shown in parentheses are still being used, with Δ indicating the position of the double bond in the heterocyclic ring. The triazoline nomenclature is used in this article.

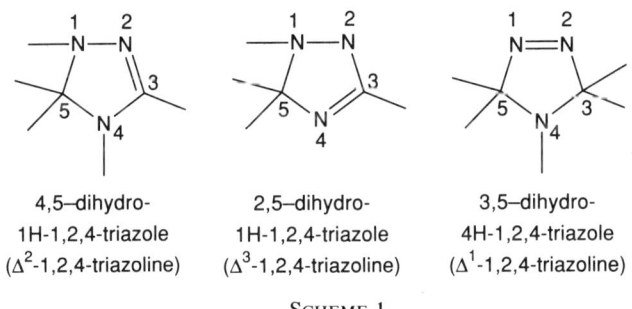

4,5–dihydro-
1H-1,2,4-triazole
(Δ^2-1,2,4-triazoline)

2,5–dihydro-
1H-1,2,4-triazole
(Δ^3-1,2,4-triazoline)

3,5–dihydro-
4H-1,2,4-triazole
(Δ^1-1,2,4-triazoline)

SCHEME 1

Most 1,2,4-triazoline chemistry is associated with the Δ^2- and Δ^3-compounds, as is apparent from the numerous literature citations; the Δ^1-1,2,4-triazolines are rarely studied.

II. Synthesis of the Triazoline Ring

The 1,2,4-triazoline ring system can be constructed from reactions of acyclic compounds or from transformation of cyclic structures. In the classical approach to triazoline synthesis, the N—C—N—N skeleton of amidrazones and other hydrazone derivatives is cyclized with the one-carbon fragment of a monocarbonyl compound (10MI1). It is a highly versatile reaction and alkyl-, aryl-, or heteroaryl-substituted triazolines are readily obtained, although the possibility of ring–chain tautomeric transformations calls for caution. More recently, hydride reduction of triazolium compounds (73BCJ1250) has evolved into a synthetic procedure for 1,2,4-triazolines.

Huisgen's studies on 1,3-dipolar cycloaddition reactions (84MI1), leading to an impressive array of heterocyclic systems, provide other approaches to 1,2,4-triazoline synthesis. The union of the C—N—N and C—N fragments of the triazoline ring is accomplished by the cycloaddition reaction of nitrilimines with Schiff bases (62T3) or azomethinimines with nitriles (63AG604; 65AG(E)701). Alternatively, the combination of the C—N—C and N—N fragments may be attained by the respective 1,3-addition of nitrile and azomethine ylides to azo and diazonium compounds (73CB3421; 84TL65).

The triazolines resulting from the various reaction schemes are predominantly Δ^2-compounds; the cycloaddition reactions of azomethinimines and nitrile ylides lead to Δ^3-triazolines. The amidrazones are unique; a Δ^3- or Δ^2-system results, depending on the presence or absence of N-2 substitution (84KGS1415). The rare Δ^1-compounds are presumed to result from the nonregioselective addition of diazomethane to the imine double bond (72LA9) or by fluorination of 3,6-diamino-s-tetrazine (67USP3326889; 70USP3515603).

A. Cyclization Reactions of Hydrazones

The ease of forming C—N and C=N bonds as opposed to N—N bond formation is reflected in the extensive use of hydrazine derivatives in 1,2,4-triazoline synthesis. Pinner (1892MI1; 1895CB465; 1897LA(297)221, 1897LA(298)1) first observed that his "monosubstituted hydrazidine" compound (amidrazone)[1] (50JA2783; 65MI1; 70CRV151), from imi-

---

[1] Amidrazones have been referred to in the literature as *hydrazidines*, *amide hydrazones* (A), and *hydrazide imides* (B).

(A) $R-C{\displaystyle{\overset{N-N{<}^R_R}{\underset{N{<}_{R^1}}{}}}}{}^{R^1}$ (B) $R-C{\displaystyle{\overset{NR^1-N{<}^R_R}{\underset{NR^1}{}}}}$

Amidrazones of types **A** and **B**, in which $R^1 \neq H$, are incapable of tautomerism. When tautomerism is possible (**A** $\rightleftharpoons$ **B**, $R^1 = H$), the terms *amide hydrazone* and *hydrazide imide* cannot be strictly applied. Hence, the term *amidrazone* is used for all compounds with structure **A**, and is the least ambiguous (70CRV151).

date–hydrazine interaction, formed the starting point for the preparation of 1,2,4-triazoles along with minor amounts of 1,2,4-triazolines. Since then, cyclization reactions of amidrazones and related hydrazones have been the subject of extensive studies, particularly ring–chain tautomeric transformations (84KGS1415). This family of reactions constitutes a major route to the synthesis of a variety of 1,2,4-triazolines, including bis- and polytriazolines as well as several fused-ring systems. In principle, however, ring–chain tautomerism can exist in all of these ring systems, which makes unambiguous product identification an essential factor in triazoline synthesis. Although valuable spectral aids have been developed to distinguish the linear from the cyclic tautomers (80ZOR942), the dependency of the tautomeric equilibrium on substituent, solvent, and temperature (77LA485; 81ZOR1825; 82KGS1264; 84KGS1415) requires careful consideration. The cycloisomerization of amidrazonium salts provides a route for the synthesis of 2-methyl-substituted 1,2,4-triazolines (84KGS1415). Reagents that attack the nucleophilic terminal amino group induce cyclization of isothiosemicarbazones (84JOC1703).

1. *Cyclocondensation of Hydrazones with Monocarbonyl Compounds*

Based on NMR spectroscopic evidence (80ZOR942; 81ZOR1825; 82KGS1264; 84KGS1415), N^1-monosubstituted amidrazones, in general, are known to react with monocarbonyl compounds to give 1,2,4-triazolines. However, early methods of synthesis usually involved cyclocondensation of unsubstituted amidrazones with aldehydes, ketones (70JHC1001, 70MI1; 71JHC173, 71JHC1043; 73JHC353), or other one-carbon sources (71BCJ780; 75JCS(PI)1433). Although the products obtained may be either alkylidene (arylidene) derivatives or 1,2,4-triazolines, in most cases the linear versus the cyclic nature of the products has not been resolved. In a few cases in which product identification has been effected, the proof is based mainly on subsequent oxidation to triazoles (64CR6470; 70JHC1001), which itself is ambiguous because the open-chain alkylideneamidrazones and the cyclic triazoline tautomers can both undergo oxidation to triazoles (Scheme 2) (64CR6470; 70JHC1001).

a. *Amidrazones* 1,2,4-Triazolines may be regarded as cyclic amidrazones and the classical cyclocondensation reaction of amidrazones with aldehydes, employed by Busch (Scheme 3), is still a widely used method of synthesis. In the Busch reaction, N^1,N^3-diphenylamidrazone, obtained from imidoyl chloride and phenylhydrazine, reacted with benzaldehyde to effect the first synthesis of 1,2,4-triazoline, while the N^2,N^3-

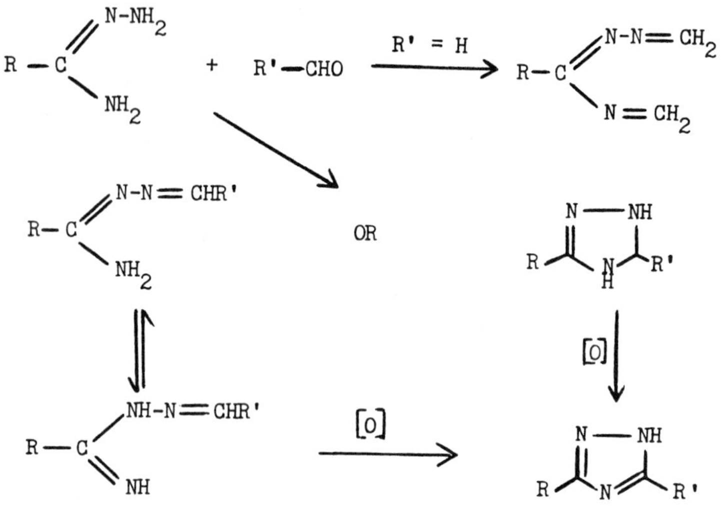

SCHEME 2

compound, unsubstituted at N^1, yielded only the arylidene derivative or Schiff base (10MI1; 14JPR310).

Δ^2-1,2,4-Triazolines are reported to result in good yields from a variety of alkyl-, aryl-, or heteroaryl-substituted amidrazones and aldehydes in refluxing ethanol (Scheme 4) (64CR6470; 70CRV151, 70JHC1001; 71JHC173). When R = alkoxy or aminocarbonyl, the resulting triazolines show herbicidal activity (86EUP189300). Aldehyde acetal has been found to effect the cyclization of N^1,N^3-diphenylamidrazone (Scheme 3) (69G69). The reaction of N^1-ethoxycarbonylbenzamidrazone with both acetaldehyde and formaldehyde requires a silica catalyst in refluxing benzene (74HCA1382). Formaldehyde fails to give a triazoline with unsub-

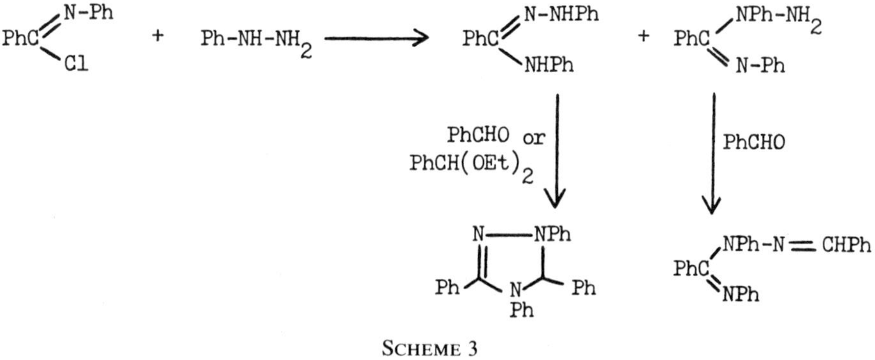

SCHEME 3

stituted amidrazones; reaction occurs at both the N^1- and N^3-amino functions (Scheme 2) (70CRV151). The reaction of an aldehyde with N^1-phenylmandelamidrazone leads directly to a 1,2,4-triazole (70CRV151).

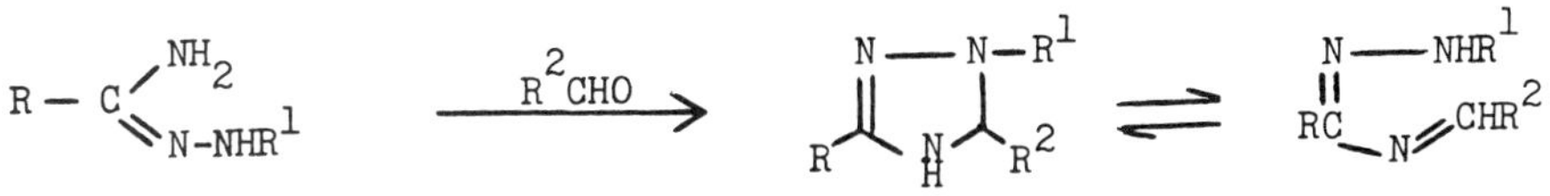

$$R-C\underset{N-NHR^1}{\overset{NH_2}{\diagup}} \xrightarrow{R^2CHO} \underset{R}{\overset{N-N-R^1}{\underset{H}{\bigwedge}R^2}} \rightleftharpoons \underset{RC}{\overset{N-NHR^1}{\underset{N}{\bigwedge}CHR^2}}$$

R and R^2 = heteroaryl, aryl or alkyl
R^1 = aryl, H, or COOEt

SCHEME 4

Several 1,3,5,5-tetraalkyl-1,2,4-triazolines (**1**) have been obtained by condensation of dialkyl ketones with N^1-monoalkylamidrazones (80ZOR942). Investigations of the spectral parameters of the simple representative triazoline structures **1** and a model alkylideneamidrazone with a fixed noncyclic structure (**2**) have led to NMR spectroscopic methods for distinguishing the cyclic structures from the linear forms (80ZOR942).

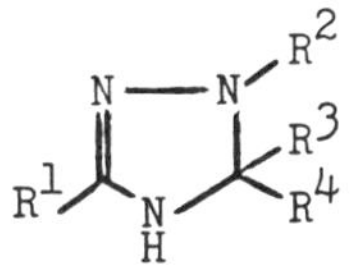

(1) 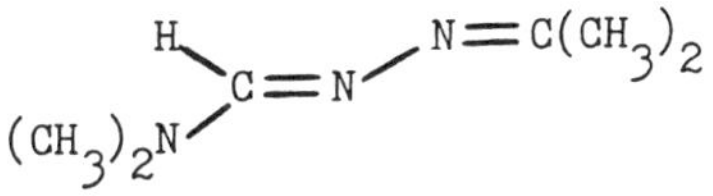 (2)

$R^1 = CH_3$, $R^2 = R^3 = R^4 = CH_3$ (a)
$R^2 = CH_3$, $R^3R^4 = (CH_2)_5$ (b)
$R^2 = CH_3$, $R^3 = H$, $R^4 = C_6H_5$ (c)
$R^2 = C_3H_7$, $R^3 = R^4 = CH_3$ (d)
$R^1 = C_6H_5$, $R^2 = R^3 = R^4 = CH_3$ (e)

Schiff base formation in the Busch reaction may be avoided by employing N-(1-chlorobenzyl)benzimidoyl chloride, in which the aldehyde one-carbon fragment is already incorporated into the N-(1-chlorobenzyl) moiety; both chlorine atoms participate in the reaction with arylhydrazine, to give modest yields of a Δ^3-1,2,4-triazoline (Scheme 5) (79ZOR1181).

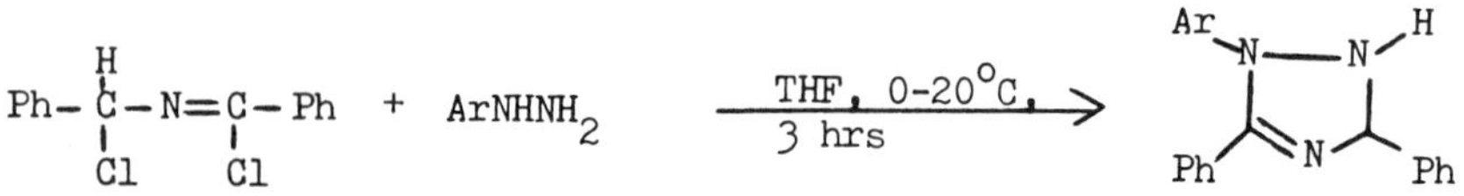

SCHEME 5

Monoketones, such as acetophenone and 2-acetylpyridine, react with unsubstituted amidrazones in refluxing ethanol in the presence of a trace of hydrochloric acid to yield 1,2,4-triazolines. Increased concentration of acid causes hydrolysis of the amidrazones leading to *N*-acylhydrazones (Scheme 6) (73JHC353).

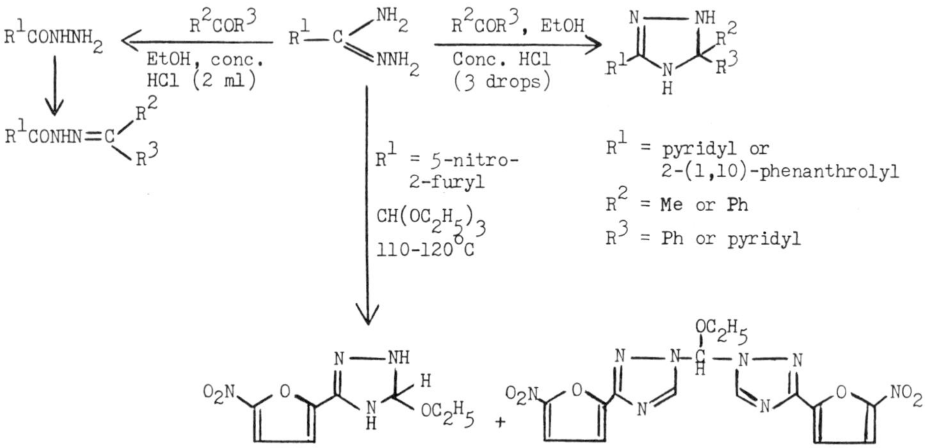

SCHEME 6

Orthoesters that are susceptible to attack by nucleophiles have been found to function as sources of the one-carbon fragment required for amidrazone cyclization; although normally 1,2,4-triazoles are the products (70MI2; 73RCR392), occasionally the intermediate triazolines have been isolated (71BCJ780). Thus the cyclization product from 5-nitro-2-furamidrazone and ethyl orthoformate has been identified as a 5-ethoxytriazoline; elimination of the alcohol fragment yields the triazole as proven by thermogravimetric (TGA) and differential thermal (DTA) analysis (Section III,D). For equimolar amounts of reactants, the triazoline is the sole product; with excess orthoformate, a mixture of the triazoline along with a bistriazole is obtained (Scheme 6). The latter is postulated to be formed directly from the triazoline by reaction with the orthoformate (Section IV,A) (71BCJ780).

Results presented in Scheme 7 are representative of the extension of the amidrazone–aldehyde (ketone) reaction to diamidrazones (71JHC1043; 75JCS(P1)1433) and dialdehydes (71JHC1043) to obtain 3,3'- and 5,5'-bistriazolines, respectively. Structural assignment is based on oxidation to the known bistriazole (**3**) (71JHC1043).

Bisalkylideneoxamidrazones, when refluxed in acetic anhydride, yield 1,4-diacetylbistriazolines; when the alkylidene chain is saturated, the product is a bistriazole. The mechanism suggested for triazoline forma-

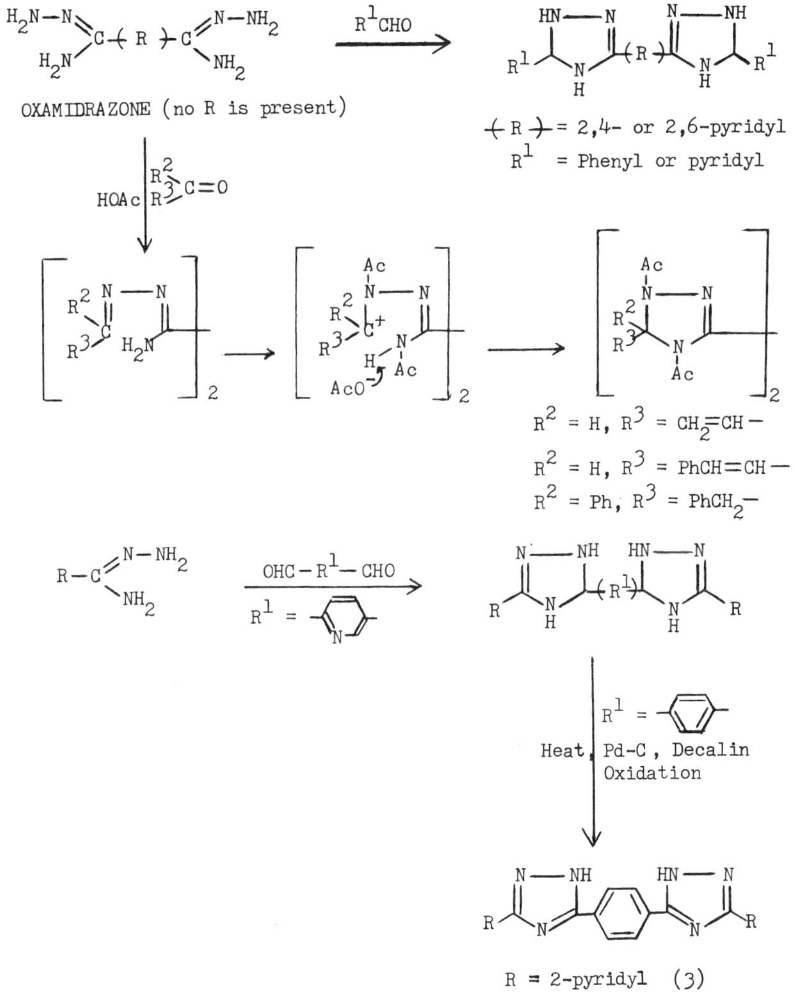

SCHEME 7

tion involves acylation of the amino group to a monoacyl derivative, with further quaternization to yield a stable carbocation, which cyclizes under the influence of acetate ion (Scheme 7) (75JCS(P1)1433).

Synthesis of soluble polytriazolines (4) of relatively high molecular weight has been achieved to a limited extent by solution polymerization of diamidrazones with certain select dialdehydes, although attempts to prepare methyl-substituted triazoline polymers have been unsuccessful (70MI1).

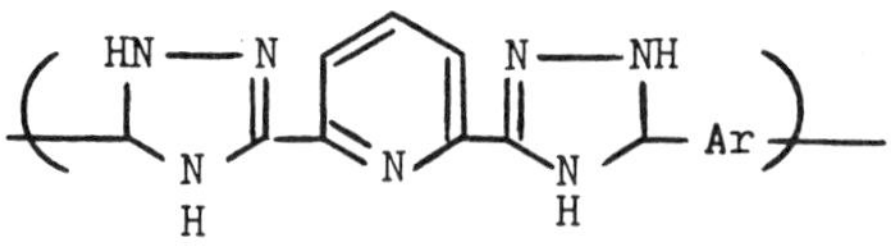

(4)

b. *N,N'-Diamidines*. The Michael reaction of aminomethylene malonates with hydrazine hydrate (Scheme 8) yields 5-amino-1,2,4-triazolines (6) (70T3069). An amidrazone derivative of the N,N'-diamidine type (5) (65MI1; 70CRV151) is postulated to be an intermediate. The aminomethylene malonate provides the one-carbon fragment to effect cyclization. In refluxing ethanol a triazole (7) is obtained, apparently by loss of amine from the triazoline precursor, as established by mass and nuclear magnetic resonance (NMR) spectra (70T3069). The amino substituent appears to be a determining factor in diethyl malonate elimination from the initial Michael adduct to give the amidrazone intermediate 5. When diethyl malonate fails to leave, an amine fragment is lost and the resulting product is a pyrazole, as shown by NMR spectra (70T3069). The preferential elimination of amine or malonate is attributed to the sp^2 or sp^3 character of the amino nitrogen in the initial adduct, the sp^3 nitrogen being a better leaving group (70T3069).

c. *Dihydroformazans*. In Scheme 9, cycloisomerization of the dihydroformazan compounds 8 leads to 5-hydroxytriazolines as the initial products; elimination of water yields the triazoles (77JHC1089).

d. α-*Aminoazo Compounds*. Thermolysis of α-aminoazo compounds 9 to afford 1,2,4-triazoles 13 has been found to proceed via a 1,2,4-triazoline intermediate (12) (Scheme 10) (82CJC285). The reaction mechanism involves the tautomeric transformation of 9 to the amidrazone 9a (65JCS3528; 82CJC285); the one-carbon fragment is provided by imine 11, derived from the free radical 10 formed by thermal decomposition of 9 (82CJC285; 83MI1). Amidrazone cyclization proceeds by elimination of

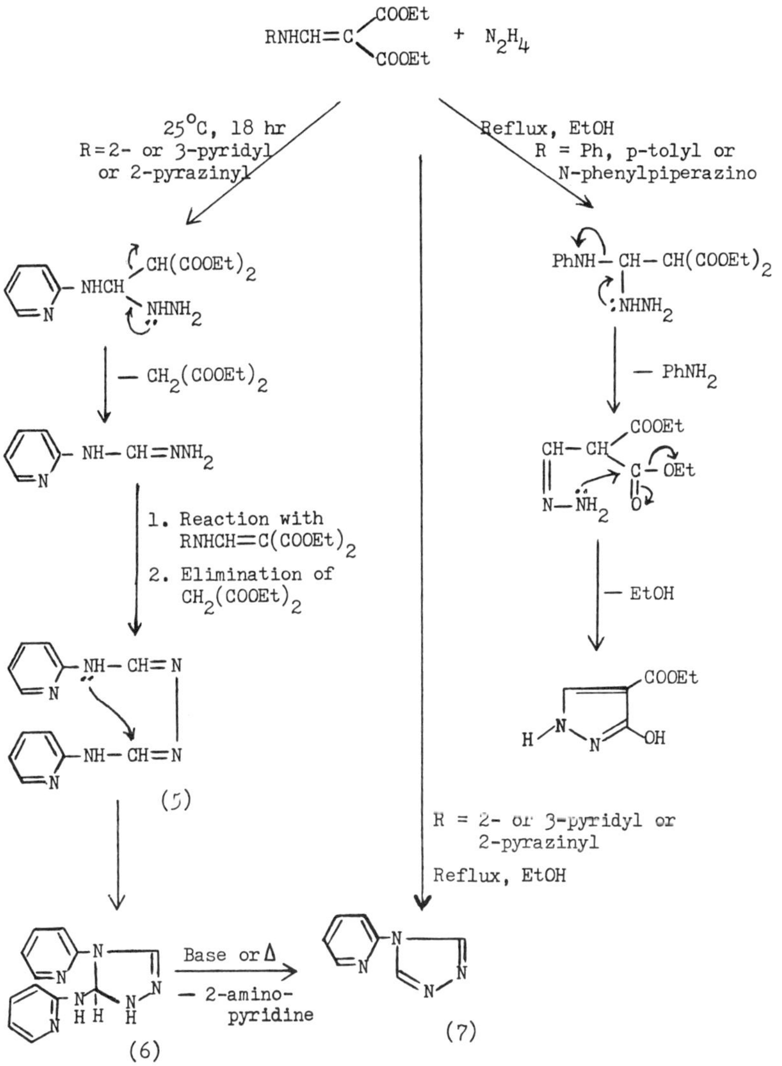

Sec. II.A] 1,2,4-TRIAZOLINES 179

SCHEME 8

ammonia, as shown in Scheme 10. Experimental evidence for the radical path rests on the failure of **9** to yield **13** in the presence of *tert*-butylcatechol, a free radical inhibitor, and the quantitative conversion of **9** to **13** in its absence (82CJC285). In addition, isolation and characterization of the stable 1,5-diaryltriazoline from the reaction of **9** with benzaldehyde provide proof for production of **9a**. The role of imines as a one-carbon source in amidrazone cyclization is further demonstrated by triazole for-

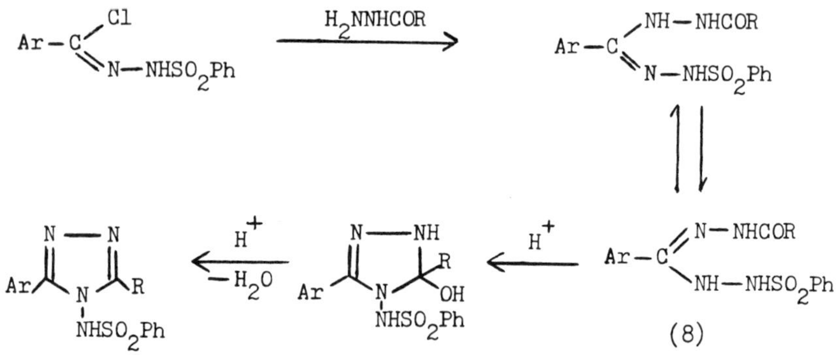

SCHEME 9

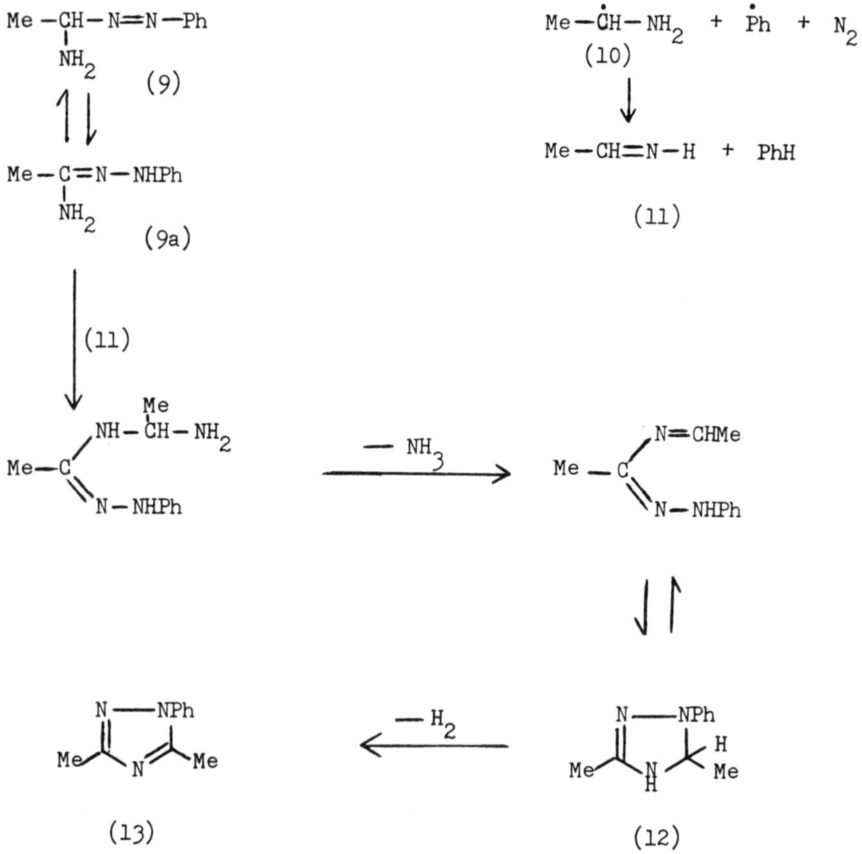

SCHEME 10

mation, in high yield, in the reaction of **9** with *N*-benzylidenemethylamine (82CJC285). The reaction sequence is similar to that of **9** with **11**; instead of ammonia, methylamine is expelled.

Likewise, in Scheme 11, ethyl thioimidate serves as the one-carbon source; the intermediate triazoline loses thiol to yield the triazole (79CCC1334).

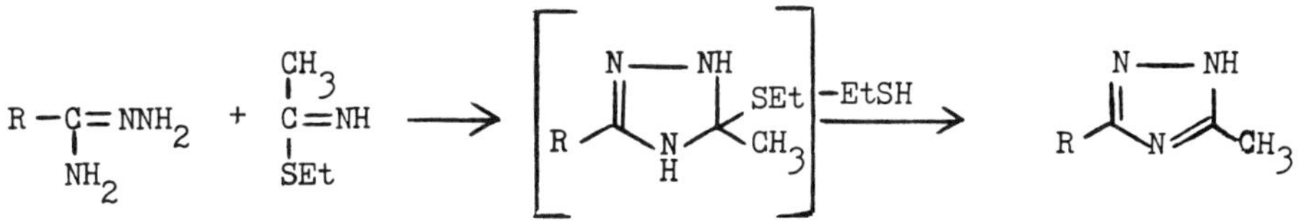

SCHEME 11

e. *Imidate–Hydrazine Interactions.* Formation of minor amounts of dihydrotriazoles was first noticed during Pinner's 1,2,4-triazole synthesis by reaction of imidates with hydrazine (1892MI1; 1895CB465; 1897LA(297)221, 1897LA(298)1). The one-carbon fragment required for amidrazone cyclization was postulated to be formed from the slow decomposition of the alkaline solution of amidrazone by exposure to air (Scheme 12) (1897LA(297)221, 1897LA(298)1).

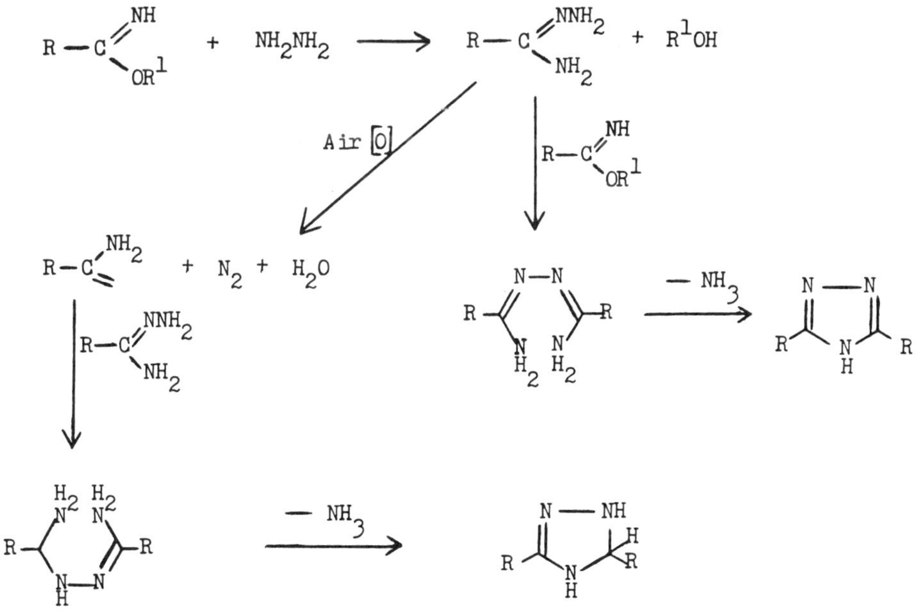

SCHEME 12

Because hydrazones are known to tautomerize to azo compounds in alkaline medium (65JCS3528), it may be more rational to interpret Pinner's R—C(NH$_2$)= fragment as a free radical similar to **10**, arising from the azo tautomer (87UP1). The reaction time of 6–8 weeks required for triazoline separation from an alkaline solution of amidrazone at room temperature in the Pinner reaction (1897LA(297)221, 1897LA(298)1) is consistent with the 2 weeks required for the tautomeric transformation of an azo compound to hydrazone at ambient temperature in a nitrogen atmosphere (82CJC285). Apparently, the low yield of triazoline in the Pinner reaction results largely from oxidative decomposition of the amidrazone due to extended exposure to air (83MI1).

f. *Hydrazino Heterocycles.* Amidrazones, in which the imine moiety is an integral part of a heterocyclic ring system, react with aldehydes and ketones to form fused-ring triazolines. The positioning of the hydrazono group on the carbon of the ring imine provides the requisite geometry for an intramolecular cycloisomerization by nucleophilic attack by the ring nitrogen on the alkylidene carbon of the hydrazono group. The reaction provides a useful route for the synthesis of different 3,4-fused triazoline ring systems.

Piperidone hydrazone **14** in the presence of silica gel in refluxing acetone yields the bicyclic triazoline **15**; while **14a** requires 4.5 hr of refluxing, the more basic **14b** affords the triazoline in 2.5 hr. In the absence of silica gel, more than 5 hr is needed to complete the reaction (76CPB3011).

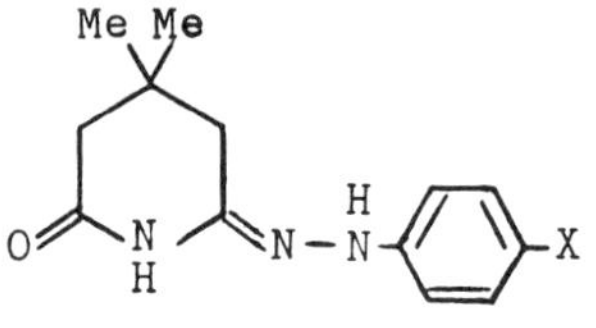

(14) a, X = H
b, X = OCH$_3$

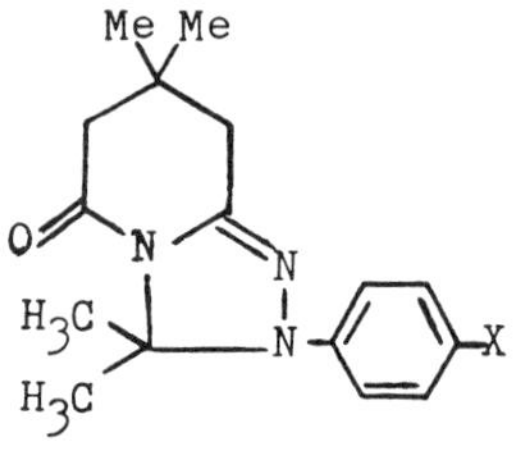

(15)

The hydrazonothiazole in Scheme 13 undergoes nitrosation with formation of a triazoline. The open-chain tautomer is ruled out, based on the single methyl resonance in the NMR spectrum (Section III) and the fact that hydrolysis of the acetonylhydrazone does not occur in dilute hydrochloric acid (Scheme 13) (78JHC401).

Likewise, acylation of pyridazinylhydrazones **15a** yields acylated dihy-

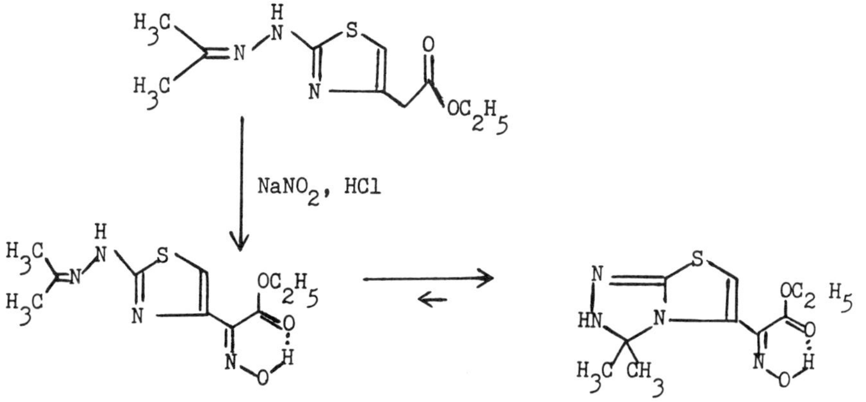

Sec. II.A] 1,2,4-TRIAZOLINES 183

SCHEME 13

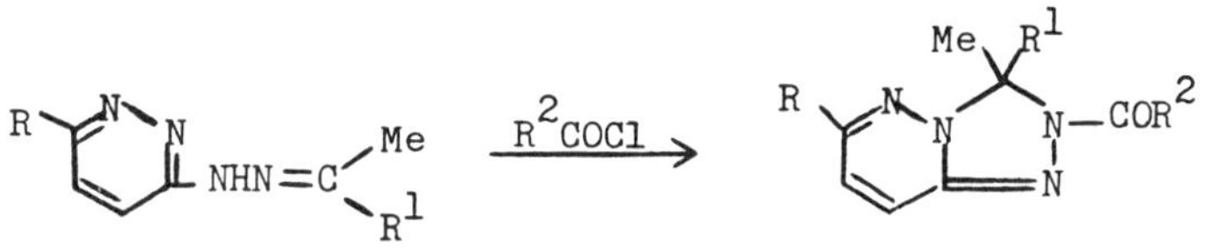

(15a) → (15b)

R^2COCl

R = Me_3C, Et_2CMe, Cl
R^1 = Me, $CH_2COOCMe_3$
R^2 = 2,6-$(MeO)_2C_6H_3$, Me_3CCH_2, $ClCH_2$, Me_2N

drotriazolopyridazines **15b**, as confirmed by X-ray analysis (85CB5009).

2-Methylhydrazino-1,4-benzodiazepine **17** reacts with acetaldehyde to yield the dihydrotriazolobenzodiazepine **18**, along with a hydrazide, **19**, suspected to be a secondary product from air oxidation of **18**. The 1-methyl isomer **(16)**, however, gives a Schiff base **(20)**, similar to the one obtained in the Busch reaction (10MI1) for triazoline synthesis (Scheme 3) (79JOC2688).

Condensation of unsubstituted 2-hydrazinobenzodiazepines with acetylacetone provides an efficient synthesis for 1-methyltriazolobenzodiazepines. A tricyclic triazoline is the postulated intermediate, in which the geometry is perfect for elimination of acetone by a cyclic mechanism.

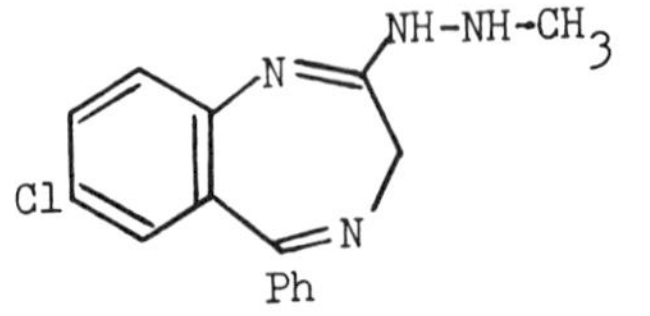

(17) 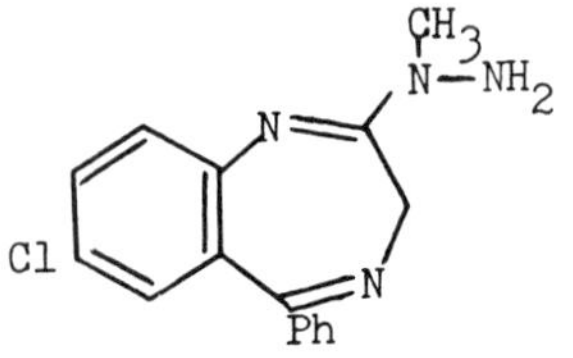 (16)

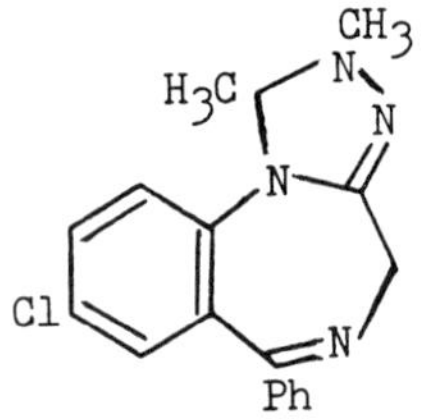

(18) 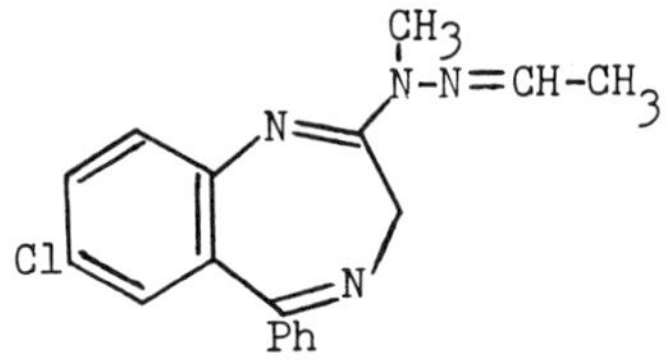 (20)

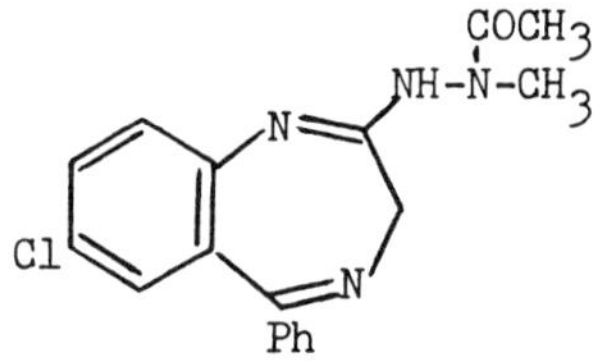

(19)

With a cyclic diketone, triazole formation occurs with simultaneous ring opening of the cyclic ketone moiety (Scheme 14) (78JHC161).

The cycloisomerization reaction of hydrazono heterocycles also provides a route for the preparation of a number of dihydrotriazolobenzazepines (21) that show good anticonvulsant activity (83EUP72029).

Fused-ring aminotriazolines are reported when hydrazino heterocycles are heated with excess dimethylformamide (DMF) in the presence of an acid. Amides, especially formamides, have proved particularly useful as a source of the one-carbon fragment in the synthesis of various nitrogen heterocycles (63JOC543). Thus, hydrazino derivatives of benzisothiazoles cyclize with phosphorus oxychloride in DMF. The resulting tricyclic aminotriazolines 22 and their physiologically acceptable salts are stable and have been found to be useful as antiinflammatory agents (79USP4140693).

Similarly, an aminotriazoline is the postulated intermediate in the reaction of hydrazinotriazinone with acetic acid and DMF, to give high yields

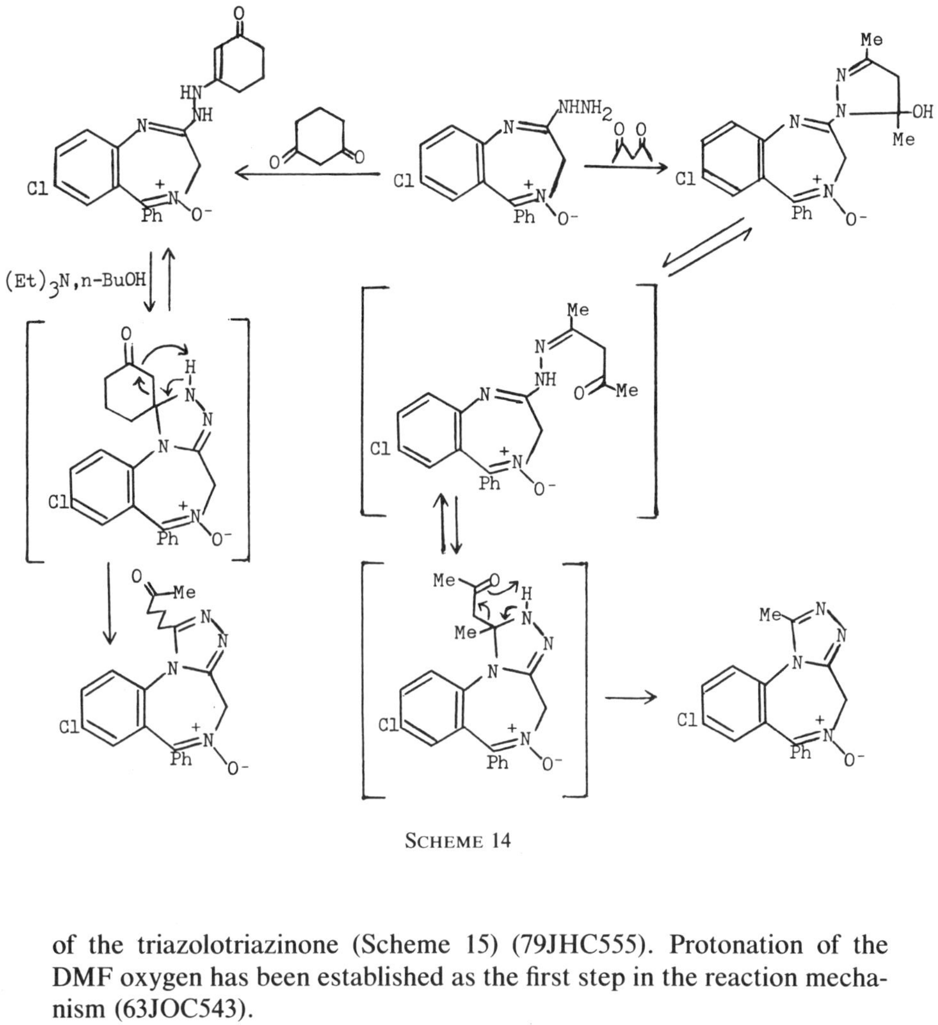

Sec. II.A] 1,2,4-TRIAZOLINES 185

SCHEME 14

of the triazolotriazinone (Scheme 15) (79JHC555). Protonation of the DMF oxygen has been established as the first step in the reaction mechanism (63JOC543).

Spectroscopic studies have shown that protonation of hydrazonophthalazines occurs at the exocyclic nitrogen leading to formation of tricyclic cations; upon deprotonation, they revert back to the original hydrazones (Scheme 16) (80TL209). The cyclic structure of the cation has been confirmed by x-ray diffraction (82KGS1100).

g. *Azo Heterocycles.* The products from the reaction of 2,2′-azopyridines 23 and 2,2′-azoquinolines 24 with diphenyldiazomethane or diazofluorene are fused-ring 1,2,4-triazolines (25–26); the structure was es-

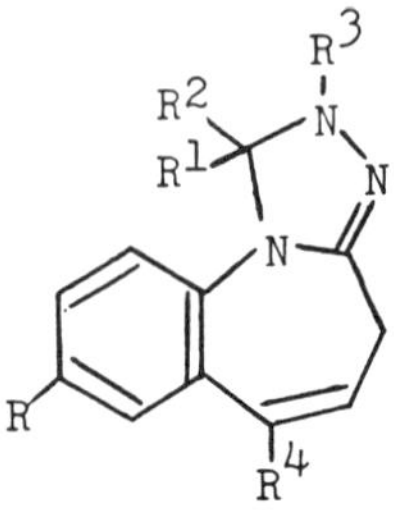

(21)

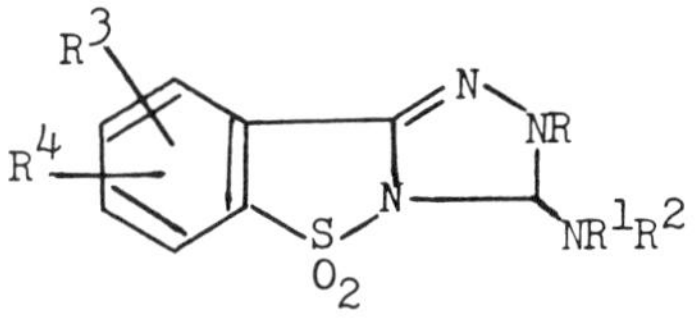

R = halo, NO_2
R^1 = H, alkyl, 4-piperidyl,
  $(CH_2)_n R^5 R^6$ ($R^5 R^6$ = H, alkyl,
  morpholino, piperidino; n= 0 or 1)
R^2 = H; $R^1 R^2$ = =O
R^3 = H, alkyl, Ph, pyridyl
R^4 = Ph, pyridyl

(22)

R = H, alkanoyl or aroyl
R^1, R^2 = alkyl
R^3 = H, halo, alkyl, alkoxy, NO_2
R^4 = H, alkoxy, halo
R^3 = R^4, when R^4 = H

tablished from chemical and UV spectral data (Scheme 17) (67TL4337). Kinetic experiments indicate that the diazoalkanes react by a carbene mechanism (67TL4337). Diazomethane itself fails to yield a triazoline; the product has been identified as N-formyl-2,2'-hydrazopyridine (29) (66JCS(C)78).

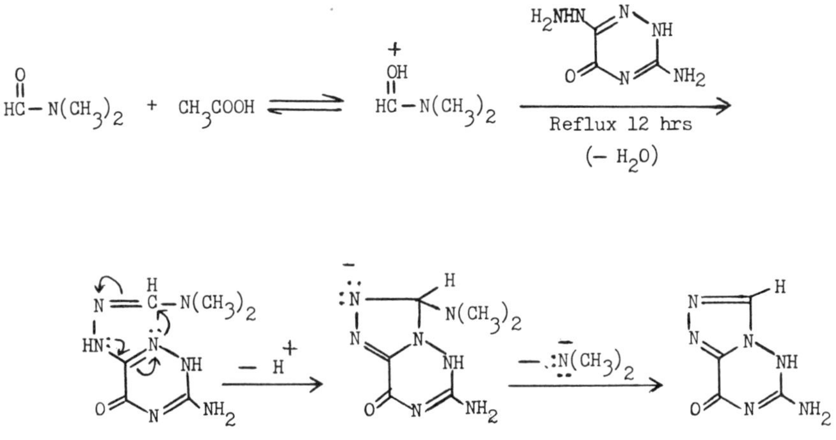

SCHEME 15

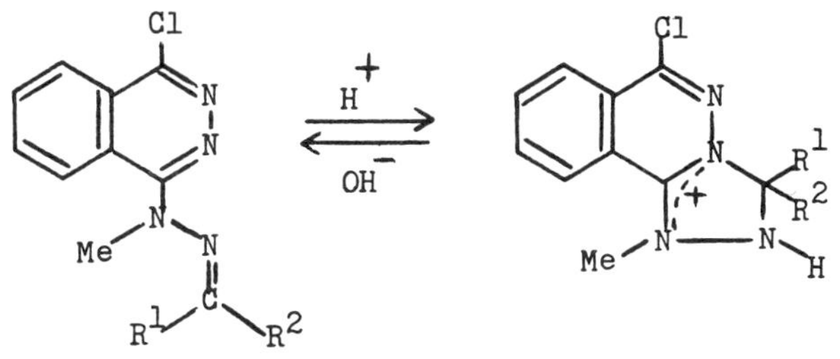

Sec. II.A]　　　　　　　　1,2,4-TRIAZOLINES　　　　　　　187

$R^1 = R^2 = H;\ R^1 = H,\ R^2 = Me;\ R^1 = R^2 = Me;\ R^1 = Me,$
$R^2 = Ph;\ R^1 = Me,\ R^2 = CH=CMe_2$

SCHEME 16

2. Tautomeric Transformation of Amidrazonium Salts

Hydrazones functionalized in the hydrazine fragment are capable, in principle, of ring–chain transformations as a result of intramolecular nucleophilic attack at the C=N bond (Scheme 18) (79MI1). Unlike acylhydrazones, which exist only in the linear form, N^1-alkylidene(arylidene) amidrazones, which may be considered nitrogenous analogues of acylhydrazones, are capable of ring–chain tautomerism (81ZOR1825; 82KGS1264, 82ZOR1613; 84KGS1415). In their free base form, they are not susceptible to tautomeric changes, and exist exclusively as the linear isomers (e.g., **35**) (80ZOR942). Their salts, however, are capable of tautomeric transition to the cyclic form (80KGS1138), and in solution exist as an equilibrium mixture with the respective dihydrotriazolium (triazolinium) compound (**34A** ⇌ **34B**) (81ZOR1825; 82KGS1264; 84KGS1415).

NMR spectroscopic studies on ring–chain tautomeric systems have led to the elucidation of factors that determine the position of the equilibrium (77LA463, 77LA485; 80KGS1138; 82KGS1264; 84KGS1415), and have provided new pathways for 1,2,4-triazoline synthesis (77LA463; 84KGS1415). Unlike amidrazone cyclocondensation reactions, which yield Δ^2-triazolines, tautomeric transformations of amidrazonium salts result in Δ^3-compounds. Previously these Δ^3 ring systems were accessible only by 1,3-cycloaddition of azomethinimines to the C≡N bond and of nitrile ylides to the N=N bond (Section II,B).

a. N^1-*Alkylidene and Arylideneamidrazones.* According to their ability to undergo ring–chain transitions, 1-alkylidene (arylidene) derivatives of amidrazonium salts occupy an intermediate position between

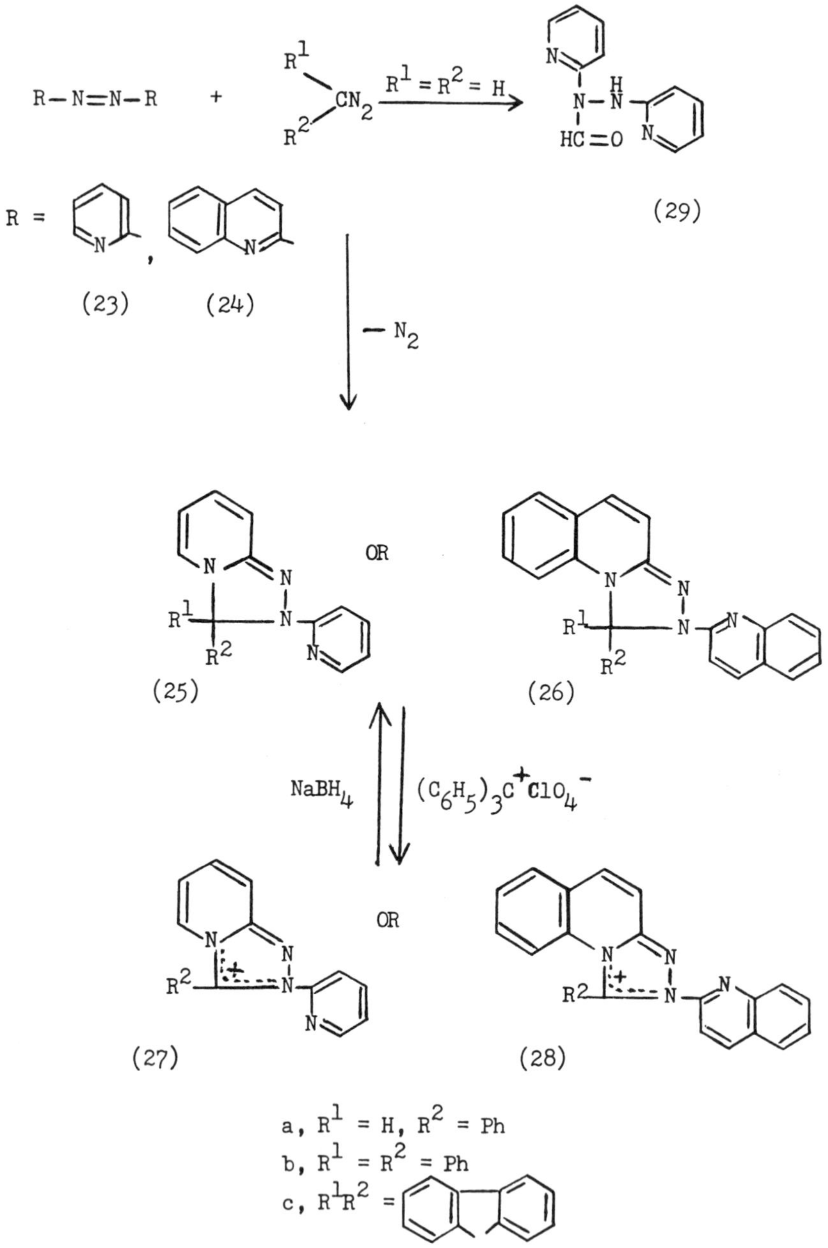

R—N≡N—R + R¹R²C(CN₂) $\xrightarrow{R^1 = R^2 = H}$ (29)

R = (23) pyridyl, (24) quinolinyl

$-N_2$ ↓

(25)   OR   (26)

⇅ NaBH₄ / $(C_6H_5)_3C^+ClO_4^-$

(27)   OR   (28)

a, $R^1 = H$, $R^2 = Ph$
b, $R^1 = R^2 = Ph$
c, $R^1R^2 =$ fluorenyl

SCHEME 17

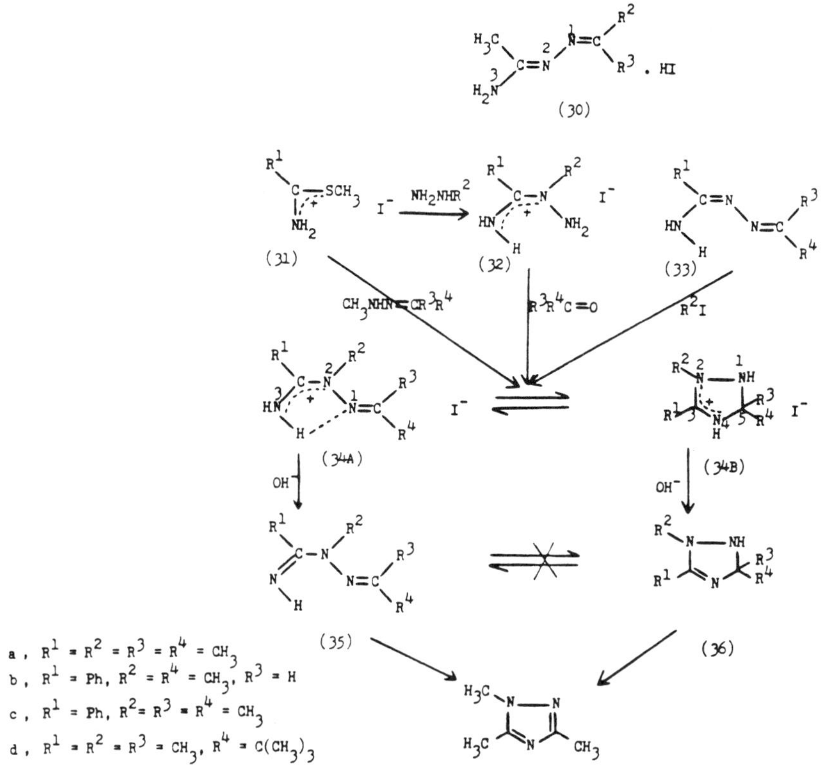

Sec. II.A] 1,2,4-TRIAZOLINES 189

SCHEME 18

a, $R^1 = R^2 = R^3 = R^4 = CH_3$
b, $R^1 = Ph$, $R^2 = R^4 = CH_3$, $R^3 = H$
c, $R^1 = Ph$, $R^2 = R^3 = R^4 = CH_3$
d, $R^1 = R^2 = R^3 = CH_3$, $R^4 = C(CH_3)_3$

acyl and thioacylhydrazones. Most important in determining their susceptibility to tautomeric transformations are structural factors. Variation of the electronic and steric properties of the substituents alters the tautomeric equilibrium over a broad range (82KGS1264).

Steric interactions of the substituents attached to the C—N^2 bond are of foremost importance. Thus acetamidrazone derivatives **30** exist exclusively in the linear form; they have an *(E)*-configuration relative to the C—N^2 bond, with no intramolecular hydrogen bonding. The unfavorable spatial orientation of the C—N^1 bond and of the amino group prevents ring–chain tautomerism (80KGS1138, 80ZOR942; 82KGS1264).

In the crystalline state, salts of benzamidrazones (**34A**, R^1 = Ph, R^2 = H) also exist in the linear form, but with a fixed *(Z)*-configuration. Increased steric interactions relative to the C—N^2 bond from phenyl substitution causes a *syn* orientation of N^1 and N^3, stabilized by intramolecular hydrogen bonding (82KGS1264). In solution, a ring–chain tautomeric equilibrium is established after 24 hr, as indicated by NMR spectroscopy. In DMSO-d_6, isopropylidenebenzamidrazone hydrochloride exists 15% in the cyclic form (80KGS1138); replacement of the anion by iodide, trifluoro acetate, or picrate decreases the cyclic tautomer to 7% (82KGS1264).

Alterations in the electronic and steric properties of substituents in the alkylidene fragment have virtually no effect on the position of the tautomeric equilibrium. However, increasing the volume of the substituents enhances the shift toward the cyclic form, apparently due to the greater sensitivity of the linear tautomer to the increase in the total steric stresses in the molecule (84KGS1415). An aromatic substituent, on the other hand, stabilizes the linear form, due to conjugation with the C=N bond (84KGS1415). A single substituent on N^3 has no effect, but when N^3 is fully substituted, ring–chain tautomerism is no longer possible (82KGS1264).

Introduction of a methyl group at N^2 has a decisive influence because the tautomeric equilibrium is extremely sensitive to steric interactions relative to the C—N^2 bond (79MI1). A substituent on N^2 favors the cyclic form, which is the only form in the crystalline state and in freshly prepared solutions; in the absence of an N^2 substituent, the linear form predominates (81ZOR1825; 82KGS1264; 84KGS1415).

The tautomeric equilibrium is also solvent and temperature dependent. At room temperature, the linear form of **34B** ($R^1 = R^2 = R^3 = R^4 = CH_3$) amounts to 32% in DMF-$d_7$, 31% in DMSO-$d_6$, and 12% in $CDCl_3$, as estimated from ^{13}C-NMR spectral data (81ZOR1825).

Knowledge of the basic principles underlying ring–chain tautomeric transformations has led to the successful synthesis of several alkyl-substituted 1,2,4-triazolines (**36a–c**, Scheme 18) (84KGS1415). Triazolinium salts **34B** may be obtained using different reaction paths for **34A**, as shown in Scheme 18 (82KGS1264, 82ZOR1613; 84KGS1415). Neutralization of the respective salts (**34B**) enables the isolation of the free bases (**36a–c**). As the substituent on N^2 is a basic requirement for tautomeric transformation to the cyclic form, the 1,2,4-triazolines resulting from this reaction scheme are Δ^3-compounds. They are stable to tautomeric changes in solution, although steric stresses caused by bulky substituents can induce rapid aromatization (e.g., **36d**). This provides a useful route for the synthesis of 1-alkyl-1,2,4-triazoles (Scheme 18) (83S483; 84KGS1415).

b. *Thiosemicarbazone S,S,S-Trioxides.* Ring–chain tautomerism between 2-(methyl)thiosemicarbazone- *S,S,S*-trioxides **37** and the respective 2-methyl-Δ^3-1,2,4-triazolinium-3-sulfonates (**38**) is also dependent on substituents, solvent, and temperature, analogous to the amidrazonium salts (Scheme 19) (77LA463, 77LA485). The acyclic form (37) increases with increasing temperature, comprising more than 50% of the equilibrium mixture at temperatures above 100°C. Unlike the alkylidene- and arylideneamidrazones, N^3-substituents have an unfavorable steric effect; in DMSO solution at 37°C, with R^3 = 2,6-diisopropylphenyl, there is only 48% of the cyclic form, and if this is combined with an isopropylidene substituent on N^1, the steric stresses in the molecule shift the equilibrium completely toward the acyclic form (77LA485).

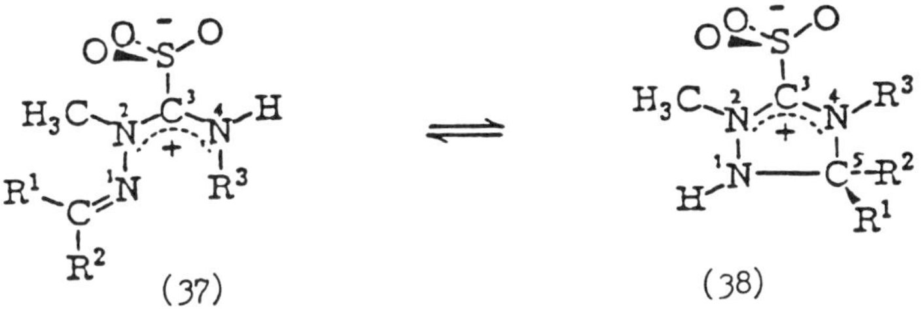

a, R^1 = H, R^2 = R^3 = CH_3

b, R^1 = H, R^2 = $CH(CH_3)_2$, R^3 = CH_3

c, R^1 = H, R^2 = C_2H_5, R^3 = $CH_2C_6H_5$

d, R^1 = H, R^2 = $CH(CH_3)_2$, R^3 = $CH_2C_6H_5$

e, R^1 = R^2 = CH_3, R^3 = $CH_2C_6H_5$

f, R^1 = H, R^2 = CH_3, R^3 = C_6H_3-2,6-$(CH_3)_2$

g, R^1 = H, R^2 = C_2H_5, R^3 = C_6H_3-2,6-$(CH_3)_2$

SCHEME 19

By appropriate selection of the thiosemicarbazones (37) and the reaction medium, complete transformation to the cyclic form can be achieved, as illustrated by the synthesis of Δ^3-triazolinium-3-sulfonates (38a–g) (77LA463).

3. *Cyclization of Hydrazone Derivatives by Reagents Susceptible to Nucleophilic Attack*

Isothiosemicarbazones, diaminomethylenehydrazones (36JA800; 37JA2077), and related compounds are polyfunctional nucleophiles (78BCJ1846, 78TL1295; 80BCJ3289). The internal nitrogen atom of the hydrazine fragment is considered to be a softer nucleophilic center than the more powerful terminal C-amino nitrogen (52JA2981). Reagents susceptible to nucleophilic attack by the terminal nitrogen have been found to induce cyclization of isothiosemicarbazones and other hydrazones, under mild reaction conditions, to afford 1,2,4-triazolines in excellent yield. The effectiveness of various agents to initiate cyclization and the possible mechanistic routes of reaction have been investigated in detail.

a. *Cyclizations Induced by Ethoxymethylenemalononitrile and Ethoxymethylene Cyanoacetate.* i. *Isothiosemicarbazones.* Cyclization of isothiosemicarbazones of aldehydes and ketones, unsubstituted on the terminal amino nitrogen (**39**), has been studied at length. Both ethoxymethylenemalononitrile (**40**, R^4 = CN) and ethoxymethylene cyanoacetate (**40**, R^4 = COOEt) are effective cyclization agents and provide a route for the one-step synthesis of novel dihydrotriazolopyrimidines (**43**) in moderate to high yield (Scheme 20) (81BCJ1767, 81JOC3956). Unambiguous product identification is effected using chemical and spectroscopic data. The triazolines derived from aldehyde isothiosemicarbazones (**43**, R^2 = H) are highly prone to oxidation; the even more susceptible dihydrotriazolopyrimidine-8-carboxylates (**43**, R^2 = H, R^4 = COOEt) undergo spontaneous oxidation during reaction to give triazolopyrimidines (**44**, R^4 = COOEt). No triazolines are found in the cyclization products; they comprise the triazoles along with 4-aminopyrimidine-5-carboxylates (**46**), competitively formed from **41** by elimination of benzonitriles.

The cyclization mechanism involves the intermediacy of **41** and not **45**; ring closure of **45** to **43** is not favored (76CC734, 76CC736). The intermediate **41** cannot be isolated when R^4 = CN (81BCJ1767); however, when R^4 = COOEt, intramolecular hydrogen bonding between the terminal amino hydrogen and the carbonyl oxygen results in a stable conformation that permits isolation of **41** (81JOC3956). Ring closure of **41** to **43** occurs by an intramolecular cycloaddition rather than a stepwise nucleophilic addition involving **45**; a 10-electron cyclic transition state (**42**) has been suggested based on spectroscopic evidence for the interconversion of **41** and **43**, a reversible process associated with hydrogen shift, similar to an electrocyclic reaction (81JOC3956).

ii. *Diaminomethylenehydrazones.* Dihydrotriazolo[1,5-*c*]-pyrimidines with a 5-amino substituent (**43a**) are obtained from ethoxymethylenemalononitrile-induced cyclization of diaminomethylenehydrazones (**39a**) (36JA800; 37JA2077; 85CPB2678).

The smaller unsubstituted N^3 atom in **39a** corresponds to the sulfur in

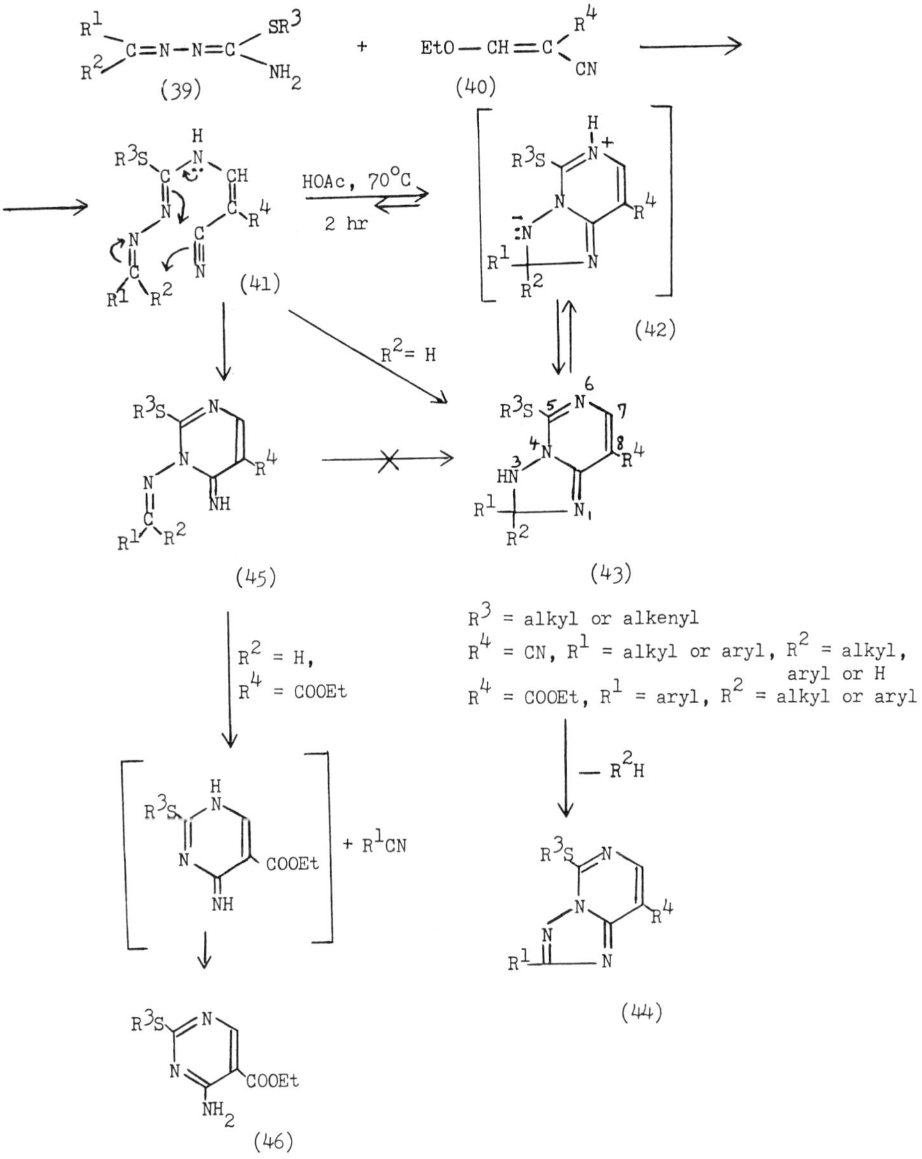

Sec. II.A] 1,2,4-TRIAZOLINES 193

SCHEME 20

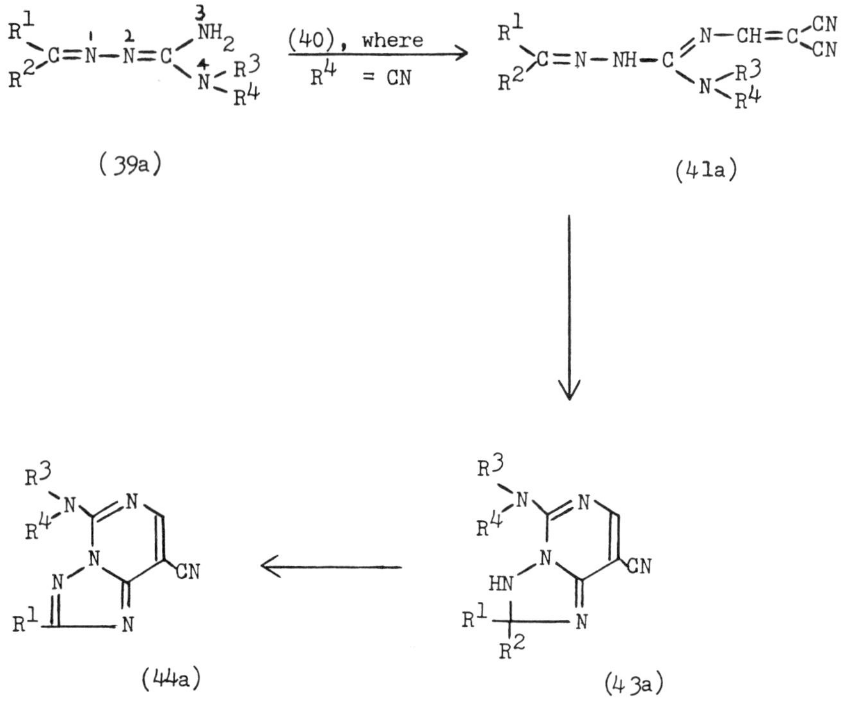

$$R^1_{R^2}C=N-N=C\underset{N-R^4}{\overset{NH_2}{\underset{R^3}{|}}} \xrightarrow{(40), \text{ where } R^4 = CN} R^1_{R^2}C=N-NH-C\underset{N-R^4}{\overset{N-CH=C\underset{CN}{CN}}{\underset{R^3}{|}}}$$

(39a)   (41a)

(44a)   (43a)

R^1 = aryl, R^2 = H or alkyl, $R^3 = R^4 = CH_3$ or $R^3R^4 = (CH_2)_4$

39; N^4-monosubstituted hydrazones were the most reactive, independent of the steric or electronic nature of the substituent group. An electrocyclic reaction mechanism, similar to the one operating in the isothiosemicarbazone cyclization (81JOC3956), has been proposed for conversion of **41a** to **43a** (85CPB2678). No intermediate analogous to **45** has been detected (85CPB2678).

b. *Isocyanate- and Isothiocyanate-Induced Cyclizations*.
Isocyanate- and isothiocyanate-initiated cyclization reactions of isothiosemicarbazones provide a novel route for the synthesis of both mono- and bicyclic triazolines in excellent yield (84ABC2913, 84JOC1703). The reactions are characterized as inter- and intramolecular double additions (84JOC1703). When a phenyl or *tert*-butyl group is present in either reagent, the isothiosemicarbazones and isocyanates in Scheme 21 react at room temperature overnight to yield 1,2,4-triazolines **47**. In the absence

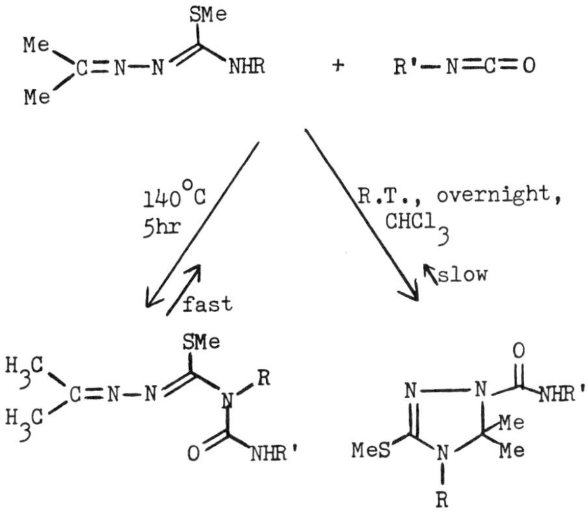

Sec. II.A]   1,2,4-TRIAZOLINES   195

$$\text{Me}_2\text{C}=\text{N}-\text{N}\overset{\text{SMe}}{=}\text{C}-\text{NHR} \quad + \quad \text{R}'-\text{N}=\text{C}=\text{O}$$

140°C, 5hr / fast   R.T., overnight, CHCl$_3$ \ slow

(48)   (47)

R = Me; Ph
R'= Me; iso-Pr; t-Bu; Ph

SCHEME 21

of a phenyl group, the products are acyclic (48); triazolines are obtained only when the reaction mixture is heated at 140°C for 5 hr in a sealed tube. At high temperature, the acyclic products yield triazolines by either a simple isomeric cyclization or by thermodynamic control of the reverse reaction to yield the more stable cyclic products 47. Triazoline formation is confirmed by the singlet signal for the *gem*-methyl groups in the NMR spectra (84JOC1703).

Scheme 22 gives the results of isothiocyanate- and isocyanate-induced cyclizations of terminal N-methoxycarbonyl-substituted isothiosemicarbazones (84JOC1703). Transformation to the fused ring system (51) is achieved through cyclization with 1,1'-carbonyldiimidazole, as established by spectroscopic data and single-crystal X-ray structure determination (84ABC2913). Several of the fused triazoline compounds 51 are reported to show considerable antifungal activity against rice blast disease, which increases with increasing bulk of the R substituent (84ABC2913; 85JAP(K)60172983).

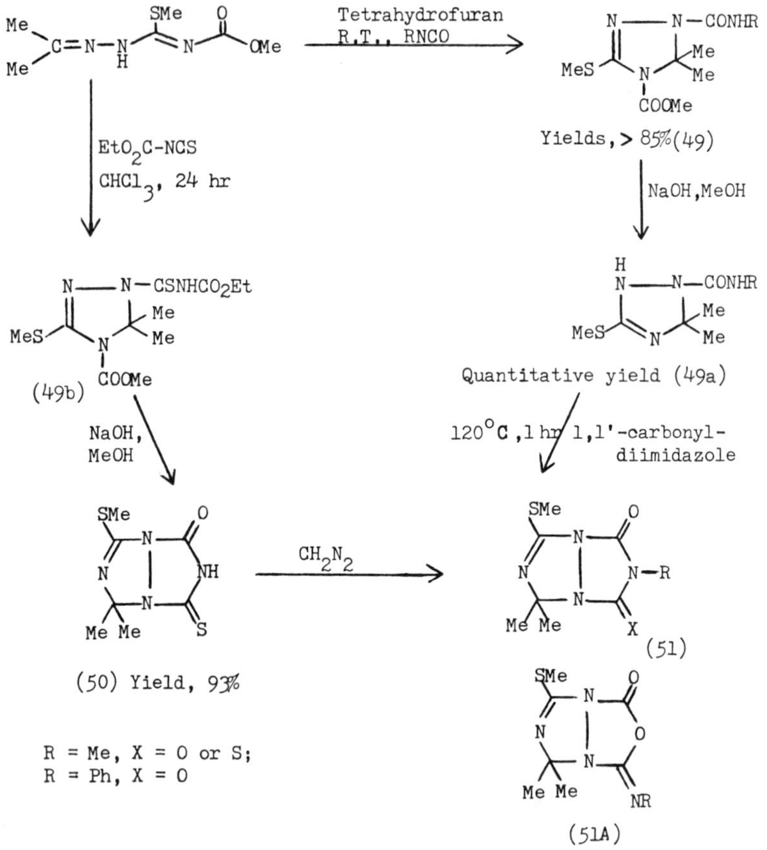

SCHEME 22

c. *Rearrangement of O-Acetyl Derivatives of 1,2-Hydroxylaminohydrazones and Thiosemicarbazones.* Acetic anhydride acylation of phenylhydrazone, semicarbazone, and thiosemicarbazones containing a hydroxylamino group at a tertiary carbon atom leads to products of O-acylation at the hydroxylamino group. These products undergo base-catalyzed rearrangement to form 4,5-dihydro-1,2,4-triazoles. An aziridine intermediate is proposed with subsequent cleavage of the C^2—C^3 bond and formation of the R^2N—C^3 bond (Scheme 22A) (86KGS352).

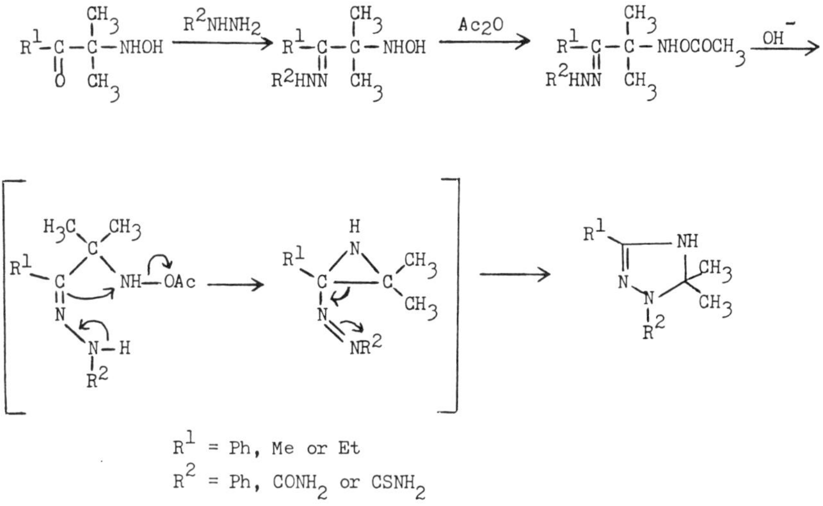

Sec. II.B] 1,2,4-TRIAZOLINES 197

$$R^1-\underset{\underset{O}{\|}}{C}-\underset{\underset{CH_3}{|}}{\overset{\overset{CH_3}{|}}{C}}-NHOH \xrightarrow{R^2NHNH_2} R^1-\underset{\underset{R^2HNN}{\|}}{C}-\underset{\underset{CH_3}{|}}{\overset{\overset{CH_3}{|}}{C}}-NHOH \xrightarrow{Ac_2O} R^1-\underset{\underset{R^2HNN}{\|}}{C}-\underset{\underset{CH_3}{|}}{\overset{\overset{CH_3}{|}}{C}}-NHOCOCH_3 \xrightarrow{OH^-}$$

$$\left[R^1-\underset{\underset{\underset{R^2}{|}}{\overset{\|}{N}}}{\overset{H_3C\diagdown\diagup CH_3}{\underset{\|}{C}}}-NH\overset{\frown}{\underset{\smile}{-}}OAc \longrightarrow R^1-\underset{\underset{\underset{NR^2}{\|}}{N}}{\overset{H\diagdown N}{\underset{\|}{C}}}\overset{CH_3}{\underset{CH_3}{\diagup}} \right] \longrightarrow \underset{\underset{R^2}{|}}{\overset{R^1\diagdown}{N}}\underset{N}{\overset{NH}{\diagup}}\overset{CH_3}{\underset{CH_3}{\diagup}}$$

$R^1 = $ Ph, Me or Et
$R^2 = $ Ph, $CONH_2$ or $CSNH_2$

SCHEME 22A

B. 1,3-DIPOLAR CYCLOADDITION REACTIONS

1,3-Dipolar cycloaddition is a fruitful synthetic pathway for the preparation of five-membered heterocyclic ring systems (64MI1; 84MI1, 84MI2). A 1,3-dipole is a species represented by a zwitterionic octet structure and undergoes a 1,3-cycloaddition to a multiple-bond system, the "dipolarophile", the formal charges being lost in the [3 + 2 → 5] cycloaddition (63AG(E)565, 63AG(E)633; 68AG(E)321, 68JOC2291). Numerous variations are available by changing the structure of both the dipolarophile and dipole. The concerted mechanism of the 1,3-cycloaddition reaction has been studied extensively and is relatively well understood (76JOC403; 84MI1, 84MI2). Molecular orbital models are also valuable aids in understanding the reactivity, regioselectivity, and stereospecificity phenomena of cycloaddition reactions and in predicting reactivities and product identities for addend pairs (76MI1; 84MI1, 84MI2).

1,3-Cycloaddition reactions provide useful routes for the synthesis of both Δ^2- and Δ^3-1,2,4-triazolines, the 1,3-dipoles contributing to three of the five members in the ring framework. The nitrilium betaines, nitrilimines, and nitrile ylides, analogous to the diazonium betaines, diazoalkanes, and azides in 1,2,3-triazoline synthesis (84AHC217), are of the octet-stabilized propargyl-allenyl type (76JOC403; 84MI1), while azomethine ylides and azomethinimines are of the allyl type (84MI1). Nitrilimines and azomethine ylides lead to Δ^2-1,2,4-triazolines; nitrile ylides and azomethinimines yield the Δ^3-isomers.

1. Nitrilimine Addition to Carbon–Nitrogen Double Bonds

A nitrilimine is commonly represented by the all-octet resonance structures **52a** and **52b** (84MI1). Nitrilimine chemistry

$$\overset{+}{C}\equiv N-\overset{-}{\underset{..}{N}}\diagdown \qquad -\overset{-}{\underset{..}{C}}=\overset{+}{N}=N\diagdown$$

(52a)   (52b)

has been reviewed (70CI(L)1216; 72RCR495; 80JHC833; 84MI1). The parent nitrilimine has been described under the name of *isodiazomethane* (34LA264), but no cycloadditions of this compound are known (68LA72, 68LA87).

Nitrilimines are unstable 1,3-dipoles, which are usually generated *in situ* by dehydrohalogenation of hydrazidoyl chlorides (64MI1; 80JHC833; 84MI1) or bromides (73T121; 83T129) under mild conditions by the action of tertiary bases (62T3). A benzene solution of the hydrazidoyl chloride and imine is treated with Et$_3$N and then refluxed after complete precipitation of the Et$_3$N·HCl (80JHC833). In the absence of dipolarophiles, dimerization to dihydrotetrazine occurs (61CB2503; 65CB1476).

High-temperature thermal elimination of nitrogen from 2,5-disubstituted tetrazoles also provides a route for the preparation of reactive nitrilimines *in situ* (61CB2503; 62T3; 64CB1085), although drastic reaction conditions make it less desirable.

Procedures have been developed for the generation of functionalized nitrilimines (75CJC3782); besides the commonly used diarylnitrilimines and C-acyl- and C-alkoxycarbonyl-N-arylnitrilimines, representatives of C-alkyl-N-aryl (72JCS(P2)44; 76IJC(B)425), C-aryl-N-alkyl (84CB1194), C-phenylazo-N-aryl (67M1618), C-aryl-N-arylsulfonyl (80T1565), and even the simple C-unsubstituted N-arylnitrilimines (76ZOR1676) have been generated from hydrazidoyl halides.

The cycloaddition of nitrilimines to the C=N double bond proceeds exclusively in a direction where only carbon–hetero bonds are formed in preference to hetero–hetero bonds, leading to Δ^2-1,2,4-triazolines (84MI1). In terms of the general molecular orbital perturbation theory (PMO) (71TL2717; 74PAC569), nitrilimines are type II 1,3-dipoles and their cycloadditions are highest occupied (HO)–lowest unoccupied (LU) (dipole) controlled. Both electron-donating (donor) and electron-withdrawing (acceptor) substituents increase the HO (LU) control, relative to the unsubstituted dipoles. The higher electronegativity of the nitrogen allows the HOMO (dipole)–LUMO (dipolarophile) interaction to dominate. The large HOMO coefficient of diphenylnitrilimine at the nitrogen atom overlaps with the carbon atom of the LUMO (dipolarophile) (Scheme 23).

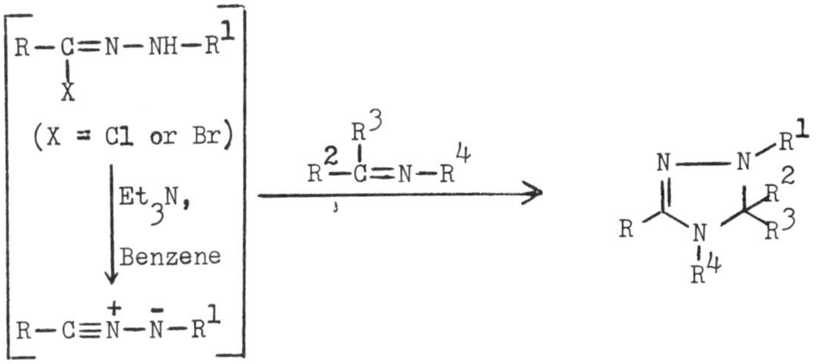

SCHEME 23

a. *Acyclic C=N Bonds. i. C=N Bonds in Schiff bases and related compounds.* The dipolarophilic character of the C=N bond is considerably more pronounced than that of the C=O bond; while acetone itself fails to add diphenylnitrilimine, acetone-*N*-isopropylimine reacts with ease to give 84% of the 1,2,4-triazoline adduct (64CB1085).

Diphenylnitrilimine has been investigated extensively; it undergoes cycloaddition to the C=N bond in aromatic, aliphatic, and heterocyclic Schiff bases to give good to excellent yields of various tetrasubstituted triazolines (53) (64CB1085; 66USP3278545). Cycloaddition to oximes, hydrazones, and imidates yields unstable triazolines 54–56 that aromatize spontaneously (65CB642).

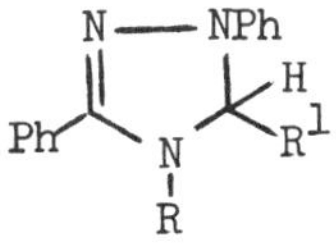

$R = R^1 = Ph$

$R = Ph, R^1 = 2\text{-Furyl}$

$R = Me \text{ or } PhCH_2, R^1 = Ph \text{ or } 2\text{-Furyl}$

$R = n\text{-Bu}, R^1 = Me$

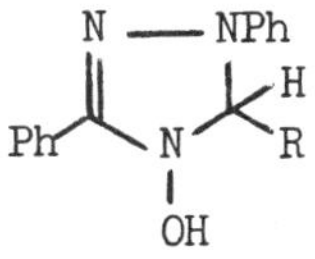

(53)

(54) 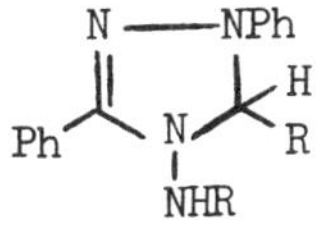 (55) 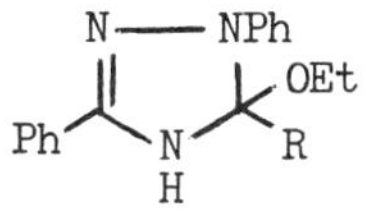 (56)

Unsymmetrical amidines can afford two cycloaddition products resulting from the two tautomeric forms; in addition, the hydrazidoyl chloride used for *in situ* generation of methoxycarbonylnitrilimine undergoes a nucleophilic attack on its carbon by the more basic amidinic nitrogen leading to imidazolinone substitution products. Although cycloaddition is the major reaction path in the case of N-aryl-substituted benzamidines, unsubstituted benzamidines favor substitution (Scheme 24) (80JHC311).

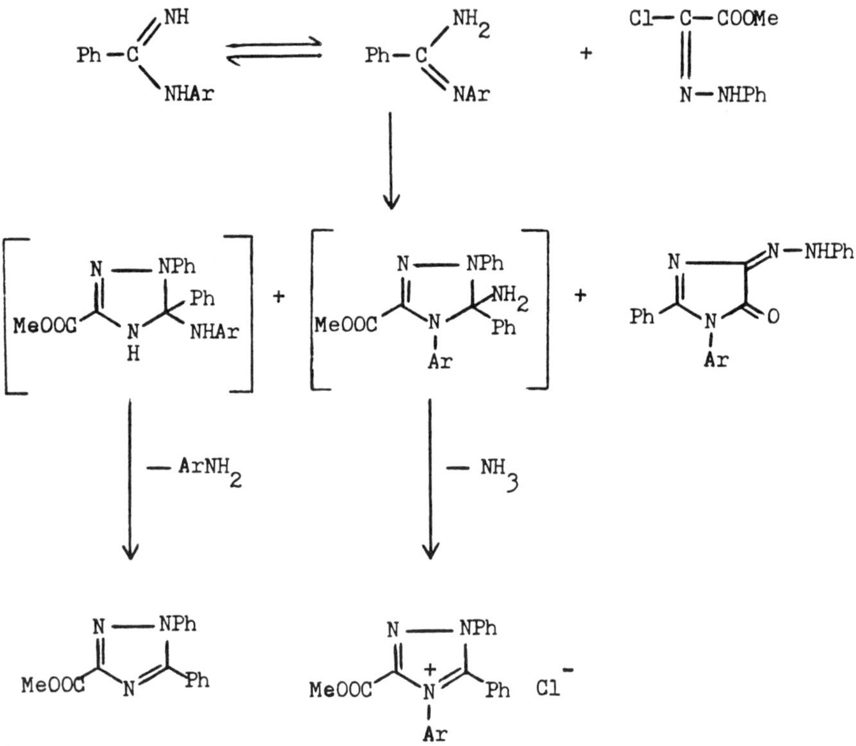

SCHEME 24

Reaction of *N*-(phenylsulfonyl)benzhydrazidoyl chloride, which fails to yield a 1,3-dipole, with N-substituted benzamidines is postulated to proceed via an intermediate formed by nucleophilic attack of the hydrazidoyl carbon by the more basic imino nitrogen of the amidine. Intramolecular cyclization of the intermediate yields a triazoline; loss of amino or phenylsulfonyl moieties from the latter leads to triazoles as final reaction products (Scheme 24A) (77BCJ2969).

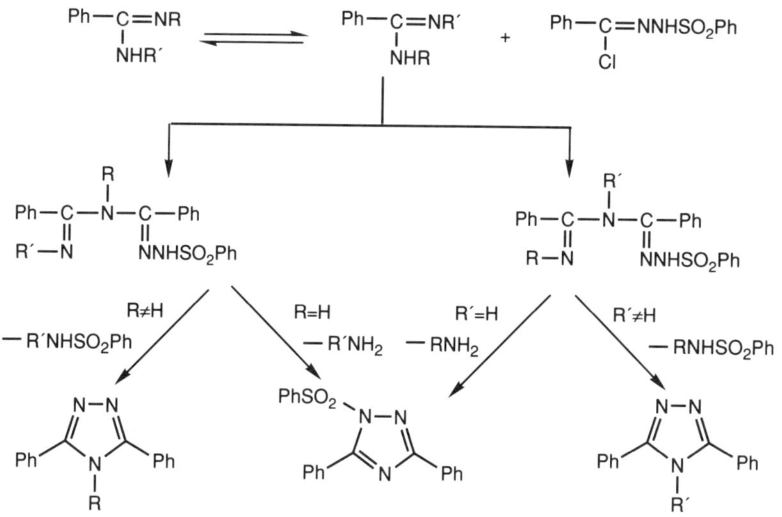

Sec. II.B] 1,2,4-TRIAZOLINES 201

SCHEME 24A

C-Benzoylimines, however, yield stable 5-benzoyl-1,2,4-triazolines (Scheme 25) (85MI1).

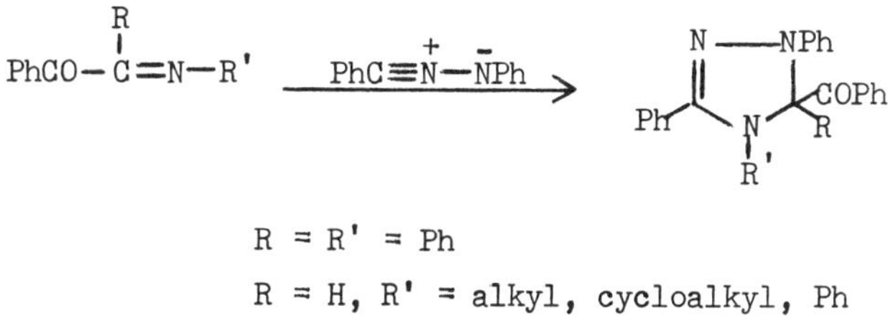

R = R' = Ph
R = H, R' = alkyl, cycloalkyl, Ph

SCHEME 25

Triazolines functionalized in the 3-position with an alkyl (76IJC(B)425; 82JHC1573; 83JIC961), acetyl, or alkoxycarbonyl group (69G69; 83T129) result by reaction of the appropriate C-substituted *N*-phenylnitrilimines with aliphatic or aromatic Schiff bases. However, *N*-phenyl formamidoxime yields the triazole *N*-oxide (**57**) (65CB642).

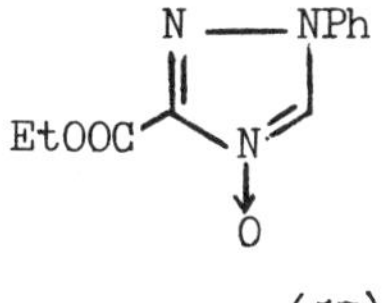

(**57**)

Diphenylnitrilimine, generated *in situ* by thermolysis of 2,5-diphenyltetrazole, reacts with benzaldehydeazine to give triazole **59** and triazoline **60** via decomposition of the mono- and bisadducts under the high-temperature reaction conditions (Scheme 26) (65CB642). However, when the dipole is generated at room temperature from hydrazidoyl chloride and triethylamine, the azine reacts to give high yield of stable 4-benzylideneamino-1,2,4-triazolines **58** (84ZOR659). Stable cycloadducts **61** are also the preferred products in the reaction of azines with C-acetyl-N-arylnitrilimines; no (4 + 3)- or "criss-cross"-type products are observed (85H1123).

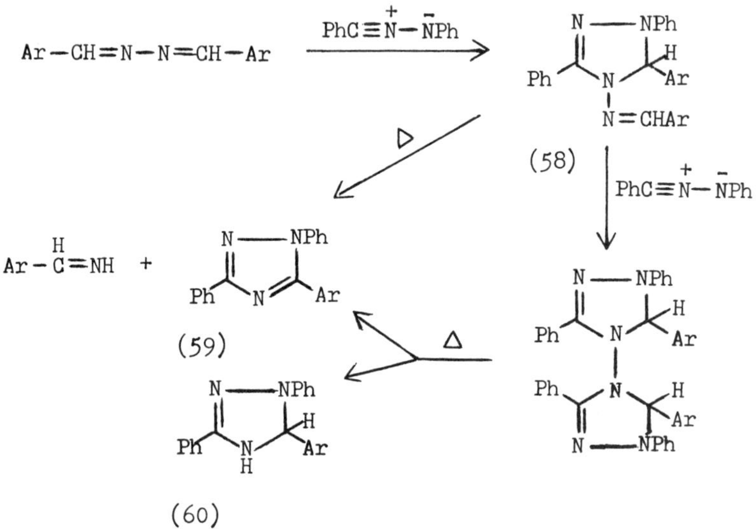

SCHEME 26

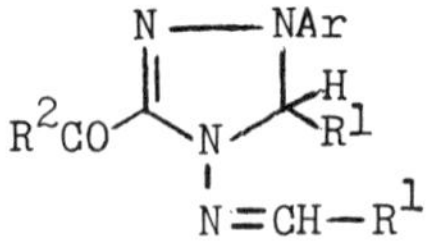

$R^1 = Ar, R^2 = Me$
$R^1 = 2\text{-furyl}, R^2 = Me \text{ or } EtO$
$R^1 = 2\text{-thienyl}, R^2 = Me$

(61)

ii. *Conjugated C=N bonds.* Site specificity in favor of the C=N bond is observed in the reaction of nitrilimines with conjugated imines, the reaction across the C=C bond being minor (84H549). Thus, 1-azabutadienes provide good building blocks for the synthesis of 5-alkenyltriazolines **62** in good to excellent yield (Scheme 27) (69CC387; 83IJC(B)1244; 84H549, 84JCR(S)56), although when R = Me and R^1 = t-Bu, the azadiene affords only the triazole (84H549).

Site-specific addition of diphenylnitrilimine to the C=N bond in 8-azaheptafulvenes yields intermediate spirotriazolines, which rearrange to isomeric triazine derivatives. Proof is provided by triazine formation from an independently generated spirotriazoline sample, as shown in Scheme 28 (79CSC341; 80T935).

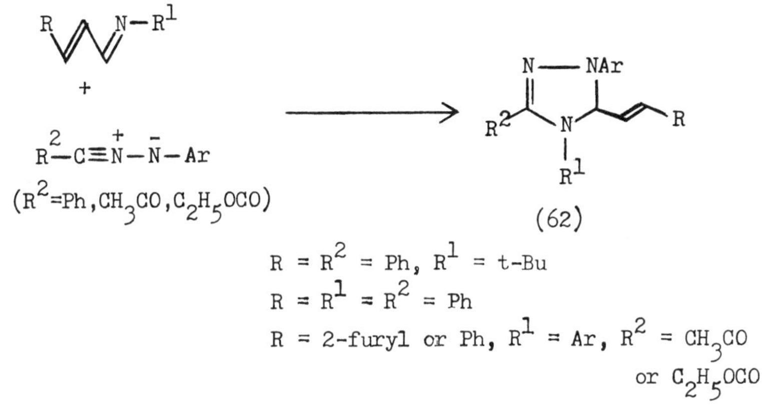

SCHEME 27

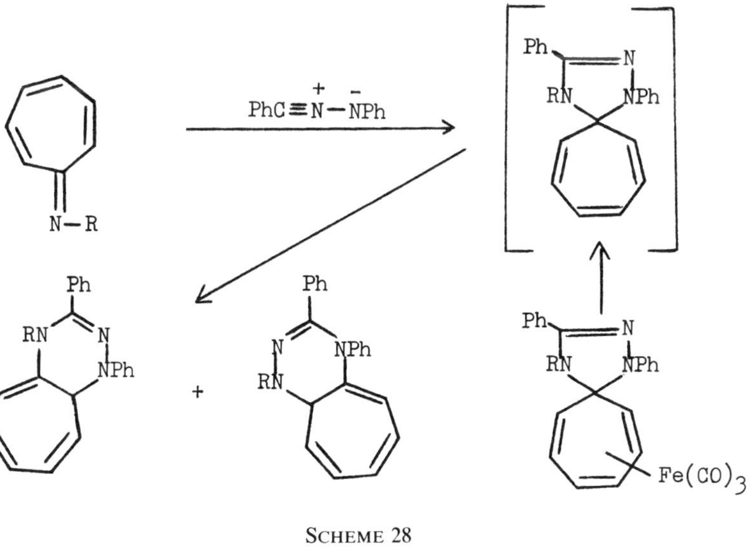

SCHEME 28

iii. *Cumulated C=N bonds.* Nitrilimine addition to carbodiimides follows the same orientation as that observed in addition reactions with imines. However, the cumulated bond system of the carbodiimide is an inferior acceptor relative to the guanidine group of the monoadduct **63**; the isolated products are thus the spiro-1,2,4-triazoline bisadducts **64** (65CB2174).

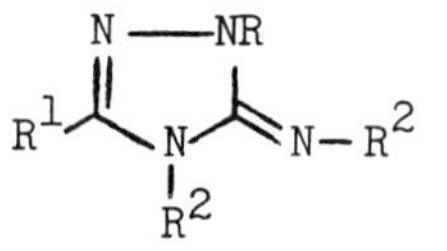

(63) 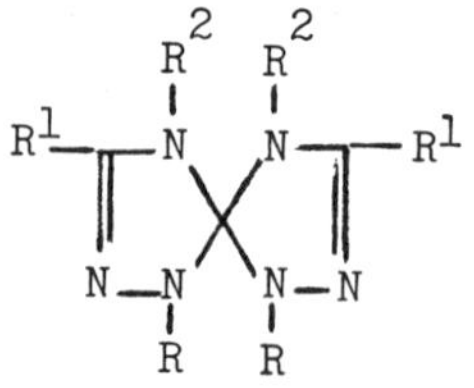   (64)

N-Phenylcyanamide reacts via its tautomeric *N*-phenylcarbodiimide form and adds *C*-carbethoxy-*N*-phenylnitrilimine on the Ph—N=C double bond (65CB2185).

iv. *Exocyclic C=N bonds.* Spirotriazoline adducts result by nitrilimine addition to exocyclic C=N bonds [Scheme 28 and formation of **(64)** from **(63)**]. Reaction of nitrilimine with phthalazinonehydrazone leads to phthalazine-1-spirotriazoline **65**, along with **66**; the structure of **65** has been confirmed by X-ray crystallography (Scheme 29) (83ZOR1069).

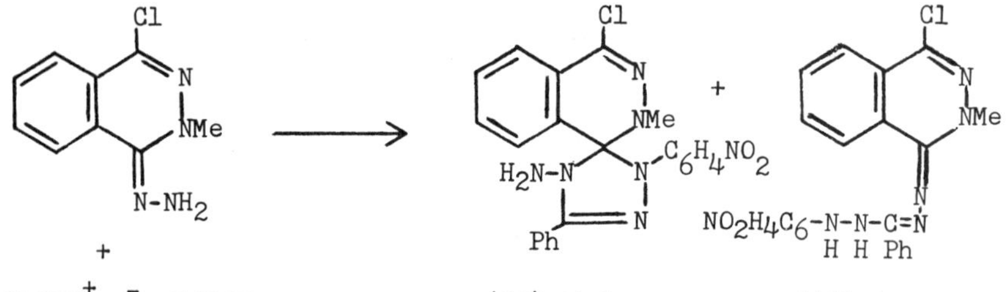

(65) 14%   (66) 60%

SCHEME 29

b. *Cyclic C=N Bonds.* The C=N bond, which is an integral part of several N-hetero ring systems, has been found to evidence dipolarophilic activity in cycloaddition reactions. Nitrilimine addition provides a promising route for the direct synthesis of several 4,5-annulated 1,2,4-triazoline ring systems, which would be tedious to synthesize by other methods. The reaction is complimentary to the cycloisomerization of hydrazono heterocycles to yield 3,4-fused triazolines (Section II,A,1,f).

i. *Pyridines, quinolines, and isoquinolines.* The products from reaction of hydrazidoyl chlorides with heterocyclic bases such as pyridine, quinoline, or isoquinoline are determined by the reaction conditions and reagents used. At low temperature (60–70°C), C-alkoxycarbonyl-N-phenylhydrazidoyl chloride undergoes nucleophilic attack at the carbon to give betaines 67 (36LA173; 37MI1); at higher temperature (160–180°C), a complex series of reactions leads to the eventual formation of N-arylcyanamides 68 (67TL3071; 68G511). In the presence of triethylamine, the product is a triazoline (69), presumably formed by cycloaddition of the nitrilimine, generated *in situ,* to the C=N bond of the heterocycle (67TL3071; 68G511).

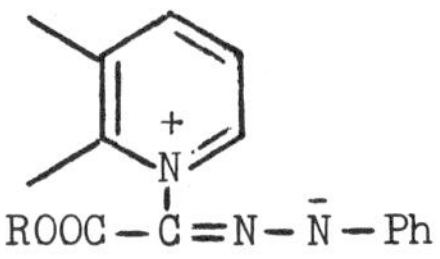

(67) 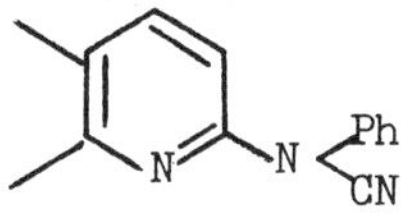   (68) 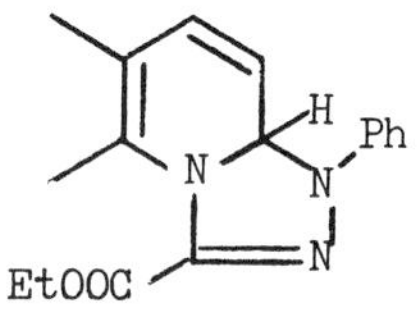   (69)

Hydrazidoyl chloridès[2] bearing strong electron-withdrawing N-phenylsulfonyl groups, from which nitrilimines cannot be generated, yield triazolines by 1,5-dipolar cyclization of the intermediate betaines (Scheme 29A) (80JHC833). The triazolines from both quinoline and isoquinoline aromatize on heating, and provide a simple route for the synthesis of 1,2,4-triazoloquinoline and -isoquinoline. The pyridinium betaine, however, fails to cyclize. The aromatic stabilization provided by the fused benzene ring in the quinoline and isoquinoline systems facilitates betaine cyclization to the respective triazolines; in the pyridinium intermediate aromatic stabilization will be lost upon cyclization (80BCJ2007). N-(Phenylsulfonyl)benzhydrazidoyl chlorides react with cyclic amidines such as 2-aminopyridines and -pyrimidines in reactions analogous to those with benzamidines (77BCJ2969) (Section II,B,1,a,i, Scheme 24A).

Dimesitylnitrilimine, on the other hand, reacts with pyridine as well as quinoline or isoquinoline to form stable cyclic 1 : 1 adducts 70 (69LA91).

---

[2] By analogy to imidoyl chlorides, RC(Cl)=NR', which are considered to be derivatives of the corresponding imidoic acids RC(OH)=NR', the compounds RC(Cl)=NNHR' are named hydrazidoyl chlorides of their corresponding hydrazidoic acids RC(OH)=NNHR' (80JHC833).

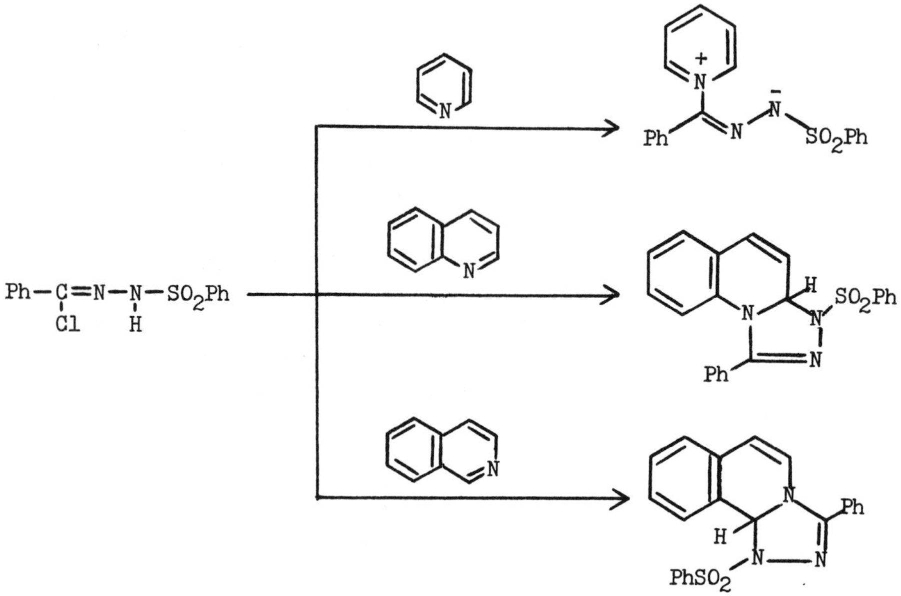

SCHEME 29A

Likewise, cycloaddition of diphenylnitrilimine to the C=N bond in dihydroisoquinoline yields **71** (85AP556).

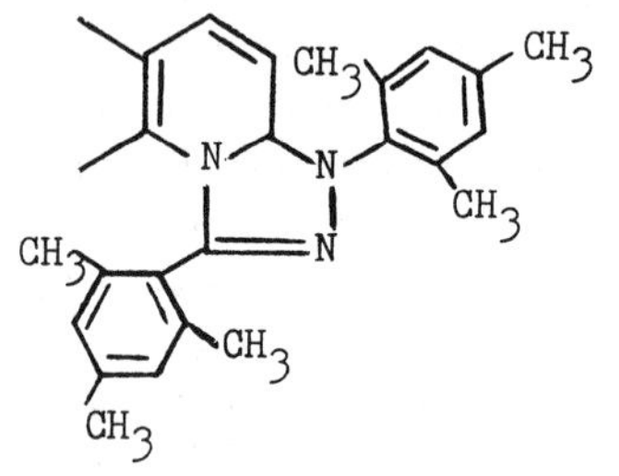

(70) 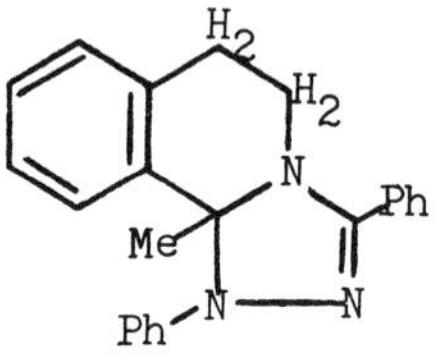   (71)

ii. *Pyrroles and pyrrolines.* The addition reaction of C-acetyl-N-phenylnitrilimine to pyrroles has been investigated (78JHC1485). In addition to the electrophilic 2-substitution product of pyrrole, 1,3-cycloaddition to the initially formed pyrroline monoadducts yields several bisadducts; two of the products (**72** and **73**) possess 1,2,4-triazoline structures as revealed by NMR spectroscopy (Scheme 30).

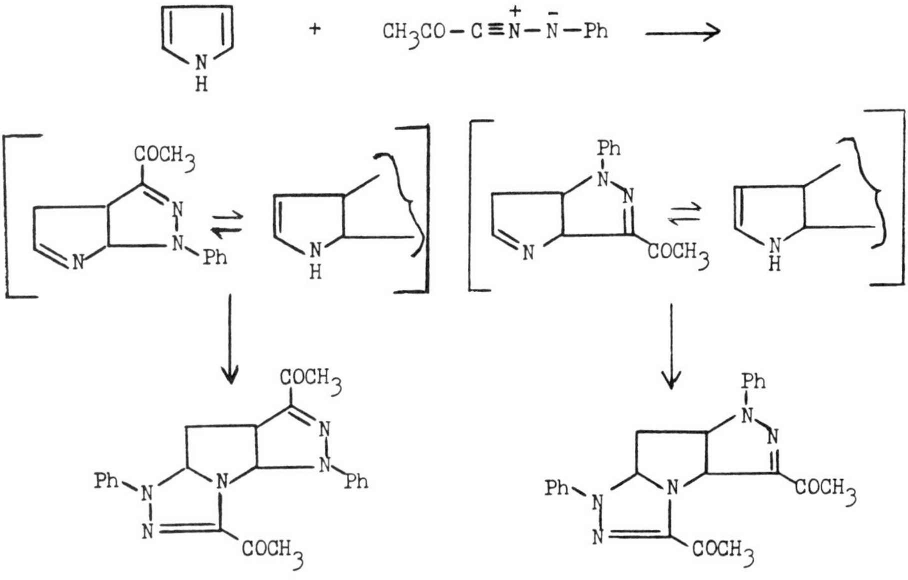

SCHEME 30

Similar cycloaddition reactions of diphenylnitrilimine with pyrrolines and benzopyrroles yield variously substituted fused-ring triazolines **74** and **75** (85AP556).

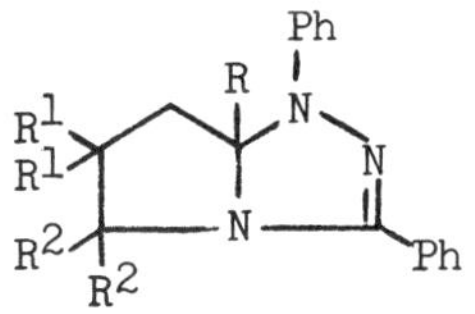

$R = Ph, PhCH_2, R^1 = R^2 = H$

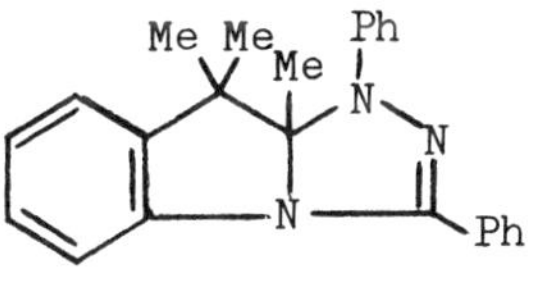

$R = R^1 = Me, R^2 = H$

$R = R^1 = H, R^2 = Me$

iii. *Imidazoles and benzimidazoles.* Reactions of diphenylnitrilimine with imidazoles, benzimidazoles, and 1-alkyl- and 1-acylbenzimidazoles have been studied. Interestingly, cycloaddition products **76** are isolated in moderate yield only from 1-acylbenzimidazoles (85H2183). In the other cases, nucleophilic substitution at the hydrazidoyl carbon is the prevailing reaction affording **77**. Unlike earlier methods of synthesis (69CB1028; 82JCS(P1)2663), this cycloaddition affords **76** in a single-step reaction.

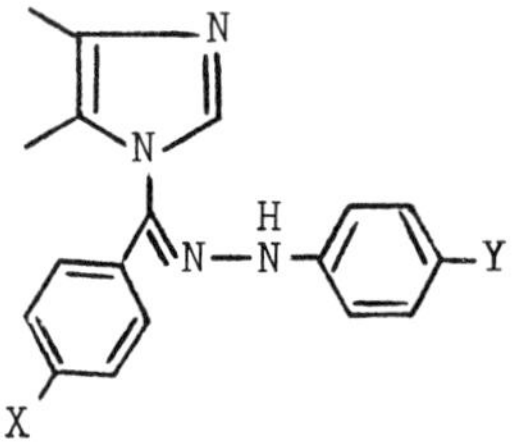

(77)

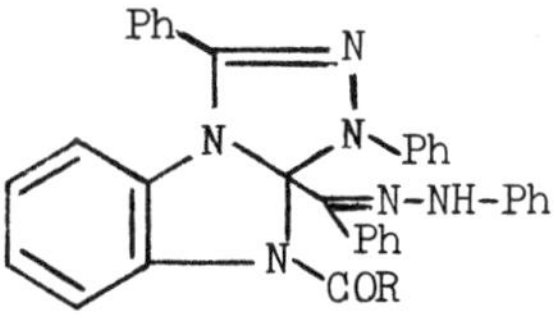

(76)

R = CH$_3$ or Ph

iv. *Benzoxazines.* The reactivity of the carbon–nitrogen double bond of benzoxazines **78** and **79** has been tested using diarylnitrilimines (86JCR(S)200). While a single cycloadduct (**80**) is obtained from **78**, two diastereoisomeric cycloadducts (**81** and **82**) in almost equimolar amounts result from **79**.

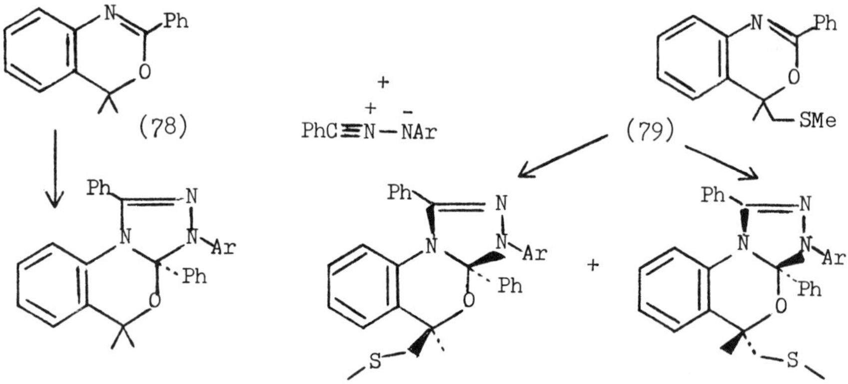

(78)   PhC≡N—NAr   (79)

(80)   (81)   (82)

v. *Norbornane/ene-fused dihydro-1,3- and 3,1-oxazines.* Diphenylnitrilimine (DPNI) cycloaddition to the C=N bond of norbornane-fused dihydro-3,1- and 1,3-oxazines **83–85** leads to tetracyclic oxazino-1,2,4-triazolines **86–88**. The stereostructures of the adducts have been elucidated by NMR spectroscopy. No cycloadducts could be isolated from the norbornene-fused dihydrooxazines (87T1931).

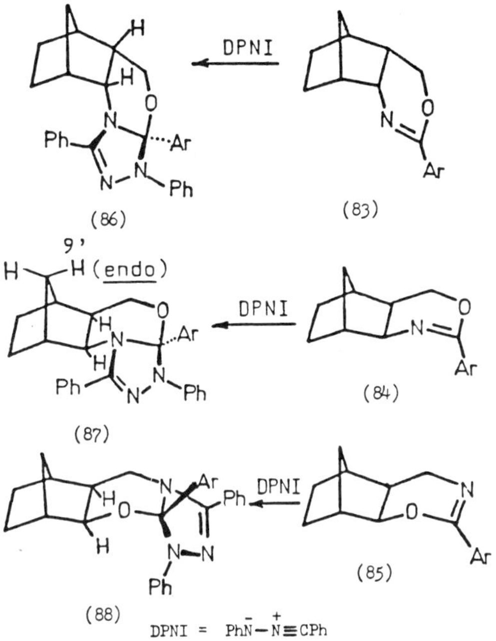

vi. *Benzothiazines*. The dipolarophilic activity of the carbon–nitrogen double bond in benzothiazines **89** and **90** has been investigated using diaryl- and C-alkoxycarbonyl-N-arylnitrilimines. The cycloaddition reaction proceeds smoothly to give excellent yields of dihydrotriazolobenzothiazines **91** and **92** (84H537).

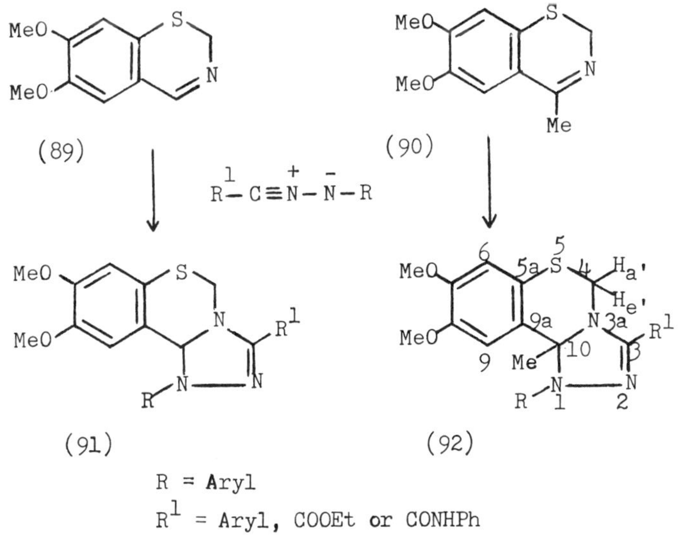

R = Aryl
R^1 = Aryl, COOEt or CONHPh

vii. *Benzodiazepines*. 1,3-Dipolar cycloaddition of suitable *C*-phenyl-*N*-arylnitrilimines to the C=N bond of the 1,4-benzodiazepine ring (93) permits synthesis of novel annulated 1,4-benzodiazepines (94). The additional triazoline nucleus dramatically reduces the conformational mobility of the seven-membered ring (85H2051), which may have interesting implications in relation to the pharmacological activity of the benzodiazepines.

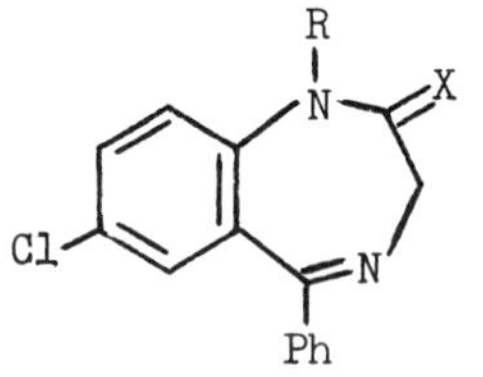

(93) 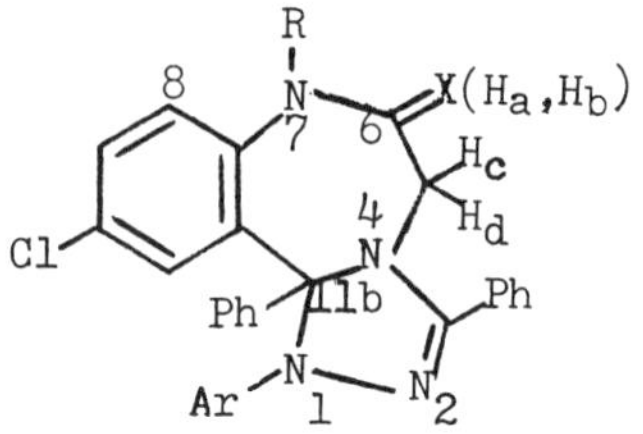   (94)

R = H or CH_3

X = H_2 or O

Similarly, cycloaddition of nitrilimines to 1,5-benzodiazepines 95 and 96 leads to good yields of previously unknown fused ring systems 97 and 98 (Scheme 31) (86S230).

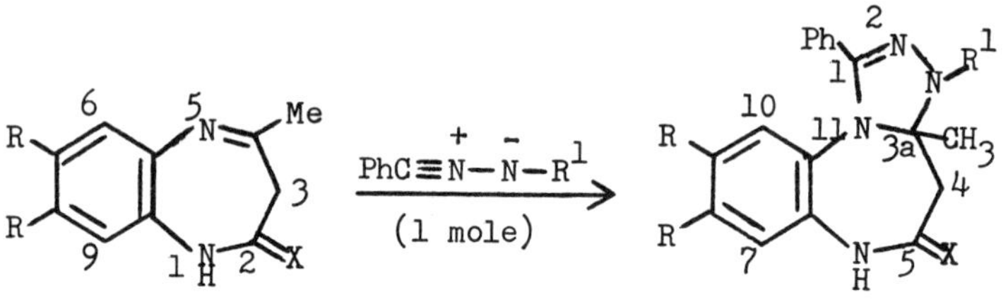

(95)   (97)

R = H or Cl
X = $2CH_3$ or O
R^1 = Ph or COOEt

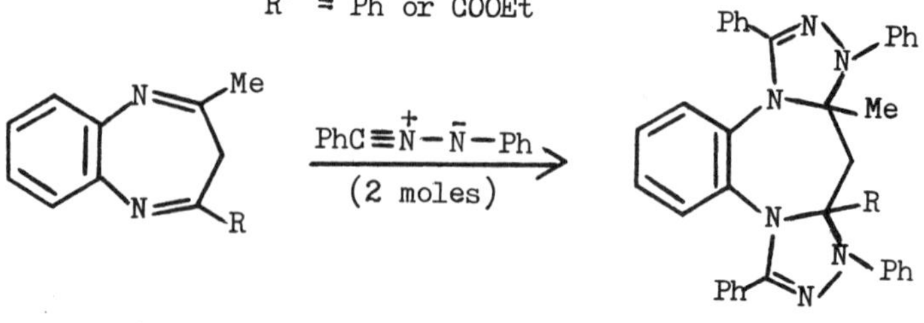

(96)   (98)

R = Me or Ph

SCHEME 31

2. Nitrile Ylide Addition to Azo Compounds

Nitrile ylides are represented by the resonance structures **99a** and **99b**, specifically named 2-azonia-1-allenide and 2-azonia-1-propynide, respectively, according to IUPAC rules (84MI1). Compared to nitrilimines, ni-

$$\overset{R}{\underset{..}{C}}{=}\overset{+}{N}{=}C\overset{R^1}{\underset{R^2}{<}} \quad \longleftrightarrow \quad R-C{\equiv}\overset{+}{N}-\overset{..}{\underset{..}{C}}\overset{R^1}{\underset{R^2}{<}}$$

(99 a)   (99 b)

trile ylides are less stable and hence more reactive; like nitrilimines, they are usually generated *in situ* in the presence of the dipolarophile (76MI2; 77H143; 84MI1).

Dehydrochlorination of *N*-(4-nitrobenzyl)benzimidoyl chloride in the presence of triethylamine provided the first route to a nitrile ylide (62AG(E)50; 72CB1258, 72CB1279, 72CB1307). Other methods of generating the dipole include thermal cycloelimination of carbon dioxide (71CB3816) or of alkyl phosphates (71AG(E)728, 71AG(E)729; 72CB3814; 73CB3421) at 140°C from oxazolin-5-ones and oxazaphospholines, respectively; thiazaphospholines lose alkyl thiophosphate at room temperature (78MI1). However, photochemically induced electrocyclic ring opening of 2*H*-azirines or their precursors is the most important method for nitrile ylide generation (72HCA745; 75S483; 76JA1048; 84HCA534).

Frontier MO calculations for structures **99a** and **99b** (76JA6397; 77JA385; 79MI2) indicate that the bent structure **99a** is energetically more favorable than the linear structure **99b** and that the HOMO/LUMO distance is independent of the nature of substituents at the nitrile carbon atom. When R = H, alkyl, aryl, aryloxy, alkoxy, or dialkylamino and R^1 and R^2 = H, alkyl, or aryl, the bent structure has the greatest HOMO coefficient at C-1; thus cycloadditions of these nitrile ylides with strongly polarized multiple bonds are HOMO (dipole)–LUMO (dipolarophile) controlled interactions and are highly regioselective as in type I dipoles. However, one or two trifluoromethyl substituents on the ylide carbon favor the linear structure **99b**; reactivity then becomes more closely related to type II dipoles such as nitrilimines, and in cycloaddition reactions both regioisomeric cycloadducts are obtained (84HCA534, 84MI1).

The full synthetic potential of nitrile ylides in 1,2,4-triazoline synthesis has not yet been explored; the few available studies are confined to reactions with symmetrical dipolarophiles. Substituent effects of unsymmetrical dipolarophiles on the regiochemistry of triazoline formation are not known.

Azo compounds undergo facile addition to nitrile ylides; the symmetrically disposed diethyl azodicarboxylate cycloadds to a variety of nitrile ylides to give good yields of Δ^3-1,2,4-triazoline-1,2,-dicarboxylates (Scheme 32) (74HCA1382; 80H929). The Δ^3-compounds can easily be saponified and decarboxylated to give the Δ^2-triazoline-1-carboxylates or the 1,2,4-triazoles by concomitant dehydrogenation (74HCA1382).

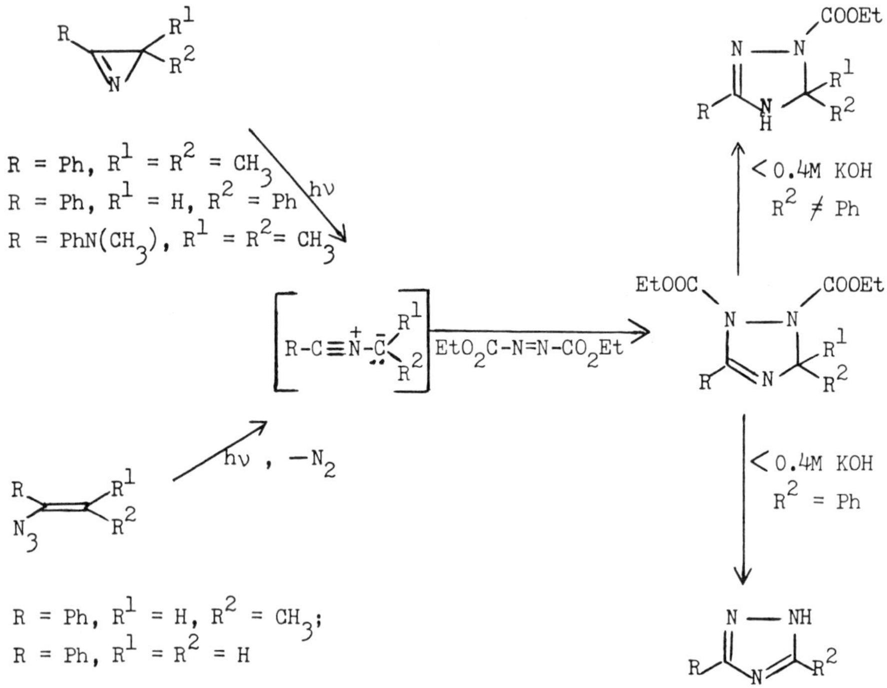

SCHEME 32

Nitrilio hexafluoro-2-propynides, generated from oxazaphospholines, have been trapped using dimethylazodicarboxylate (73CB3421) or azobenzenes (77ZN(B)607) (Scheme 33); the ylide can also be captured by the azo bond in benzo[c]cinnoline (81JHC247) to give **100**.

Functionalization at C-7 of cephalosporin has been achieved via the C-7 nitrile ylide intermediate; spirotriazoline **101** is obtained by reaction with ethylazodicarboxylate. A strong organic base such as 1,5-diazabicyclo[4,3,0]nonane (DBN) is required for nitrile ylide generation from the imidoyl chloride, which causes double bond isomerization of the Δ^3- to the Δ^2-compound (Scheme 34) (76TL1303; 77GEP152193); Et$_3$N is not effective because the hydrogen atoms at C-7 are not sufficiently acidic to effect dehydrochlorination (62AG(E)50).

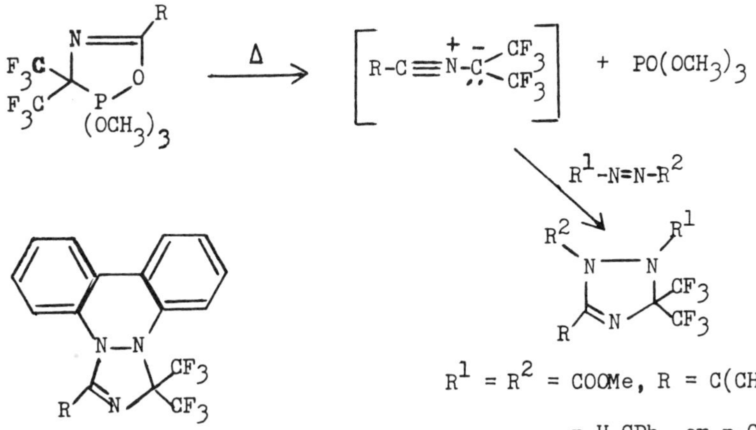

Sec. II.B] 1,2,4-TRIAZOLINES 213

$$\underset{\underset{(OCH_3)_3}{F_3C}}{\overset{R}{\underset{|}{F_3C}}}\overset{N}{\underset{P}{\bigcirc}} \quad \xrightarrow{\Delta} \quad \left[R-C\equiv\overset{+}{N}-\overset{-}{C}\overset{CF_3}{\underset{CF_3}{\diagdown}} \right] + PO(OCH_3)_3$$

$$\downarrow R^1-N=N-R^2$$

(100)

$R^1 = R^2 = $ COOMe, $R = C(CH_3)_3$, Ph, p-H$_3$CPh, or p-ClPh

$R^1 = R^2 = $ Ph, $R = C(CH_3)_3$, Ph, p-H$_3$CPh, p-ClPh, or p-FPh

SCHEME 33

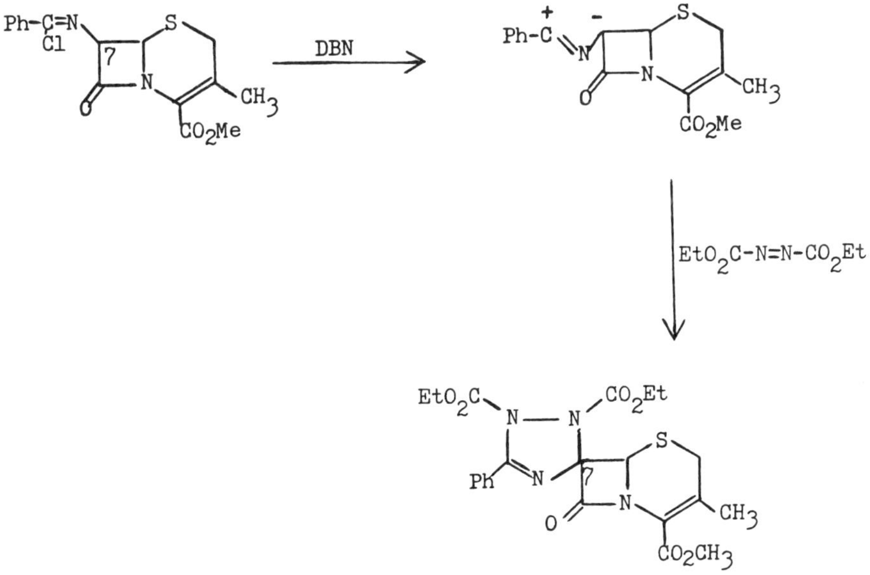

[Scheme 34 showing cephalosporin-type structure with Ph-C(Cl)=N reacting with DBN to give Ph-C=N⁺ ylide, then with EtO$_2$C-N=N-CO$_2$Et to give (101) 38 % yield]

(101) 38 % yield

SCHEME 34

3. Azomethinimine–Nitrile Interactions

Azomethinimines are better represented by their octet structures **102a** and **102b**, which clearly show the allyl anion stabilization of these 1,3-dipoles; the sextet structures **102c** and **102d** contribute little to the electron distribution of the resonance hybrid (84MI1). Resonance formula **102a** is more important as a result of the higher charge density at the nitrogen relative to carbon.

$$>C=\overset{+}{N}-\overset{-}{N}- \longleftrightarrow >\overset{-}{C}-\overset{+}{N}=N- \longleftrightarrow >\overset{+}{C}-\overset{|}{N}-\overset{-}{N}- \longleftrightarrow >\overset{-}{C}-\overset{|}{N}-\overset{+}{N}-$$

a b c d

(102)

Azomethinimines are type II HO–LU 1,3-dipoles, and resemble phenyl azides in their bidirectional behavior, although their cycloadditions are several orders of magnitude faster and the range of applicable dipolarophiles is larger (84MI1). In general, azomethinimines are too reactive to be isolated in pure form. Although stabilization of the negative charge on the nitrogen (**102a**) can be achieved by electron-withdrawing substituents (77CB500), this also reduces the charge density and the reactivity of the dipole (63CPB781, 63CPB1089; 73JA7287). Often, they are generated *in situ* and trapped with a dipolarophile (84MI1); cycloaddition to nitriles yields Δ^3-1,2,4-triazolines.

Azomethinimine dipoles could be noncyclic and not an integral part of a heterocyclic ring system, or of the cyclic type that are more or less integrated into a heterocyclic framework. The cyclic dipoles provide routes for the synthesis of a variety of annulated triazoline ring systems.

a. *C-Aryl-N,N'-dialkylazomethinimines.* Condensation of aldehydes with N,N'-disubstituted hydrazines is a frequently employed method for generating azomethinimines **103**. As a rule, the condensation products are hexahydrotetrazines **104**; these are "head-to-tail" dimers of the respective azomethinimines and provide a convenient stable source of the dipole (Scheme 35) (63AG604; 65AG(E)701, 65JOC74; 77AG(E)10).

Isolation of the hexahydrotetrazine is not essential. The 1,2,4-triazoline adduct may be formed directly by heating all three components—aromatic aldehyde, dimethylhydrazine, and dipolarophile. However, side reactions become more pronounced in this procedure and poor yields of triazolines often result.

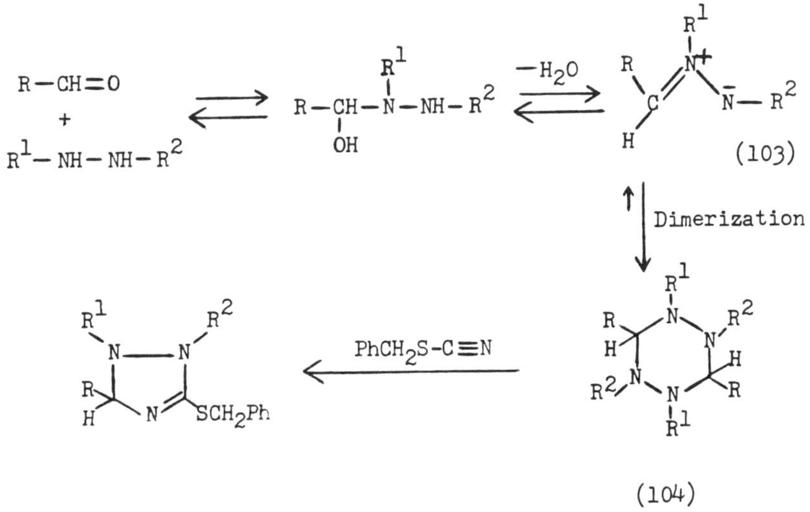

Sec. II.B]   1,2,4-TRIAZOLINES   215

SCHEME 35

b. *3,4-Dihydroisoquinolinium-N-imides.* Removal of the hydrazone proton in **105** by a base such as triethylamine or pyridine (or an excess of the hydrazine component, if R = H or alkyl) generates the deep-colored dihydroisoquinolinium-*N*-imide system (**106**). The dipole is not isolable, but displays extraordinary 1,3-dipolar activity (60AG416) and undergoes facile *in situ* additions.

In the absence of dipolarophiles, dimerization occurs, and yellow crystals of the hexahydrotetrazine derivative **107** precipitate in a few seconds (58CB1495). Solutions of 107 exhibit reversible thermochromism above 60°C, due to dissociation to the monomeric azomethinimine **106**. Likewise, the alcohol adduct **108** also serves as a stable, convenient, neutral source of the dipole at moderate temperature (Scheme 36) (62AG491; 84MI1). Most cycloadditions are usually carried out in an inert solvent at reflux temperature; the end of the reaction can often be recognized by the disappearance of the typical deep color associated with the dipole. Cycloadditions of azomethinimines are distinguished by the mild reaction conditions and the absence of side products (84MI1).

Although nitriles are highly inferior to azomethines as dipolarophiles, useful yields of dihydrotriazoloisoquinolines (**109**) are obtained by reaction of the methanol adduct (**108**) with aryl nitriles. Acetonitrile and analogous alkyl cyanides do not yield cycloadducts, but thiocyanates are superior to nitriles (Scheme 36) (62AG491).

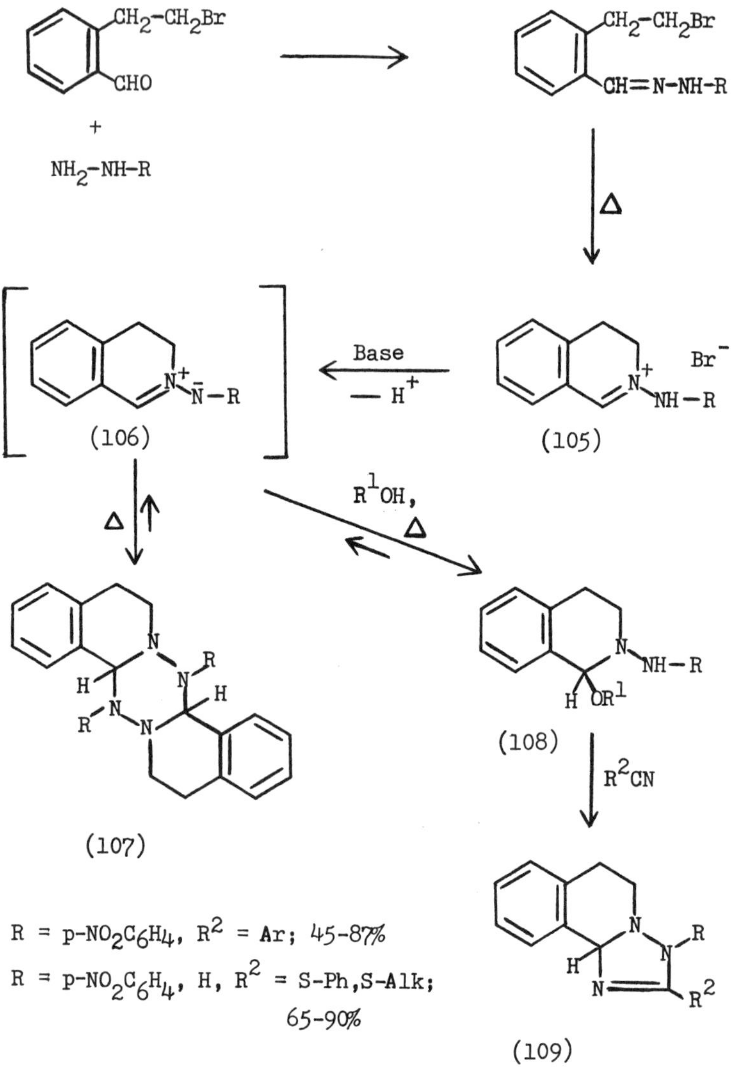

SCHEME 36

c. *Heteroaromatic N-Imides.* In this class of azomethinimines, the formal C=N bond of the 1,3-dipole is incorporated into a heteroaromatic ring system. Cycloaddition to a dipolarophile is accompanied by loss of aromaticity in the nitrogenous ring, which reduces the 1,3-activity as well as the stability of the triazoline adduct (77LA498, 77LA506).

Pyridinium-N-(imidoyl)imides **110**, prepared from the base-catalyzed reaction of N-aminopyridinium salts and N-ethoxycarbonyl imidates, undergo initial intramolecular cycloaddition in refluxing xylene to afford unstable triazoline intermediates **111**, which rearomatize to triazoles in a second step (77JOC443).

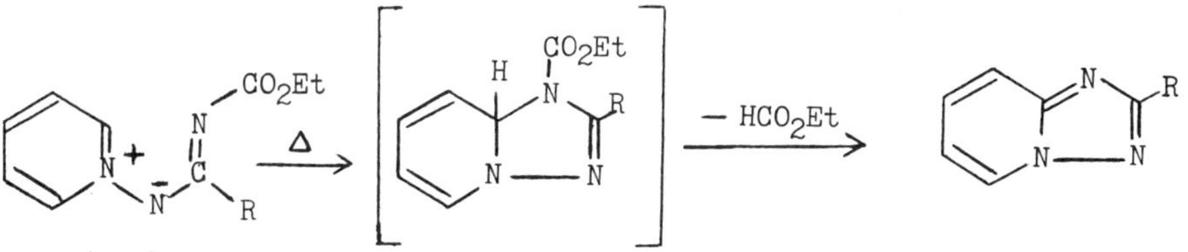

Likewise, the fused-ring triazoline adducts **112**, obtained from 1-methyl-1,2,4-triazolium-4-(acyl)imides and propiolic ester, undergo rupture at the N—N or C—N bond to yield triazole or pyrazole compounds (Scheme 37) (76CPB2568).

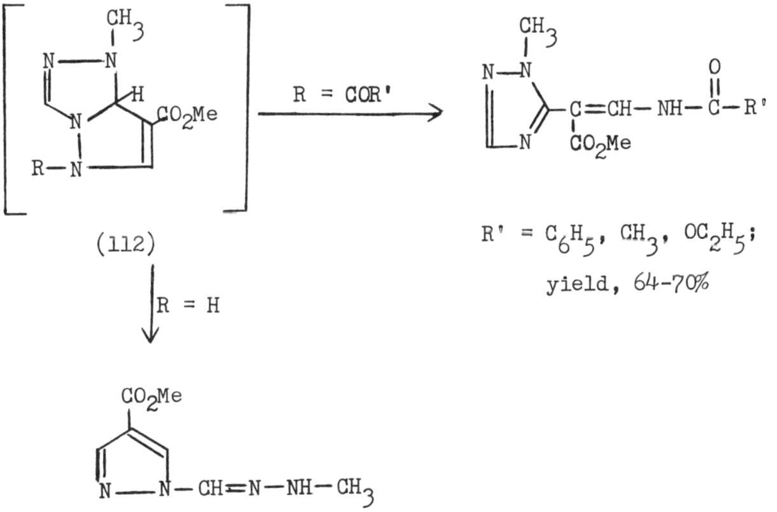

SCHEME 37

With aromatic isothiocyanates, triazolium-4-(acetyl)imide displays both 1,3- and 1,5-dipolar activity, depending on reaction conditions, mainly solvent polarity. In nonpolar solvents such as benzene, concerted 1,3-additions occur across both the C=N (**113**) and C=S (**114**) bonds of the isothiocyanate (84CCC1713); this is unlike the reaction of isothiocyanates with azomethinimines of the dihydroisoquinoline series, in which 1,3-addition is confined to the C=N bond (60AG416; 62AG292). In the polar solvent DMF, the dipole acts in its extended 1,5-dipolar form and stepwise addition leads to a single product (Scheme 38). Although theoretically a 1,5-dipole should be less stable than a 1,3-dipole, it is suggested that charge delocalization on both ends of the triazolium betaine lessens the energy requirements for stabilization of the 1,5-dipole. Thus, even in nonpolar chloroform, the triazolium betaine reacts with phenyl isocyanate in its extended form (69JPR897).

Novel 1,2,4-triazolines **114a** have been identified as characteristic photoproducts of the dimeric form of quinazolinium-*N*-acylimides, when the

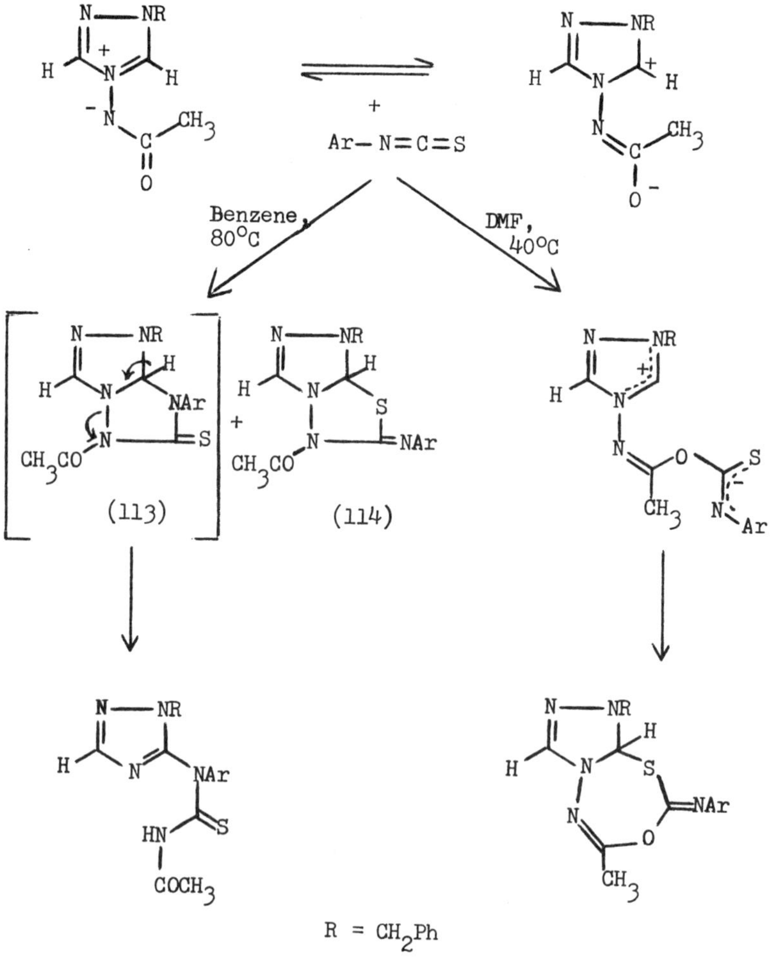

SCHEME 38

latter are irradiated in acetone in which substantial amounts of dimer are present (Scheme 38A) (83JCS(P1)2003). The triazoline structure is established by X-ray diffraction (83JCS(P1)2003).

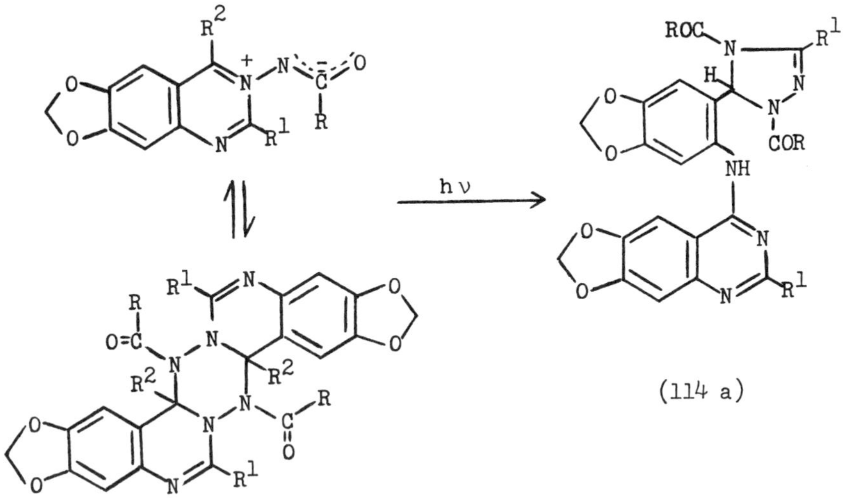

SCHEME 38A

d. *N-β-Cyanoazomethinimines*. The relatively stable, but reactive, brightly colored azomethinimines 115 are obtained by treating aromatic diazo cyanides with diaryldiazoalkanes (60TL1; 77CB500). The readily accessible prototype 115a does not react with arylnitriles (77CB571); the fluorenyl residue and the cyanamide system provide such good stabilization for the positive and negative charges, respectively, that this azomethinimine makes only minor use of its possibilities for charge cancellation. However, the more reactive ethyl cyanoformate produces 89% of the expected spirotriazoline 116 (77CB571).

Although 115a and 115b are thermally stable, the diphenyl compounds 115c and 115d trimerize in solution, particularly in refluxing acetic acid (77CB514). The trimer 117 is possibly formed by three consecutive dipolar cycloadditions, the third of which is intramolecular; its structure has been confirmed by X-ray analysis (80AG936, 80AG(E)906).

e. *Pyrazolidineazomethinimines*. Azomethinimine 118, derived from the criss-cross addition reaction of hexafluoroacetoneazine with isobutene (74AG481, 74AG482, 74AG(E)474, 74AG(E)475; 75CB1460), reacts with the nitrile group of ethyl cyanoformate to yield the expected bicyclic triazoline 119. However, nitrile adduct 120 also results from tetracyanoethylene by preferential addition to the nitrile group (76LA30).

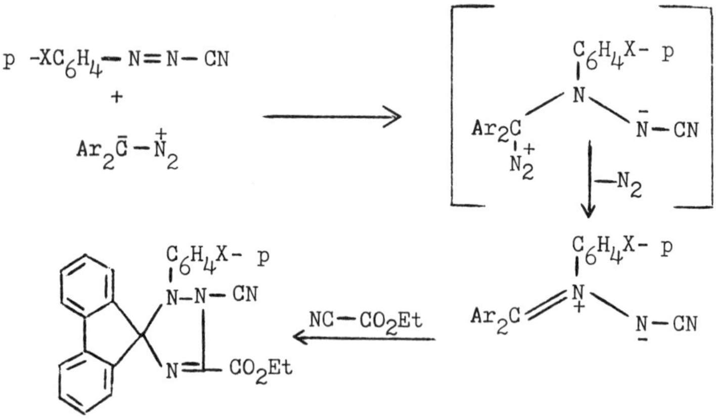

p -XC$_6$H$_4$—N=N—CN
+
Ar$_2\bar{C}$—$\overset{+}{N}_2$

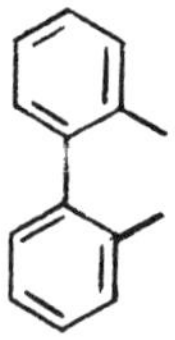

$\longrightarrow$

$\left[\begin{array}{c} \text{C}_6\text{H}_4\text{X-p} \\ | \\ \text{N} \\ \text{Ar}_2\text{C} \diagup \diagdown \bar{\text{N}}\text{—CN} \\ \overset{+}{|}\\ \text{N}_2 \end{array}\right] \downarrow -\text{N}_2$

(116) ← NC—CO$_2$Et ← Ar$_2$C=$\overset{\text{C}_6\text{H}_4\text{X-p}}{\overset{|}{\overset{+}{N}}}$—$\bar{\text{N}}$—CN

(116) [fluorene spiro structure with N–N—CN and N=C—CO$_2$Et]

(115)

a, Ar$_2$ = [o-tolyl biphenyl], X = Cl
b, Ar$_2$ = [o-tolyl biphenyl], X = I
c, Ar$_2$ = (C$_6$H$_5$)$_2$, X = Cl
d, Ar$_2$ = (C$_6$H$_5$)$_2$, X = NO$_2$

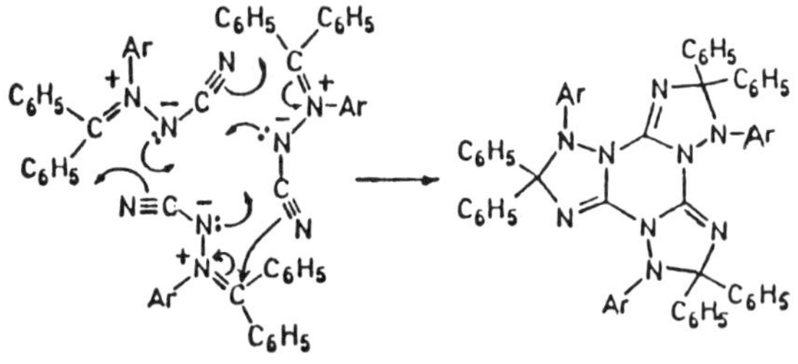

(115c) → (117)

Ar = p-Chlorophenyl

4. Reactions of Azomethine Ylides

Azomethine ylides, like azomethinimines, lack a double bond in the sextet structure but have internal octet stabilization and belong to the allyl type of dipoles (**121a** and **121b**). In terms of the PMO theory, azo-

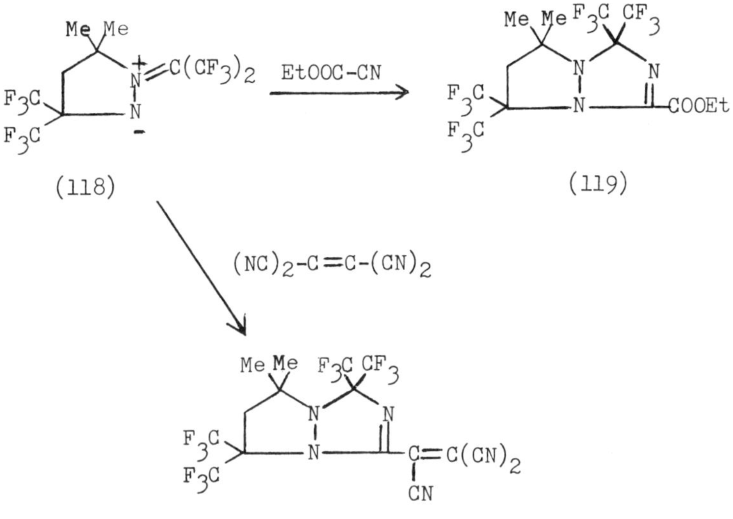

Sec. II.B] 1,2,4-TRIAZOLINES

(118) → (119)

(120)

(121) a ↔ b

methine ylides are type I dipoles and their cycloadditions are HOMO (dipole)–LUMO (dipolarophile) controlled; they are known to react most rapidly with electron-deficient alkenes (83BSB811; 84MI1).

The chemistry of these dipoles blossomed during the 1970s only after the discovery that thermal and photochemical cleavage of aziridines leads to azomethine ylides (76T2165; 79CR(C)265; 83BSB811; 84MI1). The possible utility of these dipoles in 1,2,4-triazoline synthesis has been demonstrated only recently employing heterocyclic ylides prepared by the dehydrohalogenation of immonium salts (84T369).

a. *3,4-Dihydroisoquinolinium N-Ylides.* The N≡N triple bond of benzenediazonium salts **122** acts as an electron-deficient dipolarophile in the cycloaddition reaction of 3,4-dihydroisoquinolinium N-ylides (**123**) (Scheme 39) (84TL65). The primary cycloadduct is regarded to be the unstable **124** (84TL65), based on the greater nucleophilicity of the exocyclic ylide carbon (84T369) as well as the regiochemistry of addition in **126**, which shows a triazolium proton in the 3-position.

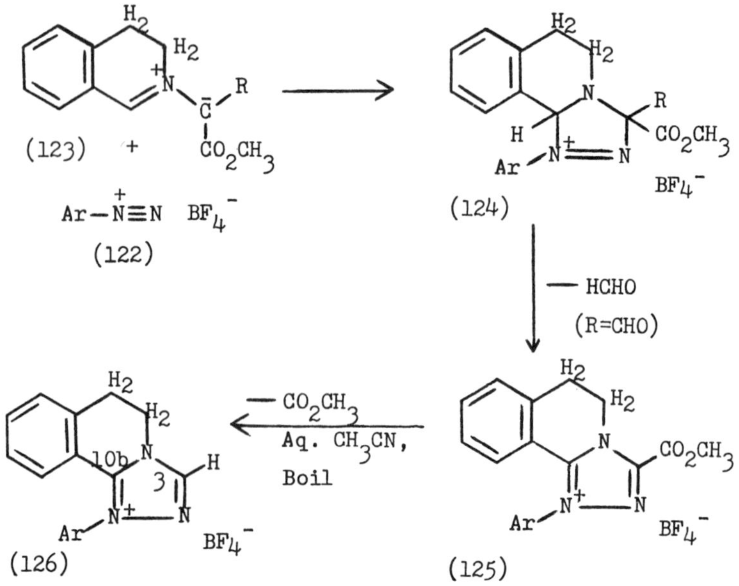

SCHEME 39

b. *1,2,4-Triazolium N-Ylides.*

Azomethine ylides that are incorporated into a triazole ring system yield bicyclic triazolines by ring annulation reactions with electron-deficient alkenes and alkynes.

Reaction of triazolium N-dicyanomethylides **127** (82JOC4409) with dimethylacetylenedicarboxylate affords cycloadducts **128**, which readily rearrange to yield triazolopyridines **129** (Section IV). Using unsymmetrical dipolarophiles, only one of the two possible regioisomers has been obtained. The results are in agreement with a second-order perturbational treatment of the orientation problem (83JCS(P2)1317). The nucleophilic extreme of the dipole is the carbon atom that carries the two cyano groups, an example of the usual polarization of azomethine ylides with electron-withdrawing substituents (73BSF2871) (Scheme 40).

Reaction of the azomethine ylide structural moiety in triazolium phenacylides is both regioselective (79RRC1053) and stereospecific (79RRC733), and yields isolable bicyclic triazolines **130** with unsymmetrically substituted ethylenes; none of the isomeric compounds **130a** is formed. The cycloadducts from alkynes, however, are unstable and undergo C—N bond fission to yield 5-substituted triazoles (Scheme 41) (81T2805, 81T2811).

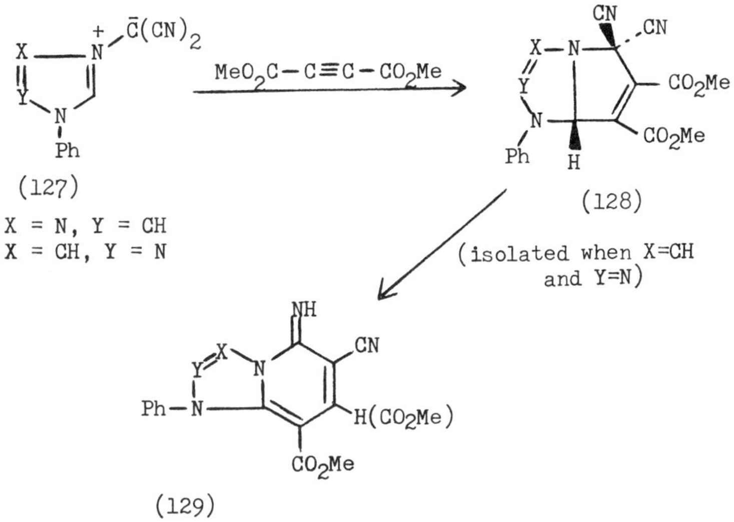

SCHEME 40

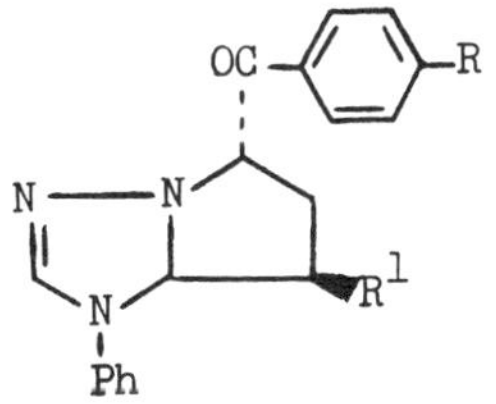

(127)
X = N, Y = CH
X = CH, Y = N

(128) (isolated when X=CH and Y=N)

(129)

(130) 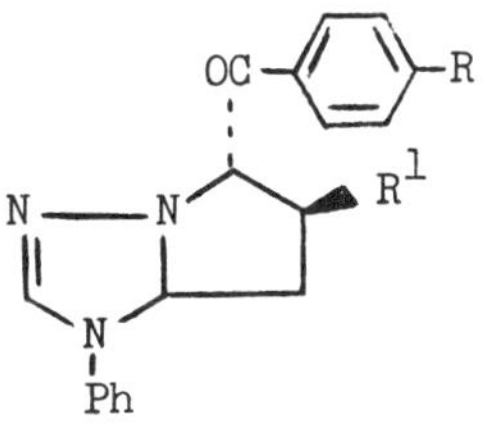   (130a)

5. Thiocarbonyl Ylides

Ring-fused mesoionic anhydro triazolium hydroxide systems **131** (84MI1) that incorporate the thiocarbonyl ylide dipole undergo cycloaddition with both olefinic and acetylenic dipolarophiles; the unstable 1 : 1 cycloadducts **132** and **133** yield ring-fused α-pyridinones **134** by elimination of hydrogen sulfide or sulfur, respectively (Scheme 42) (79JOC3803).

6. Diazoalkane Addition to Imines

The 1,3-dipolar cycloaddition of diazoalkanes to the carbon–nitrogen double bond of imines is regioselective and usually affords a single 1,2,3-triazoline adduct (61JOC2331; 66T2453; 75JHC143; 84AHC217). However, in one example of a vinylimine of hexafluoroacetone, substantial amounts of a 1,2,4-triazoline have been noticed (Scheme 43) (72LA9).

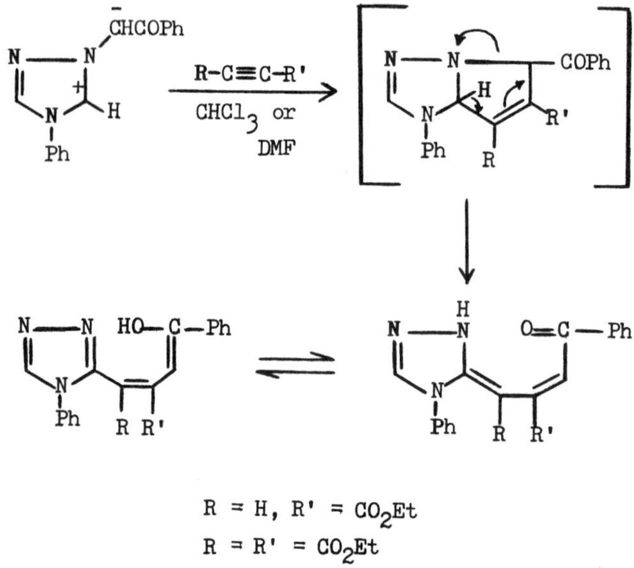

$R = H, R' = CO_2Et$
$R = R' = CO_2Et$

SCHEME 41

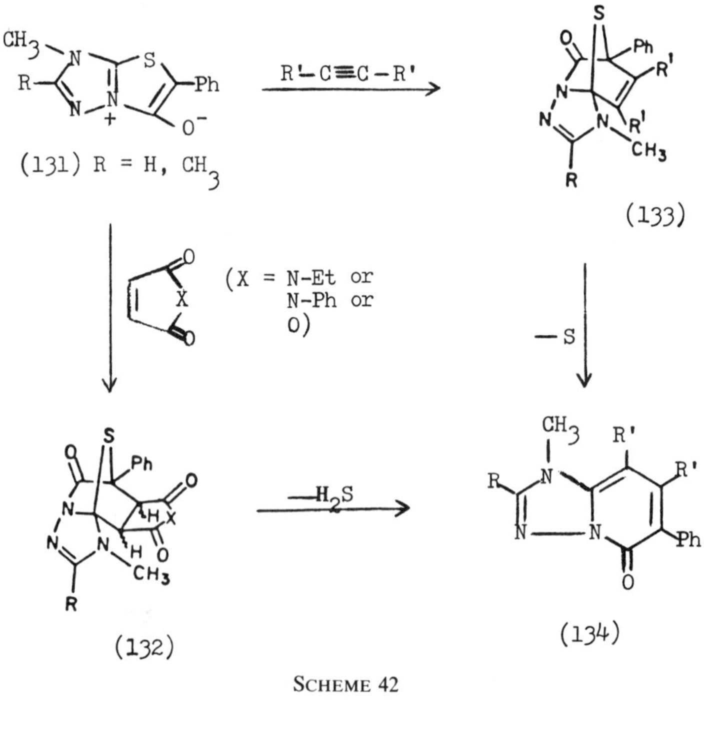

(131) $R = H, CH_3$

($X = $ N-Et or N-Ph or O)

(133)

(132)

$-H_2S$

(134)

SCHEME 42

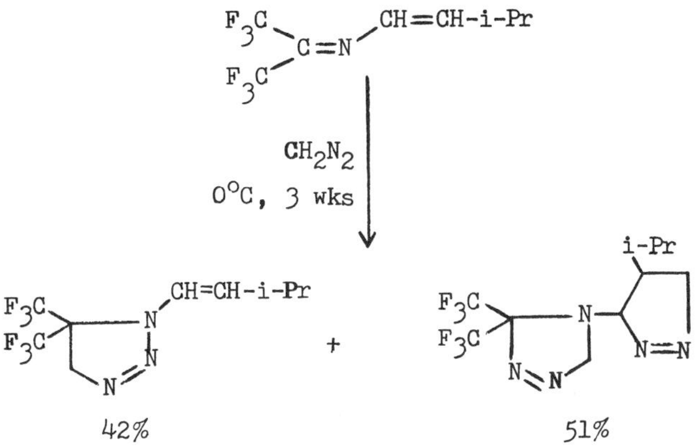

SCHEME 43

C. From Triazolium and Mesoionic Triazoles

1,2,4-Triazolium salts **135** may be represented as cationic centers, in which the positive charge is delocalized over the atoms of the ring (57QR15). The ionic centers in mesoionic compounds **136a**, on the other hand, are internally compensated; they can be named "anhydrobases" **136b** by the imaginary addition of a molecule of water to give triazolium hydroxides **137** (71JCS(B)1648). The mesoionic structures are deeply colored triazolium *C*-ylides of varying stability; because charge separation in organic molecules is generally unfavorable, the dipolar *C*-ylides **136a** would be less stable than the isomeric covalent anhydrobases or methylenetriazolines **136b**, and this is borne out by the calculated π-electron energies of both forms (71JCS(B)1648). Various methods have become available for conversion of triazolium salts and mesoionic compounds to triazolines, and this constitutes a major route for the synthesis of triazolines.

1. Sodium Borohydride Reduction of Triazolium Salts

Sodium borohydride reduction of triazolium salts **138** is well established as a useful approach to the synthesis of various tri- and tetra-substituted 1,2,4-triazolines (**141**) (Scheme 44) in good to excellent yield (Table I). The reduction mechanism involves selective nucleophilic attack by hydride ion, donated by sodium borohydride, at the electron-deficient C-5 carbon of the immonium moiety in the triazolium salt (73BCJ1250; 76JHC835, 76T2549). The selectivity of reduction is high; halogen substituents, nitro and ester groups, C=C, C=N, and even C=O double bonds all remain unchanged. Interchange of C-3 and C-5 substituents does not alter the site of hydride attack, as evidenced by the reduction of 5-methylthiotriazolium iodide **138a** to yield triazoline **141a** (76T2549).

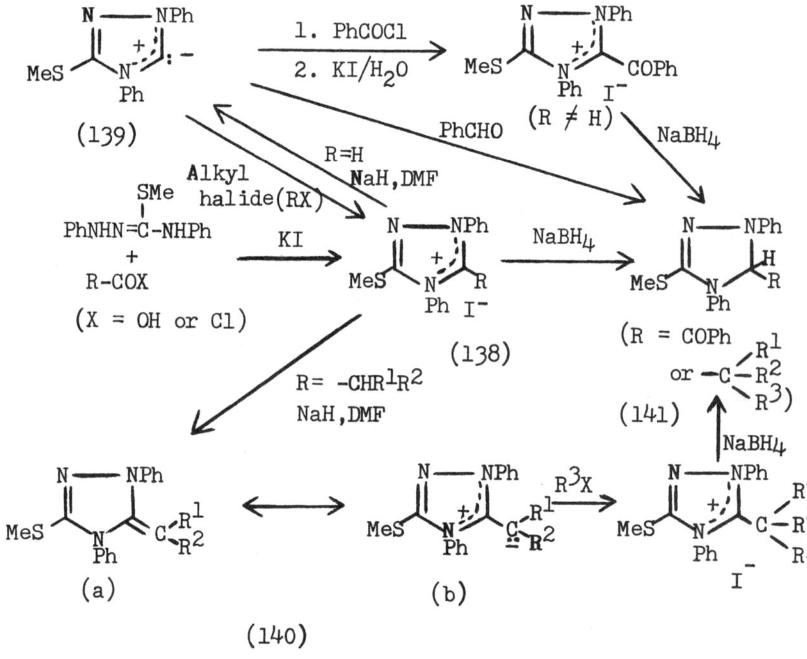

SCHEME 44

Convenient methods have been developed for the synthesis of triazolium salts. The 3-methylthio-substituted triazolium compounds **138** are readily accessible by the reaction of S-methyl-1,4-diphenylisothiosemicarbazide with acid chlorides, or (in the presence of phosphorus oxychloride) with carboxylic acids (74TL2649). Deprotonation of **138** (R = H) yields the nucleophilic carbene **139**, which is readily alkylated to yield

TABLE I
NaBH$_4$ Reduction of Triazolium Salts

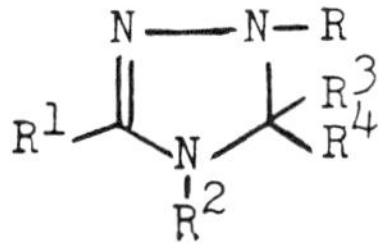

$$\underset{\underset{R^2}{|}}{N}\overset{N-R}{\underset{N}{\parallel}}\overset{R^3}{\underset{R^4}{\diagdown}}$$

R	R^1	R^2	R^3	R^4	Yield (%)	References
Ph	SMe	Ph	H	Me	94, 97	74TL2649, 76T2549
Ph	SMe	Ph	H	Et	90	75TL681, 76T2549
Ph	SMe	Ph	H	Pr	89	75TL681, 76T2549
Ph	SMe	Ph	H	t-Bu	76	75TL681, 76T2549
Ph	SMe	Ph	H	cyclo-Pr	98	75TL681, 76T2549
Ph	SMe	Ph	H	MeOOC(CH$_2$)$_4$-	85	76T2549
Ph	SMe	Ph	H	Ph	95, 99	74TL2649, 76T2549
Ph	SMe	Ph	H	4-ClC$_6$H$_4$-	99	76T2549
Ph	SMe	Ph	H	4-O$_2$NC$_6$H$_4$-	97	76T2549
Ph	SMe	Ph	H	4-MeOOCC$_6$H$_4$-	96, 95	76T2549, 74TL2649
Ph	SMe	Ph	H	3,4,5,-(MeO)$_3$C$_6$H$_2$-	90	76T2549
Ph	SMe	Ph	H	p-Phenylene	80	76T2549
Ph	SMe	Ph	H	PhCH$_2$-	98	76T2549
Ph	SMe	Ph	H	½ p-C$_6$H$_4$<	89	74TL2649
Ph	SMe	Ph	H	Ph—CH=CH—	95	74TL2649
Ph	SMe	Ph	H	PhCO-	39	75TL1889
Ph	SMe	Ph	H	3-Pyridyl	97	76T2549
Ph	CH$_3$	CH$_3$	H	CH$_3$	70	73BCJ1250
CH$_3$	CH$_3$	Ph	H	CH$_3$	86	73BCJ1250
Ph	Ph	Ph	H	Ph	97	69CB3176
Ph	Me	Ph	H	H	82	76T2549

138 (R ≠ H) (75TL1889). Anhydro bases **140a**, resulting from deprotonation of **138** (R = R^{1}R^{2}CH—), act as C-ylides (**140b**), and readily undergo C-alkylation to give **138** (R = —C—R^1) (75TL681) or **138** (R = long chain
$|\diagdown R^2$
R^3
alkyl groups) (Scheme 44) (80TL4183).

Triazolium tetrafluoroborates, obtained from the cycloaddition reac-

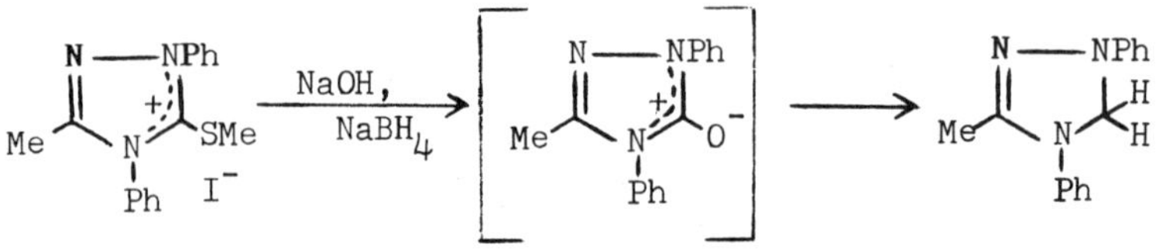

(138a) (141a)

tion of alkoxydiazenium salts with benzalanilines (69CB3176) and aldehydeazines (75JCS(P1)2474), respectively, undergo selective sodium borohydride reduction to yield **142** and **143** (Scheme 45); the methyleneaminotriazolines **143** are stable compounds and their stability, in comparison with that of the triazolium salts, is suggested to result from the conjugated azomethine bond (Scheme 45) (75JCS(P1)2474).

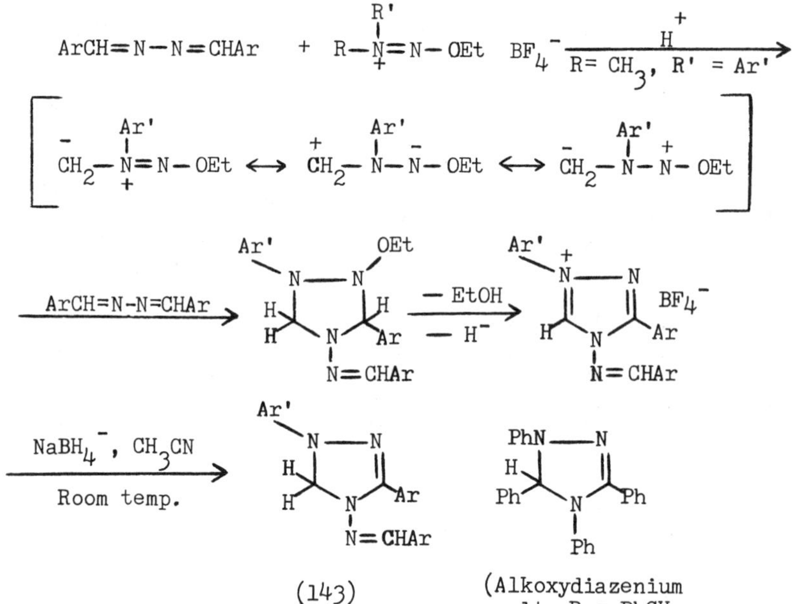

(143) (Alkoxydiazenium salt, R = $PhCH_2$, R' = Ph) (142)

SCHEME 45

Interaction of *o*-dialkylamino-substituted benzalaniline [**144**, RR = $(CH_2)_5$] with alkoxydiazenium salt [RR^1 = $(CH_2)_5$], followed by borohydride reduction, yields the rare 1,5-fused triazoline **145;** the course of reaction is not influenced by the *o*-dialkylamino group. However, with the more reactive diazenium compound (R = CH$_3$, R^1 = Ph), only tar is obtained (75JCS(P1)2474).

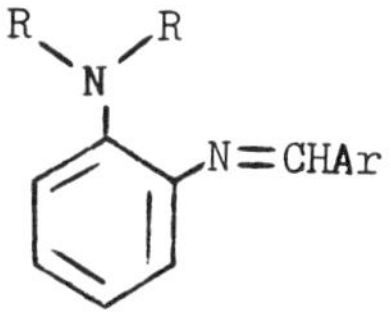

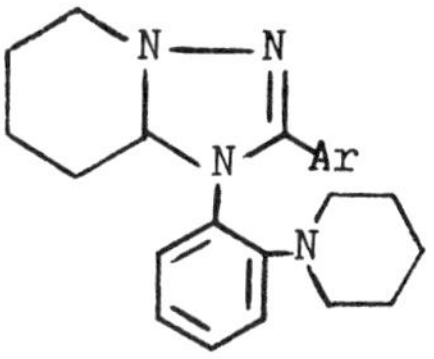

3,4-Annulated triazoline ring systems **146–148** (Scheme 46) are obtained by borohydride reduction of annulated triazolium salts. These salts result from cycloaddition of alkoxydiazenium compounds to the azomethine bond in heteroaromatic compounds such as pyridine, quinoline, and isoquinoline (69CB3176). A tetrahydrotriazoloisoquinoline (**149**) results from reduction of the triazolium salt **126** (Scheme 39) (84TL65); similarly, dihydrotriazolobenzodiazepine **18** results from reduction of the respective triazolobenzodiazepinium bromide (79JOC2688).

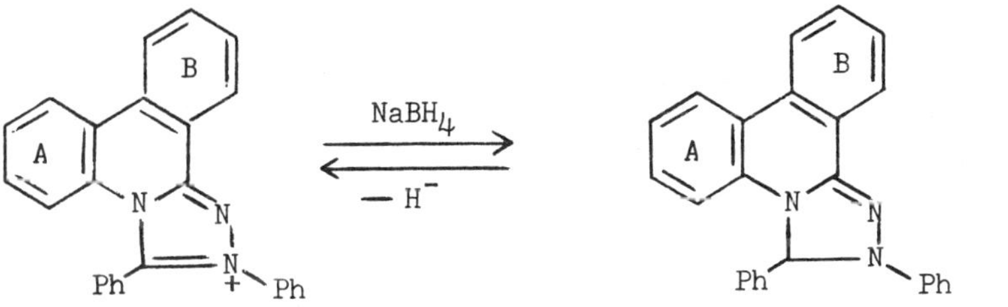

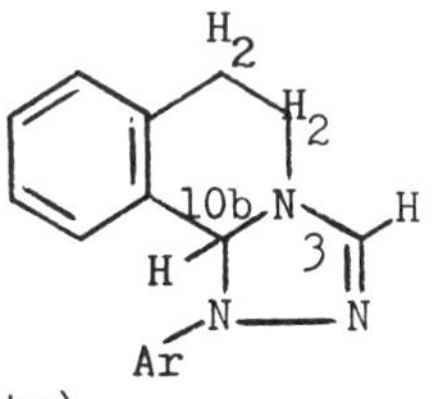

(146), Rings A and B absent (pyrido-)
(147), Ring B absent (quinolino-)
(148), Ring A absent (isoquinolino-)

SCHEME 46

Likewise, triazoloquinazolinium iodides, prepared from **150** as outlined in Scheme 47, lead to tetrahydrotriazoloquinazolines (e.g., **151**); the latter are generally difficult to isolate, probably because of the ring–chain tautomeric Schiff base structures **152** (76ACH419).

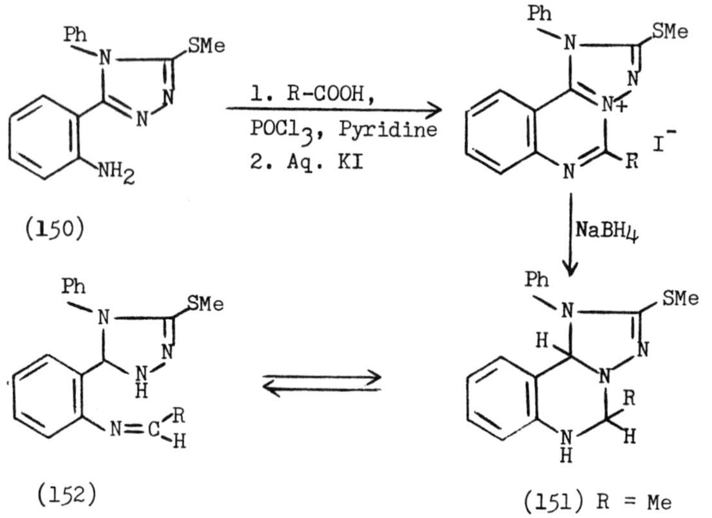

SCHEME 47

2. Reduction of Mesoionic Triazolium Compounds

1,2,4-Triazolium-3-aminides **153** belong to the general class of mesoionic compounds; they are reduced to 3-amino-1,2,4-triazolines by lithium aluminum hydride in hot dioxane (Scheme 48) (74JCS(P1)638).

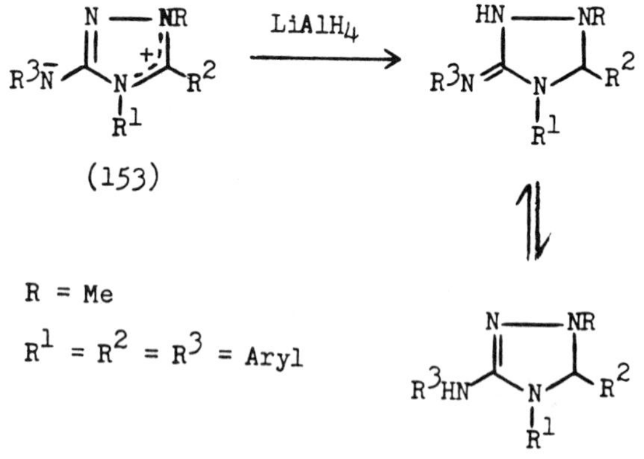

$R = Me$
$R^1 = R^2 = R^3 = Aryl$

SCHEME 48

3. *Reaction of Triazolium Salts with Nucleophiles*

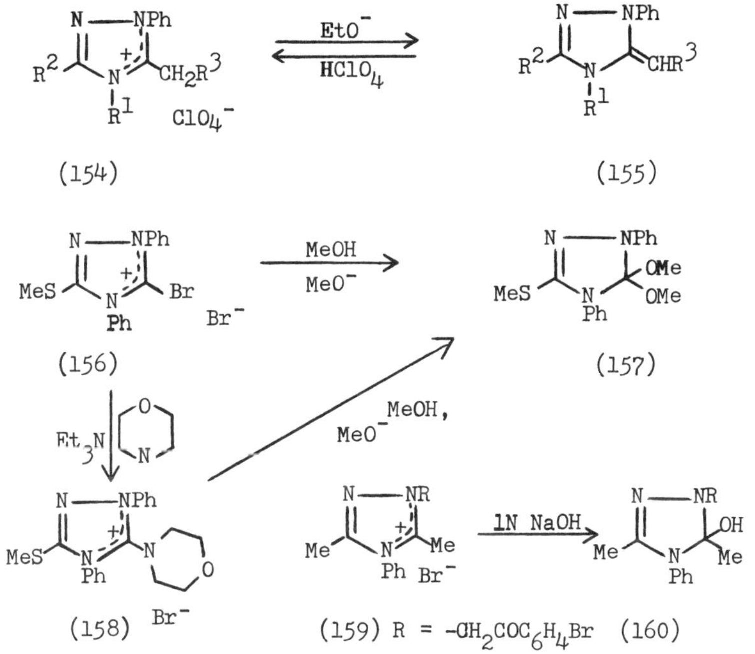

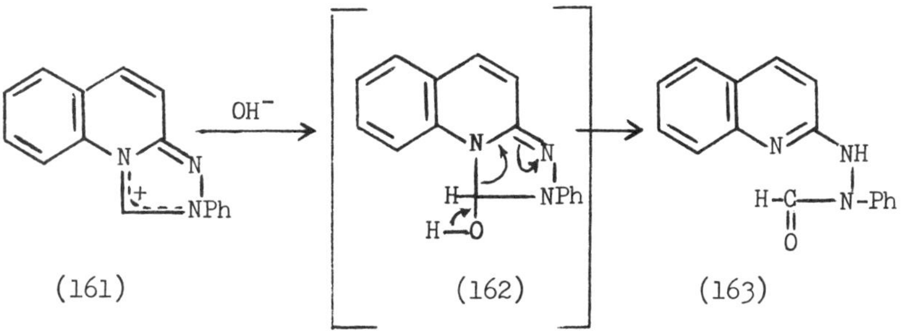

Triazolium salts are susceptible to attack by nucleophiles, similar to their reaction with hydride ions. When an activated 5-methyl function is present **(154)**, loss of a proton leads to 5-methylenetriazolines **155** (71JCS(B)1648). In the case of the reactive 5-bromotriazolium salt **156**, methoxide ion displaces the bromo group to yield the 5,5-dimethoxytriazoline **157**. The latter also results from the sodium methoxide-catalyzed methanolysis of the 5-morpholinotriazolium bromide **158** (80T1649). Likewise, the 5-methylthiotriazolium iodide **138a** furnishes the triazoline **141a**

(154) (155) (156) (157) (158) (159) R = $-CH_2COC_6H_4Br$ (160)

(161) (162) (163)

via the 5-olate compound when treated with sodium hydroxide and sodium borohydride (76T2549).

In the absence of an activating group, 5-hydroxytriazolines are formed, as illustrated by **159–160** (73S414) and **161–162** (69CB3176, 69G69). The hydroxytriazoline **162**, however, is unstable and undergoes bond cleavage and rearrangement to yield **163** (see Section IV,B,2).

4. Reaction of Triazolium Salts with Electron-Rich Multiple Bonds

Triazolium salts **164** react with ynamine to give products that are determined by the substitution pattern of the salts and the reaction conditions; [3 + 2]-cycloaddition and cycloreversion, as well as [4 + 2]-cycloaddition, are observed (83CB186). The diphenyl compound **164a** yields an un-

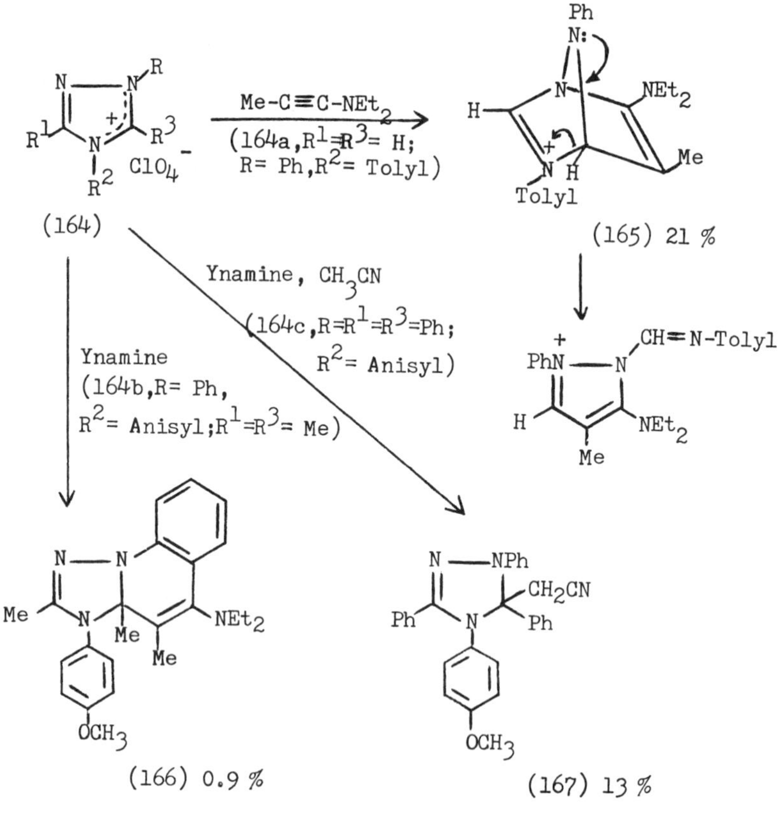

SCHEME 49

stable triazoline **165**, while a stable ring system (**166**) results from the diphenyldimethyl-substituted **164b**; ring closure in **166** does not occur on the methoxyphenyl group because of steric hindrance from the two adjacent methyl groups that disrupts its coplanarity. The fully phenyl-substituted compound **164c** undergoes cycloaddition–cycloreversion reactions in DMF, but yields triazoline **167** in acetonitrile (Scheme 49).

D. Other Routes to Triazoline Synthesis

Reduction of oxo- and thioxotriazolines has been investigated only to a limited extent. Reduction of triazolin-5-ones **168** with lithium aluminum hydride gives triazolines **169** when the ring double bond is not conjugated with the exocyclic double bond of the oxo function; the conjugated isomer **170** yields a triazolidin-5-one (**171**) (71BSF3296).

Similarly, triazolin-5-thione has been reduced using Raney nickel in ethanol (Scheme 50) (68RC247).

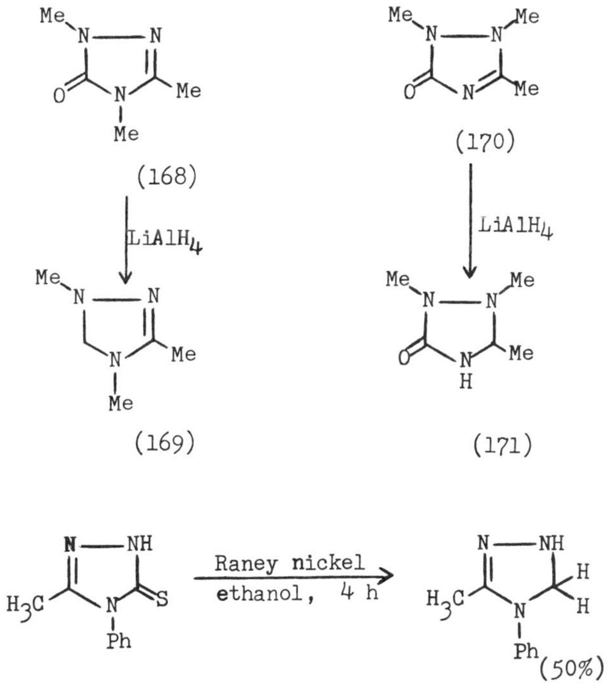

Scheme 50

E. Synthesis of Δ^1-1,2,4-Triazolines

In contrast to the various methods that have been developed for the synthesis of Δ^2- and Δ^3-triazolines, there appears to be no general synthetic routes to Δ^1-1,2,4-triazolines (**172**). Of late, the latter have generated interest as potential sources of azomethine ylides, in analogy with oxa- and thiadiazolines that yield carbonyl (thiocarbonyl) ylides upon thermolysis under mild conditions (Scheme 51) (76T2165; 85MI2).

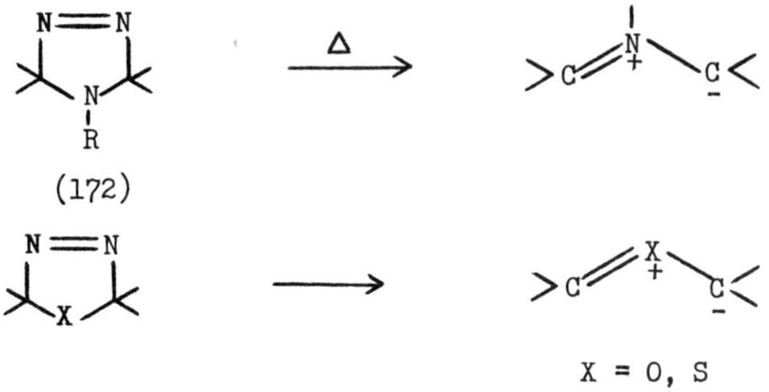

SCHEME 51

The only existing report on the synthesis of Δ^1-triazolines pertains to two patents (67USP3326889; 70USP3515603) that claim the formation of unusual perfluoroamino-1,2,4-triazolines as components of a mixture resulting from the fluorination of 3,6-diamino-s-tetrazine (Scheme 52). Numerous attempts to carry out the logical synthesis of the Δ^1-triazoline ring system as outlined in Scheme 53, using various aromatic and aliphatic amines, have resulted only in monosubstitution reactions without subsequent closure to the heterocycle (76T2165).

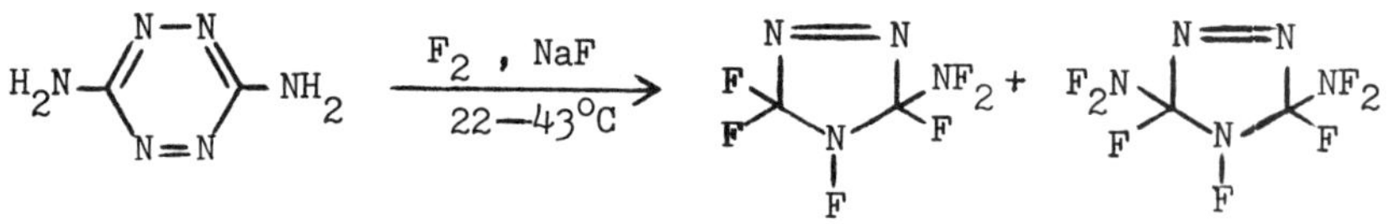

SCHEME 52

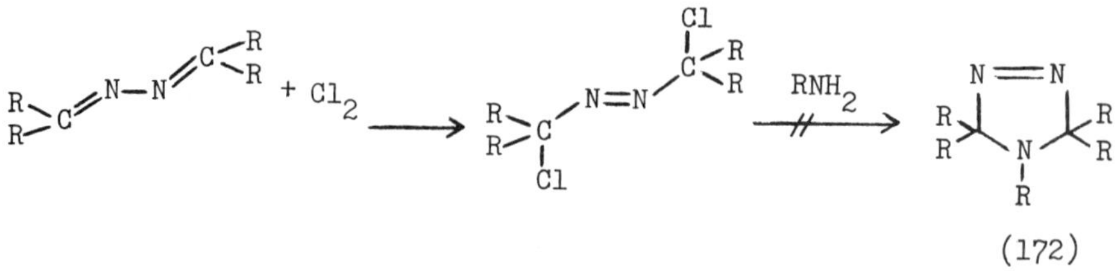

SCHEME 53

Only very recently has the accessibility of the first alkyl-substituted Δ^1-triazoline been reported, utilizing lead tetraacetate oxidation of an amidrazone (173) (85MI2). Oxidation in methylene chloride or benzene over solid potassium carbonate affords the 3-methylene-Δ^1-triazoline 174. The latter, when introduced into anhydrous methanol-d containing a trace of p-toluenesulfonic acid, equilibrates with the 3-methyl-3-methoxy-Δ^1-triazoline 175, as shown by NMR spectroscopy. The methylene protons of 174 are exchanged for ^{2}H, as are the corresponding protons of 175, as expected for an enamine system.

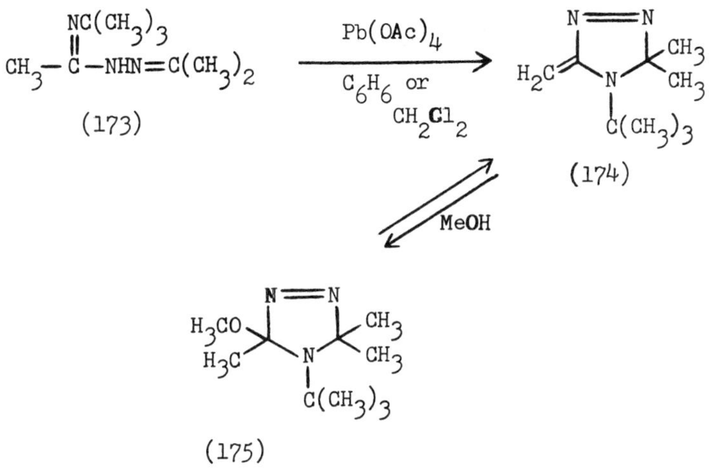

III. Structure and Physical Properties

A. STRUCTURE

1,2,4-Triazolines exist mainly in the 1H-form. The parent compound is not known.

Although the free bases do not seem to undergo tautomeric transformations, A ⇌ B, triazolinium salts are known to exhibit ring–chain tautomerism in solution (80KGS1138; 81ZOR1825; 82KGS1264, 82ZOR1613; 84KGS1415) (Section II,A,2), where the thermodynamic stabilities of the linear and cyclic forms are found to be very close (77LA463, 77LA485). The equilibrium and activation thermodynamic parameters for the ring–chain tautomerism of 2-(methyl)thiosemicarbazone S,S,S-trioxide (37) and 2-methyl-Δ^3-1,2,4-triazolinium-3-sulfonate (38) are presented in Table II.

TABLE II

ENERGY PARAMETERS FOR THE TAUTOMERISM 37 ⇌ 38

R^1	R^2	R^3	Solvent	Temp. (°C)	Conc. (mol %)	$\Delta G^\ddagger$ (kJ/mol)		E_A (kJ/mol)		$\Delta H^\ddagger$ (kJ/mol)		$\Delta S^\ddagger$ (J/Grad. mol)	
						37 38	38 37	37 38	38 37	37 38	38 37	37 38	38 37
H	Me	H	DMSO	28–50 (4)[a]	8	93.7	97.1	69.8	69.0	67.3	66.5	−86.2	−99.1
H	Et	H	DMSO	38 (1)	8	94.5	99.6	—	—	—	—	—	—
H	i-Pr	H	DMSO	29–53 (5)	8	94.5	99.9	87.8	99.6	85.3	97.0	−29.3	−9.2
H			Cl$_2$CHCOOH	39 (1)	8	97.1	97.1	—	—	—	—	—	—
Me	Me	H	DMSO	37 (1)	8	91.7	93.7	—	—	—	—	—	—
Me	Et	H	DMSO	37 (1)	8	92.1	95.8	—	—	—	—	—	—

[a] The number of experiments is given in parentheses.

B. Spectroscopic Properties

1. Nuclear Magnetic Resonance Spectra

Both ^{1}H- and ^{13}C-NMR spectra have proved to be powerful tools in structure determination and characterization of 1,2,4-triazolines. NMR spectral characteristics are used to distinguish unambiguously the linear or cyclic form (Scheme 4) of the reaction products from amidrazones and monocarbonyl compounds (Section II,A). The ^{1}H-NMR spectrum (81ZOR1825) of a freshly prepared solution of tetramethyltriazolinium iodide (**34B**, Scheme 18, $R^1 = R^2 = R^3 = R^4 = CH_3$) in DMF-$d_7$ exhibits three signals for the methyl groups at δ 3.35, 2.42, and 1.48 ppm with intensity ratio 3:3:6. The signals at δ 3.35 and 2.42 ppm clearly belong to the methyl groups at N-2 and C-3, respectively, while the two enantiotopic methyl groups at C-5 give a singlet signal in the upfield region with an intensity corresponding to 6H. A broad singlet at δ 6.87 ppm which integrates to 1H belongs to the proton at N-1. The signal for the proton at the positively charged N-4 atom is not localized at room temperature due to intermolecular exchange; however, at −40°C, a singlet appears at δ 10.46 ppm. A similar NMR pattern is observed for the triazoline free bases, both Δ^2-**1** (80ZOR942) and Δ^3-**36** (Scheme 18) (81ZOR1825) (Table III).

In the corresponding amidrazonium salt (**34A**, Scheme 18), on the other

TABLE III

^{1}H-NMR Spectra^a of Tri- and Tetrasubstituted Δ^2- and Δ^3-1,2,4-Triazolines

Compound	R^1	R^2	R^3		R^4	NH	Solvent
36a	2.05 (3H)	3.03 (3H)		1.40 (6H)		4.73 (1H)	DMSO-d_6
1a	1.78 (3H)	2.48 (3H)		1.23 (6H)		5.91 (1H)	CCl$_4$
1b	1.83 (3H)	2.53 (3H)		1.60 m (10H)		5.91 (1H)	CCl$_4$
1c	1.80 (3H)	2.48 (3H)	4.93 (1H)		7.20–7.60 m (5H)	5.31 (1H)	CDCl$_3$
1d	1.78 (3H)	0.93 t 1.80 m 2.55 t		1.21 (6H)		5.10 (1H)	CCl$_4$
1e	7.30–7.90 m (5H)	2.77 (3H)		1.34 (6H)		4.85 (1H)	CDCl$_3$

^{a}Chemical shift δ in ppm.

hand, the four methyl groups appear as four singlets of equal intensity at δ 3.43 (N^2—CH_3), 2.59 (C—CH_3), 2.13, and 2.06 ppm (*syn-* and *anti*-CH_3 groups in the alkylidene fragment). The diastereotopic protons at N^3 give two signals at δ 8.10 and 8.98 ppm, due to restricted rotation about the C=N^3 bond. The insignificant temperature dependence of these N^3-proton chemical shifts indicates a *(Z)*-configuration in relation to the C=N^1 bond for the linear tautomer (**34A**), stabilized by intramolecular hydrogen bonding (79ZOR2280; 81ZOR1825).

Significant differences also exist in the ^{13}C-NMR spectra of the linear (**34A** and **2**) and cyclic structures (**34B, 36,** and **1**) (79ZOR2280; 80ZOR942; 81ZOR1825; 82KGS1264; 84KGS1415). In the spectra of the hydrazones, there are two signals in the region 145–160 ppm, corresponding to the carbon atoms of the C=N bonds, while in the triazoline spectrum, only one signal is present in this region; the characteristic signal of C-5 lies upfield between δ 81 and 87 ppm (Table IV).

TABLE IV
^{13}C-NMR SPECTRA OF TRI- AND TETRA-SUBSTITUTED Δ^2- AND Δ^3-1,2,4-TRIAZOLINES[a]

Compound	C-3	C-5	C-R^1	C-R^2	C-R^3, R^4	Solvent	Refs.
1a	148.7 $^2J_{CCH}$ 7.7	81.2 $^2J_{CCH}$ 5.6	11.2 $^1J_{CH}$ 125.5	34.2 $^1J_{CH}$ 131.7	22.2 $^1J_{CH}$ 126.2 $^3J_{CCCH}$ 4.0	$CDCl_3$	80ZOR942
1b	148.2 $^2J_{CCH}$ 6.9	82.5 $^2J_{CCH}$ 5.5	11.4 $^1J_{CH}$ 127.6	34.7 $^1J_{CH}$ 133.9	31.9 24.3 22.0	$CDCl_3$	80ZOR942
1c	150.2 $^2J_{CCH}$ 7.9	87.3 $^1J_{CH}$ 141.7 $^3J_{CCCH}\approx{^3J_{CNCH}}\approx$ 5.0	11.3 $^1J_{CH}$ 128.2	41.0 $^1J_{CH}$ 133.9 $^3J_{CNCH}$ 6.9	139.2, 129.1, 127.9, 127.6, 127.4, 127.0	$CDCl_3$	80ZOR942
1d	148.2 $^2J_{CCH}$ 6.9	81.6 $^2J_{CCH}$ 6.0	11.8 $^1J_{CH}$ 130.1	49.8 21.6 10.8	22.9 $^1J_{CH}$ 126.0 $^3J_{CNCH}$ 3.8	$CDCl_3$	80ZOR942
1e	149.3 $^3J_{CCCH}$ 3.8	82.5 $^2J_{CCH}$ 5.0	125.0, 126.9, 127.9, 128.6, 128.7, 128.9	35.0 $^1J_{CH}$ 133.9	23.5 $^1J_{CH}$ 127.6 $^3J_{CCCH}$ 3.5	$CDCl_3$	80ZOR942
34Ba	182.4	77.3	11.5	35.2	26.3	DMSO-d_6	81ZOR1825
36a	163.2 $^2J_{CCH}$ 4.5	84.6 $^2J_{CCH}$ 3.0	14.5 $^1J_{CH}$ 126.0	39.3 $^1J_{CH}$ 134.8	29.6 $^1J_{CH}$ 126.0 $^3J_{CCCH}$ 3.0	DMSO-d_6	81ZOR1825

[a] Chemical shift δ in ppm and coupling constant *J* in Hz.

The distinct differences in the NMR pattern of the linear versus the cyclic forms provide a reliable criterion for structure determination. Based on their ^{1}H- and ^{13}C-NMR spectra, products of reactions of N^1-monosubstituted amidrazones with aldehydes and ketones are 1,2,4-triazolines (1), while the unsubstituted amidrazones form alkylidene derivatives (2) (80ZOR942). In light of these findings, the earlier assignment of triazoline structures to products resulting from unsubstituted amidrazones (70CRV151, 70JHC1001; 71JHC173; 73JHC353) should be reexamined.

NMR techniques are important diagnostic tools to follow ring–chain tautomeric transformations in solutions of triazolinium salts (**34A** ⇌ **34B** in Scheme 18 and **37** ⇌ **38** in Scheme 19), including the effect of structure on the position of the tautomeric equilibrium (Section II,A,2); such studies have helped to ascertain the principles of ring–chain tautomerism and an N^2 substituent is found to favor the cyclic form (77LA463, 77LA485; 81ZOR1825; 82KGS1264; 84KGS1415). Evidence for the establishment of a true tautomeric equilibrium in solution, for **34a** in Scheme 18, is provided by the presence of ^{13}C-NMR signals at δ 158.5 and 159.9 ppm (for C=N^1- and C=N^3- respectively, of the linear form) and at δ 77.3 and 182.4 ppm (for the respective C-5 and C-3 atoms of the cyclic triazoline) in the ^{13}C-NMR spectrum (81ZOR1825). Similar ^{1}H-NMR signals are seen for the ring–chain tautomerism of 2-methyl-Δ^3-1,2,4-triazolinium 3-sulfonates (**38**) with the thiosemicarbazone S,S,S-trioxides (**37**) in Scheme 19 (R^1 and R^2 at δ 1.83 and 2.05 for the open chain and at δ 1.34 for the cyclic) (77LA463).

The structural assignment of the dihydrotriazolopyrimidines **43** in Scheme 20, formed by cyclization of isothiosemicarbazones **39**, is confirmed by NMR spectroscopy (81BCJ1767, 81JOC3956; 85CPB2678). The H-2 and H-3 protons on the dihydrotriazole ring of **43** (R^2 = H) appear as two AB-type doublets (δ 6–8 ppm) with coupling constants ranging between 9.1 and 10.0 Hz. In the 3-deuterated compounds, the H-3 resonance disappears with the simultaneous conversion of the upfield H-2 signal from a doublet to a singlet. The chemical shift values for H-2 in triazolines **43** are ~2.0 ppm higher than those for the azomethine proton in **39**, reflecting rehybridization of the C-2 from sp^2 orbitals in **39** to sp^3 in **43**. Also, in line with the rehybridization of C-2, the 2-methyl protons (**43**, R^1 = CH_3, R^2 = H) resonate at 0.6–0.7 ppm higher than those of the corresponding compound **39**; furthermore, the anisotropic deshielding of phenyl protons ortho to the azomethine double bond in **39** (0.43 ppm in DMSO-d_6) disappears in **43** (R^1 = Ph, R^2 = H, R^4 = CN), exhibiting only a single signal for the phenyl protons at δ 7.36 ppm. The effect reappears in the triazoles **44**, in which the ortho protons of the 2-phenyl group are again deshielded by 0.49 ppm (CF$_3$COOH-d) relative to the meta and para protons, due to resonance interaction with the heteroaromatic ring system of **44**. The upfield shift of the H-7 proton resonance by 0.60–0.72 ppm in **43** relative to that in **44** is also characteristic of the triazoline structure.

The transition from the open to the cyclic structure in Scheme 16 is accompanied by a downfield shift of the aromatic protons and N—CH_3 signals, due to the deshielding effect of the positive charge, which is typical of cations. The N=CH_2 signal ($R^1 = R^2 = H$) undergoes the characteristic upfield shift and instead of two doublets [at δ 6.89 and 6.59 ppm for R^1 (Z) and R^2 (E)], a singlet signal (5.23 ppm) is obtained reflecting the C-sp^2 → C-sp^3 hybridization. Likewise, when $R^1 = R^2 = CH_3$, the two methyl signals (2.25 and 2.10 ppm) are transformed to a singlet in the upfield region (1.72 ppm). With progressive addition of weak acid, the proton signals of both the hydrazone and the cyclic cation are observed in the NMR spectrum (80TL209).

The characteristic upfield singlet signal for the geminal methyl groups at the sp^3-hybridized C-5 has been used widely in the assignment of cyclic structures for triazoline-1-mono- and -1,2-dicarboxylates (Scheme 32) (δ 1.75 and 1.63 ppm, respectively) (74HCA1382), 5,5-dimethoxytriazoline **(157)** (δ 3.20) (80T1649), 5,5-dimethyltriazolines (Scheme 22A) (δ 1.57–1.94) (86KGS352), dihydrotriazolopiperidones **(15)** (δ 1.08–1.11 ppm) (76CPB3011), 3,3-dimethyl-5-ethyloxalyl-2H-1,2,4-triazolo[3,4-b]thiazole oxime (Scheme 13) (δ 2.05) (78JHC401), 1-carbamyltriazolines **(47,** δ 1.61–1.72; **49,** δ 1.94–1.96; **49a,** δ 1.68–1.72), and triazoline-1,2-dicarboximides **(50,** δ 1.84; **51,** δ 1.68–1.75) (84JOC1703).

The 5-alkenyltriazoline structure (**62** in Scheme 27, R = R^2 = Ph, R^1 = t-Bu) has been confirmed from its ^{13}C-NMR, which gives a singlet at δ 153.0 for the C-3 and an upfield doublet at δ 78.4 for the C-5 carbons (84H549); likewise, the single upfield signal for the C-5 carbon has helped to ascribe the structures of the 5,5-dimethyltriazolines in Scheme 22A (δ 79.3–83.1 ppm) (86KGS352) and the dihydrotriazolobenzimidazoles **76** (δ 56.0 ppm) (85H2183).

Similar NMR techniques are used to identify structures **18** (79JOC2688), **6** (Scheme 8) (70T3069), and **101** (Scheme 34) (76TL1303). The regiochemistry of addition of 3,4-dihydroisoquinolinium N-ylide **(123)** to benzene diazonium salts (Scheme 39) is determined from the two singlets at δ 6.76 and 6.18 ppm for the H-3 and H-10b of the reduction product **(149)** of **126**; the opposite regioisomer would have given a single signal at higher field for the 3-methylene protons (84TL65).

NMR measurements are used for structural as well as conformational analysis of complex tri- and tetracyclic fused triazoline ring systems. The ^{1}H- and ^{13}C-NMR spectra of angularly condensed triazolobenzothiazine derivatives **91** and **92** show that these compounds exist in two conformational types, depending on the substituents (84H537, 84OMR720).

The 7- and 8-methoxy groups in **92** yield two singlets in the intervals δ 3.74–3.80 and 3.23–3.37 ppm, respectively, for the 3-ethoxycarbonyl and carboxamido compounds; the 3-aryl analogues of **92** as well as all compounds with no 10-methyl substituent (**91**) show the singlets between 3.57 and 3.97 ppm. A strikingly similar pattern is also observed for the H-6 and H-9 protons in **92**; when R^1 = CO_2Et or CONHPh, the two singlets appear at δ 6.49–6.54 and 5.93–6.07 ppm, respectively, but the signals are present in the region 6.62–7.36 ppm for the 3-aryl compounds (**92**) and for all compounds **91**.

The spectral data clearly show that the H-9 and 8-OMe protons in the 3-ethoxycarbonyl- and 3-carboxamido-10-methyl compounds (**92**) are more shielded than in the others; the average values of the measured differences for H-9 and 8-OMe are 0.84 and 0.44 ppm, respectively, compared to 0.32 and 0.11 ppm for the H-6 and 7-OMe protons. This is attributed to a probable difference in the conformation of the 3-ethoxycarbonyl/carboxamido compounds (**92**) (conformation **A**) relative to the others (conformation **B**) (Fig. 1). In **A**, the 10-methyl-3-ethoxycarbonyl derivatives are *cis*-annulated, conformationally flexible systems where H-9 and 8-OMe of the benzothiazine skeleton lie close to the 1-aryl ring of the 1,2,4-triazoline nucleus, in the preferred conformer. By assuming this more strained conformation, the molecules avoid the steric hindrance of the 10-methyl and 1-aryl groups, the shortest distance between the two groups being 2.4 Å in **A** and 1.2 Å in **B**. All compounds unsubstituted on C-10 (**91**) and the 3-aryl-substituted 10-methyl analogues (**92**) possibly occur in the **B** conformation, in which the 1-aryl ring is well removed from the H-9 and 8-OMe groups.

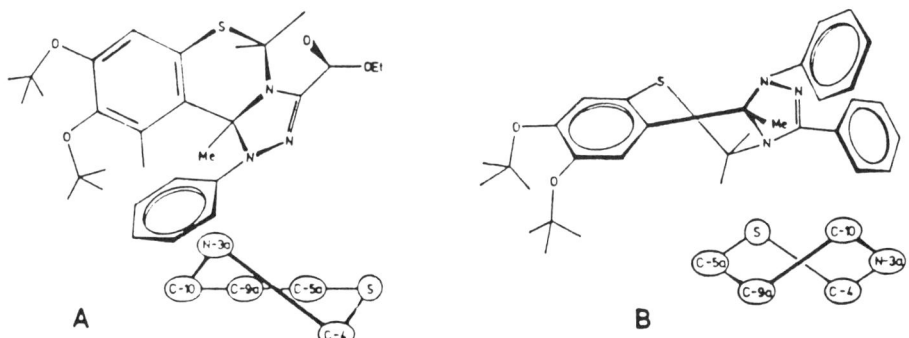

FIG. 1. Probable conformations for (**A**) 3-ethoxycarbonyl-10-methyl- and (**B**) 3-phenyl-10-methyldihydrotriazolobenzothiazines **92** (84H537, 84OMR720).

The nonequivalent protons of the 4-methylene group in **91** and **92** give an AB multiplet with chemical shifts at δ 4.05–4.80 and 4.40–5.80 ppm, respectively, and J_{AB} = 11.5–13.5 Hz. The average chemical shift difference of 1.02 ppm for the 4-methylene protons ($\Delta\delta_{4e',a'}$) in the 3-ethoxycarbonyl (carboxamido) compounds **92** compared to only 0.26 ppm in all other compounds is also commensurate with conformation **A**; the shielding of H-4e', coplanar with the carbonyl group in **A**, is diminished both by the hydrogen bonding and by the anisotropic effect of the carbonyl group.

The H-10 and 10-Me protons in compounds **91** and the 3-aryl compounds **92** are more shielded in the 1-phenyl derivatives than in the *p*-nitrophenyl analogues; this effect is not present in the 3-ethoxycarbonyl compounds **92,** in which the opposite anisotropic effect is observed. This indicates that in **A** the 1-aryl group is roughly coplanar with the 10-Me group, whereas in **B** the latter is situated above the plane of the aromatic ring.

Conformations **A** and **B** are also supported by ^{13}C-NMR data. The strained conformation **A** of the more crowded *cis*-annulated structure is indicated by the increased shielding of the C-5a (3.4–5.1 ppm), C-6 (1.3–2.5 ppm), C-8 (0.6–2.5 ppm), and C-9 (0.6–2.8 ppm) carbon atoms in the 3-ethoxycarbonyl-10-methyl compounds **92** compared to that in the 3-aryl-10-methyl analogues **92** and compounds unsubstituted on C-10 (**91**); this is also true of the field effect or steric compression shift in the shielding of carbon atoms bearing sterically hindered substituents. On the other hand, the signal for C-10 bonded to the unhindered methyl group in **A** shows a significant downfield shift of 3.6 ppm, unlike that in **B**.

The structures of the tetracyclic norbornane-fused oxazino-1,2,4-triazolines **86–88** have been proved by ^{1}H- and ^{13}C-NMR spectroscopic measurements and the various potentially stable conformations of the flexible hetero rings have been evaluated (87MRC635). Starting with the preferred conformations of the oxazine rings as deduced from the coupling constants between the H-4a and the 4-methylene protons, and discarding all structures that can be ruled out for steric reasons including those resulting from the *trans*-annulation of the hetero rings, in each case only two diastereomeric structures (**86a** and **b**; **87c** and **d**; **88e** and **f**), differing in the steric position of the aryl group attached to C-2 (to facilitate comparison of the NMR data, C-2 denotes the carbon atom between the oxygen and nitrogen atoms), remain to be distinguished.

In structure **86a,** the dihedral angles between H-4a and the 4-CH$_2$ protons are both close to 60°, which is in disagreement with the differing values of the corresponding coupling constants ($J_{4(e'),4a}$ = 1 and $J_{4'(a'),4a}$ = 5 Hz). There is also considerable steric hindrance between H-7 (endo)

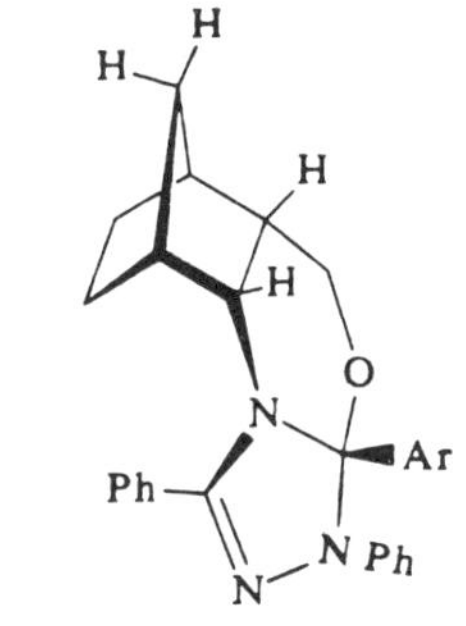

(86a)   (86 b)

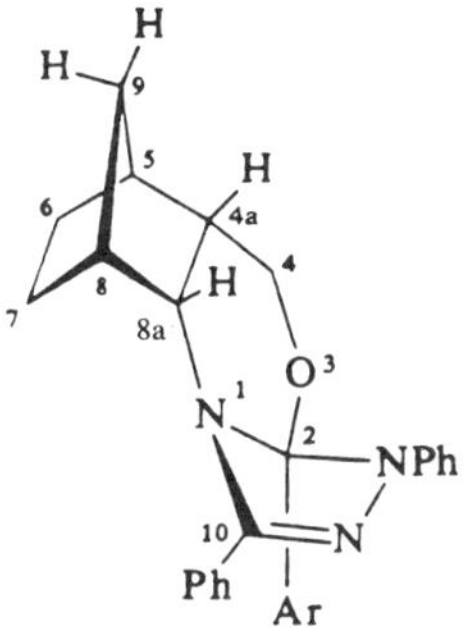

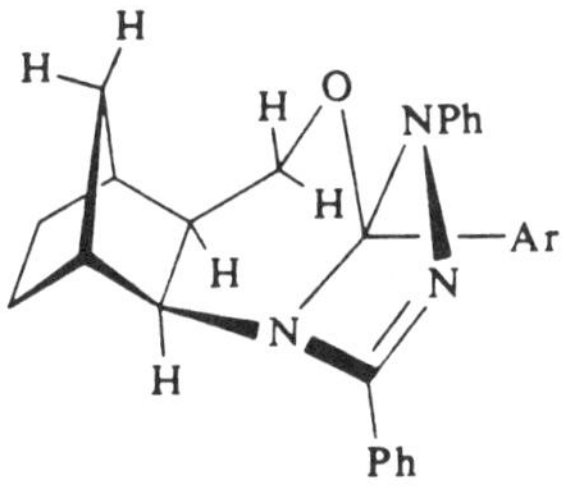

(87 d)   (87 c)

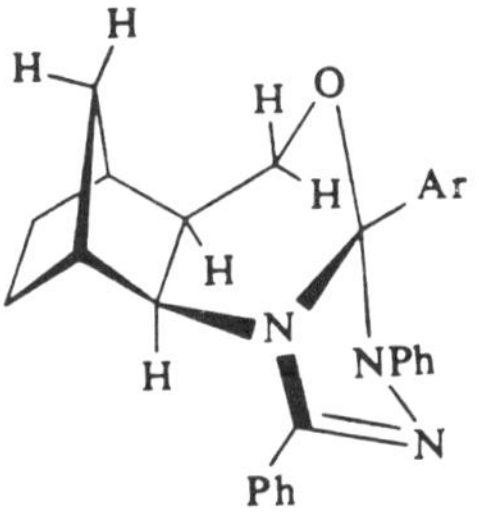

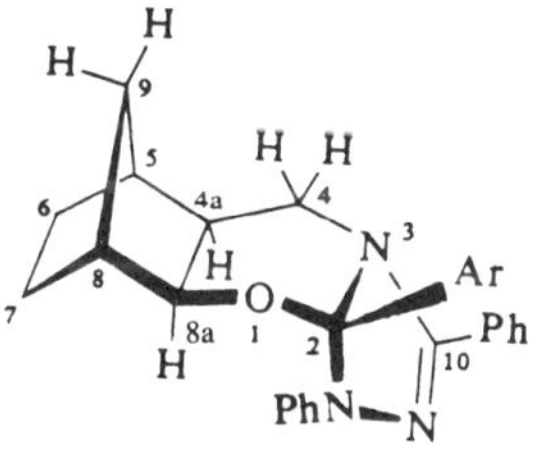

(88 e)   (88 f)

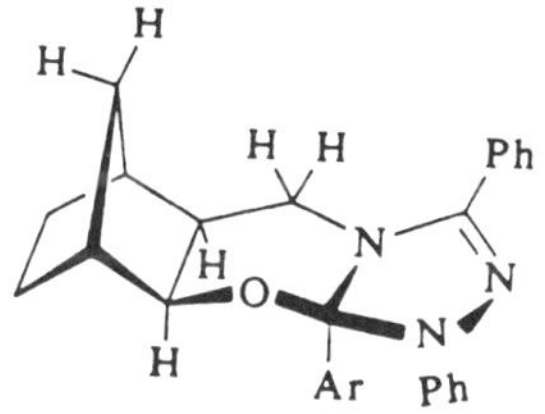

Ar = C_6H_4Cl-p

and the phenyl-substituted nitrogen atom of the triazoline ring. The more probable structure is **86b**; C-2 has the R configuration, indicating *cis*-annulation of the hetero rings and the endo position of the C-2 aryl located trans to H-4a, -8a. In this conformationally flexible stereostructure, there is no appreciable steric hindrance and, in the chair-like conformation of the oxazine ring, the dihedral angles between H-4a and the 4-CH_2 protons are nearly 30 and 90°, which are in excellent agreement with the coupling constants.

The observed ^{1}H-NMR signals for **86**, which hardly differ from those of the dipolarophile **83**, also support the **b** structure. The anomalous relative shielding of the 4-CH$_2$ protons [δ H-4'(a') (4.31 ppm) > δ H-4(e') (3.98 ppm)] in **86**, along with the doubling of the shift difference compared to **83**, can be explained by the field effect due to the 1,3-diaxial heteroatoms. In the ^{13}C-NMR spectrum, the C-4 signal shows an upfield shift difference of 6.2 ppm; in addition, the downfield shift of the C-5, C-6, C-7, C-8a signals and the upfield shift of the C-8 and C-9 signals by only a few tenths of a ppm in **86** relative to **83** indicates a nonstrained system. In the NMR spectra of **86a,** a steric compression shift would have been observed giving an upfield shift of the C-7 signal by several ppm.

DNOE measurements further support structure **b**. Saturation of the H-4(e') and H-4'(a') multiplets does not cause any increase in the signal intensities of the ortho hydrogen atoms of the aryl group, providing evidence of their distant mutual positions.

Similarly, in the case of **87**, the coupling constants of H-4a and the 4-CH$_2$ protons (4.2 and <1 Hz, respectively) indicate the conformation of the oxazine ring as similar to that in **86**. In **87d,** the aryl group on C-2 is in the cis (endo) position to H-4a,-8a; thus the probable configuration of C-2 is *R*. If it were reversed (structure **c**), a significant shift difference of the H-4'(a') signal would result, against the measured difference of zero. In accordance with configuration **d,** there is also a considerable downfield shift of the H-9' (endo) doublet in **87** compared to **84** (Δδ 1.45 ppm), resulting from the anisotropic effect of the nearby heteroatom. Likewise, the upfield shift of H-4'(a') (0.07 ppm) compared to **84** is caused by the 1,3-diaxial position of this proton and the aryl group.

Based on similar reasoning, the oxazine ring in **88e** has a boat conformation with the O-1 and C-4 atoms in exo positions; the dihedral angles between the 4-CH$_2$ protons and H-4a (180 and 60°) are in agreement with the corresponding coupling constants (12.2 and 7.1 Hz, respectively). The upfield shift of the H-8a signal by 0.35 ppm in **88** relative to **85** is also indicative of the **e** structure, where C-2 has the *R* configuration and the H-8a is appreciably shielded by the N-phenyl ring. The field effects in the ^{13}C-NMR, showing high upfield shifts of the C-8a and C-4a signals compared to **85** (7.2 and 5 ppm, respectively), again provide evidence for stereostructure **e**; in the latter there is strong steric hindrance between H-4a, -8a, and the triazoline ring, whereas in structure **f** the aryl groups can rotate about the C-2—C-Ar bond to avoid steric compression. DNOE measurements point to the remote position of H-8a relative to the C-2 aryl ring as in **e**. The signals of the ortho aryl protons are not sensitive to saturation of the H-8a doublet; in contrast, a significant increase of the

TABLE V

PROTON NMR DATA FOR TRIAZOLOBENZODIAZEPINES **94**[a]

X	Ar	R	H$_a$	H$_b$	H$_c$	H$_d$
H$_2$	Ph	H	δ3.14 J_{ab} −13.0 J_{ac} 7.8 J_{ad} 5.1	δ3.17 J_{bc} 4.2 J_{bd} 2.1	δ3.62 J_{cd} −12.4	δ 3.68
H$_2$	Ph	CH$_3$	δ2.85 J_{ab} −13.0 J_{ac} 8.7 J_{ad} 2.5	δ2.65 J_{bc} 5.8 J_{bd} 3.3	δ3.35 J_{cd} −15.3	δ 3.49
H$_2$	p−NO$_2$Ph	CH$_3$	δ2.75 J_{ab} −12.4 J_{ac} 5.6 J_{ad} 3.1	δ2.89 J_{bc} 4.1 J_{bd} 3.1	δ3.38 J_{cd} −13.4	δ 3.49
O	p−NO$_2$Ph	H	—	—	δ3.94 J_{cd} −15.2	δ 4.05
O	p−NO$_2$Ph	CH$_3$	—	—	δ3.46 J_{cd} −12.5	δ 4.15

[a] Chemical shifts δ in ppm and coupling constants J in Hz. From (85H2051).

intensity should occur in **f**, as the minimum distance between H-8a and the ortho hydrogens would be only 0.5 Å in **f** (87MRC635, 87T1931).

The structure and conformation of tetrahydro-1*H*-s-triazolo[4,3-*d*]-1,4-benzodiazepines and -azepinones (**94**, X = H$_2$ or O) have been distinguished by comparison of their NMR spectra with that of the starting 2,3-dihydro-1*H*-1,4-benzodiazepines (**93**) (85H2051). Annulation of the triazoline ring has been found to exert a profound influence on the conformational mobility of the diazepine ring system. The starting compounds (**93**) exist in CDCl$_3$ solution as two pseudo-boat conformers that are rapidly interconverting at room temperature; the alicyclic region in the ^{1}H-NMR spectrum is characterized by an AA'BB' system nearly symmetrical about its middle point. The ethylene protons in the triazolobenzodiazepines **94**, on the other hand, are nonequivalent and resonate as an ABCD spin system giving rise to two complex unsymmetrical groups of peaks (Table V); these spectral features suggest that compounds **94** exist in solution as one conformer that does not interconvert at room temperature. Using temperature-dependent NMR analysis, the ring-inversion barriers have been calculated to be 18.7 and 19.4 kcal/mol, respectively, for compounds **94** when X = H$_2$ (R = CH$_3$) and X = O (R = H). The 11b-phenyl group in **94** may be responsible for the blocked configuration of the seven-membered ring at room temperature (85H2051).

Similarly, the tetrahydrotriazolo-1,5-benzodiazepines **97** and **98** have been identified by comparison of their NMR data with those of the benzodiazepine compounds **95** and **96**. The proton resonance of the —N=C(CH$_3$)— moiety in **95** and **96** is shifted to higher field for the saturated N—C(CH$_3$) moiety in **97** and **98**; an analogous shift of the C-4 resonance in **95** for the C-3a resonance in **97** is also observed due to saturation of the C=N double bond. In addition, although the signals of the methylene protons in **95** and **96** appear as a singlet, they form an AB system in the spectra of **97** and **98**. Also, the single sharp peak between δ 1.34 and 1.33 ppm signalling the geminal methyl groups in **95** (X = 2CH$_3$) appears as two distinct singlets for **97** in the regions 1.38–1.46 and 1.13–1.21 ppm, respectively. These NMR data are indicative of a fixed conformation for the annulated products **97** and **98**; temperature-dependent ^{1}H-NMR analysis in the range −88 to +150°C indicates no change in the conformation of the seven-membered ring (86S230).

The NMR spectra of the tricyclic triazolines **72** and **73** show the pyrrolidine protons as an ABMPX system, which can be analyzed on an approximate first-order basis (Table VI). The comparable chemical shift values of H$_A$, H$_B$, and H$_P$ indicate the presence of the Z moiety in both compounds (Fig. 2). Furthermore, in **72**, the ortho protons of the phenyl group at N-5 are shifted downfield due to the deshielding effect of the carbonyl group at C-3. This effect, absent in **73**, along with the chemical shifts of

TABLE VI

PROTON NMR DATA FOR TRIAZOLINES **72** AND **73**[a]

Compound	δ (ppm)				
	H$_A$	H$_B$	H$_M$	H$_P$	H$_X$
72	1.98 ddd	2.60 ddd	3.70 td	5.46 dd	7.12
73	2.00 ddd	2.62 dd	4.55 dd	5.50 dd	6.70 d

	Coupling Constants (Hz)					
	J_{AB}	J_{AM}	J_{AP}	J_{BM}	J_{BP}	J_{MX}
72	13.5	10.1	9.0	1.2	5.2	10.1
73	13.0	7.8	8.6	0	5.2	10.1

[a] Data from (78JHC1485).

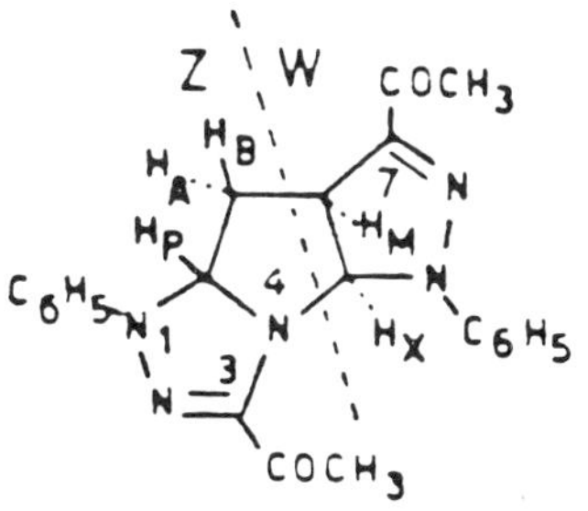

Sec. III.B] 1,2,4-TRIAZOLINES 247

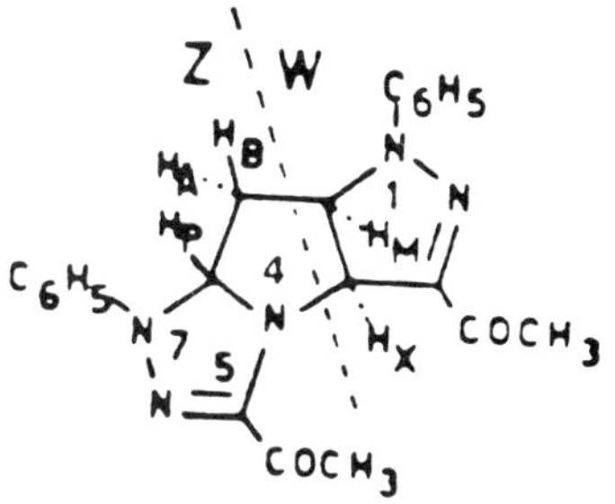

FIG. 2. Structure of tricyclic triazolines 72 and 73, based on proton NMR data.

H_M and H_X at higher and lower field, respectively, compared to that in 72, provides evidence for the junction of the two heterocyclic rings in the W moiety (78JHC1485).

2. Ultraviolet and Infrared Spectra

Tetrasubstituted triazolines bearing ethoxycarbonyl (Scheme 32) (74HCA1382), carbamyl or thiocarbamyl (Scheme 22A) (86KGS352), and aryl or alkyl (53) (64CB1085) substituent groups absorb in the UV region 359–228 nm (log ϵ = 3.44–4.29), with the 1,3,4,5-tetraphenyltriazoline showing an absorption maximum at 359 nm (64CB1085). The characteristic UV spectra of the tetraaryl compounds are used to distinguish these from the triaryl-5-triazolinones; replacement of the 5-aryl with an oxo group significantly lowers the absorption maximum from 359 to 275 nm (64CB1085).

Unlike NMR, UV spectroscopic measurements have been used only to a very limited extent for structure identification purposes. Consequently, relevant data are not available for the evaluation of substituent effects on the triazoline chromophore. However, UV data on certain annulated triazoline ring systems indicate that both the position and intensity of the absorption maximum can be affected by annulation; for example, dihydrotriazolobenzodiazepines 18, λ_{max} = 350–214 nm (log ϵ = 1.8–3.4) (79JOC2688), dihydrotriazolopyridines 25, λ_{max} = 450–430 nm (log ϵ = 1.1–2.7), and dihydrotriazoloquinolines 26, λ_{max} = 463–440 nm (log ϵ = 7.3–9.2) (67TL4337). Similarly, the metal complex resulting from [SbCl$_6$]$^-$ and (1,4-dimethyl-1,2,4-triazolin-3-yl)-4-azo-N,N-diethylaniline has a maximum absorption at 546–554 nm (70ZAK500).

IR spectra of triazolines show characteristic absorption bands for the ring C=N; some selected examples include tri- (**60**) and tetraaryltriazolines (**53**) (1550–1570 cm^{-1}) (64CB1085; 65CB642), 1-, 1,2-, and 3-alkoxycarbonyltriazolines (Schemes 32 and 33) (1530–1645 cm^{-1}) (73CB3421; 74HCA1382; 83T129), and tetrahydrotriazolobenzodiazepines (**94, 97,** and **98**) (1580–1605 cm^{-1}) (85H2051; 86S230). The NH stretching vibrations of a ring NH (**60**, Schemes 21, 22, and 32) appear between 3195 and 3590 cm^{-1} (65CB642; 74HCA1382; 78JHC401; 84JOC1703; 85H2051); its absence at 3370 cm^{-1} has been used to assign the structure for 4-benzylideneaminotriazoline **58** (84ZOR659).

Carbamyl- and thiocarbamyl-substituted triazolines (Schemes 21, 22, and 22A) display both the NH/NH$_2$ and CO bands of the substituent in the ranges 3330–3430 and 1630–1670 cm^{-1}, respectively (84JOC1703; 86KGS352). Similarly, the alkoxycarbonyl C=O (Schemes 32 and 33) gives absorption peaks in the frequency range 1670–1775 cm^{-1} (73CB3421; 74HCA1382; 83T129). The triazolo[3,4-*b*]thiazole oxime (Scheme 13) gives, in addition to the NH peak at 3330 cm^{-1}, a very broad band between 3200 and 3780 cm^{-1} for the N—OH group, indicating the existence of hydrogen bonding in the compound (78JHC401). Likewise, the CN substituent in the dihydrotriazolopyrimidines **43** (Scheme 20) can be clearly detected by its absorption band at 2220 cm^{-1} (81BCJ1767). In the case of 5-alkenyltriazolines **62**, the presence of the trans CH=CH bond is indicated by a set of absorption bands between 877 and 975 cm^{-1}, the one at 975 cm^{-1} being the strongest (69CC387).

3. *Mass Spectral Fragmentations*

Unlike 1,2,3-triazolines (84AHC217), most 1,2,4-triazolines show molecular ion peaks in their mass spectra (MS) with varying peak intensity. This property has enabled molecular weight determinations of a wide range of triazolines, as exemplified by 5-(2-pyridyl)aminotriazoline (**6**) (M$^+$, 240) (70T3069), dihydrotriazoloquinoline **167** (M$^+$, 390, 2%) (83CB186), triazolothiazole oxime in Scheme 13 (M$^+$, 270) (78JHC401), 4-benzylideneaminotriazolines **61** (M$^+$, 362–442) (85H1123), 5-styryltriazoline **62** (M$^+$, 381) (84H549), 1- and 1,2-alkoxycarbonyltriazolines (Scheme 32) (M$^+$, 219–247) (74HCA1382), various carbamyl- and thiocarbamyl-substituted ring systems (Schemes 21 and 22) (M$^+$, 216–314, 3–27%) (84JOC1703), dihydrotriazolo-1,4-benzodiazepine **18** (M$^+$, 324) (79JOC2688), and tetrahydrotriazolo-1,4- and 1,5-benzodiazepines **94** (M$^+$, 450–523) and **97** (M$^+$, 382, 20%) (85H2051; 86S230).

In several instances, mass spectral data along with NMR studies have been used in structure determination. In the MS of dihydrotriazolopyrimidines **43** (Scheme 20) molecular ions are obtained for all compounds with an intensity (5–41%) that is dependent upon the steric crowding at C-2; the abundance of the ions $M^+ - R^1$ (1–100%) and $M^+ - R^2$ (1–100%) further supports structure **43**. The fragmentation pathways have been confirmed by the MS of deuterated compounds in which $R^2 = D$ or CD_3. Additionally, triazolopyrimidines **44** (Scheme 20), obtained by aromatization of **43**, give intense molecular ion peaks (24–100%), indicating their extended conjugation systems compared to the respective triazolines. The fragment ion R^1CN (5–38%) appears to be characteristic of the triazole compounds **44**, thus providing a means of distinguishing the latter from the dihydro structures **43** (81BCJ1767, 81JOC3956; 85CPB2678).

The 3,3-bis(trifluoromethyl)triazolines in Scheme 33 give characteristic mass spectral fragmentations: M^+, $(M^+ - F)$, $(M^+ - CF_3)$, $(M^+ - CF_3 - CF_3CN)$, $(M^+ - CF_3 - RCN)$, $(M^+ - C_{12}H_{10}N_2)$, $(RC{\equiv}N{-}C_6H_5)^+$, $(CF_3C{\equiv}N{-}C_6H_5)^+$, $RC{\equiv}N^+$, $C_6H_5N^+$, R^+, and CF_3^+ (77ZN(B)607).

The bicyclic triazoline **114** gives, besides the molecular ion peak at *m/e* 351, a peak at *m/e* 248 corresponding to the 1,4-substituted 1,2,4-triazole-5-thione, formed by loss of phenylisonitrile (*m/e* 103). In contrast, the ring-opened product from **113** decomposes differently in the MS, first splitting off its side chain in a stepwise process; this observation provides conclusive evidence in the structure determination of **114** (84CCC1713). Likewise, in the case of 5-(2-pyridyl)aminotriazoline **6** (Scheme 8), the fragment ion peaks at *m/e* 146 (M^+ – 2-aminopyridine), *m/e* 119 (M^+ – 2-aminopyridine – HCN), and *m/e* 94 (2-aminopyridine). These data, along with the MS data for the 4-(2-pyridyl)-1,2,4-triazole, have led to the elucidation of structure **6** (70T3069). Similarly, a base peak at *m/e* 401 (M^+) and the fragment-ion peaks at *m/e* 324 ($M^+ - C_6H_5$), 310 ($M^+ - C_6H_5N$), and 298 ($M^+ - C_6H_5{-}CH{=}CH{-}$) have enabled the identification of the 1,3,4-triphenyl-5-styryltriazoline **62** (69CC387).

C. X-Ray Crystallography

The X-ray crystal structure of 4-benzamido-1-benzoyl-4,5-dihydro-1,2,4-triazoline **(176)** (87AX(C)168) shows the triazoline ring to be essentially planar with a maximum deviation from a least-squares plane

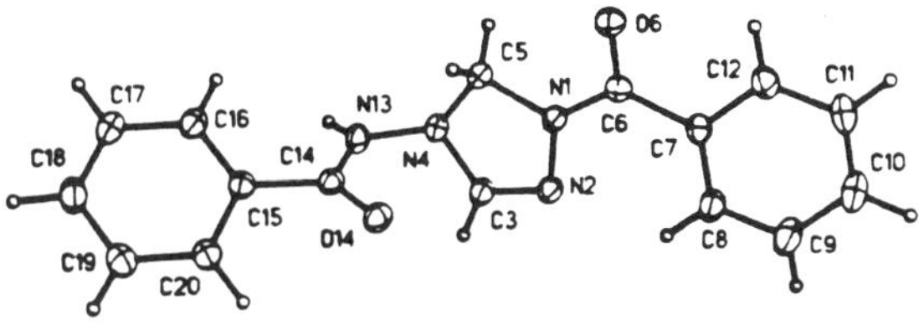

(**176**)

through the ring of 0.04 Å; however, the ring nitrogen atom with the benzamido substituent is pyramidal. Bond distances and angles are normal; an intermolecular hydrogen bond (2.93 Å) occurs between the secondary amido group and the benzoyl oxygen atom (87AX(C)168).

X-Ray analysis has been used widely along with NMR spectroscopy in the absolute structure determination of 1,2,4-triazolines. In Scheme 16, the anomalous ring closure of the nonplanar hydrazone structure upon protonation to yield the dihydrotriazolo[3,4-a]phthalazinium cation (bisulfate salt) (80TL209) has been investigated by X-ray diffraction (Fig. 3) to gain information on the precise geometry of the two molecules (82KGS1100). The length of the N-2—C-10 bond [1.474(15) Å], the change in the hybridization of the C-10 atom (the sum of the angles at C-10 is close to the value for the tetrahedral C atom, 544.9°), and the increase in the length of the N-4—C-10 bond from 1.273(3) Å in the hydrazone (Fig. 3a) to 1.474(3) Å in the cation (Fig. 3b) all constitute evidence that the amidrazone fragment N-2—C-8—N-3—N-4—C-10 in the hydrazone, upon protonation, forms a triazolinium ring rather than undergoing a simple change in its conformation with random drawing together of the N-2 and C-10 atoms. The N-3 atom in the cation assumes a planar configuration while N-4 is pyramidal, the sum of the angles being 359.2 and 316.6°, respectively. Conversion of the hydrazone to the cation is also accompanied by a decrease in the angle at N-4.

Comparison of the phthalazine ring system itself in the two molecules indicates only small changes; the N-1—N-2—C-8 angle increases slightiy, while the N-2—C-8—C-7 angle decreases and the N-2—C-8 bond becomes only slightly longer. Although the phthalazine system in the cation is virtually planar, the cation as a whole is not, due to the bent envelope-like structure of the triazolinium ring along the N-2----N-4 line. The cation and the bisulfate anion in the crystal are connected by hydrogen bonds N-4—H···O-1 [2.03(3) Å] and N-4···O-1 [2.95(3) Å] with the angle N-4—H—O-1 equalling 17.1(2)°. The bisulfate anions themselves are joined together to form centrosymmetric dimers by means of two hydrogen bonds, O-4—H···O-2′ and O-2···H—O-4′.

Single-crystal X-ray structure determination of **51** has confirmed the compound to be 7-(methylthio)-2,5,5-trimethyl-1H,5H—1,2,4-triazolo [1,2-a]-1,2,4-triazole-1,3(2H)-dione. A bent structure is manifested by the C-3—N-4—N-8—C-7 torsional angle (−135.7°). Distinction between structure **51** and the alternative **51A** by NMR spectroscopy has been unsuccessful (84JOC1703). Likewise, X-ray analysis of **15b** unequivocally identifies the structure as a dihydrotriazolo[4,3-b]pyridazine with the acyl group at the N-2 position (85CB5009). X-Ray diffraction has also helped to confirm the structure of the 5-(quinazolinylaminophenyl)-Δ^2-1,2,4-triazoline-1,4-dicarboxylate (R = OEt) in Scheme 38A (83JCS(P1)2003).

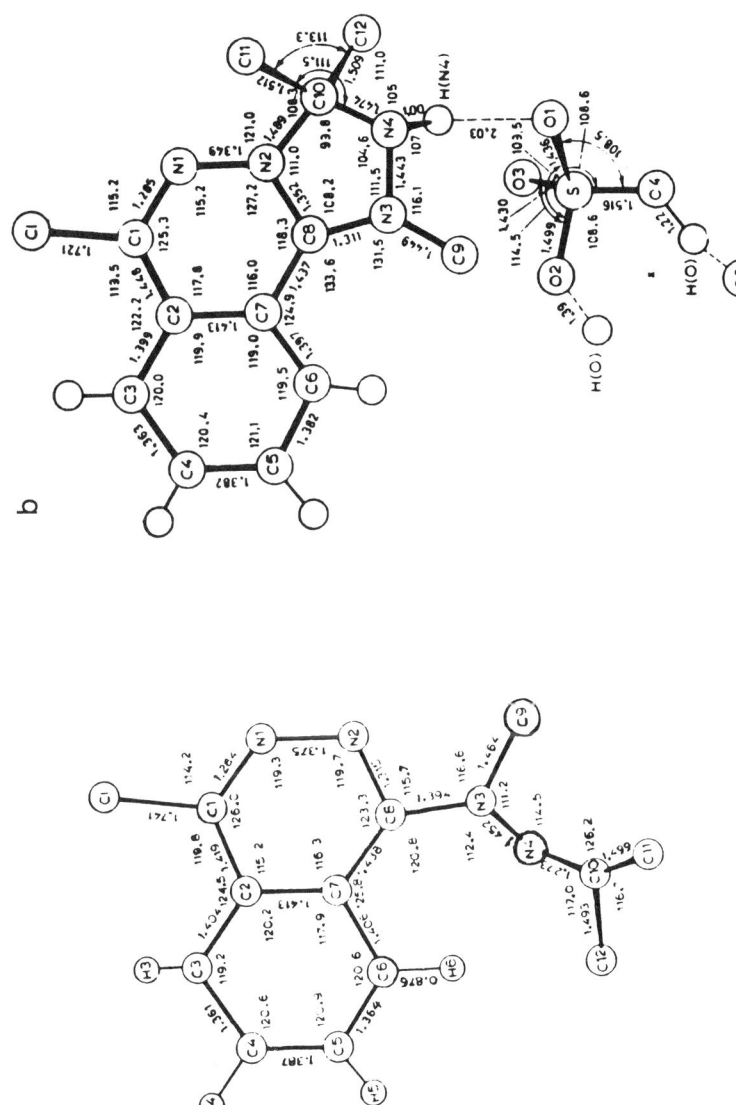

FIG 3. Bond lengths and bond angles for (a) nonplanar hydrazonophthalazines and (b) dihydrotriazolo[3,4-a]phthalazinium cation from Scheme 16 (82KGS1100).

X-Ray determination of the stereochemistry and regiochemistry of diastereomer **81** shows that the CH_2—S—Me group and the five-membered triazoline ring lie on the same side with respect to the main plane of the molecule. As the reaction is regiospecific, the stereo- and regiochemistry of **82** and **80** can also be deduced (86JCR(S)200).

Using single-crystal X-ray analysis of tetrahydrotriazolo[1,5]benzodiazepine **97**, the NMR-based structural assignments have been confirmed. In addition, the influence of the *[a]*- fusion of the triazoline ring on the conformation of the benzodiazepine system was determined. The seven-membered ring adopts a chair conformation, which can be described in terms of the dihedral angles between four least-squares planes. Moreover, the angle of 5.2° between plane 3 (three-atom moiety of the seven-membered ring, C-3a, C-4, and C-5) and plane 4 [eight atoms (fused benzene ring and N-6 and N-11 atoms)] is indicative of a flattening of the chair conformation, which alleviates the unfavorable synaxial interference between the 3a-methyl and the 5-methyl groups. The interatomic distance between the two groups in the crystal equals 3.23 Å, compared to ~2.5 Å in a Dreiding model of the ideal chair structure. Owing to steric inhibition caused by the fused benzene nucleus, the C-1 phenyl ring, while rotated from the mean plane of the triazoline ring, the N-3 phenyl is nearly coplanar to it. The N-3 atom has an almost planar configuration, whereas N-11 shows a flattened pyramidal structure (86JHC1431).

X-Ray crystal analysis has resolved the ambiguities of interpretation of the ^{1}H-NMR spectral data in the conformational analysis of structures **91** and **92** (85JST271). X-Ray analysis of three 10-methyl compounds (**92a**, R = *p*-NO$_2$Ph, R^1 = *p*-ClPh; **92b**, R = Ph, R^1 = *p*-ClPh; **92c**, R = *p*-ClPh, R^1 = COOEt) basically substantiates the NMR observations. Although **92a** and **92b** assume conformation **B**, **92c** is characterized by conformer **A**, both conformations with *cis*-X/Y ring junctions (Fig. 1) (Section III,B,1). The thiazine ring has only one of the cis inverse forms, as shown by the synclinal H-4 (axial)—C-4—S-5—C-5A torsion angles (Fig. 4). The rigidity of the X/Y ring annulation hinders inversion of the sp^3 nitrogen at 3A; the pronounced pyramidality of N-3A is not enough to hinder its inversion.

In accordance with the characteristics of conformer **B**, the thiazine ring in **92a** assumes a skew-boat shape shifted somewhat toward the twist-boat form, whereas in **92b** it has an almost perfect skew-boat conformation. The Cremer–Pople puckering parameters (ϕ = 89–96° and θ = 69–77°) and the asymmetry factors for these conformations indicate a less (**92a**) and a more (**92b**) perfect symmetry axis bisecting the C-4—S-5 bond. Thus, the *cis*-fused triazoline ring with its N-2—C-3 double bond has an envelope conformation in **92a** with the C-10 on the flap, but a half-chair conformation in **92b** with a twofold axis bisecting C-3. Together,

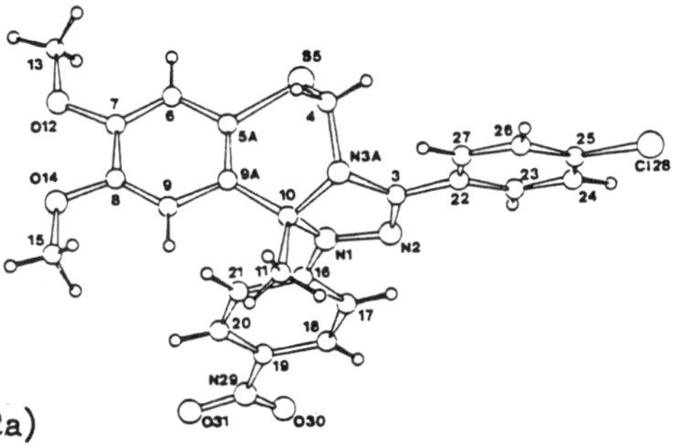

(92a)

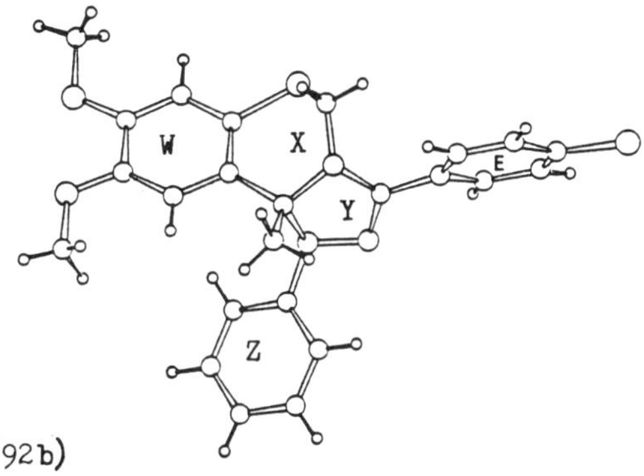

(92b) 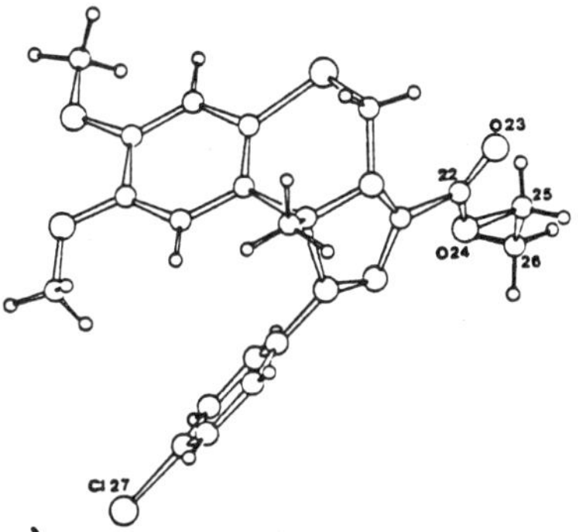 (92c)

FIG. 4. Conformations for 3-(p-chlorophenyl)-N^1-(p-nitrophenyl)- (**92a**), 3-(p-chlorophenyl)-N^1-phenyl- (**92b**), and 3-ethoxycarbonyl-N^1-(p-chlorophenyl)-10-methyldihydrotriazolobenzothiazines (**92c**) deduced from X-ray analysis (85JST271).

these differences result in the best plane of the triazoline ring being perpendicular to that of the benzothiazine moiety [89.3(2)°] in **92a**, but slightly deviating in **92b** [76.7(1)°]. Consequently, the 10-methyl group is nearer to an equatorial orientation to the thiazine ring in **92a** than in **92b**, the exocyclic torsion angles, C-11—C-10—C-9A—C-5A, being 172.3(11)° and 153.4(4)°, respectively. Also, the H···H interactions between the 10-Me and one of the ortho hydrogen atom of the 1-aryl moiety are much weaker (2.69 Å for **92a** and 2.91 Å for **92b**) and those between H-9 and the 10-Me are much stronger (2.11 and 2.07 Å, respectively) than that expected from the NMR-based model **B**.

When the 3-aryl substituent is replaced by a COOEt group as in **92c**, the flexible thiazine ring moves via pseudorotation from the skew-boat/twist-boat form to an intermediate shape between a half-chair and an envelope conformation ($\phi = 45.5°$ and $\theta = 52.8°$) with the lowest puckering amplitude (θ = 0.53Å) among all three structures (**92a–c**). This is accompanied by an inversion of N-1 together with the triazoline ring, *cis*-annulated to ring X (Fig. 4). The inverse triazoline ring maintains its envelope shape with C-10 on the flap, but the endocyclic torsion angles bear the opposite sign to those of **92a**. As a result, C-11 (of the 10-Me group) gains an increased pseudoaxial → axial orientation, while N-1 is shifted into a pseudoequatorial → equatorial position as seen from the torsion angles and Newman projections. Thus, as expected from model A (Fig. 1), the H-4 (equatorial) comes close to the 3-COOEt with a measured H···O distance of 2.44 Å and not 1.5 Å as calculated from the NMR model. Simultaneously, the H-4 (axial) approaches a proton of the quasi-axial 10-Me group (H-4a···H—C-11 = 2.34 Å), which moves away from H-9, the shortest H···H being 2.37 Å. This is also concomitant with an approach between the 1-aryl, H-9, and terminal part of the 8-OMe; the distance from the center of the 1-phenyl ring to H-9 (4.86 and 3.93 Å) and to C-15 (of 8-OMe) (6.57 and 5.37 Å) in **92a** and **92b,** respectively, comes within 2.89 and 3.97 Å in **92c.**

D. Other Physical Properties

The thermostability of 3-(2-furyl)-5-ethoxy-1,2,4-triazoline (Scheme 6) has been investigated by thermogravimetric (TGA) and differential thermal (DTA) analysis (71BCJ780). An endothermic peak at 128°C in the DTA curve coincides with the melting point of the triazoline, whereas two exothermic peaks, at 151 and 254°C, respectively, correspond to the two stages of weight reduction in the TGA curve. The initial 20% weight loss corresponds to the loss of an ethanol molecule from the triazoline (theoretical, 20.3%); the second stage of weight reduction agrees with the decomposition point of the triazole during pyrolysis.

TGA of polytriazolines **4** of low thermal stability indicates weight losses in air and in nitrogen, commencing at 280 and 400°C, respectively (70MI1).

IV. Reactivity of 1,2,4-Triazolines

Reactivity is discussed in terms of reaction types; no attempt is made to subdivide reactivity between ring atoms and substituent groups, because in many instances reactions of substituent groups are intimately associated with changes in the ring itself.

A. AROMATIZATION TO TRIAZOLES

The partially reduced, nonaromatic 1,2,4-triazolines are relatively unstable molecules. In this respect they resemble the 1,2,3-triazolines (84AHC217), and, as such, are readily aromatized to the stable triazoles by oxidation, isomerization reactions, or elimination of stable molecular fragments. Oxidative dehydrogenation reactions leading to triazoles usually require two free hydrogen atoms in adjacent positions on the ring; when they are not adjacent or when only one hydrogen atom is present, acidic oxidizing agents cause hydride ion abstraction and triazolium salt formation. For other aromatization reactions, at least one hydrogen in a position adjacent to that of the expelled fragment is required. Triazoline aromatization is a highly recognized synthetic route for the preparation of 1,2,4-triazoles.

1. *Oxidative Dehydrogenation Reactions*

Several oxidizing agents including air are known to effect triazoline aromatization to triazoles. Triazolines **36** in Scheme 18 are oxidized with elimination of water (**36b**) or methanol (**36a** and **c**), slowly in air but quantitatively upon heating for 4 hr at 100°C, to yield the 1,3,5-substituted triazoles. Steric stresses caused by the *tert*-butyl group in **36d** lead to rapid aromatization with elimination of *tert*-butyl alcohol, which makes impossible the isolation of the free triazoline base (84KGS1415). Similarly, air oxidation of 4-benzylideneaminotriazoline **58** (Scheme 26) at 185°C results in 1,3,5-triphenyltriazole **59** (84ZOR659). The dihydrotriazolobenzodiazepine **18** becomes susceptible to air oxidation and yields the triazole when the N—CH$_3$ group is replaced by hydrogen atom (79JOC2688). The tricyclic triazolinium salt **124** in Scheme 39 is likewise converted to the triazolium salt **125** by autoxidation when R = H (84TL65).

The dihydrotriazolopyrimidine-8-carboxylates **43** (R^2 = H, R^4 = COOEt) (Scheme 20) undergo spontaneous dehydrogenation during synthesis which requires heating the reaction mixture in 1-butanol, DMF, dioxane, or pyridine. Moderate yields of the triazolopyrimidines **44** are obtained (R^4 = COOEt) (81JOC3956). Likewise, dihydrotriazolopyrimidines **43a** (R^2 = H) with a 5-amino substituent lead to 5-aminotriazolo compounds **44a** in the presence of triethylamine (85CPB2678). DMSO functions as an exceptionally mild oxidizing agent in the conversion of the 8-cyano compounds **43** (R^2 = H, R^4 = CN). The oxidation can be performed by simply leaving a solution of the triazoline in DMSO to stand in an open vessel at ambient temperature, whereupon the relatively insoluble triazole will crystallize (81BCJ1767). Dehydrogenation of **43** and **43a** could also be achieved with iron(III) chloride in aqueous acetic acid (81BCJ1767; 85CPB2678) or iodine in ethanol (85CPB2678), but with yields varying over a wide range. The oxidation reaction provides a one-step synthesis for the preparation of 2,5-disubstituted 8-cyano(ethoxycarbonyl)-1,2,4-triazolo[1,5-c]pyrimidines.

Potassium permanganate oxidation in dilute acetic acid has been used to provide proof of formation of a cyclic product in the amidrazone–aldehyde cyclocondensation reactions in Scheme 4 (64CR6470). Dehydrogenation of both mono- (Scheme 4) (70JHC1001) and bistriazolines (Scheme 7) (71JHC1043) has been effected using 10% palladium on carbon and, despite the low yields, used as proof in structure determinations. Oxidation of 5-styryltriazolines **62** (R = Ph, R^1 = t- Bu) with loss of tert-butyl groups occurs upon chromatographic treatment on a silica column and leads to 5-styryltriazole, but alumina has no effect (84H549). The aromatization is also effected by acid hydrolysis, although the triazole is mixed with the aldehyde, PhCH=CHCHO, formed from the accompanying ring cleavage reaction (84H549) (Section IV,B,1).

Quinones are efficient oxidizing agents. Several Δ^3-1,2,4-triazolinium-3-sulfonates **38** (R^1 = H, Scheme 19) can be converted to the respective triazolium compounds **(177)** by p-benzoquinone (77LA463). Chloranil (2,3,5,6-tetrachloro-p-benzoquinone) and DDQ (2,3-dichloro-5,6-dicyano-p-benzoquinone) effect the oxidation of dihydrotriazoloquinolines and -isoquinolines in Scheme 29A; these do not have a free hydrogen on the nitrogen atom and yet good yields of triazoles **178** and **179** are obtained. Apparently, the phenylsulfonyl group is a good leaving group and provides the driving force for the oxidation. Benzenesulfonates of the reduced forms of DDQ and chloranil are formed and isolated which indicates benzenesulfinic acid is generated during the oxidation reaction (80BCJ2007).

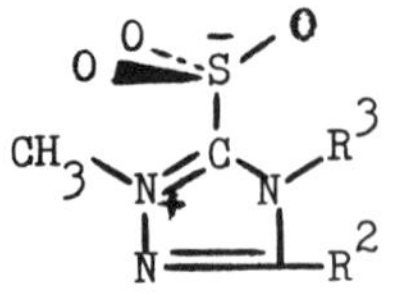

(177) 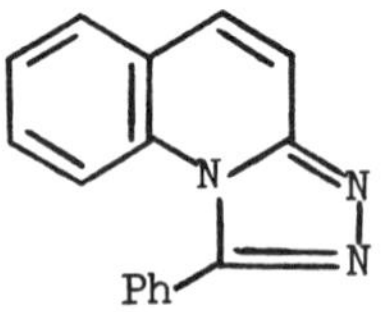   (178) 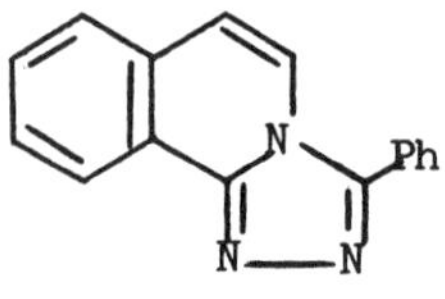   (179)

Triazoline ring systems bearing only a single hydrogen on the ring, when oxidized under acidic conditions at low temperature, lead to triazolium salts by hydride abstraction. Higher temperatures lead to ring cleavage (Section IV,B) (67TL3071; 68G511). Thus, perchloric acid oxidation of mono- and bistriazolines results in the respective triazolium perchlorates (e.g., **180**, Scheme 54) (69G69). Similar results are obtained in the chromium trioxide–acetic acid oxidation of fused ring triazolines (**181**, Scheme 54) (67TL3071; 68G511). Likewise, tropilium tetrafluoroborate treatment yields triazolium tetrafluoroborates **182** (Scheme 54) (69CB3176), and hydride abstraction from triazolines **25** and **26** (R^1 = H) by triphenylmethyl perchlorate leads to the triazolium salts **27** and **28** (Scheme 17) (67TL4337).

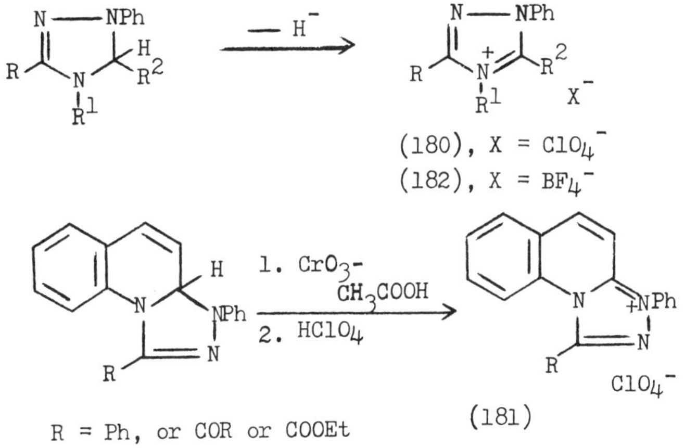

SCHEME 54

2. *Elimination of Stable Fragments*

Aromatization to triazoles by the elimination of stable molecules is common among the 1,2,4-triazolines. The reaction occurs spontaneously or under the influence of heat, acid, or base, which invariably forms part of the reaction conditions under which the triazoline synthesis usually takes place. Thus, in many cases, the only reaction products are triazoles.

Cycloaddition of diphenylnitrilimine to oximes, hydrazones, and imidates seldom yields the 4-hydroxy- (**54**), 4-amino- (**55**), or 5-alkoxytriazolines (**56**). The latter aromatize by the spontaneous loss of a molecule of water, amine, or ethanol, and the isolated reaction products are triazoles (65CB642). Loss of ethanol from the 3-(5-nitro-2-furyl)-5-ethoxy-1,2,4-triazoline in Scheme 6 occurs at 128–154°C or in dilute sulfuric acid, but not under the refluxing temperature (110–120°C) of ethyl orthoformate or 2-methoxyethanol. However, when heated in an excess of ethyl orthoformate, the triazoline yields the bistriazole (Scheme 6) (71BCJ780). The triazole in Scheme 11 is formed by loss of a molecule of thiol (79CCC1334), whereas that in Scheme 9 is obtained by the acid-induced dehydration of the intermediate 5-hydroxytriazoline (77JHC1089). Acid treatment of 5,5-dimethoxytriazoline **157** likewise yields the triazole-5-one (Scheme 55) (80T1649). The tetra- and tricyclic triazolines **132** and **133** lose H_2S or S, respectively, in refluxing xylene to yield triazolopyridinones **134** in reasonably good yield (Scheme 42) (79JOC3803).

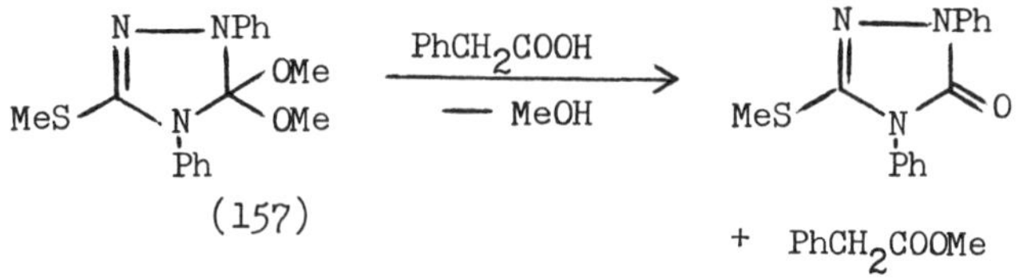

SCHEME 55

Examples of elimination of tertiary alkyl- and arylamine fragments from 5-amino-substituted triazoline intermediates are outlined in Scheme 15 (79JHC555) and Scheme 24 (80JHC311), respectively. In Scheme 8, a heteroarylamine molecule (e.g., 2-aminopyridine) is lost from triazoline **6** in refluxing ethanol (70T3069). The simultaneous formation of triazole **59** and triazoline **60** in Scheme 26 is a unique case of aromatization of a bistriazoline by the thermally induced expulsion of a molecule of triazoline (65CB642) and may be likened to the elimination of an amine fragment in triazole formation. Similarly, triazoline **111** loses propionic acid in refluxing xylene (77JOC443). The thermally induced elimination of benzenesulfinic acid from dihydrotriazoloquinoline and dihydrotriazoloisoquinoline (Scheme 29A) is effected similarly by refluxing the triazolines in dioxane or CCl_4; the resulting products are identical to triazoles **178** and **179** obtained in the DDQ or chloranil oxidations (Section IV,A,1) (80BCJ2007). On the other hand, hydrolytic removal of the formyl group in triazolinium salt **124** is readily achieved during aqueous work-up of the reaction mixture in the cold (Scheme 39) (84TL65).

An interesting case of double fragment elimination of benzenesulfinic acid along with an amine molecular moiety is seen in the reaction of N-substituted benzamidines and 2-aminopyridines (or -pyrimidines), respectively, with N-(phenylsulfonyl)benzhydrazidoyl chloride (Scheme 24A). The fragment elimination pattern is determined by the nature of the R substituent on the ring nitrogen of the transient triazoline intermediate (Scheme 56). An analogous reaction takes place with benzimidates, an alcohol fragment being expelled in place of the amine molecule. This reaction scheme serves as a simple synthetic procedure for the preparation of

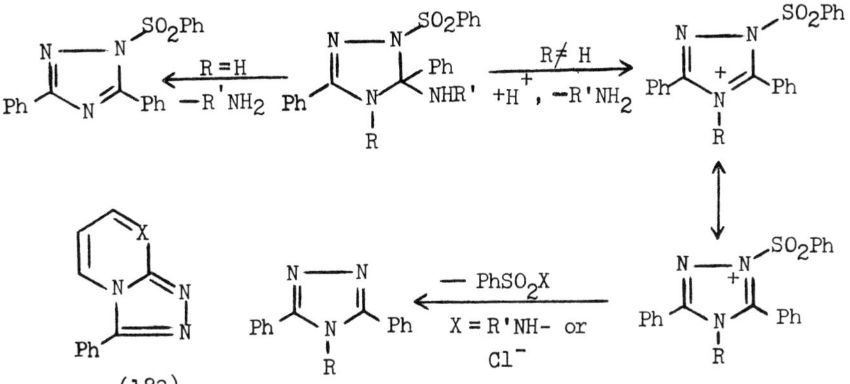

Sec. IV.A]   1,2,4-TRIAZOLINES   259

SCHEME 56

1,2,4-triazolopyridines and -pyrimidines (**183**), which are reported to be useful as anticonvulsants and as tranquilizers (77BCJ2969).

Aromatization by deacetylation has been observed under the influence of acetic anhydride, according to the mechanism outlined in Scheme 57

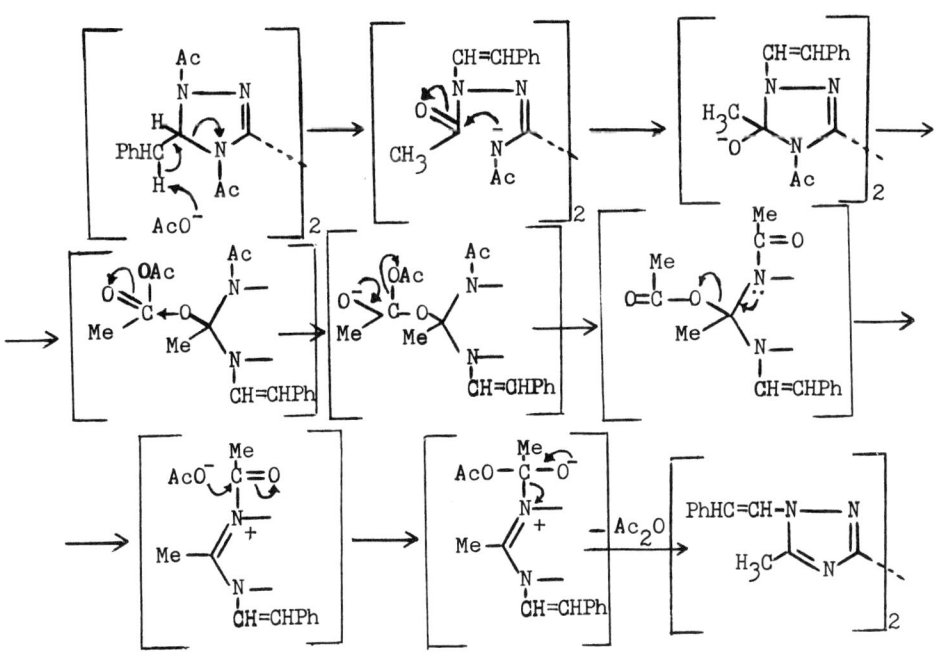

SCHEME 57

(75JCS(P1)1433). Another interesting case of aromatization involves the formation of triazolethiones **184** by loss of phenylisonitrile from the fused-ring triazolines **114** in Scheme 38 when subjected to acid hydrolysis (84CCC1713).

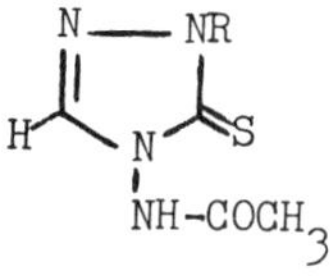

$$\underset{\underset{NH-COCH_3}{|}}{\overset{N \mathrel{=\!=} NR}{\underset{N}{H\diagdown}}\diagup}S$$

(184)

A special case of aromatization of a tricyclic triazoline resulting from elimination of an acetone molecule is outlined in Scheme 14 (78JHC161). Another unusual aromatization reaction by pyrolytic elimination of methane occurs among the dihydrotriazolopyrimidine-8-carboxylates **43** when $R^2 = CH_3$. The reaction is facilitated by hot pyridine and even the loss of a benzene molecule could be induced when $R^2 = Ph$. The driving force of elimination of R^2H may be establishment of an extended conjugation system over two heteroaromatic rings and/or release from steric crowding at C-2 (81JOC3956).

3. Isomerization Reactions

Closely related to aromatization by elimination of stable molecular fragments are the isomerization reactions of fused-ring triazolines leading to triazoles. The bicyclic triazolines *112* in Scheme 37 undergo isomeric ring opening at the N—N bond to yield the 5-substituted triazoles in 64–70% yield (76CPB2568). Compounds **113** and **114**, although both primary cycloadducts, tend to aromatize. This is assisted in **113** by the steric congestion on N-1 and N-6. The triazole structure is established from NMR spectroscopic and chemical hydrolysis data (84CCC1713). Cycloadducts **128** in Scheme 40 readily rearrange to yield triazolopyridines **129**; an initial ring opening with subsequent six-membered ring closure has been postulated (83JCS(P2)1317). Similarly, the unstable bicyclic triazoline in Scheme 41 yields the 5-substituted triazoles by C—N bond fission (81T2805, 81T2811).

4. Retro Diels–Alder Reactions

Diels–Alder type cycloreversions, although frequently observed among the bi- and tricyclic 1,2,3-triazolines (84AHC217), are not common among the 1,2,4-ring systems. In one recorded case, azomethine ylides of the mesoionic oxazolone type (185) react with benzenediazonium fluoroborate at room temperature with evolution of carbon dioxide. The intermediate cycloadduct 186 undergoes a Diels–Alder cycloreversion, which is finished in 2 min. The product has been identified as the triazolium salt 187 (84TL65).

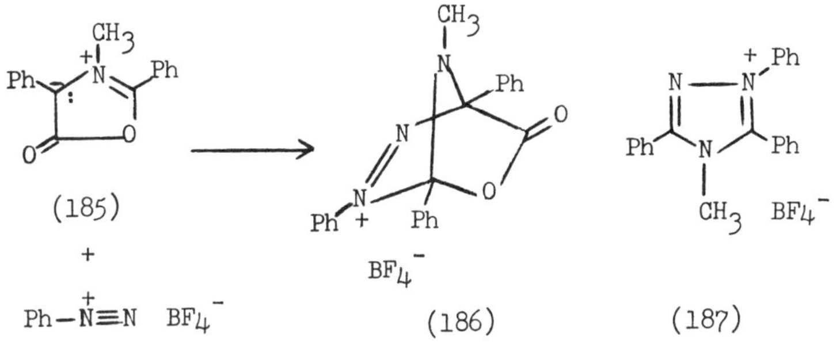

B. Ring Cleavage Reactions

1. *Hydrolytic Cleavage by Acids: A Synthetic Route to Aldehydes*

1,2,4-Triazolines are not stable under acidic conditions: Acid hydrolysis causes decomposition of the triazoline ring system to yield aldehydes. The reaction has been developed into a synthetic route for aldehydes. There is no need to isolate or purify the triazoline. The reaction mixture resulting from sodium borohydride reduction of the triazolium salt (see Scheme 44) (Section II,C,1) is usually decomposed directly by acid to yield the aldehyde. Several convenient procedures have been developed for the synthesis of the triazolium salts themselves, and because the 5-R substituent is introduced by employing carboxylic acids, acid chlorides, or alkyl halides (see Scheme 44), the reaction affords a very simple, efficient, and selective method for the transformation of these functionalities to aldehydes. Generally, the yields are in the range 40–80%, based on the acid, acid chloride, or alkyl halide. Because of the high selectivity of the borohydride reduction of triazolium salts, aldehydes that contain double bonds; halogen, nitro, keto, or ester substituents; or one or two side chains on the α-carbon; or aldehydes that are alicyclic, have all been synthesized. The method is also suitable for the conversion of dicarboxylic acids to aldehydic acids (see Table VII) (76T2549).

Aldehyde synthesis by triazoline acid hydrolysis has advantages over the Meyers procedure (72JA3243; 73JOC36; 75JOC2021), in view of the

TABLE VII

DECOMPOSITION OF TRIAZOLINES 141 TO ALDEHYDES

R	Yield (%)[a]	References
Me	>55	74TL2649
	45	75TL1889
Et	52	75TL1889
	90	76T2549
	47	75TL681
Pr	72	76T2549
	31	75TL681
i−Pr	87	76T2549
	34	75TL681
n−Bu	79	75TL1889
t−Bu	75	76T2549
	32	75TL681
Cyclopropyl	63	76T2549
$Cl(CH_2)_2$−	30	75TL681
$Cl(CH_2)_3$−	61	76T2549
$MeOOCC_2H_4$−	29	75TL681
$MeOOCC_4H_8$−	64	76T2549
$EtOOC(CH_2)_5$−	46	74TL2649
$ClCH_2CH_2COOMe$−	30	75TL1889
$Br(CH_2)_3Cl$−	26	75TL1889
Ph	90	74TL2649
$PhCH_2Cl$−	44	75TL1889
2−MeC_6H_4−	78	76T2549
4−ClC_6H_4−	81	76T2549
4−$NO_2C_6H_4$−	91	76T2549
4−$MeOOCC_6H_4$−	66	74TL2649
	77	76T2549
3,4,5−$(MeO)_3C_6H_2$−	91	76T2549
PhCH=CH−	83	74TL2649
$\frac{1}{2}$ 4−C_6H_4<	55	74TL2649

[a]Based on the carboxylic acid, acid chloride, or alkyl halide.

simplicity of both the preparation and reduction of the triazolium salts (Scheme 44) (76T2549). Decomposition by warm aqueous sulfuric acid or cold aqueous formaldehyde–hydrochloric acid yields the desired aldehyde (74TL2649). Based on the ability of the easily accessible anhydro bases 140 to undergo ready C-alkylation, a novel method for the synthesis of aldehydes (RCHO, with R = $R^1R^2R^3C$-) has been devised, starting with carboxylic acids, acid chlorides, or alkyl halides. In general, the C-alkylated triazolium salts can be reduced, without purification, with aqueous sodium borohydride and the resulting triazolines **141** (Scheme 44) cleaved with acid to yield the aldehydes (75TL681). Similarly, alkylation of the nucleophilic carbene **139** (Scheme 44) with alkyl halides provides the triazolium salts **138**, which are reduced and hydrolyzed to the corresponding aldehydes (75TL1889). This procedure, which is an alternative to the Corey aldehyde synthesis (65AG(E)1075, 65AG(E)1077), does not require an inert gas atmosphere, and in the final step the aldehyde is liberated by simple acid treatment.

In addition, this synthetic scheme offers a mild reaction route for the degradation of α-hydroxy acids to the corresponding aldehydes containing one less carbon atom or for their conversion to aldehydes with the same number of carbon atoms. For example, reaction of 1,4-diphenylthiosemicarbazide with α-acetoxyacyl chlorides yields 5-(α-hydroxyalkyl)-1,2,4-triazolium salts, which, upon sodium hydride reduction and acid treatment, give benzaldehyde along with **138** (R = H) (Scheme 58) (75TL1889). Moreover, the 5-(α-hydroxyalkyl)-1,2,4-triazolium salts can be easily converted to the 5-alkyl compounds (Scheme 58) and borohydride reduction of the latter followed by acid hydrolysis provides a method for the transformation of α-hydroxy acids to aldehydes with the

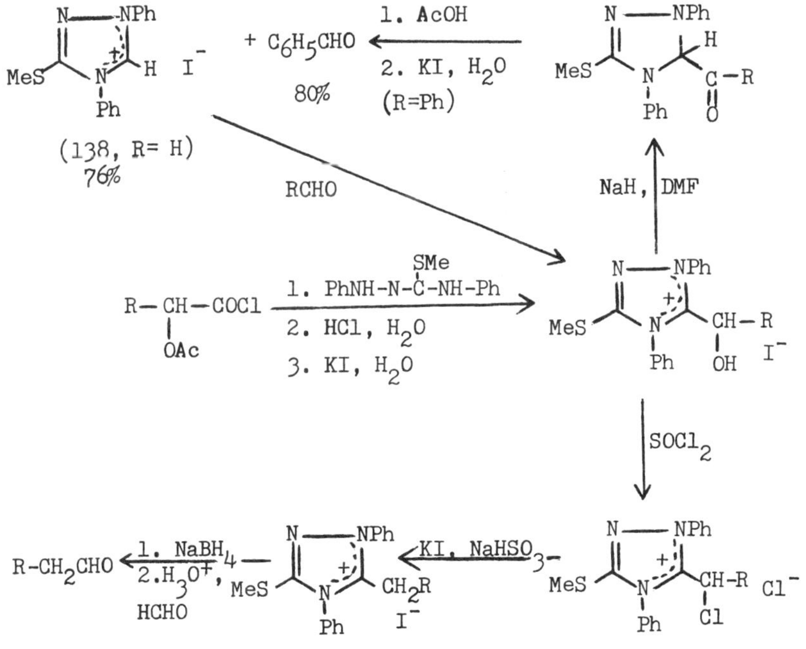

SCHEME 58

same number of carbon atoms (80TL4183). On the other hand, the reaction of triazolium salt **138** (R = H) with aliphatic aldehydes also yields the 5-(α-hydroxyalkyl)-1,2,4-triazolium salts, and, based on this reaction, a method has been devised for lengthening the carbon chain of the resulting aldehydes by one CH_2 group as compared to the starting aldehydes (Table VIII) (Scheme 58) (80TL4183).

2. Oxidative Ring Cleavage

Oxidation of triazolines could lead to aromatization or ring cleavage, depending on reaction conditions, ring substitution, and steric stresses of the molecule. Oxidation of triazolines bearing two free hydrogen atoms or a hydrogen atom along with an easily displaceable alkyl group in adjacent positions usually results in aromatization to triazoles by loss of water or alcohol (Section IV,A,1). However, when the triazoline contains only a single hydrogen atom in the 5-position or when the second hydrogen atom is not adjacent, oxidation leads to ring cleavage reactions through the intermediacy of unstable 5-hydroxytriazolines that cannot be stabilized by aromatization or through the formation of triazolium salts susceptible to nucleophilic attack. The type of ring opening is determined by the reaction conditions and the nature and position of the substituent groups, particularly those in the 3-position of the triazoline ring. (To avoid confusion in referring to the 3-position in the different fused ring systems, the numbering takes into account only the triazoline nucleus.)

TABLE VIII

REACTION OF TRIAZOLIUM SALT **138** WITH ALIPHATIC ALDEHYDES[a]

RCH_2CHO	Yield (%)	RCH_2COOH	Yield (%)
 (isobutyl)CHO	72	 (isobutyl)COOH	75
 (n-pentyl)CHO	53	 (n-pentyl)COOH	70
 (branched)CHO	57	 (OMe-branched)COOH	71
 (long chain with double bond)CHO	41	 (long chain with double bond)COOH	37

[a] See scheme 58. Data from (80TL4183).

a. *Oxidation by Oxygen in Organic Solvents: C—N Bond Cleavage via 5-Hydroxytriazolines.* Triazoline oxidation with oxygen in organic solvents yields 5-hydroxytriazolines that subsequently undergo ring-opening reactions at the C-5—N-4 or N-1 bond to give N-acylamidrazones. Under neutral oxidation conditions, 5-hydroxylation is the favored reaction, regardless of the 3-substituent group. This is exemplified by the oxidation of 1,4,5-substituted 3-acetyltriazolines as outlined in Scheme 59 (69CB1028), as well as by the air oxidation of dihydrotriazolobenzodiazepine 18 in wet ethyl acetate–chloroform to give 38% yield of the *N*-acylamidrazone 19 (79JOC2688).

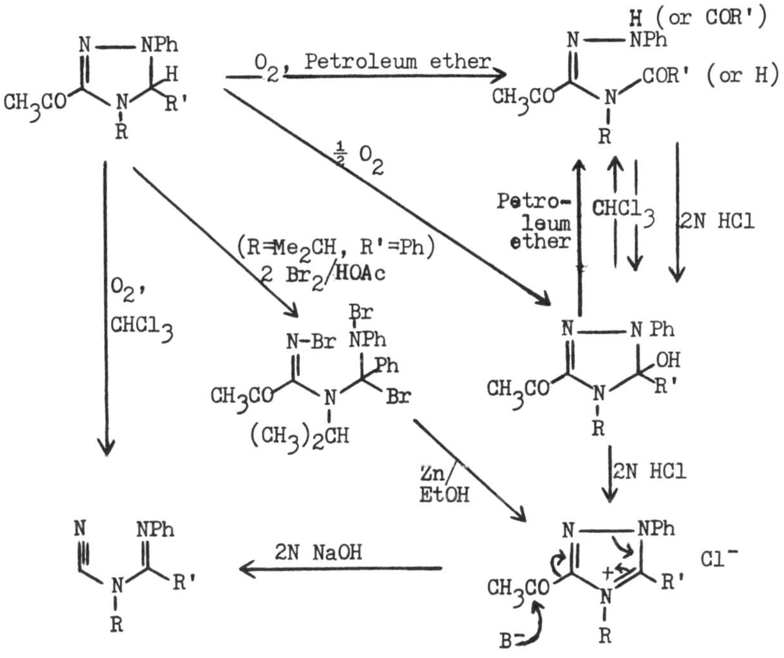

SCHEME 59

b. *Oxidation by Acidic Oxidizing Agents Followed by Nucleophilic Attack.* Triazolium salts are obtained from triazolines by oxidation with acidic oxidizing agents (Section IV,A,1). Subsequent treatment with alkali leads to either C—N or N—N bond cleavage, depending on the 3-substituent. While a 3-aryl or -alkyl group causes C-5—N-1 or —N-4 bond opening, the presence of a 3-acyl or 3-alkoxycarbonyl or the absence of 3-substitution brings about N—N bond cleavage.

i. *C—N Bond cleavage via 3-aryl-substituted triazolium salts.* The triazolium salts resulting from oxidation of 3-aryltriazolines are highly susceptible to nucleophilic attack by hydroxide ion on the heterocyclic ring carbon. The 5-hydroxytriazoline intermediates thus formed undergo C—N bond cleavage similar to that of the 3-acetyl-5-hydroxytriazolines shown in Scheme 59. Three typical cases are illustrated in Scheme 60, leading to *N*-acyl- **(188–189)** or *N*-formyl- **(190)** amidrazones (69CB3176, 69G69). *N*-Formylhydrazinoazines **191** are the products of oxidation of 5-unsubstituted 4-arylideneaminotriazolines **58,** also by way of the 5-hydroxytriazolines (75JCS(P1)2474). The *N*-formylamidrazones **190** provide convenient building blocks for the synthesis of 2-arylazo heterocycles in 80–100% yield (69CB3176).

ii. *N—N Bond cleavage via 3-acyl, 3-alkoxycarbonyl, or 3-unsubstituted triazolium salts: N-Cyanamidine synthesis.* When the 3-substituent is an acyl or alkoxycarbonyl or when there is no 3-substitution, the N—N bond is cleaved giving rise to *N*-cyanamidines **192** and **192a** (Scheme 61) (68G511). Unlike the 3-aryltriazolium salts in Scheme 60, the compounds in Scheme 61 do not yield hydroxytriazoline intermediates; while proton abstraction prevails in the 3-unsubstituted salts, in the 3-acyl and 3-alkoxycarbonyl compounds, nucleophilic attack by hydroxide ion occurs preferentially on the carbonyl carbon of the 3-substituent groups, leading to their elimination (69G69). Thus, treatment of the appropriately substituted triazolium salts with dilute sodium hydroxide followed by heating provides a high-yield route for the synthesis of *N*-cyanamidines, both simple **(192)** as well as those incorporating a heterocyclic ring system **(192a)** (69G69).

As shown in Scheme 59, in the absence of a nucleophilic reagent, the 3-acetyl-5-hydroxytriazolines undergo ring opening at the C—N bond, similar to the 3-aryl analogues in Scheme 60. Under acidic conditions, however, the triazolium salts formed from the 3-acetyl-5-hydroxy compounds yield *N*-cyanamidines by N—N bond cleavage (Scheme 59) (69CB1028), similar to **180** (R = COOEt or COMe) in Scheme 61. *N*-Cyanamidines also result directly when the 3-acetyltriazolines in Scheme 59 are reacted with oxygen in chloroform solution or indirectly via N—N ring opening to a tribromo compound and reduction of the latter to the triazolium salt (69CB1028).

Similar reaction paths have been suggested to rationalize the formation of heterocyclic substituted arylcyanamides **194** in the high-temperature reactions of pyridine, quinoline, and isoquinoline, respectively, with *C*-

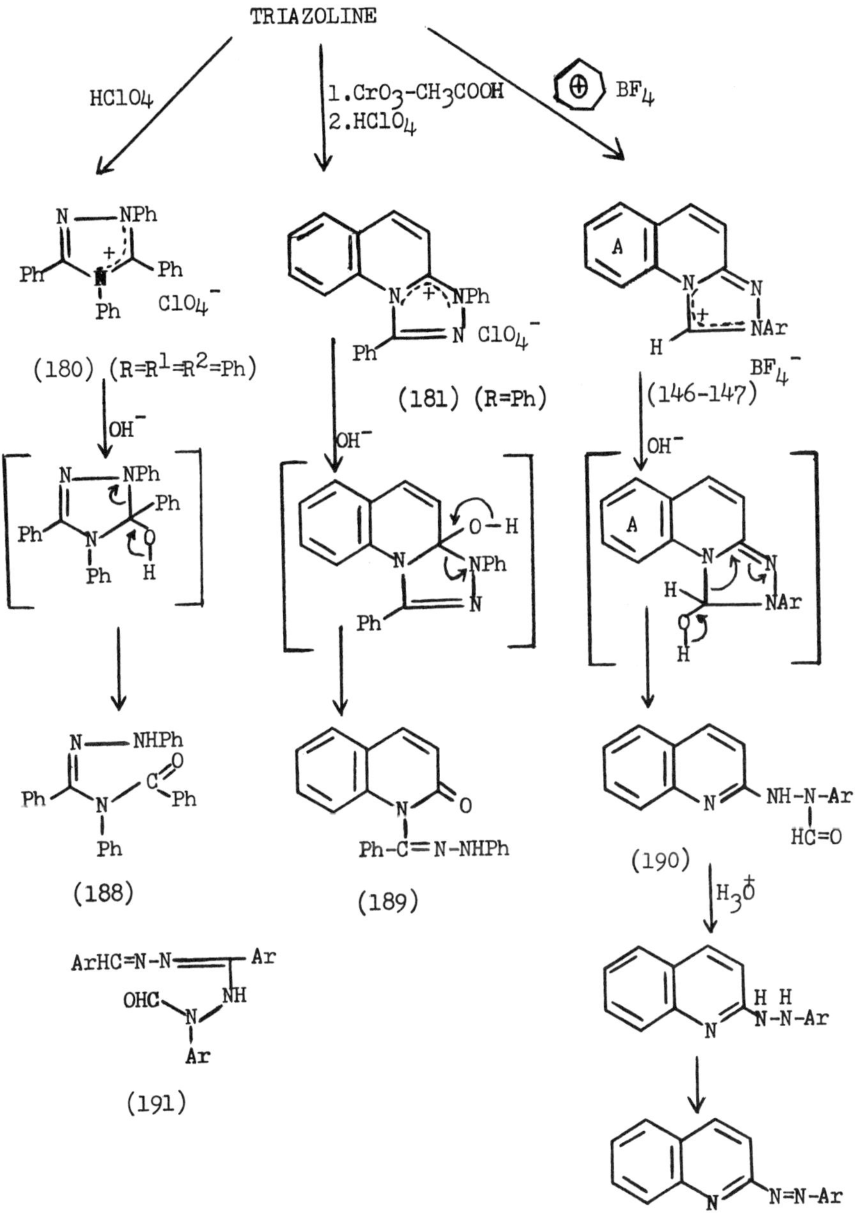
TRIAZOLINE
HClO4
1.CrO3-CH3COOH
2.HClO4
BF4
(180) (R=R1=R2=Ph)
(181) (R=Ph)
(146-147)
OH⁻
(188)
(189)
(190)
H3O+
(191)

SCHEME 60

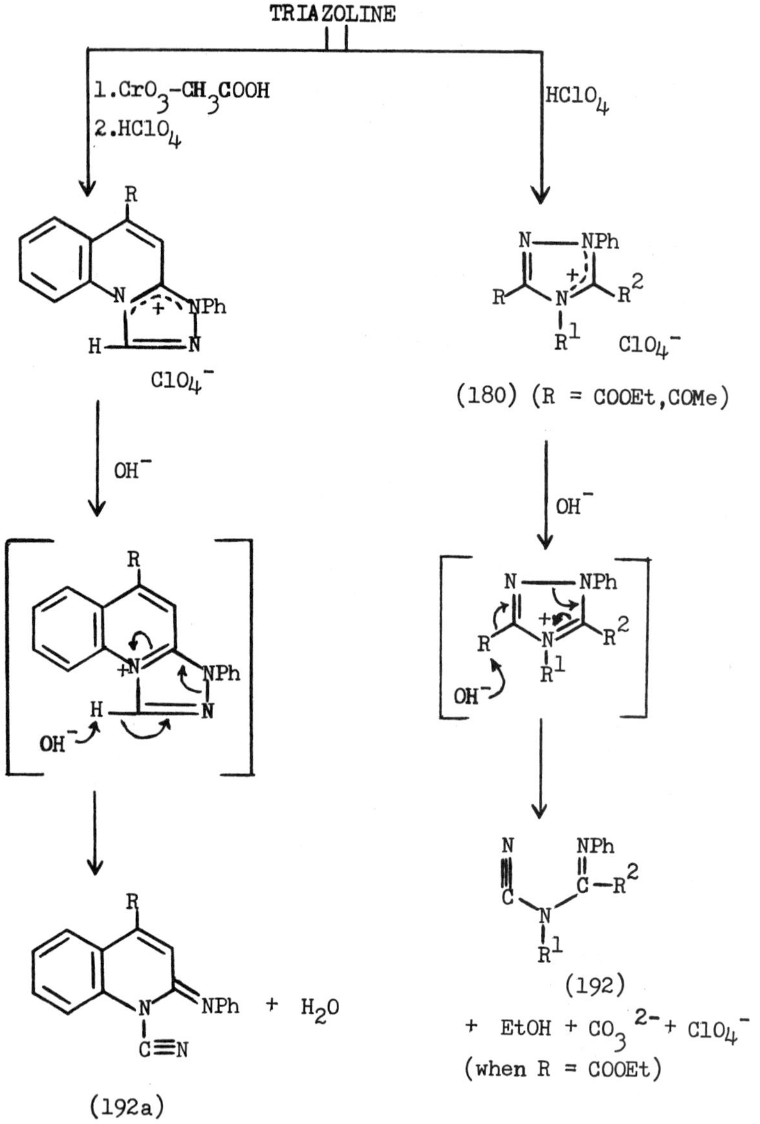

SCHEME 61

ethoxycarbonyl-*N*-arylhydrazidoyl chloride. Elimination of the ethoxycarbonyl group from the triazolium salt is effected by the heterocyclic base in place of the hydroxide ion, and the exocyclic cyano derivative **194** is obtained by thermal isomerization of the *N*-cyanoquinoline compound **193,** as outlined in Scheme 62 (67TL3071; 68G511).

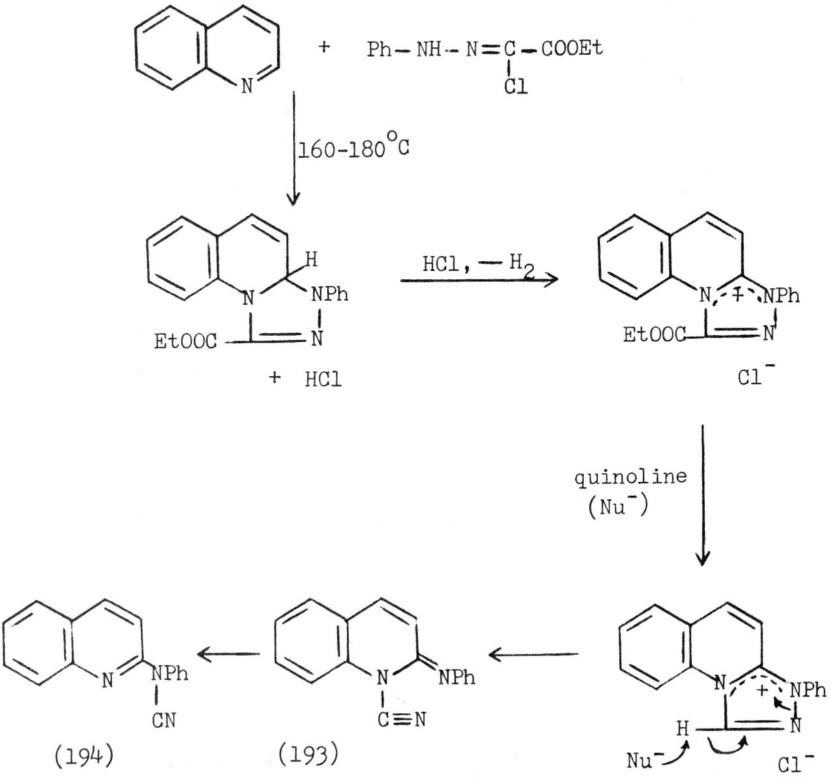

SCHEME 62

C. OTHER REACTIONS

1. Oxidation by Acidic Oxidizing Agents Followed by Electrophilic Attack

The CH hydrogen adjacent to the positively charged nitrogen in quaternary triazolium salts is known to exchange for deuterium (66CB2017), thus allowing triazolium salts to react with electrophiles. Unlike the reactions of nucleophiles, electrophilic attack does not lead to ring cleavage. The triazolium salt resulting from the oxidation of dihydrotriazoloisoquinoline reacts with phenyl isothiocyanate or benzothiazolium tetrafluoroborate, in the presence of triethylamine, as outlined in Scheme 63 (69CB3176).

A methyl group in place of the CH hydrogen is also sufficiently reactive and undergoes electrophilic attack by quinolinium salts as shown in

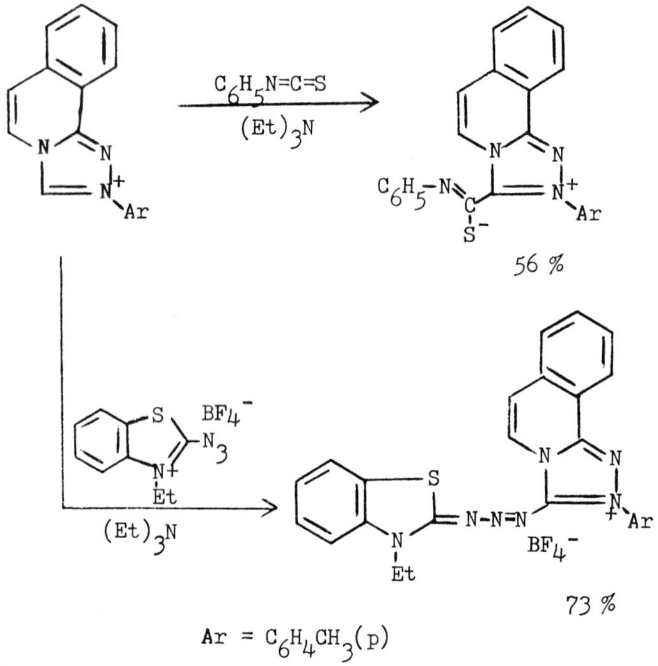

SCHEME 63

Scheme 64 to afford enamine compounds. The latter are isolated as the perchlorate salts by treating the reaction mixture with lithium perchlorate and water (69CB3176).

2. Alkali-Induced Reactions

a. *Dealkoxycarbonylations with Double Bond Isomerization.* 1,2-Ethoxycarbonyl-Δ^3-1,2,4-triazolines containing no phenyl substituent in the 5-position, when refluxed in 0.2–0.4 M aqueous ethanolic potassium hydroxide, undergo saponification with decarboxylation to give the isomeric Δ^2-triazoline-1-carboxylates. On the other hand, under the same conditions, the 5-phenyl-1,2-ethoxycarbonyltriazolines yield the triazoles. In 10 M aqueous potassium hydroxide, compounds with either type of 5-substitution yield the triazole (Scheme 32) (74HCA1382). Likewise, 10% aqueous methanolic solution of sodium hydroxide effects demethoxycarbonylation with simultaneous isomerization of 4-methoxycarbonyl-Δ^2-triazoline **49** to the Δ^3-compound (**49a**) in quantitative yield (Scheme 22) (84ABC2913, 84JOC1703).

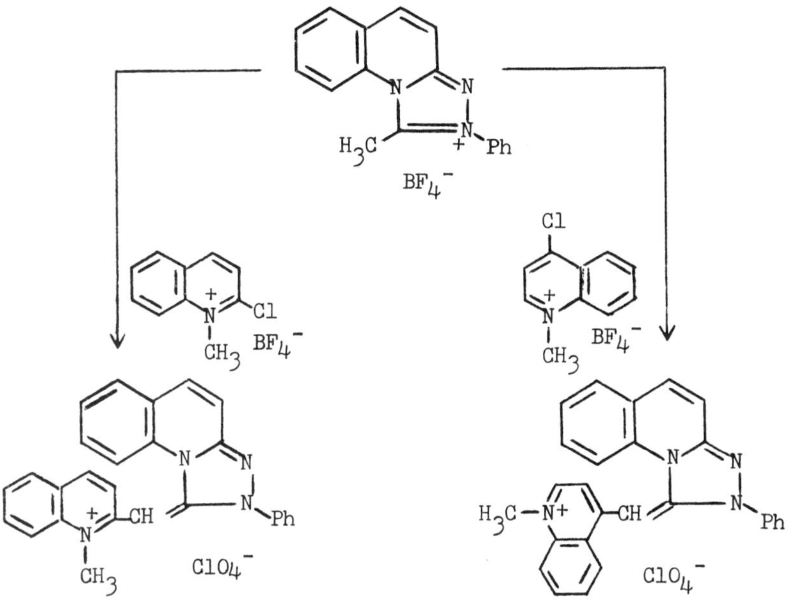

SCHEME 64

b. *Intramolecular Cyclizations to Fused-Ring Systems.*
Reaction of 4-methoxycarbonyltriazoline **49b**, unlike that of **49**, with 10% aqueous methanolic sodium hydroxide effects not only the removal of the methoxycarbonyl group but also the simultaneous intramolecular cyclization of the triazoline to the fused heterocycle **50** in 93% yield (Scheme 22) (84JOC1703). Treatment of 3-acetyl-4-(2-acetylaminophenyl)triazoline **195** with boiling 2 N aqueous methanolic sodium hydroxide results in loss of the N-acetyl group with subsequent intramolecular cyclization to a tricyclic system **(196)**. Oxidation of the latter by oxygen yields the rearranged dihydrotriazolobenzimidazole **198** via the intermediate hydroxytriazoline **197**. The primary amino group resulting from the hydrolytic cleavage of the C=N bond of the pyrazine ring attacks the C-1 position in the hydroxytriazoline intermediate with elimination of a molecule of water (Scheme 65) (69CB1028).

3. *1,1'-Carbonyldiimidazole-Induced Cyclization Reactions*

Transformation of 1-carbamyltriazolines **49a** to fused-ring systems **(51)** has been achieved through cyclization reaction with 1,1'-carbonyldiimidazole at 120°C for 1 hr (Scheme 22) (84ABC2913, 84JOC1073).

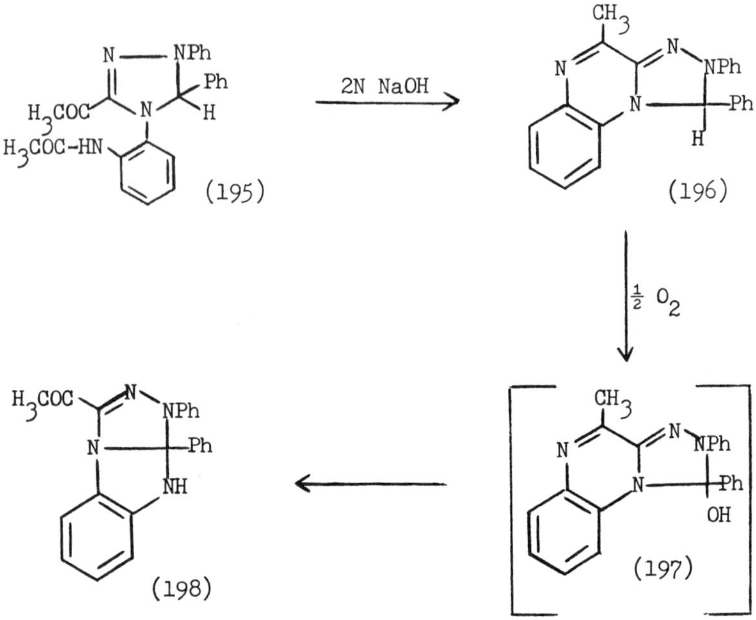

SCHEME 65

4. Acid Hydrolysis

Although acid treatment of triazolines generally leads to aromatization to triazoles or ring cleavage with aldehyde formation (Sections IV,A,1 and IV,B,1), in one recorded case hydrolysis of the fused-ring compound **76** by a 1 : 1 mixture of phosphoric acid and acetic acid left the triazoline ring unaffected; the reaction is confined to the substituent group as evidenced by product analysis **(199)** (85H2183).

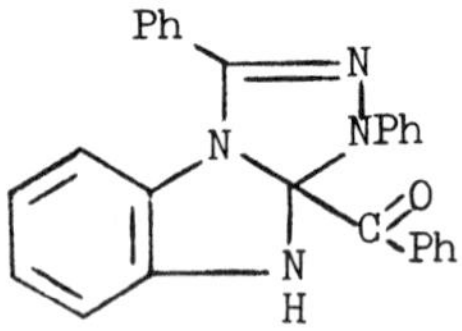

5. Photolytic Reaction with Trifluoro Acetate

Photolysis of 1-ethoxycarbonyl-5, 5-dimethyl-3-phenyl-Δ^2-1,2,4-triazoline (Scheme 32) in benzene in the presence of oxygen and methyl trifluoroacetate gives the 5-methoxy-2,2-dimethyl-4-phenyl-5-trifluoromethyl-

3-oxazoline (**200**). 5,5-Dimethyl-3-phenyl-1,2,4-triazole seems to be the intermediate, which upon losing nitrogen gives the benzonitrileisopropylide (Scheme 66) (74HCA1382).

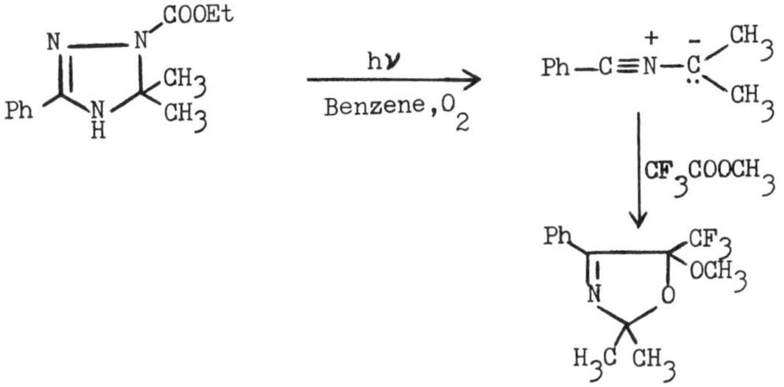

SCHEME 66

6. Ring Alkylation Reactions

Regioselective N-(2)-methylation of triazoline **50** has been attained using a solution of diazomethane in chloroform; the N-methyl derivative (**51**) is obtained in 84% yield (Scheme 22) (84JOC1703).

Likewise, alkylation of 2,3,5,5-tetramethyltriazoline **36a** with methyl iodide occurs exclusively at the N-4 position to yield the pentamethyltriazolinium iodide **201** (84KGS1415).

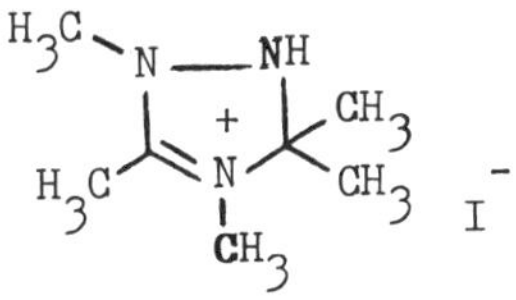

(201)

Aldehydes react with triazolinium-3-sulfonates **38** to afford 1-hydroxyalkyltriazolinium-3-sulfonates **202** (77LA463).

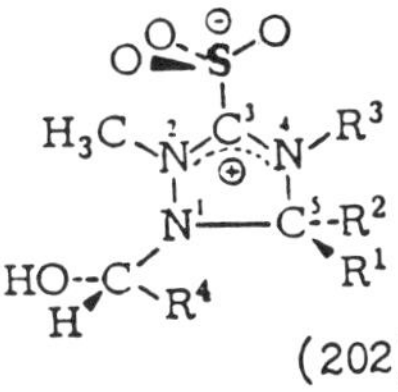

(202)

ACKNOWLEDGMENTS

The author thanks Dr. Terence Tita for reviewing the manuscript, Mrs. Janet Ritchie for typing, and NINCDS (NIH) for research grant award NS-24750.

References

1892MI1	A. Pinner, "Die Imidoäther und ihre Derivate." Oppenheim, Berlin, 1892.
1895CB465	A. Pinner and N. Caro, *Chem. Ber.* **28**, 465 (1895).
1897LA(297)221	A. Pinner, *Justus Liebigs Ann. Chem.* **297**, 221 (1897).
1897LA(298)1	A. Pinner, *Justus Liebigs Ann. Chem.* **298**, 1 (1897).
10MI1	M. Busch and R. Ruppenthal, *Ber. Dtsch. Chem. Ges.* **43**, 3001 (1910).
14JPR310	M. Busch and C. Schneider, *J. Prakt. Chem.* **89**, 310 (1914).
34LA264	E. Müller and W. Kreutzmann, *Justus Liebigs Ann. Chem.* **512**, 264 (1934).
36JA800	G. W. Kirsten and G. B. L. Smith, *J. Am. Chem. Soc.* **58**, 800 (1936).
36LA173	P. W. Neber and H. Worner, *Justus Liebigs Ann. Chem.* **526**, 173 (1936).
37JA2077	G. B. L. Smith and E. P. Shoub, *J. Am. Chem. Soc.* **59**, 2077 (1937).
37MI1	F. Krollpfeiffer and E. Braun, *Ber. Dtsch. Chem. Ges.* **70**, 89 (1937).
50JA2783	H. Rapoport and R. M. Bonner, *J. Am. Chem. Soc.* **72**, 2783 (1950).
52JA2981	W. G. Finnegan, R. A. Henry, and G. B. L. Smith, *J. Am. Chem. Soc.* **74**, 2981 (1952).
57QR15	W. Baker and W. D. Ollis, *Q. Rev., Chem. Soc.* **11**, 15 (1957).
58CB1495	E. Schmitz, *Chem. Ber.* **91**, 1495 (1958).
60AG416	R. Huisgen, R. Grashey, P. Laur, and H. Leitermann, *Angew. Chem.* **72**, 416 (1960).
60TL1	R. Huisgen, R. Fleischmann, and A. Eckell, *Tetrahedron Lett.*, 1 (1960).
61CB2503	R. Huisgen, J. Sauer, and M. Seidel, *Chem. Ber.* **94**, 2503 (1961).
61JOC2331	P. K. Kadaba and J. O. Edwards, *J. Org. Chem.* **26**, 2331 (1961).
62AG292	R. Grashey and K. Adelsberger, *Angew. Chem.* **74**, 292 (1962)
62AG491	R. Grashey, H. Leitermann, R. Schmidt, and K. Adelsberger, *Angew. Chem.* **74**, 491 (1962).
62AG(E)50	R. Huisgen, H. Stangl, H. J. Sturm, and H. Wagenhofer, *Angew. Chem., Int. Ed. Engl.* **1**, 50 (1962).
62T3	R. Huisgen, M. Seidel, G. Wallbillich, and H. Knupfer, *Tetrahedron* **17**, 3 (1962).
63AG604	R. Huisgen, *Angew. Chem.* **75**, 604 (1963).
63AG(E)565	R. Huisgen, *Angew, Chem., Int. Ed. Engl.* **2**, 565 (1963).
63AG(E)633	R. Huisgen, *Angew. Chem. Int. Ed. Engl.* **2**, 633 (1963).
63CPB781	T. Okamoto, M. Hirobe, C. Mizushima, and A. Ohsawa, *Chem. Pharm. Bull.* **11**, 781 (1963).
63CPB1089	T. Okamoto, M. Hirobe, and Y. Tamai, *Chem. Pharm. Bull.* **11**, 1089 (1963).
63JOC543	K. T. Potts, *J. Org. Chem.* **28**, 543 (1963).
64CB1085	R. Huisgen, R. Grashey, H. Knupfer, R. Kunz, and M. Seidel, *Chem. Ber.* **97**, 1085 (1964).

64CR6470	B. G. Baccar and F. Mathis, *C. R. Hebd. Seances Acad. Sci* **258**, 6470 (1964).
64MI1	R. Huisgen, R. Grashey, and J. Sauer, *in* "The Chemistry of Alkenes" (S. Patai, ed), pp. 806–878. Wiley, New York, 1964.
65AG(E)701	R. Grashey, *Angew. Chem., Int. Ed. Engl.* **4**, 701 (1965).
65AG(E)1075	E. J. Corey and D. Seebach, *Angew. Chem., Int. Ed. Engl.* **4**, 1075 (1965)
65AG(E)1077	E. J. Corey and D. Seebach, *Angew Chem., Int. Ed. Engl.* **4**, 1077 (1965).
65CB642	R. Huisgen, R. Grashey, E. Aufderhaar, and R. Kunz, *Chem. Ber.* **98**, 642 (1965).
65CB1476	R. Huisgen, E. Aufderhaar, and G. Wallbillich, *Chem. Ber.* **98**, 1476 (1965).
65CB2174	R. Huisgen, R. Grashey, R. Kunz, G. Wallbillich, and E. Aufderhaar, *Chem. Ber.* **98**, 2174 (1965).
65CB2185	R. Huisgen and E. Aufderhaar, *Chem. Ber.* **98**, 2185 (1965).
65JCS3528	A. J. Bellamy and R. D. Guthrie, *J. Chem. Soc.*, 3528 (1965).
65JOC74	R. Grashey, R. Huisgen, K. K. Sun, and R. M. Moriarty, *J. Org. Chem.* **30**, 74 (1965).
65MI1	"IUPAC, Nomenclature of Organic Chemistry," Sect. C, p. 221. Butterworth, London, 1965.
66CB2017	H. Quast and S. Hunig, *Chem. Ber.* **99**, 2017 (1966).
66JCS(C)78	A. R. Katritzky and S. Musierowicz, *J. Chem. Soc. C*, 78 (1966).
66T2453	P. K. Kadaba, *Tetrahedron* **22**, 2453 (1966).
66USP3278545	R. Huisgen, R. W. Grashey, and R. J. Crabtree, U.S. Pat. 3,278,545 (1966).
67M1618	R. Huisgen, K. Adelsberger, E. Aufderhaar, H. Knupfer, and G. Wallbillich, *Monatsh. Chem.* **98**, 1618 (1967).
67TL3071	R. Fusco, P. Dalla Croce, and A. Salvi, *Tetrahedron Lett.*, 3071 (1967).
67TL4337	J. Markert and E. Fahr, *Tetrahedron Lett.*, 4337 (1967).
67USP3326889	H. A. Brown, U.S. Pat. 3,326,889 (1967).
68AG(E)321	R. Huisgen, *Angew. Chem., Int. Ed. Engl.* **7**, 321 (1968).
68G511	R. Fusco, P. Dalla Croce, and A. Salvi, *Gazz. Chim. Ital.* **98**, 511 (1968).
68JOC2291	R. Huisgen, *J. Org. Chem.* **33**, 2291 (1968).
68LA72	E. Müller, R. Beutler, and B. Zeeh, *Justus Liebigs Ann. Chem.* **719**, 72 (1968).
68LA87	E. Müller, P. Kästner, R. Beutler, W. Rundel, H. Suhr, and B. Zeeh, *Justus Liebigs Ann. Chem.* **713**, 87 (1968).
68RC247	T. Bany, *Rocz. Chem.* **42**, 247 (1968) [(*CA* **69**, 67340)].
69CB1028	H. Daniel, *Chem. Ber.* **102**, 1028 (1969).
69CB3176	T. Eicher, S. Hunig, and P. Nikolaus, *Chem. Ber.* **102**, 3176 (1969).
69CC387	N. Singh, S. Mohan, and J. S. Sandhu, *J. C. S. Chem. Commun.*, 387 (1969).
69G69	R. Fusco and P. Dalla Croce, *Gazz. Chim. Ital.* **99**, 69 (1969).
69JPR897	H. G. O. Becker, N. Sauder, and H.-J. Timpe, *J. Prakt. Chem.* **311**, 897 (1969).
69LA91	C. Grundmann and K. Flory, *Justus Liebigs Ann. Chem.* **721**, 91 (1969).
70CI(L)1216	R. N. Butler and F. L. Scott, *Chem. Ind. (London)*, 1216 (1970).
70CRV151	D. G. Neilson, R. Roger, J. W. M. Heatlie, and L. R. Newlands, *Chem. Rev.* **70**, 151 (1970).

70JHC1001	F. H. Case, *J. Heterocycl. Chem.* **7**, 1001 (1970).
70MI1	P. M. Hergenrother and L. A. Carlson, *J. Polym. Sci.* A-1, **8**, 1003 (1970).
70MI2	R. H. De Wolfe, "Carboxylic Ortho Acid Derivatives." Academic Press, New York, 1970.
70T3069	C. M. Gupta, A. P. Bhaduri, and N. M. Khanna, *Tetrahedron* **26**, 3069 (1970).
70USP3515603	H. A. Brown, J. G. Erickson, D. R. Husted, and C. D. Wright, U.S. Pat. 3,515,603 (1970).
70ZAK500	P. P. Kish and Yu. K. Onishchenko, *Zh. Anal Khim.* **25**, 500 (1970).
71AG(E)728	K. Burger and J. Fehn, *Angew. Chem., Int. Ed. Engl.* **10**, 728 (1971).
71AG(E)729	K. Burger and J. Fehn, *Angew. Chem., Int. Ed. Engl.* **10**, 729 (1971).
71BCJ780	I. Hirao, Y. Kato, T. Hayakawa, and H. Tateishi, *Bull. Chem. Soc. Jpn.* **44**, 780 (1971).
71BSF3296	J. Daunis, Y. Guindo, R. Jacquier, and P. Viallefont, *Bull. Soc. Chim. Fr.*, 3296 (1971).
71CB3816	W. Steglich, P. Gruber, H.-U. Heininger, and F. Kneidl, *Chem. Ber.* **104**, 3816 (1971).
71JCS(B)1648	G. V. Boyd and A. J. H. Summers, *J. Chem. Soc. B*, 1648 (1971)
71JHC173	F. H. Case, *J. Heterocycl. Chem.* **8**, 173 (1971).
71JHC1043	F. H. Case, *J. Heterocycl. Chem.* **8**, 1043 (1971).
71TL2717	R. Sustmann, *Tetrahedron Lett.*, 2717 (1971).
72CB1258	R. Huisgen, H. Stangl, H. J. Sturm, R. Raab, and K. Bunge, *Chem. Ber.* **105**, 1258 (1972).
72CB1279	K. Bunge, R. Huisgen, R. Raab, and H. Stangl, *Chem. Ber.* **105**, 1279 (1972).
72CB1307	K. Bunge, R. Huisgen, R. Raab and H. J. Sturm, *Chem. Ber.* **105**, 1307 (1972).
72CB3814	K. Burger and J. Fehn, *Chem. Ber.* **105**, 3814 (1972).
72HCA745	H. Giezendanner, M. Märky, B. Jackson, H.-J. Hansen, and H. Schmid, *Helv. Chim. Acta* **55**, 745 (1972).
72JA3243	A. I. Meyers and N. Nazarenko, *J. Am. Chem. Soc.* **94**, 3243 (1972).
72JCS(P2)44	A. F. Hegarty, M. P. Cashman, and F. L. Scott, *J. C. S. Perkin* **2**, 44 (1972).
72LA9	K. Burger, J. Fehn, and A. Gieren, *Justus Liebigs Ann. Chem.* **757**, 9 (1972).
72RCR495	Yu. P. Kitaev and B. I. Buzykin, *Russ. Chem. Rev. (Engl. Transl.)* **41**, 495 (1972).
73BCJ1250	T. Isida, T. Akiyama, N. Mihara, S. Kozima, and K. Sisido, *Bull. Chem. Soc. Jpn.* **46**, 1250 (1973).
73BSF2871	J. Bastide, J. Hamelin, F. Texier, and Y. V. Quang, *Bull. Soc. Chim. Fr.*, 2871 (1973).
73CB3421	K. Burger and K. Einhellig, *Chem. Ber.* **106**, 3421 (1973).
73JA7287	K. N. Houk, J. Sims, R. E. Duke, Jr., R. W. Strozier, and J. K. George, *J. Am. Chem. Soc.* **95**, 7287 (1973).
73JHC353	F. H. Case, *J. Heterocycl. Chem.* **10**, 353 (1973).
73JOC36	A. I. Meyers, A. Nabeya, H. W. Adickes, I. R. Politzer, G. R. Malone, A. C. Kovelevsky, R. L. Nolen, and R. C. Portnoy, *J. Org. Chem.* **38**, 36 (1973).

73RCR392	V. V. Mezheritskii, E. P. Olekhnovich, and G. N. Dorofeenko, *Russ. Chem. Rev. (Engl. Transl.)* **42**, 392 (1973).
73S414	H. G. O. Becker, H. D. Steinleitner, and H. J. Timpe, *Synthesis*, 414 (1973).
73T121	A. S. Shawali and H. M. Hassaneen, *Tetrahedron* **29**, 121 (1973).
74AG481	K. Burger, W. Thenn, and A. Gieren, *Angew. Chem.* **86**, 481 (1974).
74AG482	A. Gieren, P. Narayanan, K. Burger, and W. Thenn, *Angew. Chem.* **86**, 482 (1974).
74AG(E)474	K. Burger, W. Thenn, and A. Gieren, *Angew. Chem., Int. Ed. Engl.* **13**, 474 (1974).
74AG(E)475	A. Gieren, P. Narayanan, K. Burger, and W. Thenn, *Angew. Chem., Int. Ed. Engl.* **13**, 475 (1974).
74HCA1382	P. Gilgen, H. Heimgartner, and H. Schmid, *Helv. Chim. Acta* **57**, 1382 (1974).
74JCS(P1)638	W. D. Ollis and C. A. Ramsden, *J. C. S. Perkin I*, 638 (1974)
74PAC569	R. Sustmann, *Pure Appl. Chem.* **40**, 569 (1974).
74TL2649	G. Doleschall, *Tetrahedron Lett.*, 2649 (1974).
75CB1460	K. Burger, W. Thenn, R. Rauh, H. Schickaneder, and A. Gieren, *Chem. Ber.* **108**, 1460 (1975).
75CJC3782	J. W. Lown and B. E. Landberg, *Can. J. Chem.* **53**, 3782 (1975).
75JCS(P1)1433	M. J. Cooper, R. Hull, and M. Wardleworth, *J. C. S. Perkin I*, 1433 (1975).
75JCS(P1)2474	S. S. Mathur and H. Suschitzky, *J. C. S. Perkin I*, 2474 (1975).
75JHC143	P. K. Kadaba, *J. Heterocycl. Chem.* **12**, 143 (1975).
75JOC2021	A. I. Meyers and J. C. Durandetta, *J. Org. Chem.* **40**, 2021 (1975).
75MI1	K. M. Watson and D. G. Neilson, in "The Chemistry of Amidines and Imidates" (S. Patai, ed.), p. 491. Wiley (Interscience), New York, 1975.
75S483	D. J. Anderson and A. Hassner, *Synthesis*, 483 (1975).
75TL681	G. Doleschall, *Tetrahedron Lett.*, 681 (1975).
75TL1889	G. Doleschall, *Tetrahedron Lett.*, 1889 (1975).
76ACH419	G. Doleschall, *Acta Chim. Acad. Sci. Hung.* **90**, 419 (1976).
76CC734	J. E. Baldwin, *J. C. S. Chem. Commun.*, 734 (1976).
76CC736	J. E. Baldwin, *J. C. C Chem. Commun.*, 736 (1976).
76CPB2568	Y. Tamura, H. Hayashi, and M. Ikeda, *Chem. Pharm. Bull.* **24**, 2568 (1976).
76CPB3011	T. Yamazaki, K. Matoba, S. Imoto, and M. Terashima, *Chem. Pharm. Bull.* **24**, 3011 (1976).
76IJC(B)425	A. S. Shawali and H. M. Hassaneen, *Indian J. Chem., Sect. B* **14B**, 425 (1976).
76JA1048	A. Padwa, A. Ku, A. Mazzu, and S. I. Wetmore, Jr., *J. Am. Chem. Soc.* **98**, 1048 (1976).
76JA6397	P. Caramella and K. N. Houk, *J. Am. Chem. Soc.* **98**, 6397 (1976).
76JHC835	P. K. Kadaba, B. Stanovnik, and M. Tisler, *J. Heterocycl. Chem.* **13**, 835 (1976).
76JOC403	R. Huisgen, *J. Org. Chem.* **41**, 403 (1976).
76LA30	K. Burger, H. Schickaneder, and M. Pinzel, *Liebigs Ann. Chem.*, 30 (1976).
76MI1	I. Fleming, "Frontier Orbitals and Organic Chemical Reactions," Wiley, New York, 1976.

76MI2	I. Zugravescu and M. Petrovanu, "N-Ylid Chemistry," p. 315. McGraw-Hill, New York, 1976.
76T2165	R. M. Kellogg, *Tetrahedron* **32**, 2165 (1976).
76T2549	G. Doleschall, *Tetrahedron* **32**, 2549 (1976).
76TL1303	K. Hirai, Y. Iwano, T. Saito, T. Hiraoka, and Y. Kishida, *Tetrahedron Lett.*, 1303 (1976).
76ZOR1676	B. I. Buzykin, L. P. Sysoeva, and Y. P. Kitaev, *Zh. Org. Khim.* **12**, 1676 (1976).
77AG(E)10	W. Oppolzer, *Angew. Chem., Int. Ed. Engl.* **16**, 10 (1977).
77BJC969	S. Ito, Y. Tanaka, A. Kakehi, and H. Miyazawa, *Bull. Chem. Soc. Jpn.* **50**, 2969 (1977).
77CB500	R. Huisgen, R. Fleischmann, and A. Eckell, *Chem. Ber.* **110**, 500 (1977).
77CB514	R. Huisgen, R. Fleischmann, and A. Eckell, *Chem. Ber.* **110**, 514 (1977).
77CB571	A. Eckell and R. Huisgen, *Chem. Ber.* **110**, 571 (1977).
77GEP152193	K. Hirai, Y. Iwano, T. Saito, T. Hiraoka, Y. Kishida, and T. Nishimura, *Ger. Pat.* 152193 (1977).
77H143	P. Gilgen, H. Heimgartner, H. Schmid, and H.-J. Hansen, *Heterocycles* **6**, 143 (1977).
77JA385	P. Caramella, R. W. Gandour, J. A. Hall, C. G. Deville, and K. N. Houk, *J. Am. Chem. Soc.* **99**, 385 (1977).
77JHC1089	A. S. Shawali and A.-G. A. Fahmi, *J. Heterocycl. Chem.* **14**, 1089 (1977).
77JOC443	A. Kakehi, S. Ito, K. Uchiyama, Y. Konno, and K. Kondo, *J. Org. Chem.* **42**, 443 (1977).
77LA463	W. Walter and C. Rohloff, *Liebigs Ann. Chem.*, 463 (1977)
77LA485	W. Walter and C. Rohloff, *Liebigs Ann. Chem.*, 485 (1977).
77LA498	R. Huisgen, R. Grashey, and R. Krischke, *Liebigs Ann. Chem.* 498 (1977).
77LA506	R. Huisgen, R. Grashey, and R. Krischke, *Liebigs Ann. Chem.* 506 (1977).
77ZN(B)607	K. Burger, H. Goth, and W.-D. Roth, *Z. Naturforsch., B. Anorg. Chem., Org. Chem.* **32B**, 607 (1977).
78BCJ1846	C. Yamazaki, *Bull. Chem. Soc. Jpn.* **51**, 1846 (1978).
78JHC161	A. Walser and G. Zenchoff, *J. Heterocycl. Chem.* **15**, 161 (1978).
78JHC401	E. Campaigne and T. P. Selby, *J. Heterocycl. Chem.* **15**, 401 (1978).
78JHC1485	M. Ruccia, N. Vivona, G. Cusmano, and G. Macaluso, *J. Heterocycl. Chem.* **15**, 1485 (1978).
78MI1	K. Burger and R. Ottlinger, *J. Fluorine Chem.* **12**, 519 (1978).
78TL1295	C. Yamazaki, *Tetrahedron Lett.*, 1295 (1978).
79CCC1334	T. Vanek, J. Farkas, and J. Gut, *Collect. Czech. Chem. Commun.* **44**, 1334 (1979).
79CR(C)265	J. Vebrel, E. Cerutti, and R. Carrie, *C. R. Hebd. Seances Acad. Sci., Ser. C* **288**, 265 (1979).
79CSC341	R. Oberti, M. C. Domeneghetti, and R. Gandolfi, *Cryst. Struct. Commun.* **8**, 341 (1979).
79JHC555	C. A. Lovelette, *J. Heterocycl. Chem.* **16**, 555 (1979).
79JOC2688	J. B. Hester, Jr., C. G. Chidester, and J. Szmuszkovicz, *J. Org. Chem.* **44**, 2688 (1979).
79JOC3803	K. T. Potts and S. Kanemasa, *J. Org. Chem.* **44**, 3803 (1979).
79MI1	R. E. Valter, "Ring-Chain Isomerism in Organic Chemistry" (in Russian). Zinatne. Riga, 1979.

79MI2	K. N. Houk, *Top. Curr. Chem.* **79,** 1 (1979).
79RRC733	M. Petrovanu, C. Luchian, G. Surpateanu, and V. Barboiu, *Rev. Roum. Chim.* **24,** 733 (1979).
79RRC1053	M. Petrovanu, C. Luchian, G. Surpateanu, and V. Barboiu, *Rev. Roum. Chim.* **24,** 1053 (1979).
79USP4140693	P. C. Wade, B. R. Vogt, and T. P. Kissick, U.S. Pat. 4,140,693 (1979).
79ZOR1181	B. S. Drach and T. P. Popovich, *Zh. Org. Khim.* **15,** 1181 (1979).
79ZOR2280	V. A. Khrustalev, K. N. Zelenin, and V. P. Sergutina, *Zh. Org. Khim.* **15,** 2280 (1979).
80AG936	J. L. Flippen-Anderson, I. Karle, R. Huisgen, and H. U. Reissig, *Angew. Chem.* **92,** 936 (1980).
80AG(E)906	J. L. Flippen-Anderson, I. Karle, R. Huisgen, and H. U. Reissig, *Angew. Chem., Int. Ed. Engl.* **19,** 906 (1980).
80BCJ2007	S. Ito, A. Kakehi, T. Matsuno, and J. Yoshida, *Bull. Chem. Soc. Jpn.* **53,** 2007 (1980).
80BCJ3289	C. Yamazaki, *Bull. Chem. Soc. Jpn.* **53,** 3289 (1980).
80H929	K. Dietliker, W. Stegmann, and H. Heimgartner, *Heterocycles* **14,** 929 (1980).
80JHC311	F. Anzani and P. Dalla Croce, *J. Heterocycl. Chem.* **17,** 311 (1980).
80JHC833	A. S. Shawali and C. Parkanyi, *J. Heterocycl. Chem.* **17,** 833 (1980).
80KGS1138	V. A. Khrustalev, K. N. Zelenin, V. P. Sergutina, and V. V. Pinson, *Khim. Geterotsikl. Soedin.,* 1138 (1980).
80T935	R. Gandolfi and L. Toma, *Tetrahedron* **36,** 935 (1980).
80T1565	T. Sasaki, S. Eguchi, and Y. Tanaka, *Tetrahedron* **36,** 1565 (1980).
80T1649	G. Doleschall and G. Toth, *Tetrahedron* **36,** 1649 (1980).
80TL209	B. I. Buzykin, A. P. Stolyarov, and N. N. Bystrykh, *Tetrahedron Lett.* **21,** 209 (1980).
80TL4183	G. Doleschall, *Tetrahedron Lett.* **21,** 4183 (1980).
80ZOR942	K. N. Zelenin, V. A. Khrustalev, and V. P. Sergutina, *Zh. Org. Khim.* **16,** 942 (1980).
81BCJ1767	C. Yamazaki, *Bull. Chem. Soc. Jpn.* **54,** 1767 (1981).
81JHC247	K. Burger, S. Tremmel, W.-D. Roth, and H. Goth, *J. Heterocycl. Chem.* **18,** 247 (1981).
81JOC3956	C. Yamazaki, *J. Org. Chem.* **46,** 3956 (1981).
81MI1	C. Temple, Jr., "Triazoles-1,2,4." Wiley, New York, 1981.
81T2805	M. Petrovanu, C. Luchian, G. Surpateanu, and V. Barboiu, *Tetrahedron* **37,** 2805 (1981).
81T2811	M. Petrovanu, C. Luchian, G. Surpateanu, and V. Barboiu, *Tetrahedron* **37,** 2811 (1981).
81ZOR1825	K. N. Zelenin, V. A. Khrustalev, V. P. Sergutina, and V. V. Pinson, *Zh. Org. Khim.* **17,** 1825 (1981).
82CJC285	P. Métra and J. Hamelin, *Can. J. Chem.* **60,** 285 (1982).
82JCS(P1)2663	H. A. Elfahham, K. U. Sadek, G. E. H. Elgemeie, and M. H. Elnagdi, *J. C. S. Perkin I,* 2663 (1982).
82JHC1573	R. S. Tewari, P. D. Dixit, and P. Parihar, *J. Heterocycl. Chem.* **19,** 1573 (1982).
82JOC4409	E. Diez-Barra, M. C. Pardo, and J. Elguero, *J. Org. Chem.* **47,** 4409 (1982).
82KGS1100	I. A. Litvinov, Yu. T. Struchkov, N. N. Bystrykh, and B. I. Buzykin, *Khim. Geterotsikl. Soedin.,* 1100 (1982).

82KGS1264	V. A. Khrustalev, V. P. Sergutina, K. N. Zelenin, and V. V. Pinson, *Khim. Geterotsikl. Soedin.*, 1264 (1982).
82ZOR1613	K. N. Zelenin, V. V. Pinson, and V. A. Khrustalev, *Zh. Org. Khim.* **18**, 1613 (1982).
83BSB811	A. Padwa, Y.-Y. Chen, K. F. Koehler, and M. Tomas, *Bull. Soc. Chim. Belg.* **92**, 811 (1983).
83CB186	G. Scherowsky and J. Pickardt, *Chem. Ber.* **116**, 186 (1983).
83EUP72029	R. Borer, M. Gerecke, and E. Kyburz, Eur. Pat. Appl. EP72,029 (1983).
83IJC(B)1244	D. Prajapati, J. S. Sandhu, and J. N. Baruah, *Indian J. Chem., Sect. B* **22B**, 1244 (1983).
83JCS(P1)2003	G. Barta-Szalai, J. Fetter, K. Lempert, J. Moller, and L. Parkanyi, *J. C. S. Perkin 1*, 2003 (1983).
83JCS(P2)1317	E. Diez-Barra, M. C. Pardo, J. Elguero, and J. Arriau, *J. C. S. Perkin 2*, 1317 (1983).
83JIC961	R. S. Tewari and P. Parihar, *J. Indian Chem. Soc.* **60**, 961 (1983).
83MI1	P. A. S. Smith, "Derivatives of Hydrazine and other Hydro-nitrogens having N-N Bonds." Benjamin/Cummings, Reading, Massachusetts, 1983.
83S483	M. Perez, C. Dorado, and J. Soto, *Synthesis*, 483 (1983).
83T129	R. S. Tewari and P. Parihar, *Tetrahedron* **39**, 129 (1983).
83ZOR1069	B. I. Buzykin and N. N. Bystrykh, *Zh. Org. Khim.* **19**, 1069 (1983).
84ABC2913	Y. Sanemitsu, Y. Nakayama, and S. Inoue, *Agric. Biol. Chem.* **48**, 2913 (1984).
84AHC217	P. K. Kadaba, B. Stanovnik, and M. Tisler, *Adv. Heterocycl. Chem.* **37**, 217 (1984).
84CB1194	W. Fliege, R. Grashey, and R. Huisgen, *Chem. Ber.* **117**, 1194 (1984).
84CCC1713	P. Zalupsky and A. Martvon, *Collect. Czech. Chem. Commun.* **49**, 1713 (1984).
84H537	L. Fodor, M-S. El-Gharib, J. Szabo, G. Bernath, and P. Sohar, *Heterocycles* **22**, 537 (1984).
84H549	Y. Ohshiro, M. Komatsu, M. Uesaka, and T. Agawa,*Heterocycles* **22**, 549 (1984).
84HCA534	T. Büchel, R. Prewo, J. H. Bieri, and H. Heimgartner, *Helv. Chim. Acta* **67**, 534 (1984).
84JCR(S)56	D. Prajapati, J. S. Sandhu, and J. N. Baruah, *J. Chem. Res., Synop.*, 56 (1984).
84JOC1703	Y. Nakayama and Y. Sanemitsu, *J. Org. Chem.* **49**, 1703 (1984).
84KGS1415	V. V. Pinson, V. A. Khrustalev, K. N. Zelenin, and Z. M. Matveeva, *Khim. Geterotsikl. Soedin.*, 1415 (1984).
84MI1	A. Padwa (Ed.), "1,3-Dipolar Cycloaddition Chemistry," Vol. 1. Wiley, New York, 1984.
84MI2	A. Padwa (Ed.), "1,3-Dipolar Cycloaddition Chemistry," Vol. 2. Wiley, New York, 1984.
84MI3	J. B. Polya, *Compr. Heterocycl. Chem.* **5**, 733 (1984).
84OMR720	P. Sohar, L. Fodor, J. Szabo, and G. Bernath, *Org. Magn. Reson.* **22**, 720 (1984).
84T369	Z. Bende, L. Toke, L. Weber, G. Toth, F. Janke, and G. Csonka, *Tetrahedron* **40**, 369 (1984)
84TL65	F. Bronberger and R. Huisgen, *Tetrahedron Lett.* **25**, 65 (1984).
84ZOR659	L. P. Vasil'eva, G. S. Akimova, and V. N. Chistokletov, *Zh. Org. Khim.* **20**, 659 (1984).

85AP556	G. Dannhardt and I. Sommer, *Arch. Pharm. (Weinheim, Ger.)* **318**, 556 (1985).
85CB5009	R. F. Abdulla, N. D. Jones, and J. K. Swartzendruber, *Chem. Ber.* **118**, 5009 (1985).
85CPB2678	Y. Miyamoto, *Chem. Pharm. Bull.* **33**, 2678 (1985).
85H1123	D. Prajapati, J. S. Sandhu, T. Kametani, H. Nagase, K. Kawai, and T. Honda, *Heterocycles* **23**, 1123 (1985).
85H2051	G. Capozzi, A. Chimirri, S. Grasso, G. Romeo, and G. Zappia, *Heterocycles* **23**, 2051 (1985).
85H2183	E. Sayanna, R. V. Venkataratnam, and G. Thyagarajan, *Heterocycles* **23**, 2183 (1985).
85JAP(K)60172983	Y. Nakayama, M. Sanemitsu, K. Maeda, and S. Inoue, *Jpn. Kokai Tokkyo Koho* JP 60, 172, 983 [85, 172, 983], (1985).
85JST271	A. Kalman, G. Y. Argay, P. Sohar, L. Fodor, J. Szabo, and G. Bernath, *J. Mol. Struct.* **129**, 271 (1985).
85MI1	B. Alcaide, G. Escobar, R. Perez-Ossorio, and J. Plumet, *An. Quim., Ser. C* **81**, 85 (1985) [*CA* **106**, 196329 (1987)].
85MI2	A. Schwan and J. Warkentin, *Proc. Int. Congr. Heterocycl. Chem., 10th, 1985* Abstr., G2-12 (1985).
86EUP189300	T. Shida, T. Watanabe, S. Yamazaki, H. Shinkawa, and K. Satake, Eur. Pat. Appl. EP 189,300 (1986).
86JCR(S)200	G. Capozzi, R. Ottana, G. Romeo, and G. Valle, *J. Chem. Res., Synop.*, 200 (1986).
86JHC1431	M. C. Aversa, A. Ferlazzo, P. Giannetto, F. H. Kohnke, A. M. Z. Slawin, and D. J. Williams, *J. Heterocycl. Chem.* **23**, 1431 (1986).
86KGS352	N. V. Belova, L. B. Volodarskii, and A. Ya. Tikhonov, *Khim. Geterotsikl. Soedin.*, 352 (1986).
86S230	M. C. Aversa, A. Ferlazzo, P. Giannetto, and F. H. Kohnke, *Synthesis*, 230 (1986).
87AX(C)168	J. L. Flippen-Anderson, *Acta Crystallogr., Sect. C* **C43**, 168 (1987).
87MRC635	P. Sohar, G. Stajer, and G. Bernath, *Magn. Reson. Chem.* **25**, 635 (1987).
87T1931	G. Stajer, A. E. Szabo, G. Bernath, and P. Sohar, *Tetrahedron* **43**, 1931 (1987).
87UP1	P. K. Kadaba, unpublished observations.

Cumulative Index of Authors, Volumes 1–45

Abboud, J. L. M., see Catalan, J., **41**, 187.
Abramovitch, R. A., Saha, J. G., *Substitution in the Pyridine Series: Effect of Substituents*, **6**, 229.
Abramovitch, R. A., Spenser, I. D., *The Carbolines*, **3**, 79.
Acheson, R. M., *Reactions of Acetylenecarboxylic Acids and Their Esters with Nitrogen-Containing Heterocyclic Compounds*, **1**, 125.
Acheson, R. M., Elmore, N. F., *Reactions of Acetylenecarboxylic Esters with Nitrogen-Containing Heterocycles*, **23**, 263.
Adam, W., *The Chemistry of 1,2-Dioxetanes*, **21**, 437.
Albert, A., *4-Amino-1,2,3-triazoles*, **40**, The Chemistry of 8-Azapurines (1,2,3-Triazolo[4,5-d] pyrimidines), **39**, 117; *Annelation of a Pyrimidine Ring to an Existing Ring*, **32**, 1; *Covalent Hydration in Nitrogen Heterocycles*, **20**, 117.
Albert, A., Armarego, W. L. F., *Covalent Hydration in Nitrogen-Containing Heteroaromatic Compounds. I. Qualitative Aspects*, **4**, 1.
Albert, A., Yamamoto, H., *Heterocyclic Oligomers*, **15**, 1.
Anastassiou, A. G., Kasmai, H. S., *Medium-Large and Large π-Excessive Heteroannulenes*, **23**, 55.
Anderson, P. S., see Lyle, R. L., **6**, 45.
ApSimon, J. W., see Paré, J. R. J., **42**, 335.
Arán, V. J., Goya, P., Ochoa, C., *Heterocycles Containing the Sulfamide Moiety*, **44**, 81.
Armarego W. L. F., *Quinazolines*, **1**, 253; **24**, 1.
Armarego W. L. F., see Albert, A., **4**, 1.

Ashby, J., Cook, C. C., *Recent Advances in the Chemistry of Dibenzothiophenes*, **16**, 181.
Avendano Lopez, C., Gonzalez Trigo, G., *The Chemistry of Hydantoins*, **38**, 177.
Badger, G. M., Sasse, W. H. F., *The Action of Metal Catalysts on Pyridines*, **2**, 179.
Balaban, A. T., Dinculescu, A., Dorofeenko, G. N., Fischer, G., Koblik, A. V., Mezheritskii, V. V., Schroth, W., *Pyrylium Salts: Syntheses, Reactions and Physical Properties*, S2.
Balaban, A. T., Schroth, W., Fischer, G., *Pyrylium Salts, Part I. Syntheses*, **10**, 241.
Bapat, J. B., Black, D. St. C., Brown, R. F. C., *Cyclic Hydroxamic Acids*, **10**, 199.
Baram, S. G., see Mamaev, V. P., **42**, 1
Barker, J. M., *gem-Dithienylalkanes and Their Derivatives*, **32**, 83; *The Thienopyridines*, **21**, 65.
Barton, H. J., see Bojarski, J. T., **38**, 229.
Barton, J. W., *Benzo[c]cinnolines*, **24**, 151.
Beke, D., *Heterocyclic Pseudobases*, **1**, 167.
Belen'kii, L. I., *The Literature of Heterocyclic Chemistry, Part III*, **44**, 269.
Benassi, R., Folli, U., Schenetti, L., Taddei, F., The *Conformations of Acyl Groups in Heterocyclic Compounds*, **41**, 75.
Berg, U., see Gallo, R., **43**, 173.
Bhatt, M. V., see Shirwaiker, G. S., **37**, 67.
Black, D. St. C., Doyle J. E., *1-Azabicyclo[3.1.0]hexanes and Analogs with Further Heteroatom Substitution*, **27**, 1.

Black, D. St. C., see Bapat, J. B., **10**, 199.
Blaha, K., Červinka, O., *Cyclic Enamines and Imines*, **6**, 147.
Bobbitt, J. M., *The Chemistry of 4-Oxy- and 4-Keto-1,2,3,4-tetrahydroisoquinolines*, **15**, 99.
Bodea, C., Silberg, I., *Recent Advances in the Chemistry of Phenothiazines*, **9**, 321.
Bojarski, J. T., Mokrosz, J. L., Barton, H. J., Paluchowska, M. H., *Recent Progress in Barbituric Acid Chemistry*, **38**, 229.
Bonnett, R., North, S. A., *The Chemistry of the Isoindoles*, **29**, 341.
Bosshard, P., Eugster, C. H., *The Development of the Chemistry of Furans, 1952–1963*, **7**, 377.
Boulton, A. J., Ghosh, P. B., *Benzofuroxans*, **10**, 1.
Boulton, A. J., see Gasco, A., **29**, 251; Wuensch, K. H., **8**, 277.
Bradsher, C. K., *Cationic Polar Cycloaddition*, **16**, 289; **19**, xi.
Brown, C., Davidson, R. M., *1,4-Benzothiazines, Dihydro-1,4-benzothiazines, and Related Compounds*, **38**, 135.
Brown, R. F. C., see Bapat, J. B., **10**, 199.
Broy, W., see Mayer, R., **8**, 219.
Bryce, M. R., Vernon, J. M., *Reactions of Benzyne with Heterocyclic Compounds*, **28**, 183.
Bulka, E., *The Present State of Selenazole Chemistry*, **2**, 343.
Bunting, J. W., *Heterocyclic Pseudobases*, **25**, 1.
Butler, R. N., *Recent Advances in Tetrazole Chemistry*, **21**, 323.
Cagniant, P., Cagniant, D., *Recent Advances in the Chemistry of Benzo[b]furan and Its Derivatives. Part I. Occurrence and Synthesis*, **18**, 337.
Calf, G. E., Garnett, J. L., *Isotopic Hydrogen Labeling of Heterocyclic Compounds by One-Step Methods*, **15**, 137.
Catala Noble, A., see Popp, F. D., **8**, 21.
Catalan, J., Abboud, J. L. M., Elguero, J., *Basicity and Acidity of Azoles*, **41**, 187.

Červinka, O., see Blaha, K., **6**, 147.
Chambers, R. D., Sargent, C. R., *Polyfluoroheteroaromatic Compounds*, **28**, 1.
Charushin, V. N., Chupakhin, O. N., van der Plas, H. C., *Reactions of Azines with Bifunctional Nucleophiles: Cyclizations and Ring Transformations*, **43**, 301.
Cheeseman, G. W. H., *Recent Advances in Quinoxaline Chemistry*, **2**, 203.
Cheeseman, G. W. H., Werstiuk, E. S. G., *Quinoxaline Chemistry: Developments 1963–1975*, **22**, 367; *Recent Advances in Pyrazine Chemistry*, **14**, 99.
Chupakhin, O. N., see Charushin, V. N., **43**, 301.
Clapp, L. B., *1,2,4-Oxadiazoles*, **20**, 65.
Claramunt, R. M., see Elguero, J., **22**, 183.
Cleghorn, H. P., see Lloyd, D., **17**, 27.
Comins, D. L., O'Connor, S., *Regioselective Substitution in Aromatic Six-Membered Nitrogen Heterocycles*, **44**, 199
Cook, C. C., see Ashby, J., **16**, 181.
Cook, M. J., Katritzky, A. R., Linda, P., *Aromaticity of Heterocycles*, **17**, 255.
Crabb, T. A., Katritzky, A. R., *Conformational Equilibria in Nitrogen-Containing Saturated Six-Membered Rings*, **36**, 1.
Daltrozzo, E., see Scheibe, G., **7**, 153.
Davidson, J. L., Preston, P. N., *Use of Transition Organometallic Compounds in Heterocyclic Synthesis*, **30**, 319.
Davidson, R. M., see Brown, C., **38**, 135.
Davis, M., *Benzisothiazoles*, **14**, 43; *Recent Advances in the Chemistry of Benzisothiazoles and Other Polycyclic Isothiazoles*, **38**, 105; *Sulfur Transfer Reagents in Heterocyclic Synthesis*, **30**, 47.
Deady, L. W., see Zoltewicz, J. A., **22**, 71.
Dean, F. M., *Recent Advances in Furan Chemistry, Part I*, **30**, 167; *Part II*, **31**, 237.
den Hertog, H. J., van der Plas, H. C., *Hetarynes*, **4**, 121.

Dinculescu, A., see Balaban, A. T., S2.
Donald, D. S., Webster, O. W., *Synthesis of Heterocycles from Hydrogen Cyanide Derivatives*, **41**, 1.
Dorofeenko, G. N., see Balaban, A. T., S2.
Dou, H. J. M., see Gallo, R. J., **36**, 175.
Doyle, J. E., see Black, D. St. C., **27**, 1.
Drum, C., see Katritzky, A. R., **40**, 1.
Duffin, G. F., *The Quaternization of Heterocyclic Compounds*, **3**, 1.
Dyke, S. F., *1,2-Dihydroisoquinolines*, **14**, 279.
Eckstein, Z., Urbański, T., *1,3-Oxazine Derivatives*, **2**, 311; **23**, 1.
Eisch, J. J., *Halogenation of Heterocyclic Compounds*, **7**, 1.
Elgemeie, G. E. H., see Elnagdi, M. H., **41**, 319.
Elguero, J., see Catalan, J., **41**, 187.
Elguero, J., Claramunt, R. M., Summers, A. J. H., *The Chemistry of Aromatic Azapentalenes*, **22**, 183.
Elguero, J., Marzin, C., Katritzky, A. R., Linda, P., *The Tautomerism of Heterocycles*, S1.
Elmoghayar, M. R. H., see Elnagdi, M. H., **41**, 319.
Elmore, N. F., see Acheson, R. M., **23**, 263.
Elnagdi, M. H., Elgemeie, G. E. H., Elmoghayar, M. R. H., *Chemistry of Pyrazolopyrimidines*, **41**, 319.
El'tsov, A. V., see Timpe, H. J., **33**, 185.
Elvidge, J. A., Jones, J. R., O'Brien, C., Evans, E. A., Sheppard, H. C., *Based-Catalyzed Hydrogen Exchange*, **16**, 1.
Eugster, C. H., see Bosshard, P., **7**, 377.
Evans, E. A., see Elvidge, J. A., **16**, 1.
Fedrick, J. L., see Shepherd, R. G., **4**, 145.
Ferles, M., Pliml, J., *3-Piperideines (1,2,3,6-Tetrahydropyridines)*, **12**, 43.
Filler, R., *Recent Advances in Oxazolone Chemistry*, **4**, 75.
Filler, R., Rao, Y. S., *New Developments in the Chemistry of Oxazolones*, **21**, 175.
Fischer, G. W., see Balaban, A. T., **10**, 241; S2.
Fletcher, I. J., Siegrist, A. E., *Olefin Synthesis with Anils*, **23**, 171.
Flitsch, W., *The Chemistry of 4-Azaazulenes*, **43**, 35; *Hydrogenated Porphyrin Derivatives: Hydroporphyrins*, **43**, 73.
Flitsch, W., Jones, G., *The Chemistry of Pyrrolizines*, **37**, 1.
Flitsch, W., Kraemer, U., *Cyclazines and Related N-Bridged Annulenes*, **22**, 321.
Folli, U., see Benassi, R., **41**, 75.
Fowler, F. W., *Synthesis and Reactions of 1-Azirines*, **13**, 45.
Freeman, F., *The Chemistry of 1-Pyrindines*, **15**, 187.
Friedrichsen, W., *Benzo[c]furans*, **26**, 135.
Fringuelli, F., Marino, G., Taticchi, A., *Tellurophene and Related Compounds*, **21**, 119.
Fujita, E., Nagao, Y., *Chiral Induction Using Heterocycles*, **45**, 1.
Furusaki, F., see Takeuchi, Y., **21**, 207.
Gallo, R. J., Makosza, M., Dou, H. J. M., Hassanaly, P., *Applications of Phase Transfer Catalysts in Heterocyclic Chemistry*, **36**, 175.
Gallo, R. J., Roussel, C., Berg, U., *The Quantitative Analysis of Steric Effects in Heteroaromatics*, **43**, 173.
Gardini, G. P., *The Oxidation of Monocyclic Pyrroles*, **15**, 67.
Garnett, J. L., see Calf, G. E., **15**, 137.
Gasco, A., Boulton, A. J., *Furoxans and Benzofuroxans*, **29**, 251.
George, M. V., Khetan, S. K., Gupta, R. K., *Synthesis of Heterocycles through Nucleophilic Addition to Acetylenic Esters*, **19**, 279.
Ghosh, P. B., see Boulton, A. J., **10**, 1.
Gilchrist, T. L., *Ring-Opening of Five-Membered Heteroaromatic Anions*, **41**, 41.
Gilchrist, T. L., Gymer, G. E., *1,2,3-Triazoles*, **16**, 33.
Glushkov, R. G., Granik, V. G., *The Chemistry of Lactim Ethers*, **12**, 185.
Gol'dfarb, Ya. L., see Litvinov, V. P., **19**, 123.

Gompper, R., *The Reactions of Diazomethane with Heterocyclic Compounds*, **2**, 245.
Gonzalez Trigo, G., see Avendano Lopez, C., **38**, 177.
Goya, P., see Arán V. J., **44**, 81.
Grandberg, I. I., see Kost, A. N., **6**, 347.
Granik, V. G., see Glushkov, R. G., **12**, 185.
Griffin, T. S., Woods, T. S., and Klayman, D. L., *Thioureas in the Synthesis of Heterocycles*, **18**, 99.
Grimmett, M. R., *Advances in Imidazole Chemistry*, **12**, 103; **27**, 241.
Grimmett, M. R., Keene, B. R. T., *Reactions of Annular Nitrogens of Azines with Electrophiles*, **43**, 127.
Gronowitz, S., *Recent Advances in the Chemistry of Thiophenes*, **1**, 1.
Gupta, R. K., see George, M. V., **19**, 279.
Gut, J., *Aza Analogs of Pyrimidine and Purine Bases of Nucleic Acids*, **1**, 189.
Gymer, G. E., see Gilchrist, T. L., **16**, 33.
Hanson, P., *Heteroaromatic Radicals, Part I: General Properties; Radicals with Group V Ring Heteroatoms*, **25**, 205; *Part II: Radicals with Group VI and Groups V and VI Ring Heteroatoms*, **27**, 31.
Hardy, C. R., *The Chemistry of Pyrazolopyridines*, **36**, 343.
Heacock, R. A., *The Aminochromes*, **5**, 205.
Heacock, R. A., Kasparek, S., *The Indole Grignard Reagents*, **10**, 43.
Heinz, B., See Ried, W., **35**, 199.
Hermecz, I., *Chemistry of Diazabicycloundecene (DBU) and Other Pyrimidoazepines*, **42**, 83.
Hermecz, I., Vasvari-Debreczy, L., *Tricyclic Compounds with a Central Pyrimidine Ring and One Bridgehead Nitrogen*, **39**, 281.
Hermecz, I. Meszaros, Z., *Chemistry of Pyrido[1,2-a]pyrimidines*, **33**, 241.
Hettler, H., *3-Oxo-2,3-dihydrobenz[d]isothiazole-1,1-dioxide (Saccharin) and Derivatives* **15**, 233.
Hetzheim, A., Moeckel, K., *Recent Advances in 1,3,4-Oxadiazole Chemistry*, **7**, 183.
Hewitt, D., *The Chemistry of Azaphosphorines*, **43**, 1.
Hibino, S., see Kametani, T., **42**, 245.
Hiremath, S. P., Hosmane, R. S., *Applications of Nuclear Magnetic Spectroscopy to Heterocyclic Chemistry: Indole and Its Derivatives*, **15**, 277.
Hiremath, S. P., Hooper, M., *Isatogens and Indolones*, **22**, 123.
Holm, A., *1,2,3,4-Thiatriazoles*, **20**, 145.
Honda, T., see Kametani, T., **39**, 181.
Hooper, M., see Hiremath, S. P. **22**, 123.
Hörnefeldt, A. B., *Selenophenes*, **30**, 127.
Hosmane, R. S., see Hiremath, S. P., **15**, 277.
Hunt, J. H., see Swinbourne, F. J., **23**, 103.
Iddon, B., *Benzol[c]thiophenes* **14**, 331.
Iddon, B., Scrowston, R. M., *Recent Advances in the Chemistry of Benzo[b]thiophenes*, **11**, 177.
Ikeda, M., see Tamura Y., **29** 71.
Illuminati, G., *Nucleophilic Heteroaromatic Substitution*, **3**, 285.
Illuminati, G., Stegel, F., *The Formation of Anionic σ-Adducts from Heteroaromatic Compounds: Structures, Rates, and Equilibra*, **34**, 305.
Ionescu, M., Mantsch, H., *Phenoxazines*, **8**, 83.
Irwin, W. J., Wibberley, D. G., *Pyridopyrimidines: 1,3,5-, 1,3,6-, 1,3,7-, and 1,3,8-Triazanaphthalenes*, **10**, 149.
Jaffé, H. H., Jones, H. L., *Applicants of the Hammett Equation to Heterocyclic Compounds*, **3**, 209.
Jankowski, K., see Paré, J. R. J., **42**, 335.
Jankowski, K., Paré, J. R. J., Wightman, R. H., *Mass Spectrometry of Nucleic Acids*, **39**, 79.
Jensen, K. A., Pedersen, C., *1,2,3,4-Thiatriazoles*, **3**, 263.
Johnson, C. D., see Tomasik, P., **20**, 1.
Johnson, F., Madroñero, R., *Heterocyclic Syntheses Involving Nitrilium Salts and Nitriles under Acidic Conditions*, **6**, 95.
Jones, G., *Aromatic Quinolizines*, **31**, 1.
Jones, G., Sliskovic, D. R., *The Chemistry of the Triazolopyridines*, **34**, 79.
Jones, G., see Flitsch, W., **37**, 1.

Jones, H. L., see Jaffé, H. H., **3**, 209.
Jones, J. R., see Elvidge, J. A., **16**, 1.
Jones, P. M., see Katritzky, A. R., **25**, 303.
Jones, R. A., *Physicochemical Properties of Pyrroles*, **11**, 383.
Joule, J. A., *Recent Advances in the Chemistry of 9H-Carbazoles*, **35**, 83.
Kadaba, P. K., Δ^3-*and* Δ^4-*1,2,3-Triazolines*, **37**, 351.
Kadaba, P. K., Stanovnik, B., Tišler M., Δ^2-*1,2,3-Triazolines*, **37**, 217.
Kametani, T., Hibino, S., *The Synthesis of Natural Heterocyclic Products by Hetero Diels-Alder Cycloaddition Reactions*, **42**, 245.
Kametani, T., Honda, T., *The Application of Aziridines to the Synthesis of Natural Products*, **39**, 181.
Kanemasa, S., see Tsuge, O., **45**, 231.
Kappe, T., Stadlbauer, W., *Isatoic Anhydrides and Their Uses in Heterocyclic Synthesis*, **28**, 127.
Kasmai, H. S.., see Anastassiou, A. G., **23**, 55.
Kasparek, S., *1-, 2- and 3-Benzazepines*, **17**, 45.
Kasparek, S., see Heacock, R. A., **10**, 43.
Katritzky, A. R., Drum, C., *Advances in Heterocyclic Chemistry: Prospect and Retrospect*, **40**, 1.
Katritzky, A. R., Jones, P. M., *The Literature of Heterocyclic Chemistry, Part II*, **25**, 303.
Katritzky, A. R., Lagowski, J. M., *Prototropic Tautomerism of Heteroaromatic Compounds. I. General Discussion and Methods of Study*, **1**, 311; *II. Six-Membered Rings*, **1**, 339; *III. Five-Membered Rings and One Hetero Atom*, **2**, 1; *IV. Five-Membered Rings with Two or More Hetero Atoms*, **2**, 27.
Katritzky, A. R., Weeds, S. M., *The Literature of Heterocyclic Chemistry*, **7**, 225.
Katritzky, A. R., see Cook, M. J., **17**, 255; Crabb, T. A., **36**, 1; Elguero, J., S1; Sammes M. P., **32**, 233; **34**, 1, 53; **35**, 375, 413.

Keay, J. G., *The Reduction of Nitrogen Heterocycles with Complex Metal Hydrides*, **39**, 1.
Keene, B. R. T., see Grimmett, M. R., **43**, 127.
Keene, B. R. T., Tissington, P., *Recent Developments in Phenanthridine Chemistry*, **13**, 315.
Khetan, S. K., see George, M. V., **19**, 279.
Kirschke, K., see Schulz M., **8**, 165.
Klayman, D. L., see Griffin, T. S., **18**, 99.
Klemm, L. H., *Syntheses of Tetracyclic and Pentacyclic Condensed Thiophene Systems*, **32**, 127.
Klinkert, G., see Swinbourne, F. J., **23**, 103.
Knabe, J., *1,2-Dihydroisoquinolines and Related Compounds*, **40**, 105.
Kobayashi, Y., Kumadaki, I., *Dewar Heterocycles and Related Compounds*, **31**, 169.
Koblik, A. V., see Balaban, A. T., S2.
Kobylecki, R. J., McKillop, A., *1,2,3-Triazines*, **19**, 215.
Kochetkov, N. K., Likhosherstov, A. M., *Advances in Pyrrolizidine Chemistry*, **5**, 315.
Kochetkov, N. K., Sokolov, S. D., *Recent Developments in Isoxazole Chemistry*, **2**, 365.
Kost, A. N., Grandberg, I. I., *Progress in Pyrazole Chemistry*, **6**, 347.
Koutecký, J. see Zahradnik. R., **5**, 69.
Kraemer, U., see Flitsch, W. **22**, 321.
Kress, T. J., see Paudler, W. W., **11**, 123.
Kricka, L. J., Vernon, J. M., *Nitrogen-Bridged Six-Membered Ring Systems: 7-Azabicyclo-[2.2.1]hepta-2,5-dienes, Naphthalen-1,4-imines, and Anthracen-9,10-imines*, **16**, 87.
Kuhla, D. E., Lombardino, J. G., *Pyrrolodiazines with a Bridgehead Nitrogen*, **21**, 1.
Kuhla, D. E., see Lombardino, J. G., **28**, 73.
Kumadaki, I., see Kobayashi, Y., **31**, 169.
Kurzer, F., *1,2,4-Thiadiazoles*, **32**, 285.
Kuthan, J., *Pyrans, Thiopyrans, and Selenopyrans*, **34**, 145.

Lagowski, J. M., see Katritzky, A. R., **1**, 311, 339; **2**, 1, 27.
Lakhan, R., Ternai, B., *Advances in Oxazole Chemistry*, **17**, 99.
Lalezari, I., Shafiee, A., Yalpani, M., *Selenium-Nitrogen Heterocycles*, **24**, 109.
Likhosherstov, A. M., see Kochetkov, N. K., **5**, 315.
Linda, P., see Cook, M. J., **17**, 255; Elguero, J., **S1**.
Lindner, E., *Metallacyclo-alkanes and -alkenes*, **39**, 237.
Lister, J. H., *Current Views on Some Physicochemical Aspects of Purines*, **24**, 215; *Physicochemical Aspects of the Chemistry of Purines*, **6**, 1.
Litvinov, V. P. Gol'dfarb, Ya, L., *The Chemistry of Thienothiophenes and Related Systems*, **19**, 123.
Lloyd, D., Cleghorn, H. P., *1,5-Benzodiazepines*, **17**, 27.
Lloyd, D., Cleghorn, H. P., Marshall, D. R., *2,3-Dihydro-1,4-diazepines*, **17**, 1.
Lombardino, J. G., Kuhla, D. E., *1,2-and 2,1-Benzothiazines and Related Compounds*, **28**, 73.
Lombardino, J. G., see Kuhla, D. E., **21**, 1.
Lozac'h, N., *1,6,6aS^{IV}-Trithiapentalenes and Related Structures*, **13**, 161.
Lozac'h, N., Stavaux, M., *The 1,2- and 1,3-Dithiolium Ions*, **27**, 151.
Lund, H., *Electrolyis of N-Heterocyclic Compounds*, **12**, 213.
Lund, H., Tabakovic, I., *Electrolysis of N-Heterocyclic Compounds, Part II*, **36**, 235.
Lyle, R. E., Anderson, P. S., *The Reduction of Nitrogen Heterocycles with Complex Metal Hydrides*, **6**, 45.
Madroñero, R., see Johnson, F., **6**, 95.
Magdesieva, N. N., *Advances in Selenophene Chemistry*, **12**, 1.
Mamaev, V. P., Shkurko, O. P., Baram, S. G., *Electron Effects of Heteroaromatic and Substituted Heteroaromatic Groups*, **42**, 1.
Mann, M. E., see White, J. D., **10**, 113.
Mantsch, H., see Ionescu, M., **8**, 83.
Marino, G., *Electrophilic Substitutions of Five-Membered Rings*, **13**, 235.
Marino, G., see Fringuelli, F., **21**, 119.
Marshall, D. R., see Lloyd, D., **17**, 1.
Marzin, C., see Elguero, J., **S1**.
Mayer, R., Broy, W., Zahradnik, R., *Monocyclic Sulfur-Containing Pyrones*, **8**, 219.
McGill, C. K., Rappa, A., *Advances in the Chichibabin Reaction*, **44**, 1.
McKillop, A., see Kobylecki, R. J., **19**, 215.
McNaught, A., *The Nomenclature of Heterocycles*, **20**, 175.
Merlini, L., *Advances in the Chemistry of Chrom-3-enes*, **18**, 159.
Meszaros, Z., see Hermecz, I., **33**, 241.
Meth-Cohn, O., Suschitzky, H., *Heterocycles by Ring-Closure of Ortho-Substituted t-Anilines—The t-Amino Effect*, **14**, 211.
Meth-Cohn, O., Tarnowski, B., *Cyclizations under Vilsmeier Conditions*, **31**, 207; *Thiocoumarins*, **26**, 115.
Mezheritskii, V. V., see Balaban, A. T., **S2**.
Minisci, F., Porta, O., *Advances in Homolytic Substitution of Heteroaromatic Compounds*, **16**, 123.
Moeckel, K., see Hetzheim, A., **7**, 183.
Mokrosz, J. L., see Bojarski, J. T., **38**,
Moody, C. J., *Azodicarbonyl Compounds in Heterocyclic Synthesis*, **30**, 1.
Moody, C. J., *Claisen Rearrangements in Heteroaromatic Systems*, **42**, 203.
Moynahan, E. B., see Popp, F. D., **13**, 1.
Nagao, Y., see Fujita, E., **45**, 1.
Nair, M. D., see Rajappa, S., **25**, 113.
Nayak, A., see Newkome, G. R., **25**, 83.
Newkome, G. R., Nayak, A., *4-Thiazolidinones*, **25**, 1.
Norman, R. O. C., Radda, G. K., *Free-Radical Substitution of Heteroaromatic Compounds*, **2**, 131.
North, S. A., see Bonnett, R., **29**, 341.
O'Brien, C., see Elvidge J. A., **16**, 1.
Ochoa, C., see Arán, V. J., **44**, 81.
O'Connor, S., see Comins, D. L., **44**, 199.
Ollis, W. D., Ramsden, C. A., *Meso-ionic Compounds*, **19**, 1.
Paluchowska, M. H., see Bojarski, J. T., **38**, 229.
Paré, J. R. J., see Jankowski, K., **39**, 79

Paré, J. R. J., Jankowski, K., ApSimon, J. W., *Mass Spectral Techniques in Heterocyclic Chemistry: Applications and Stereochemical Considerations in Carbohydrates and Other Oxygen Heterocycles,* **42,** 335.
Paudler, W. W., Kress, T. J., *The Naphthyridines,* **11,** 123.
Paudler, W. W., Sheets, R. M., *Recent Developments in Naphthyridine Chemistry,* **33,** 147.
Pedersen, C., see Jensen, K. A., **3,** 263.
Pedersen, C. Th., *1,2-Dithiole-3-thiones and 1,2-Dithiol-3-ones,* **31,** 63.
Perlmutter, H. D., *1,4-Diazocines,* **45,** 185.
Perlmutter, H. D., Trattner, R. B., *Azocines,* **31,** 115.
Perrin, D. D., *Covalent Hydration in Nitrogen Heteroaromatic Compounds. II. Quantitative Aspects,* **4,** 43.
Pliml, J., Prystas, M., *The Hilbert-Johnson Reaction of 2,4-Dialkoxypyrimidines with Halogenoses,* **8,** 115.
Pliml, J., see Ferles, M., **12,** 43.
Popp, F. D., *Developments in the Chemistry of Reissert Compounds, 1968–1978,* **24,** 187; *Reissert Compounds,* **9,** 1; *The Chemistry of Isatin,* **18,** 1.
Popp, F. D., Catala Noble, A., *The Chemistry of Diazepines,* **8,** 21.
Popp, F. D., Moynahan, E. B., *Heterocyclic Ferrocenes,* **13,** 1.
Porta, O., see Minisci, F., **16,** 123.
Porter, A. E. A., *The Chemistry of Thiophenium Salts and Thiophenium Ylids,* **45,** 151.
Preston, P. N., see Davidson, J. L., **30,** 321.
Prinzbach, H., Futterer, E., *The 1,2- and 1,3-Dithiolium Ions,* **7,** 39.
Prystas, M., see Pliml, J., **8,** 115.
Pullman, A., Pullman, B., *Electronic Aspects of Purine Tautomerism,* **13,** 77.
Pullman, B., see Kwiatkowski, J. S., **18,** 199.
Radda, G. K., see Norman, R. O. C., **2,** 131.
Rajappa, S., Nair, M. D., *Ring Synthesis of Heteroaromatic Nitro Compounds,* **25,** 113.

Ramsden, C. A., *Heterocyclic Betaine Derivatives of Alternant Hydrocarbons,* **26,** 1.
Ramsden, C. A., see Ollis, W. D., **19,** 1.
Rao, Y. S., see Filler, R., **21,** 175.
Rappa, A., see McGill, C. K., **44,** 1.
Rees, C. W., Smithen, C. E., *The Reactions of Heterocyclic Compounds with Carbenes,* **3,** 57.
Reid, S. T., *The Photochemistry of Heterocycles,* **11,** 1; *The Photochemistry of Oxygen- and Sulfur-Containing Heterocycles,* **33,** 1; *Photochemistry of Nitrogen-Containing Heterocycles,* **30,** 239.
Reinhoudt, D. N., *(2 + 2)-Cycloaddition and (2 + 2)-Cycloreversion Reactions of Heterocyclic Compounds,* **21,** 253.
Ried, W., Heinz, B., *Four-Membered Rings Containing One Sulfur Atom,* **35,** 199.
Robins, D. J., *Advances in Pyrrolizidine Chemistry,* **24,** 247.
Roussel, C., see Gallo, R., **43,** 173.
Ruccia, M., Vivona, N., Spinelli, D., *Mononuclear Heterocyclic Rearrangements,* **29,** 141.
Saha, J. G., see Abramovitch, R. A., **6,** 229.
Sammes, M. P., Katritzky, A. R., *The 2H-Imidazoles,* **35,** 375; *The 4H-Imidazoles,* **35,** 413; *The 3H-Pyrazoles,* **34,** 1; *The 4H-Pyrazoles,* **34,** 53; *The 2H- and 3H-Pyrroles,* **32,** 233.
Sandström, J., *Recent Advances in the Chemistry of 1,3,4-Thiadiazoles,* **9,** 165.
Sargent, C. R., see Chambers, R. D., **28,** 1.
Sargent, M. V., Stransky, P. O., *Dibenzofurans,* **35,** 1.
Sasaki, T., *Heteroadamantanes,* **30,** 79.
Sasse, W. H. F., see Badger, G. M., **2,** 179.
Scheibe, G., Daltrozzo, E., *Diquinolylmethane and Its Analogs,* **7,** 153.
Schenetti, L., see Benassi, R., **41,** 75.
Schmitz, E., *Three-Membered Rings with Two Hetero Atoms,* **2,** 83; **24,** 63.
Schneller, S. W., *Thiochromanones and Related Compounds,* **18,** 59.

Schroth, W., see Balaban, A. T., **10**, 241; S2.

Schulz, M., Kirschke K., *Cyclic Peroxides*, **8**, 165.

Scrowston, R. M., *Recent Advances in the Chemistry of Benzo[b]thiophenes*, **29**, 171.

Scrowston, R. M., see Iddon, B., **11**, 177.

Shafiee, A., see Lalezari, I., **24**, 109.

Shepherd, R. G., Fedrick, J. L., *Reactivity of Azine, Benzoazine, and Azinoazine Derivatives with Simple Nucleophiles*, **4**, 145.

Sheppard, H. C., see Elvidge, J. A., **16**, 1.

Shirwaiker, G. S., Bhatt, M. V., *Chemistry of Arene Oxides*, **37**, 67.

Shkurko, O. P., see Mamaev, V. P., **42**, 1.

Siegrist, A. E., see Fletcher, I. J., **23**, 171.

Silberg, I., see Bodea, C., **9**, 321.

Slack, R., Wooldridge, K. R. H., *Isothiazoles*, **4**, 107.

Sliskovic, D. R., see Jones, G., **34**, 79.

Smalley, R. K., *The Chemistry of Indoxazenes and Anthranils, 1966-1979*, **29**, 1.

Smith, G. F., *The Acid-Catalyzed Polymerization of Pyrroles and Indoles*, **2**, 287.

Smithen, C. E., see Rees, C. W., **3**, 57.

Spenser, I. D., see Abramovitch, R. A., **3**, 79.

Speranza, M., *The Reactivity of Heteroaromatic Compounds in the Gas Phase*, **40**, 25.

Spinelli, D., see Ruccia, M., **29**, 141.

Spiteller, G., *Mass Spectrometry of Heterocyclic Compounds*, **7**, 301.

Stadlbauer, W., see Kappe, T., **28**, 127.

Stanovnik, B., see Kadaba, P. K., **37**, 217; Tisler, M., **9**, 211; **24**, 363.

Stavaux, M., see Lozac'h, N., **27**, 151.

Stegel, F., see Illuminati, G., **34**, 305.

Stoodley, R. J., *1,4-Thiazines and Their Dihydro Derivatives*, **24**, 293.

Stransky, P. O., see Sargent, M. V., **35**, 1.

Summers, A. J. H., see Elguero, J., **22**, 183.

Summers, L. A., *The Bipyridines*, **35**, 281; *The Phenanthrolines*, **22**, 1.

Suschitzky, H., see Meth-Cohn, O., **14**, 211.

Swinbourne, F. J., Hunt, J. H., Klinkert, G., *Advances in Indolizine Chemistry*, **23**, 103.

Tabakovic, I., see Lund, H., **36**, 235.

Taddei, F., see Benassi, R., **41**, 75.

Takeuchi, Y., Furusaki, F., *The Chemistry of Isoxazolidines*, **21**, 207.

Tamura, Y. Ikeda, M., *Advances in the Chemistry of Heteroaromatic N-Imines and N-Aminoazonium Salts*, **29**, 71.

Tarnowski, B., see Meth-Cohn, O., **26**, 115; **31**, 207.

Taticchi, A., see Fringuelli, F., **21**, 119.

Tedder, J. M., *Heterocyclic Diazo Compounds*, **8**, 1.

Ternai, B., see Lakhan, R., **17**, 99.

Thyagarajan, B. S., *Aromatic Quinolizines*, **5**, 291; *Claisen Rearrangements in Nitrogen Heterocyclic Systems*, **8**, 143.

Timpe, H. J., *Heteroaromatic N-Imines*, **17**, 213.

Timpe, H. J., El'tsov, A. V., *Pseudoazulenes*, **33**, 185.

Tišler, M., *Heterocyclic Quinones*, **45**, 37.

Tišler, M., see Kadaba, P. K., **37**, 217.

Tišler, M., Stanovnik, B., *Pyridazines*, **9**, 211; *Recent Advances in Pyridazine Chemistry*, **24**, 363.

Tissington, P., see Keene, B. R. T., **13**, 315.

Tomasik, P., Johnson, C. D., *Applications of the Hammett Equation to Heterocyclic Compounds*, **20**, 1.

Toomey, J. E., Jr., *Synthesis of Pyridines by Electrochemical Methods*, **37**, 167.

Tsuge, O., Kanemasa, S., *Recent Advances in Azomethine Ylide Chemistry*, **45**, 231.

Trattner, R. B., see Perlmutter, H. D., **31**, 115.

Ugi, I., *Pentazoles*, **3**, 373.

Urbański, T., see Eckstein, Z., **2**, 311; **23**, 1.

van den Haak, H. J., see van der Plas, H. C., **33**, 95.

van der Plas, H. C., Wozniak, M., van den Haak, H. J., *Reactivity of Naphthyridines toward Nitrogen Nucleophiles*, **33**, 95.

van der Plas, H. C., see Charushin, V. N., **43**, 301.
van der Plas, H. C., see den Hertog, H. J., **4**, 121.
Vasvari-Debreczy, L., see Hermecz, I., **39**, 281.
Vernon, J. M., see Bryce, M. R., **28**, 183; Kricka, L. J., **16**, 87.
Vivona, N., see Ruccia, M., **29**, 141.
Wakefield, B. J., Wright, D. J., *Isoxazole Chemistry 1963*, **25**, 147.
Wamhoff, H., *Heterocyclic β-Enamino Esters, Versatile Synthons in Heterocyclic Synthesis*, **38**, 299.
Weber, H., *Oxidative Transformations of Heteroaromatic Iminium Salts*, **41**, 275.
Weeds, S. M., see Katritzky, A. R., **7**, 225.
Weinstock, L. M., Pollak, P. I., *The 1,2,5-Thiadiazoles*, **9**, 107.
Weis, A. L., *Recent Advances in the Chemistry of Dihydroazines*, **38**, 1.
Wentrup, C., *Carbenes and Nitrenes in Heterocyclic Chemistry: Intramolecular Reactions*, **28**, 231.
Werstiuk, E. S. G., see Cheeseman, G. W. H., **14**, 99; **22**, 367.
White, J. D., Mann, M. E., *Isoindoles*, **10**, 113.

Wightman, R. H., see Jankowski, K., **39**, 79.
Willette, R. E., *Monoazaindoles: The Pyrrolopyridines*, **9**, 27.
Woods, T. S., see Griffin, T. S. **18**, 99.
Wooldridge, K. R. H., *Recent Advances in the Chemistry of Mononuclear Isothiazoles*, **14**, 1.
Wooldridge, K. R. H., see Slack, R., **4**, 107.
Wozniak, M., see van der Plas, H. C., **33**, 95.
Wright, D. J., see Wakefield, B. J., **25**, 147.
Wünsch, K. H., Boulton, A. J., *Indoxazenes and Anthranils*, **8**, 277.
Yakhontov, L. N., *Quinuclidine Chemistry*, **11**, 473.
Yalpani, M., see Lalezari, I., **24**, 109.
Zahradnik, R., *Electronic Structure of Heterocyclic Sulfur Compounds*, **5**, 1.
Zahradnik, R., Koutecky, J., *Theoretical Studies of Physico-chemical Properties and Reactivity of Azines*, **5**, 69.
Zahradnik, R., see Mayer, R., **8**, 219.
Zoltewicz, J. A., Deady, L. W., *Quaternization of Heteroaromatic Compounds: Quantitative Aspects*, **22**, 71.

Cumulative Index of Titles, Volumes 1–45

A

Acetylenecarboxylic acids and esters, reactions with N-heterocyclic compounds, **1**, 125
Acetylenecarboxylic esters, reactions with nitrogen-containing heterocycles, **23**, 263
Acetylenic esters, synthesis of heterocycles through nucleophilic additions to, **19**, 297
Acid-catalyzed polymerization of pyrroles and indoles, **2**, 287
Acidity of azoles, basicity and, **41**, 187
Acyl groups in heterocyclic compounds, conformations of, **41**, 75
Advances
 in the Chichibabin reaction, **44**, 1
 in chrom-3-ene chemistry, **18**, 159
 in heterocyclic chemistry, prospect and retrospect, **40**, 1
 in homolytic substitution of heteroaromatic compounds, **16**, 123
 in imidazole chemistry, **12**, 103; **27**, 241
 in indolizine chemistry, **23**, 103
 in oxazole chemistry, **17**, 99
 in pyrrolizidine chemistry, **5**, 315; **24**, 247
 in selenophene chemistry, **12**, 1
t-Amino effect, **14**, 211
N-Aminoazonium salts, N-imines and, **29**, 71
Aminochromes, **5**, 205
4-Amino-1,2,3-triazoles, **40**, 129
Anils, olefin synthesis with, **23**, 171
Anionic σ-adducts of heterocycles, **34**, 305
Anions, ring-opening of five-membered heteroaromatic, **41**, 41
Annelation of a pyrimidine ring to an existing ring, **32**, 1
Annular nitrogens of azines with electrophiles, reactions of, **43**, 127
Annulenes, N-bridged, cyclazines and, **22**, 321
Anthracen-1,4-imines, **16**, 87
Anthranils, **8**, 277; **29**, 1
Applications
 of the Hammett equation to heterocyclic compounds, **3**, 209; **20**, 1
 of mass spectral techniques and stereochemical considerations in carbohydrates and other oxygen heterocycles, **42**, 335
 of NMR spectroscopy to indole and its derivatives, **15**, 277
 of phase-transfer catalysis to heterocyclic chemistry, **36**, 175
Arene oxides, chemistry of, **37**, 67
Aromatic azapentalenes, **22**, 183
Aromatic quinolizines, **5**, 291; **31**, 1
Aromatic six-membered nitrogen heterocycles, regioselective substitution in **44**, 199
Aromaticity of heterocycles, **17**, 255
Aza analogs of pyrimidine and purine bases, **1**, 189
4-Azaazulenes, chemistry of, **43**, 35
7-Azabicyclo[2.2.1]hepta-2,5-dienes, **16**, 87
1-Azabicyclo[3.1.0]hexanes and analogs with further heteroatom substitution, **27**, 1
Azapentalenes, aromatic, chemistry of, **22**, 183
Azaphosphorines, chemistry of **43**, 1
8-Azapurines, chemistry of, **39**, 117
Azines
 reactions of annular nitrogens of, with electrophiles, **43**, 127
 reactivity with nucleophiles, **4**, 145

theoretical studies of, physicochemical properties and reactivity of, **5**, 69
Azinoazines, reactivity with nucleophiles, **4**, 145
Aziridine intermediates, synthesis of natural products via, **39**, 181
1-Azirines, synthesis and reactions of, **13**, 45
Azocines, **31**, 115
Azodicarbonyl compounds in heterocyclic synthesis, **30**, 1
Azoles, basicity and acidity of, **41**, 187
Azomethine ylide chemistry, recent advances in, **45**, 231

B

Barbituric acid, recent progress in chemistry of, **38**, 229
Base-catalyzed hydrogen exchange, **16**, 1
Basicity and acidity of azoles, **41**, 187
1-, 2-, and 3-Benzazepines, **17**, 45
Benzisothiazoles, **14**, 43; **38**, 105
Benzisoxazoles, **8**, 277; **29**, 1
Benzoazines, reactivity with nucleophiles, **4**, 145
Benzo[c]cinnolines, **24**, 151
1,5-Benzodiazepines, **17**, 27
Benzo[b]furan and derivatives, recent advances in chemistry of, Part I, occurrence and synthesis, **18**, 337
Benzo[c]furans, **26**, 135
Benzofuroxans, **10**, 1; **29**, 251
2H-1-Benzopyrans (chrom-3-enes), **18**, 159
1,2- and 2,1-Benzothiazines and related compounds, **28**, 73
1,4-Benzothiazines and related compounds, **38**, 135
Benzo[b]thiophene chemistry, recent advances in, **11**, 177; **29**, 171
Benzo[c]thiophenes, **14**, 331
1,2,3-Benzotriazines, **19**, 215
Benzyne, reactions with heterocyclic compounds, **28**, 183
Betaines, heterocyclic, derivatives of alternant hydrocarbons, **26**, 1
Bifunctional nucleophiles: cyclizations and ring transformations on reaction of azines with, **43**, 301

Biological pyrimidines, tautomerism and electronic structure of, **18**, 199
Bipyridines, **35**, 281
Bridgehead nitrogen, tricyclic compounds with a central pyrimidine ring and, **39**, 281

C

9H-Carbazoles, recent advances in, **35**, 83
Carbenes
  and nitrenes, intramolecular reactions, **28**, 231
  reactions with heterocyclic compounds, **3**, 57
Carbohydrates and other oxygen heterocycles, applications of mass spectral techniques and stereochemical considerations in, **42**, 335
Carbolines, **3**, 79
Cationic polar cycloaddition, **16**, 289 (**19**, xi)
Chemistry
  and rearrangements of 1,2-dihydroisoquinolines, **40**, 105
  of arene oxides, **37**, 67
  of aromatic azapentalenes, **22**, 183
  of 4-azaazulenes, **43**, 35
  of azaphosphorines, **43**, 1
  of 8-azapurines, **39**, 117
  of azomethine ylides, recent advances in, **45**, 231
  of barbituric acid, recent progress in, **38**, 229
  of benzo[b]furan, Part I, occurrence and synthesis, **18**, 337
  of benzo[b]thiophenes, **11**, 177; **29**, 171
  of chrom-3-enes, **18**, 159
  of diazabicycloundecene (DBU) and other pyrimidoazepines, **42**, 83
  of diazepines, **8**, 21
  of dibenzothiophenes, **16**, 181
  of dihydroazines, **38**, 1
  of 1,2-dioxetanes, **21**, 437
  of furans, **7**, 377
  of hydantoins, **38**, 177
  of isatin, **18**, 1
  of isoindoles, **29**, 341
  of isoxazolidines, **21**, 207
  of lactim ethers, **12**, 185

of mononuclear isothiazoles, **14,** 1
of 4-oxy- and 4-keto-1,2,3,4-tetrahydroisoquinolines, **15,** 99
of phenanthridines, **13,** 315
of phenothiazines, **9,** 321
of polycyclic isothiazoles, **38,** 1
of pyrazolopyridines, **36,** 343
of pyrazolopyrimidines, **41,** 319
of pyrido[1,2-*a*]pyrimidines, **33,** 241
of 1-pyrindines, **15,** 197
of pyrrolizines, **37,** 1
of tetrazoles, **21,** 323
of 1,3,4-thiadiazoles, **9,** 165
of thienothiophenes, **19,** 123
of thiophenes, **1,** 1
of thiophenium salts and thiophenium ylids, **45,** 151
of triazolopyridines, **34,** 79
Chichibabin reaction, advances in, **44,** 1
Chiral induction using heterocycles, **45,** 1
Chrom-3-ene chemistry, advances in, **18,** 159
Claisen rearrangements
 in heteroaromatic systems, **42,** 203
 in nitrogen heterocyclic systems, **8,** 143
Complex metal hydrides, reduction of nitrogen heterocycles with, **6,** 45; **39,** 1
Condensed thiophene systems, tetra- and pentacyclic, **32,** 127
Conformational equilibria in nitrogen-containing saturated six-membered rings, **36,** 1
Conformations of acyl groups in heterocyclic compounds, **41,** 75
Covalent hydration
 in heteroaromatic compounds, **4,** 1, 43
 in nitrogen heterocycles, **20,** 117
Current views on some physicochemical aspects of purines, **24,** 215
Cyclazines, and related N-bridged annulenes, **22,** 321
Cyclic enamines and imines, **6,** 147
Cyclic hydroxamic acids, **10,** 199
Cyclic peroxides, **8,** 165
Cyclizations and ring transformations on reaction of azines with bifunctional nucleophiles, **43,** 301
Cyclizations under Vilsmeier conditions, **31,** 207
Cycloaddition, cationic polar, **16,** 289 (**19,** xi)

(2 + 2)-Cycloaddition and (2 + 2)-cycloreversion reactions of heterocyclic compounds, **21,** 253

D

Developments in the chemistry
 of furans (1952–1963), **7,** 377
 of Reissert compounds (1968–1978), **24,** 187
Dewar heterocycles and related compounds, **31,** 169
2,4-Dialkoxypyrimidines, Hilbert-Johnson reaction of, **8,** 115
Diazabicycloundecene (DBU) and other pyrimidoazepines, chemistry of, **42,** 83
Diazepines, chemistry of, **8,** 21
1,4-Diazepines, 2,3-dihydro-, **17,** 1
Diazirines, diaziridines, **2,** 83; **24,** 63
1,4-Diazocines, **45,** 185
Diazo compounds, heterocyclic, **8,** 1
Diazomethane, reactions with heterocyclic compounds, **2,** 245
Dibenzofurans, **35,** 1
Dibenzothiophenes, chemistry of, **16,** 181
Dihydroazines, recent advances in chemistry of, **38,** 1
Dihydro-1,4-benzothiazines, and related compounds, **38,** 135
2,3-Dihydro-1,4-diazepines, **17,** 1
1,2-Dihydroisoquinolines and related compounds, **14,** 279; **40,** 105
1,2-Dioxetanes, chemistry of, **21,** 437
Diquinolylmethane and its analogs, **7,** 153
gem-Dithienylalkanes and their derivatives, **32,** 83
1,2-Dithiole-3-thiones and 1,2-dithiol-3-ones, **31,** 63
1,2- and 1,3-Dithiolium ions, **7,** 39; **27,** 151

E

Electrochemical synthesis of pyridines, **37,** 167
Electrolysis of *N*-heterocyclic compounds
 Part I, **12,** 213
 Part II, **36,** 235
Electronic aspects of purine tautomerism, **13,** 77

Electronic effects of heteroaromatic and substituted heteroaromatic groups, **42**, 1
Electronic structure
  of biological pyrimidines, tautomerism and, **18**, 199
  of heterocyclic sulfur compounds, **5**, 1
Electrophiles, reactions of annular nitrogens of azines with, **43**, 127
Electrophilic substitutions of five-membered rings, **13**, 235
Enamines and imines, cyclic, **6**, 147
β-Enamino esters, heterocyclic, as heterocyclic synthons, **38**, 299
π-Excessive heteroannulenes, medium-large and large, **23**, 55

F

Ferrocenes, heterocyclic, **13**, 1
Five-membered heteroaromatic anions, ring-opening of, **41**, 41
Five-membered rings, electrophilic substitutions of, **13**, 235
Formation of anionic σ-adducts from heteroaromatic compounds, **34**, 305
Four-membered rings containing one sulfur atom, **35**, 199
Free radical substitutions of heteroaromatic compounds, **2**, 131
Furan chemistry, recent advances in, Part I, **30**, 167; Part II, **31**, 237
Furans, developments of the chemistry of (1952–1963), **7**, 377
Furans, dibenzo-, **35**, 1
Furoxans, **29**, 251

G

Gas phase reactivity of heteroaromatic compounds, **40**, 25
Grignard reagents, indole, **10**, 43

H

Halogenation of heterocyclic compounds, **7**, 1
Hammett equation, applications to heterocyclic compounds, **3**, 209; **20**, 1
Hetarynes, **4**, 121
Heteroadamantanes, **30**, 79
Heteroannulenes, medium-large and large π-excessive, **23**, 55
Heteroaromatic N-aminoazonium salts, **29**, 71
Heteroaromatic compounds
  free-radical substitutions of, **2**, 131
  homolytic substitution of, **16**, 123
  nitrogen, covalent hydration in, **4**, 1, 43
  prototropic tautomerism of, **1**, 311, 339; **2**, 1, 27; **SI**
  quaternization of, **22**, 71
  reactivity of, in gas phase, **40**, 25
Heteroaromatic N-imines, **17**, 213; **29**, 71
Heteroaromatic nitro compounds, ring synthesis of, **25**, 113
Heteroaromatic radicals, Part I, general properties; radicals with Group V ring heteroatoms, **25**, 205; Part II, radicals with Group VI and Groups V and VI ring heteroatoms, **27**, 31
Heteroaromatic and substituted heteroaromatic groups, electronic effects, **42**, 1
Heteroaromatic substitution, nucleophilic, **3**, 285
Heteroaromatic systems, Claisen rearrangements in, **42**, 203
Heteroaromatics, quantitative analysis of steric effects in, **43**, 173
Heterocycles
  aromaticity of, **17**, 255
  chiral induction using, **45**, 1
  containing the sulfamide moiety, **44**, 81
  nomenclature of, **20**, 175
  photochemistry of, **11**, 1
  by ring closure of ortho-substituted t-anilines, **14**, 211
Heterocyclic betaine derivatives of alternant hydrocarbons, **26**, 1
Heterocyclic chemistry
  applications of phase-transfer catalysis in **36**, 175
  literature of, **7**, 225; **25**, 303; **44**, 269
Heterocyclic compounds
  application of Hammett equation to, **3**, 209; **20**, 1
  (2 + 2)-cycloaddition and (2 + 2)-cycloreversion reactions of, **21**, 253
  halogenation of, **7**, 1
  isotopic hydrogen labeling of, **15**, 137
  mass spectrometry of, **7**, 301

quaternization of, **3**, 1; **22**, 71
reactions of, with carbenes, **3**, 57
reactions of diazomethane with, **2**, 245
reactions with benzyne, **28**, 183
N-Heterocyclic compounds (*see also* Nitrogen heterocycles)
electrolysis of, **12**, 213
photochemistry of, **30**, 239
reaction of acetylenecarboxylic acids and esters with, **1**, 125; **23**, 263
Heterocyclic diazo compounds, **8**, 1
Heterocyclic ferrocenes, **13**, 1
Heterocyclic iminium salts, oxidative transformations of, **41**, 275
Heterocyclic oligomers, **15**, 1
Heterocyclic products, natural, synthesis of by hetero Diels-Alder cycloaddition reactions, **42**, 245
Heterocyclic pseudobases, **1**, 167; **25**, 1
Heterocyclic quinones, **45**, 37
Heterocyclic sulphur compounds, electronic structure of, **5**, 1
Heterocyclic synthesis
azodicarbonyl compounds and, **30**, 1
heterocyclic β-enamino esters and, **38**, 299
involving nitrilium salts and nitriles under acidic conditions, **6**, 95
through nucleophilic additions to acetylenic esters, **19**, 279
sulfur transfer reagents in, **30**, 47
thioureas in, **18**, 99
uses of isatoic anhydrides in, **28**, 73
Hetero Diels-Alder cycloaddition reactions, synthesis of natural heterocyclic products by, **42**, 245
Hilbert-Johnson reaction of 2,4-dialkoxypyrimidines with halogenoses, **8**, 115
Homolytic substitution of heteroaromatic compounds, **16**, 123
Hydantoins, chemistry of, **38**, 177
Hydrogen cyanide derivatives, synthesis of heterocycles from, **41**, 1
Hydrogen exchange
base-catalyzed, **16**, 1
one-step (labeling) methods, **15**, 137
Hydrogenated porphyrin derivatives: hydroporphyrins, **43**, 73
Hydroxamic acids, cyclic, **10**, 199

I

Imidazole chemistry, advances in, **12**, 103; **27**, 241
2H-Imidazoles, **35**, 375
4H-Imidazoles, **35**, 413
N-Imines, heteroaromatic, **17**, 213; **29**, 71
Iminium salts, oxidative transformations of heterocyclic, **41**, 275
Indole Grignard reagents, **10**, 43
Indole(s)
acid-catalyzed polymerization, **2**, 287
and derivatives, application of NMR spectroscopy to, **15**, 277
Indolizine chemistry, advances in, **23**, 103
Indolones, isatogens and, **22**, 123
Indoxazenes, **8**, 277; **29**, 1
Isatin, chemistry of, **18**, 1
Isatogens and indolones, **22**, 123
Isatoic anhydrides, uses in heterocyclic synthesis, **28**, 127
Isoindoles, **10**, 113; **29**, 341
Isoquinolines
1,2-dihydro-, **14**, 279
4-oxy- and 4-keto-1,2,3,4-tetrahydro-, **15**, 99
Isothiazoles, **4**, 107
recent advances in the chemistry of monocyclic, **14**, 1
polycyclic, recent advances in chemistry of, **38**, 105
Isotopic hydrogen labeling of heterocyclic compounds, one-step methods, **15**, 137
Isoxazole chemistry, recent developments in, **2**, 365; since 1963, **25**, 147
Isoxazolidines, chemistry of, **21**, 207

L

Lactim ethers, chemistry of, **12**, 185
Literature of heterocyclic chemistry, **7**, 225; **25**, 303; **44**, 269

M

Mass spectral techniques in heterocyclic chemistry: applications and stereochemical considerations in carbohydrates and other oxygen heterocycles, **42**, 335

Mass spectrometry
  of heterocyclic compounds, **7**, 301
  of nucleic acids, **39**, 79
Medium-large and large π-excessive heteroannulenes, **23**, 55
Meso-ionic compounds, **19**, 1
Metal catalysts, action on pyridines, **2**, 179
Metallacycloalkanes and -alkenes, **39**, 237
Monoazaindoles, **9**, 27
Monocyclic pyrroles, oxidation, of, **15**, 67
Monocyclic sulfur-containing pyrones, **8**, 219
Mononuclear heterocyclic rearrangements, **29**, 141
Mononuclear isothiazoles, recent advances in chemistry of, **14**, 1

N

Naphthalen-1,4-imines, **16**, 87
Naphthyridines, **11**, 124
  reactivity of, toward nitrogen nucleophiles, **33**, 95
  recent developments in chemistry of, **33**, 147
Natural heterocyclic products by hetero Diels-Alder cycloaddition reactions, synthesis of, **42**, 245
Natural products, synthesis via aziridine intermediates, **39**, 181
New developments in the chemistry of oxazolones, **21**, 175
Nitrenes, carbenes and, intramolecular reactions of, **28**, 231
Nitriles and nitrilium salts, heterocyclic synthesis involving, **6**, 95
Nitro-compounds, heteroaromatic, ring synthesis of, **25**, 113
Nitrogen-bridged six-membered ring systems, **16**, 87
Nitrogen heterocycles (*see also* N-Heterocyclic compounds)
  aromatic six-membered, regioselective substitution in, **44**, 199
  conformational equilibria in saturated six-membered rings, **36**, 1
  covalent hydration in, **20**, 117
  photochemistry of, **30**, 239
  reactions of acetylenecarboxylic esters with, **23**, 263
  reduction of, with complex metal hydrides, **6**, 45; **39**, 1
Nitrogen heterocyclic systems, Claisen rearrangements in, **8**, 143
Nomenclature of heterocycles, **20**, 175
Nuclear magnetic resonance spectroscopy, application to indoles, **15**, 277
Nucleic acids, mass spectrometry of, **39**, 79
Nucleophiles, bifunctional, cyclisations and ring transformations on reaction of azines with, **43**, 301
Nucleophiles, reactivity of azine derivatives with, **4**, 145
Nucleophilic additions to acetylenic esters, synthesis of heterocycles through, **19**, 299
Nucleophilic heteroaromatic substitution, **3**, 285

O

Olefin synthesis with anils, **23**, 171
Oligomers, heterocyclic, **15**, 1
Organometallic compounds, transition metal, use in heterocyclic synthesis, **30**, 321
1,3,4-Oxadiazole chemistry, recent advances in, **7**, 183
1,2,4-Oxadiazoles, **20**, 65
1,2,5-Oxadiazoles, **29**, 251
1,3-Oxazine derivatives, **2**, 311; **23**, 1
Oxaziridines, **2**, 83; **24**, 63
Oxazole chemistry, advances in **17**, 99
Oxazolone chemistry
  new developments in, **21**, 175
  recent advances in, **4**, 75
Oxidation of monocyclic pyrroles, **15**, 67
Oxidative transformations of heteroaromatic iminium salts, **41**, 275
3-Oxo-2,3-dihydrobenz[*d*]isothiazole 1,1-dioxide (saccharin) and derivatives, **15**, 233
Oxygen heterocycles, applications of mass spectral techniques and stereochemical considerations in carbohydrates and others, **42**, 335
4-Oxy- and 4-keto-1,2,3,4-tetrahydroisoquinolines, chemistry of, **15**, 99

P

Pentazoles, 3, 373
Peroxides, cyclic, 8, 165 (see also 1,2-Dioxetanes)
Phase transfer catalysis, applications in heterocyclic chemistry, 36, 175
Phenanthridine chemistry, recent developments in, 13, 315
Phenanthrolines, 22, 1
Phenothiazines, chemistry of, 9, 321
Phenoxazines, 8, 83
Photochemistry
  of heterocycles, 11, 1
  of nitrogen-containing heterocycles, 30, 239
  of oxygen- and sulfur-containing heterocycles, 33, 1
Physicochemical aspects of purines, 6, 1; 24, 215
Physicochemical properties
  of azines, 5, 69
  of pyrroles, 11, 383
3-Piperideines, 12, 43
Polyfluoroheteroaromatic compounds, 28, 1
Polymerization of pyrroles and indoles, acid-catalyzed, 2, 1
Porphyrin derivatives, hydrogenated: hydroporphyrins, 43, 73
Present state of selenazole chemistry, 2, 343
Progress in pyrazole chemistry, 6, 347
Prototropic tautomerism of heteroaromatic compounds, 1, 311, 339; 2, 1,27; S1
Pseudoazulenes, 33, 185
Pseudobases, heterocyclic, 1, 167; 25, 1
Purine bases, aza analogs of, 1, 189
Purines
  physicochemical aspects of, 6, 1; 24, 215
  tautomerism, electronic aspects of, 13, 77
Pyrans, thiopyrans, and selenopyrans, 34, 145
Pyrazine chemistry, recent advances in, 14, 99
Pyrazole chemistry, progress in, 6, 347
3H-Pyrazoles, 34, 1
4H-Pyrazoles, 34, 53
Pyrazolopyridines, 36, 343
Pyrazolopyrimidines, chemistry of, 41, 319

Pyridazines, 9, 211; 24, 363
Pyridine(s)
  action of metal catalysts on, 2, 179
  effect of substituents on substitution in, 6, 229
  synthesis by electrochemical methods, 37, 167
  1,2,3,6-tetrahydro-, 12, 43
Pyridoindoles (the carbolines), 3, 79
Pyridopyrimidines, 10, 149
Pyrido[1,2-a]pyrimidines, chemistry of, 33, 241
Pyrimidine bases, aza analogs of, 1, 189
Pyrimidine ring annelation to an existing ring, 32, 1
Pyrimidine ring, tricyclic compounds with a central, 39, 281
Pyrimidines
  2,4-dialkoxy-, Hilbert-Johnson reaction of, 8, 115
  fused tricyclic, 39, 281
  tautomerism and electronic structure of biological, 18, 199
Pyrimidoazepines, chemistry of diazabicycloundecene (DBU) and other, 42, 83
1-Pyrindines, chemistry of, 15, 197
Pyrones, monocyclic sulfur-containing, 8, 219
Pyrroles
  acid-catalyzed polymerization of, 2, 287
  oxidation of monocyclic, 15, 67
  physicochemical properties of, 11, 383
2H- and 3H-Pyrroles, 32, 233
Pyrrolizidine chemistry, 5, 315; 24, 247
Pyrrolizines, chemistry of, 37, 1
Pyrrolodiazines with a bridgehead nitrogen, 21, 1
Pyrrolopyridines, 9, 27
Pyrylium salts
  syntheses, 10, 241
  syntheses, reactions, and physical properties, S2

Q

Quantitative analysis of steric effects in heteroaromatics, 43, 173
Quaternization
  of heteroaromatic compounds, 22, 71
  of heterocyclic compounds, 3, 1

Quinazolines, **1**, 253; **24**, 1
Quinolizines, aromatic, **5**, 291; **31**, 1
Quinones, heterocyclic, **45**, 37
Quinoxaline chemistry
  developments 1963–1975, **22**, 367
  recent advances in, **2**, 203
Quinuclidine chemistry, **11**, 473

R

Reactions
  of annular nitrogens of azines with electrophiles, **43**, 127
  of azines with bifunctional nucleophiles: cyclizations and ring transformations, **43**, 301
Reactivity
  of heteroaromatic compounds in the gas phase, **40**, 25
  of naphthyridines toward nitrogen nucleophiles, **33**, 95
Rearrangements, mononuclear heterocyclic, **29**, 141
Recent advances
  azomethine ylide chemistry, **45**, 231
  in benzo[*b*]thiophene chemistry, **11**, 177
  in furan chemistry
    Part I, **30**, 168
    Part II, **31**, 237
  in 1,3,4-oxadiazole chemistry, **7**, 183
  in oxazolone chemistry, **4**, 75
  in pyrazine chemistry, **14**, 99
  in pyridazine chemistry, **24**, 363
  in quinoxaline chemistry, **2**, 203
  in tetrazole chemistry, **21**, 323
  in the chemistry
    of benzisothiazoles and other polycyclic isothiazoles, **38**, 105
    of benzo[*b*]furans, occurrence and synthesis, **18**, 337
    of benzo[*b*]thiophenes, **29**, 171
    of 9*H*-carbazoles, **35**, 83
    of dibenzothiophenes, **16**, 181
    of dihydroazines, **38**, 1
    of mononuclear isothiazoles, **14**, 1
    of phenothiazines, **9**, 321
    of 1,3,4-thiadiazoles, **9**, 165
    of thiophenes, **1**, 1
Recent developments
  in naphthyridine chemistry, **33**, 147
  in isoxazole chemistry, **2**, 365
  in phenanthridine chemistry, **13**, 315
Recent progress in barbituric acid chemistry, **38**, 229
Reduction of nitrogen heterocycles with complex metal hydrides, **6**, 45; **39**, 1
Regioselective substitution in aromatic six-membered nitrogen heterocycles, **44**, 199
Reissert compounds, **9**, 1; **24**, 187
Ring closure of ortho-substituted *t*-anilines, heterocycles by, **14**, 211
Ring-opening of five-membered heteroaromatic anions, **41**, 41
Ring synthesis of heteroaromatic nitro compounds, **25**, 113
Ring transformations and cyclizations on reaction of azines with bifunctional nucleophiles, **43**, 301

S

Saccharin and derivatives, **15**, 233
Selenazole chemistry, present state of, **2**, 343
Selenium-nitrogen heterocycles, **24**, 109
Selenophene chemistry, advances in, **12**, 1
Selenophenes, **30**, 127
Selenopyrans, **34**, 145
Six-membered ring systems, nitrogen bridged, **16**, 87
Steric effects in heteroaromatics, quantitative analysis of, **43**, 173
Substitution(s)
  electrophilic, of five-membered rings, **13**, 235
  homolytic, of heteroaromatic compounds, **16**, 123
  nucleophilic heteroaromatic, **3**, 285
    in pyridines, effect of substituents, **6**, 229
  regioselective, in aromatic six-membered nitrogen heterocycles, **44**, 199
Sulfamide moiety, heterocycles containing the, **44**, 81
Sulfur compounds
  electronic structure of heterocyclic, **5**, 1
  four-membered rings, **35**, 199
Sulfur transfer reagents in heterocyclic synthesis, **30**, 47

Synthesis
  and reactions of 1-azirines, **13**, 45
  of heterocycles
  by ring-closure of *o*-substituted *t*-anilines, **14**, 211
  from hydrogen cyanide derivatives, **41**, 1
  from nitrilium salts and nitriles under acidic conditions, **6**, 95
  thioureas in, **18**, 99
  through nucleophilic additions to acetylenic esters, **19**, 279
  of natural heterocyclic products by hetero Diels-Alder cycloaddition reactions, **42**, 245
  of tetracyclic and pentacyclic condensed thiophene systems, **32**, 127
  of pyridines by electrochemical methods, **37**, 167

T

Tautomerism
  electronic aspects of purine, **13**, 77
  and electronic structure of biological pyrimidines, **18**, 199
  prototropic, of heteroaromatic compounds, **1**, 311, 339; **2**, 1, 27; S1
Tellurophene and related compounds, **21**, 119
1,2,3,4-Tetrahydroisoquinolines, 4-oxy- and 4-keto-, **15**, 99
1,2,3,6-Tetrahydropyridines, **12**, 43
Tetrazole chemistry, recent advances in, **21**, 323
Theoretical studies of physicochemical properties and reactivity of azines, **5**, 69
1,2,4-Thiadiazoles, **5**, 119; **32**, 285
1,2,5-Thiadiazoles, chemistry of, **9**, 107
1,3,4-Thiadiazoles, recent advances in the chemistry of, **9**, 165
Thiathiophthenes (1,6,6aS^{IV}-trithiapentalenes), **13**, 161
1,2,3,4-Thiatriazoles, **3**, 263; **20**, 145
1,4-Thiazines and their dihydro derivatives, **24**, 293

4-Thiazolidinones, **25**, 83
Thienopyridines, **21**, 65
Thienothiophenes and related systems, chemistry of, **19**, 123
Thiochromanones and related compounds, **18**, 59
Thiocoumarins, **26**, 115
Thiophenes, recent advances in the chemistry of, **1**, 1
Thiophenium salts and thiophenium ylids, chemistry of, **45**, 151
Thiopyrans, **34**, 145
Thiopyrones (monocyclic sulfur-containing pyrones), **8**, 219
Thioureas in synthesis of heterocycles, **18**, 99
Three-membered rings with two heteroatoms, **2**, 83; **24**, 63
Transition organometallic compounds in heterocyclic synthesis, use of, **30**, 321
1,3,5-, 1,3,6-, 1,3,7-, and 1,3,8-Triazanaphthalenes, **10**, 149
1,2,3-Triazines, **19**, 215
1,2,3-Triazoles, **16**, 33
1,2,3-Triazoles, 4-amino-, **40**, 129
Δ^2-1,2,3-Triazolines, **37**, 217
Δ^3- and Δ^4-1,2,3-Triazolines, **37**, 351
Triazolopyridines, **34**, 79
1,2,3-Triazolo[4,5-*d*]pyrimidines (8-azapurines), chemistry of **39**, 117
Tricyclic compounds with a central pyrimidine ring and one bridgehead nitrogen, **39**, 281
1,6,6aS^{IV}-Trithiapentalenes, **13**, 161

U

Use of transition organometallic compounds in heterocyclic synthesis, **30**, 321

V

Vilsmeier conditions, cyclization under, **31**, 207

Cumulative Subject Index, Volumes 41–45

A

Acamelin, structure and occurrence, **45**, 55
Acenaphtho[1,2-c][1,2,5]thiadiazole 2,2-dioxide, thermolysis, **44**, 138
Acenaphthylene, cycloaddition
  to tetrachlorothiophenium ethoxycarbonylimine, **45**, 177
  to tetrachlorothiophenium ylid, **45**, 172
Aceperimidines, Chichibabin amination, **44**, 69
Acetone, addition
  to pyridinium salts, **43**, 340
  to pyrimidines, **43**, 345
  to quinolinium salts, **43**, 341
Acetonedicarboxylic ester, addition to 5-nitropyrimidin-2-one, **43**, 336
Acetyl heterocycles, conformation, **41**, 100; **45**, 214
7-Acetylbicyclo[4.2.0]oct-3-enes, mass spectra, **42**, 340
Acetylenedicarboxylic esters reaction
  with azaazulenes and azafulvenes, **43**, 59
  with azomethine ylids, **45**, 234–239, 242, 288, 295
  with 1,3,4,6-diazadiphosphorines, **43**, 27
  with dihydropyrazines, **45**, 195, 221
  with nitrones, **45**, 289, 335
  with protoporphyrin IX ester, **43**, 102
  with 2-pyridones, **43**, 214
  with a quinone, **45**, 340
  ring-expansions using, **45**, 222
Acidity of cyclic sulfonamides, **44**, 109, 147, 160
Acidity and basicity of azoles (review), **41**, 187
  correlation of, **41**, 231
  tabulation of data, **41**, 234

Acidity functions, **41**, 208
Acridine, amination, **44**, 4, 47
Acridine N-oxides, 1,2,3,4-tetrahydro-, photochemistry, **43**, 61
Acridine quinones, **45**, 90, 91
Acridines, 1,4-diazepino-fused, **45**, 191
Acridinium salts, disproportionation of pseudobases, **41**, 298
Acridizinium ions
  cycloadditions to, **43**, 343
  steric effects in cycloadditions of, **43**, 214
Actinidine, synthesis, **42**, 317
Actinomycinol, structure and occurrence, **45**, 90
Actinorhodin
  reaction with diazomethane, **45**, 76
  structure and tautomerism, **45**, 108
Acyl groups, conformation (review), **41**, 75
N Acylazoles, rotation barriers, **43**, 251
Acylation
  of azines, **43**, 147
  of azinones, **43**, 149
N-Acylindoles, rotation barriers, **43**, 251
Acylnitroso compounds, cycloaddition of, **42**, 288
Adaline, synthesis, **42**, 321
Adamantane, phospha-analogs, **43**, 21
Addition reactions, DBU in, **42**, 116
σ-Adducts in Chichibabin reaction, **44**, 3, 4, 9, 48, 52–55, 71
Adenine, formation from HCN oligomers, **41**, 5
Akuammigine, synthesis, **42**, 326
Albidin, structure, **45**, 67
Aldol condensations, diastereoselective, **45**, 7
Alkali-induced disproportionation of pyridinium salts, **41**, 280
Alkaloid synthesis by hetero-Diels–Alder reactions, **42**, 245

303

Alkyl groups, steric effects of, on quaternization of adjacent nitrogen, **43**, 180
Alkylamides, alkali metal, in Chichibabin reactions, **44**, 31
Alkylamination of heterocycles, **44**, 32
Alkylamination, intramolecular, of pyridines, **44**, 35
Alkylation
  of azines, **43**, 131, 180
  of 1,2,6-thiadiazine 1,1-dioxides, **44**, 111
  of 1,2,4,6-thiatriazine 1,1-dioxides, **44**, 161
2-Alkylpyridinium salts, oxidation, **41**, 283
Alkynylation of pyridine, **44**, 206, 208, 212
Alloxazinium salts, reaction with 1,3-diols, **43**, 308
Alstonine, tetrahydro-, synthesis, **42**, 326
Amaryllidaceae alkaloids, synthesis, **42**, 288
Amauromine, synthesis, **42**, 206
Amidines, tautomerism of, **43**, 305
Amination
  of benzimidazoles, **44**, 57
  of isoquinolines, **44**, 41
  of naphthyridines, **44**, 52
  of pyrazines, **44**, 48, 236
  of pyridazines, **44**, 48
  of pyridines, **44**, 2, 203
  of pyrimidines, **44**, 49, 232
  of quinolines, **44**, 41
  of quinones, **45**, 87
  of 1,2,4-triazines, **44**, 241
  of 1,3,5-triazines, **44**, 70
  for Aminodehalogenation reactions, *see* Nucleophilic substitution under the appropriate ring system
Amination, Chichibabin (review), **44**, 1
N-Amination of azines, **43**, 161
Amine anions, ring-cleavage, **41**, 66
Aminoacids, racemic, resolution of, **45**, 7
Aminoacids, racemisation by aldehydes, **45**, 252
Aminomalononitrile, **41**, 3
3-Aminopyrazoles, condensation with 1,3-dicarbonyl compounds, **41**, 321
Aminopyridines, quaternization, **43**, 201
Amino-sugars, synthesis, **42**, 255, 274, 283, 284, 298
Ammonia, liquid, σ-adduct formation by amide ions in, **44**, 9

Analytical separation of diastereomers, **45**, 6
Angles of twist in biaryls
  calculated, **43**, 259, 261, 262
  observed, **43**, 258, 261
Anhydrocannabisativene, synthesis, **42**, 266
Anhydrovilangin, structure, **45**, 109
Anion cleavage of five-membered heterocycles (review), **41**, 41
Annellation effects on azole basicity and acidity, **41**, 228
ANRORC mechanism
  in Chichibabin reactions, **44**, 11
  in chloronitropyridine-amide ion reaction, **43**, 303
  in ring interconversions of six-membered heterocycles, **43**, 337, 346
Ansathiazin, structure, **45**, 117
Anthracene, cycloaddition to tetrachlorothiophenium ethoxycarbonylimine, **45**, 178, 180
Antihypertensives, **44**, 189
Aphid pigments, **45**, 106
Apo-β-erythroidine, synthesis, **43**, 51
Aqueous solution, acid-base equilibria in, **41**, 216
Aromaticity
  of five-membered heterocycles, **45**, 152
  of 1,2,6-thiadiazine 1,1-dioxides, **44**, 100
Arylamination of heterocycles, **44**, 31
Aryne mechanism for Chichibabin reaction, **44**, 72
Asteriomycin, synthesis, **42**, 256
Asymmetric induction, *see* Chiral induction
Autocatalysis in Chichibabin reaction, **44**, 8
3-Azaazulene-1-carboxylic ester, 2-ethoxy-, cycloaddition to, **43**, 59
3a-Azaazulenes and related compounds, *see* 4-Azaazulenes
4-Azaazulenes (*correctly* 3a-Azaazulenes)
  chemistry of (review), **43**, 35
  cycloadditions, **43**, 46
  natural products derived from, **43**, 68
  oxidation, **43**, 44
  pharmaceutical activity (of benzo-derivatives), **43**, 68
  protonation, **43**, 42

SUBJECT INDEX

reactions with electrophiles and nucleophiles, **43**, 44
spectra, **43**, 38
synthesis, **43**, 49
tautomerism, **43**, 40
theoretical calculations, **43**, 37
4-Azaazulenium salts, **43**, 42
4-Azaazulen-1-one, 3-methyl-
rearrangement, **43**, 62
synthesis, **43**, 62
4-Azaazulen-3-one, synthesis, **43**, 63
4-Azaazulen-5-one, synthesis, **43**, 63
4-Azaazulen-7-one, synthesis, **43**, 66
4-Azaazulen-7-one, 1,2,3-tribromo-, synthesis, **43**, 68
4-Azaazulen-9-one, synthesis, **43**, 67
Azabenz[*a*]azulenes, **43**, 36, 39, 42, 51
1-Azabutadiene systems, cycloaddition, **42**, 302
2-Azabutadiene systems, cycloaddition, **42**, 307
Aza-Cope rearrangements, **42**, 213
1,2,6-Azadiphosphorines, hexahydro-, **43**, 25
Azaphosphinanes, *see* Azaphosphorinanes
Azaphosphorinanes, **43**, 4, 6, 8
Azaphosphorines, chemistry of (review), **43**, 1
1,2-Azaphosphorines, **43**, 2, 4
1,3-Azaphosphorines, **43**, 5
1,4-Azaphosphorines, **43**, 8
Azaquinones, formation from diazidoquinones, **45**, 85
1-Azatriptycene, photolysis, **43**, 55
2-Azatriptycene, Chichibabin amination, **44**, 72
Azepino[1,2-*a*]indoles, n.m.r. spectra, **43**, 39
Azetidines, acyl, conformation, **41**, 137
Azetidinones, **42**, 94, 97, 127, 148, 172, 292
2-Azetidinones, 4-acetoxy-, chiral syntheses using, **45**, 13
Azetinones, fused, **41**, 60
Azide cyclizations, intramolecular, **43**, 54
Azidoformic ester, reaction with 1,4-naphthoquinone, **45**, 78
Azido-heterocycles, cleavage, **41**, 67, 70
Azidoquinones, photochemical reaction with dienes, **45**, 44

o-Azidostyrene derivatives, indoles from, **45**, 43
Azines, reactions
of annular nitrogens with electrophiles (review), **43**, 127
with bifunctional nucleophiles (review), **43**, 301
Azirines, photolysis to nitrile ylids, **45**, 53
Aziridines, nitrosative deimination, **45**, 192
Aziridines, acyl, conformation, **41**, 121, 136
Azodicarboxylic esters, cycloaddition to azomethine ylids, **45**, 305
Azoles
acidity of, in DMSO, **41**, 217, 228
basicity and acidity (review), **41**, 187
Azoles, *N*-acyl-, rotation barriers, **43**, 251
Azomethine *N*-oxides, *see* Nitrones
Azomethine ylids (review), **45**, 231
cycloaddition
FMO discussion of, **45**, 295
forming 2,5-dihydropyrroles, **45**, 234
intramolecular, **45**, 333
regioselectivity of, **45**, 319, 327
stereochemistry of, **45**, 235
stereoselectivity of, **45**, 314, 327
to azo compounds, **45**, 305
to carbonyl compounds, **45**, 301, 327
to imines, **45**, 301, 329
to thiocarbonyl compounds, **45**, 304
to unactivated olefins and acetylenes, **45**, 298
dimerization to pyrazine derivatives, **45**, 261
generation, **45**, 233
by decarboxylation, **45**, 269, 311
by deprotonation, **45**, 256, 310
by desilylation, **45**, 239
by proton transfer in dienamines, **45**, 294
by tautomerism of aminoacid derivatives, **45**, 249, 309
from t-amine *N*-oxides, **45**, 277
from aziridines, **45**, 234, 306
from 4-isoxazolines, **45**, 287
from metal-coordinated species, **45**, 280, 313
from 4-oxazolines, **45**, 287
geometry, **45**, 306
syn-anti isomerization, **45**, 235, 236

Azomethine ylids, (*cont*.)
  rearrangement by proton transfer, **45,** 237, 290
Azomethine ylids, *C*-chloro-, **45,** 238
Azomethine ylids, cyano-, as nitrilium ylid equivalents, **45,** 253
Azomethine ylids, *N*-phthalimido-, **45,** 238
3a-Azoniaazulenes, *see* 4-Azaazulenium salts

B

Bacteriochlorins
  structure, **43,** 75
  synthesis, **43,** 110
Bacteriophin, synthesis, **43,** 114
Barbituric acid, alkylation, **43,** 142
Barbituric acids, 1,3-dialkyl-, conformational preferences in, **43,** 231
Basagran (herbicide), **44,** 188
Basicity and acidity of azoles (review), **41,** 187
Basicity
  of azines (monocyclic), **43,** 128
  of azoles, **41,** 187
  of diazabicycloundecane (DBU), **42,** 91
  of hydroporphyrins, **43,** 87
  of imidazoles, **41,** 241
  of monocyclic azines, **43,** 128
  of organic bases, **42,** 91
  of pyridazines, **43,** 129
  of pyridines, steric effects on, **43,** 276
  of pyrroles, **41,** 234
  of 1,2,6-thiadiazine 1,1-dioxides, **44,** 109
Bentazone, **44,** 130, 188
Benz[*de*]anthracen-7-one, amination, **44,** 72
2,3-Benzazaphosphorines, **43,** 3
Benzazetinones, formation from 2,1-benzisoxazolium salts, **41,** 61
Benzene di- and tri-imines, synthesis and reactivity, **45,** 152
Benzimidazole, quaternization rates in, **43,** 197
Benzimidazole, 1-methyl-, Chichibabin amination, **44,** 5, 16, 17, 57–62, 64
Benzimidazole quinones, **45,** 79
Benzimidazole, pyridyl- and pyridylalkyl-, Chichibabin amination, **44,** 63

Benzimidazoles
  acidity and basicity, **41,** 244
  Chichibabin amination, **44,** 57
  Claisen rearrangements in, **42,** 212
  formation from dibenzo[*b,f*][1,4]diazocine derivatives, **45,** 220
Benzimidazolium salts, oxidation, **41,** 306
Benzimidazolyl group, substituent constants, **42,** 51, 54
Benz[*e*]indole-4,5-dione, 3-acetyl-1-ethoxycarbonyl-2-hydroxy-, **45,** 42
Benz[*f*]indole-4,9-dione, 2-phenyl-, **45,** 44
Benz[*g*]indole-4,5-dione, 3-ethoxycarbonyl-2-methyl-, **45,** 40
Benz[*e*]indolizines, reduced, synthesis, **42,** 304
Benz[*f*]isoindole-4,9-dione, 2-methyl-, synthesis, **45,** 54
1,2-Benzisothiazole-4,5-diones, synthesis, **45,** 78
Benzisoxazole 4,7-quinones, formation by cycloaddition, **45,** 77
2,1-Benzisoxazole-4,7-dione, 3-alkoxy-, synthesis and reactions, **45,** 78
1,2-Benzisoxazoles, cleavage of carbanions, **41,** 50
Benzobistriazole quinones, synthesis, **45,** 83
Benzo[*b*]carbazole-6,11-diones, synthesis, **45,** 50
Benzo[3,4]cyclobuta[1,2-*b*]quinoxaline
  oxidation, **45,** 195
  photolysis, **45,** 204
Benzocyclobutene-1,2-diones, condensation with 1,2-diamines, **45,** 208
Benzocyclobutenes, 5-acetyl-1,4,4a,5,6,6a-hexahydro-, CI mass spectra, **42,** 340
Benzocyclobutenes, dihydro-, ring-opening and cycloaddition, **42,** 246, 248, 259
1,3,2-Benzodiazaphosphorines, **43,** 3
1,6-Benzodiazocine, 1,6-dihydro-1,6-dimethyl-, **45,** 194, 209, 218
2,5-Benzodiazocine, 1,2,3,4-tetrahydro-, derivative, **45,** 187
1,4-Benzodiazocine-2,5-dione, 1,3,4,6-tetrahydro-, **45,** 194
1,6-Benzodiazocines, 2,5-diaryl-, controversy surrounding, **45,** 204
1,6-Benzodiazocines, 1,2,3,4,5,6-hexahydro-, synthesis, **45,** 193

2,5-Benzodiazocines, uses, **45**, 223
2,5-Benzodiazocines, 1,6-diaryl-3,4-dihydro-, synthesis, **45**, 206
2,5-Benzodiazocines, 1,2,3,4,5,6-hexahydro-, synthesis, **45**, 193
1,4-Benzodiazocin-2(1H)-one, 3,4,5,6-tetrahydro-, synthesis, **45**, 187
Benzodi[1,2-b;5,4($4,5$?)-b']di[1]benzothiophene-6,12-dione, synthesis, **45**, 73
Benzo[1,2-b;5,4-b']difuran, **45**, 58, 59
Benzo[1,2-d;4,5-d']diimidazole-4,8-diones, synthesis, **45**, 79
Benzodiimidazoles, inertness under Chichibabin conditions, **44**, 62
Benzo[1,2-c;4,5-c']diisochromene-5,7,12,14-tetrone, **45**, 114
Benzo[1,2-d;4,5-d']dioxazole-4,8-diones, synthesis, **45**, 80
1,3-Benzodioxolane quinones, **45**, 81
Benzo[1,2-b;4,5-b']diquinoline-6,7,13,14-tetrones, synthesis, **45**, 92
Benzo[1,2-g;4,5-g']diquinoline-5, 7,12,14-tetrones, synthesis, **45**, 92
1,4-Benzodithiin-5,8-diones, 2,3-dicyano-, synthesis, **45**, 115
1,3-Benzodithiolane quinones, **45**, 82
Benzo[1,2-b;4,3-b']dithiophene-4,5-dione, synthesis, **45**, 73
Benzo[1,2-c;3,4-c']dithiophene-4,8-dione, synthesis, **45**, 73
Benzo[2,1-b;3,4-b']dithiophene-4,5-dione, synthesis, **45**, 73
Benzo[1,2-b;4,5-b']dithiophene-4,8-diones, **45**, 70
Benzofuran quinones, **45**, 54
Benzofuran-4,7-dione, 3,5-dimethyl-, ^{13}C NMR, **45**, 56
Benzofurans
Claisen rearrangements in, **42**, 228
cleavage of carbanions, **41**, 44
Benzo[b]furans, acyl, conformation, **41**, 83, 95
Benzo-fusion
effect on acidity and basicity, **41**, 228
effect on Chichibabin amination, **44**, 20
steric effects of, **43**, 186
Benzo[b]naphtho[2,3-d]furan-6,11-diones, synthesis, **45**, 64, 65

Benzonitrile oxide, cycloaddition to quinones, **45**, 77
Benzo[a]phenazine-5,6-dione, **45**, 100
Benzophenone, 4-nitro-, amination under Chichibabin conditions, **44**, 72
Benzoporphyrins, **43**, 118
2-Benzopyran-5,8-diones, 3,4-dihydro-, **45**, 102
1-Benzopyran-2-ones, Claisen rearrangements in, **42**, 224, 231
1-Benzopyran-2,7,8-triones, **45**, 100
Benzo[f]quinazolines, amination, **44**, 67
Benzo[f]quinazoline-3(4H)-trione, 1-methyl-, **45**, 99
Benzo[f]quinazolinones, Chichibabin amination, **44**, 68
Benzo[g]quinoline-4,9-diones, naturally-occurring, **45**, 90
Benzoquinolines, Chichibabin amination, **44**, 45
Benzo[g]quinoxaline-5,10-diones, **45**, 99
Benzoselenophenes, cleavage of carbanions, **41**, 48
Benzotellurophenes, cleavage of carbanions, **41**, 48
3,2,4-Benzothiadiazepine 3,3-dioxides, 1,2,4,5-tetrahydro-, synthesis, **44**, 184
3H-2,1,3-Benzothiadiazine 2,2-dioxides, **44**, 108, 116
2,1,3-Benzothiadiazin-4(3H)-one 2,2-dioxides
synthesis, **44**, 130
ultraviolet spectra, **44**, 97
uses, **44**, 188
1,2,3-Benzothiadiazole-4,5-dione, 6,7-dichloro-, synthesis, **45**, 83
1,2,3-Benzothiadiazole-4,7-dione, synthesis, **45**, 83
2,1,3-Benzothiadiazole-4,5- and 4,7-diones, synthesis, **45**, 84, 123
2,1,3-Benzothiadiazoline 2,2-dioxide
acidity, **44**, 147
electrophilic substitution, **44**, 145
oxidation, **44**, 140
Benzothiazoles, Claisen rearrangements in, **42**, 213, 230
Benzothiazolyl group, substituent constants, **42**, 52
[1]Benzothieno[6,5-b][1]benzothiophene-4,10-dione, synthesis, **45**, 73

Benzo[b]thiophene-2,3-dione, use in synthesis of benzothieno-fused quinones, **45,** 73
Benzo[b]thiophene-4,5-diones, synthesis, **45,** 69
Benzo[b]thiophene-4,7-diones, occurrence and synthesis, **45,** 70
Benzo[b]thiophenes
 Claisen rearrangements in, **42,** 229
 cleavage of carbanions, **41,** 45
Benzo[b]thiophenium salts, structure, **45,** 158
Benzo[b]thiophenium S-ylids, synthesis, **45,** 161
Benzotriazole, quaternization rates in, **43,** 197
Benzotriazole-4,5-diones, synthesis, **45,** 82
Benzotriazole-4,7-diones, synthesis, **45,** 82
Benzotriazoles, acidity and basicity, **41,** 262
Benzotriazolyl groups, substituent constants, **42,** 52
3,1-Benzoxazepines, formation by photorearrangement, **43,** 61
4,1,2-Benzoxazin-5,8-diones, synthesis, **45,** 121
Benzoxazoles, Claisen rearrangements in, **42,** 213
Benzoxazolyl group, substituent constants, **42,** 51
Benzoyl heterocycles, conformation, **41,** 102
Benzoylation of uracil, **43,** 149
Berberine alkaloids, reactions, **43,** 60
Betaines of 1,2,4,6-thiatriazine dioxide series, **44,** 159
Biacridanylidenes, stereochemistry, **43,** 253
Biaryls
 chiral, **43,** 256
 restricted rotation in, **43,** 256
Bicyclo[4.2.0]oct-3-enes, 7-acetyl-, CI mass spectra, **42,** 340
Biflavanylidenes, stereochemistry, **43,** 253
Bifluorenylidenes, stereochemistry, **43,** 253
Biindoles, polybromo-, chiral, **43,** 256
Bikaverin, structure, occurrence, biological activity, **45,** 110
Bile pigments, photochemistry, **43,** 58

Binaphthoquinones, polycyclic furan quinones from, **45,** 65
Biochemical aspects of hydroporphyrins, **43,** 116
Biological activity
 of 1,4-diazocines and derivatives, **45,** 223
 of pyrazolopyrimidines, **41,** 366
 of quinoline quinones, **45,** 88
Biological properties of sulfamide-containing ring systems, **44,** 188
Biosynthesis of naphthyridinomycin, **45,** 95
Biphenyl, heterocyclic analogs, restricted rotation and energy barriers, **43,** 256
Bipyridines, Chichibabin amination, **44,** 41
Bipyridines, bridged, chirality of, **43,** 269
2,2'-Bipyridines, formation in Chichibabin reaction, **44,** 29
N,N'-Bipyrroles, chiral, **43,** 263
Borane-pyridine adducts, **43,** 185, 203
Borinic esters, dibutyl-, stereocontrol using, **45,**
Boron trichloride-pyridine adducts, **43,** 203
Boronate esters, mass spectra, **42,** 395
Boronic acids, aryl-, coupling with bromopyridines using Pd catalyst, **44,** 208
Boronic esters, pyridyl-, synthetic uses, **44,** 222
Bostrycoidin and derivatives
 structure and occurrence, **45,** 95
 synthesis, **45,** 96, 123
Brasanquinones, structure and synthesis, **45,** 64, 65
Brevicomin, synthesis, **42,** 270, 321
Bridged azaphosphorinanes, **43,** 6, 9, 11, 19, 21, 26
Bromination
 of pyridines, **44,** 217
 of 1,2,4-triazines, **44,** 242
Brönsted equations, **43,** 176
Brucine, oxidation, **45,** 41
Bryaquinone, structure, **45,** 64
Buchapsine, synthesis by Claisen rearrangement, **42,** 221
Butenolides, synthesis, **42,** 319
Buttressing effects, **43,** 235, 273, 275
Butynedioic acid/ester, *see* Acetylenedicarboxylic

C

Caldariellaquinone, occurrence and structure, **45**, 70
Cannabinoids, mass spectra, **42**, 394
Cannabinol, (-)-(9R)-hexahydro-, synthesis, **42**, 326
Caprolactam, conversion into DBU, **42**, 85
Caprolactim ethers, synthesis of pyrimidoazepines from, **42**, 147, 159, 161
Carbacyclin, chiral synthesis, **45**, 27
Carbapenems, chiral synthesis, **45**, 16
Carbazole-1,2-dione, **45**, 41
Carbazole-1,4-dione, 5,6,7,8-tetrahydro-3-hydroxy-2-methyl-, **45**, 40
Carbazole-3,4-dione, 5,6,7,8-tetrahydro-2-methyl-, **45**, 40
Carbazole-1,4-diones, naturally occurring, **45**, 49
Carbazoles, acidity and basicity, **41**, 238
Carbazoles, N-acyl-, rotation barriers, **43**, 251
Carbazolyl groups, substituent constants, **42**, 53
Carbene formation from thiophenium ylids, **45**, 170
Carbene, difluoro-, azomethine ylid generation using, **45**, 293
Carbenes, reaction with 1,2,6-thiadiazine 1,1-dioxides, **44**, 117
Carbenes and carbenoids, reaction with thiophenes, **45**, 165, 172, 174
Carbenes, heterocyclic, ring-cleavage, **41**, 67
Carbohydrates, mass spectrometry of (review), **42**, 335
Carbonyl groups, as dienophiles, **42**, 270
Carboxylation using DBU, **42**, 119
Carpanone, synthesis, **42**, 325
Cathodic reduction of phosphonium salts, **43**, 10
Cellobiose, FAB mass spectrum, **42**, 385
Cellobiose acetate, mass spectrum, **42**, 389
Cephalosporin, phosphorus analogs of, **43**, 7
Cephalotaxine and derivatives, **43**, 69
Cephalotaxinone, synthesis, **43**, 54
Cephem, chiral recognition on acylation, **45**, 5

Cervinomycins, structure, **45**, 112
Characterisation of carbohydrates by mass spectrometry, **42**, 343
Chemical ionization (CI) mass spectrometry
  of 7-acetylbicyclo[4.2.0]oct-3-enes, **42**, 340
  of cyclic diols, **42**, 341
  of saccharides, **42**, 349
  of steroid derivatives, **42**, 340
Chemiluminescence of acridinium oxidation, **41**, 299
Chichibabin amination, *see* Amination
Chichibabin reaction, advances in (review), **44**, 1
Chilenine, **43**, 42, 69
Chiral induction
  in cyclohex-4-ene-1,2-dicarboxylic acid, **45**, 27
  in 2,4-dimethylglutaric acid, **45**, 24
  in 3-methylglutaric acid, **45**, 22
  in norborn-5-ene-2,3-dicarboxylic acid, **45**, 26
  using heterocycles (review), **45**, 1
Chiral recognition in aminolysis of 3-acylthiazolidinethiones, **45**, 4
o-Chloranil
  aromatization using, **44**, 209, 210, 212, 224, 225
  cyclization using, **45**, 87
Chlorination of pyridines, **44**, 217, 228
Chlorins
  reactions, **43**, 100, 106
  structure, **43**, 74
  synthesis, **43**, 90, 96, 97, 107
1-Chloro-1-nitrosocyclohexane, cycloadditions of, **42**, 287, 293
β-Chlorophlorin, structure, **43**, 75
Chlorophyll-*a*, FAB mass spectrum, **42**, 365
Chlorosulfonyl isocyanate (CSI), use of in synthesis
  of dithiatriazines **44**, 174
  of 1,2,4,6-thiatriazine, 1,1-dioxides, **44**, 170
  of trithiatriazines, **44**, 174
Cholestanes, 3-dimethylamino-3-methyl-, mass spectra, **42**, 339
Chrom-3-ene-5,6-diones, biological activity, **45**, 102

Chrom-3-ene-5,8-diones, ring-chain tautomerism, **45**, 101
Cine-substitution on amination of 5-bromopyrimidine, **44**, 51
Cinnolines, quaternisation rates in, **43**, 197, 198
Circular dichroism (CD)
  and conformation study, **41**, 80; **43**, 218
  temperature dependence, **43**, 230
Claisen and aza-Cope rearrangements
  in benzimidazoles, **42**, 212
  in benzofurans, **42**, 228
  in benzopyranones, **42**, 224, 231
  in benzothiazoles, **42**, 213, 230
  in benzothiophenes, **42**, 229
  in benzoxazoles **42**, 213
  in coumarins, **42**, 224, 231
  in furans, **42**, 207
  in heteroaromatic systems (review), **42**, 203
  in indoles, **42**, 205, 215, 227
  in isoxazoles, **42**, 215
  in naphthalenes, **42**, 225
  in oxazoles, **42**, 211
  in pyrazoles, **42**, 211
  in pyridines, **42**, 216
  in pyrimidines, **42**, 222
  in pyrroles, **42**, 205, 213
  in quinolines, **42**, 220, 230
  in tetrazoles, **42**, 212
  in thiophenes, **42**, 207
  in triazines, **42**, 223
Cleavage
  of heterocyclic carbanions, **41**, 44
  promoted by N lone-pairs, **41**, 55
Cleistopholine, occurrence and structure, **45**, 90
σ-Complexes, *see* σ-Adducts
Concentration effects on acid-base equilibria, **41**, 218
Condensations using DBU, **42**, 122
Conformation
  of acyl groups (review), **41**, 75
  of dibenzo[*e,g*][1,4]diazocines, **45**, 212
  of 1,4-dihydro-1,4-diazocines, **45**, 211
  of 5,12-dihydrodibenzo[*b,f*][1,4]diazocine-6,11-diones, **45**, 215
  of 5,6,11,12-tetrahydrodibenzo[*b,f*][1,4]-diazocines, **45**, 214

Conformational equilibria, acyl, substituent effects, **41**, 103
δ-Coniceine, synthesis, **42**, 261
Copper catalysis
  in Grignard reactions, **44**, 210, 225
  in rearrangement of thiophenium ylids, **45**, 168
Copper complexes
  in azomethine ylid generation, **45**, 280, 313
  of 1,4-azaphosphorines, **43**, 10
Copper reagents, organo-, reactions with pyridines, **44**, 210
Copper-pyridine reagents, **44**, 222
Cornforth rearrangement of oxazoles, **41**, 57
Coronands, keto-, conformation, **41**, 113
Correlation of acidity and basicity of azoles, **41**, 231
Coumarins, Claisen rearrangements in, **42**, 224, 231
Covalent hydration of fused pyrazines, **43**, 321
Crinane, synthesis, **42**, 288
Cruciaromaticity, **43**, 79
Cryptopleurine, synthesis, **42**, 265
Cryptosporin, structure, **45**, 103
Crystal structure
  of porphyrins, **43**, 81
  see also X-ray crystallography
Cyanide ion, cleavage of pyridiniopyridines by, **44**, 214
Cyanocycline-A, structure, **45**, 95
Cyanogen, condensation with sulfamides, **44**, 149
[2.3.4]Cyclazine-3,4-quinone, 5-cyano-, **43**, 44
[2.2.4]Cyclazinium salts, **43**, 44
Cycloaddition reactions, hetero- (review), **42**, 245
Cycloadditions
  forming 3a-azaazulenes, **43**, 59
  of azomethine ylids, **45**, 231–344
  of six-membered heterocycles, **43**, 334, 342
  to 3a-azaazulenes, **43**, 46
  to vinylporphyrins
    of DMAD, **43**, 102
    of nitrosobenzenes, **43**, 104

of tetracyanoethylene, **43**, 102
Cyclobutenones, 4-hydroxy-4-aryl-, rearrangement, **45**, 56
Cyclodextrins, FAB mass spectra, **42**, 358
Cyclopenta[*gh*]perimidines, Chichibabin amination, **44**, 69
Cyclospongiaquinones, structure, **45**, 110
Cyperaquinones, occurrence and structure, **45**, 57
Cytosine, alkylation **43**, 140

D

Daunomycin, analogs of, synthesis, **45**, 51, 73
Daunosamine, synthesis, **42**, 273
Deactivation factors in substituted pyrimidines, **42**, 63
Deacylcyclindrocarpol, mass spectrum, **42**, 338
Dealkylation kinetics in pyridinium salts, **43**, 189
Decarboxylation route to azomethine ylids, **45**, 269
Decarboxylative ring-cleavage, **41**, 50, 51, 52, 54
Decker oxidation
of [1]benzopyrano[2,3-*b*]pyridinium salts, **41**, 302
of isoquinolinium salts, **41**, 299
of pyridinium salts, **41**, 276, 286
of quinolinium salts, **41**, 296
of quinoxalinium salts, **41**, 304
Deprotonation route to azomethine ylids, **45**, 256
Dequaternization of pyridinium salts, **43**, 278
Dequaternization reactions, **43**, 189, 201
Desosamine, synthesis, **42**, 298
Deuterium; size of CD bonds relative to CH, **43**, 187
Diaminofumaronitrile, **41**, 4
Diaminomaleonitrile (DAMN), **41**, 4, 15
Di-anionic σ-adducts in Chichibabin reactions, **44**, 5, 23, 71
1,2-Diarylpyridinium salts, photocyclization, **41**, 294
Diastereocontrolled aldol reactions, **45**, 7

Diazabicyclo[2.2.2]octane (DABCO) quaternization, **43**, 182
1,8-Diazabicyclo[5.4.0]undec-7-ene (DBU) and other pyrimidoazepines (review), **42**, 83
1,8-Diazabicyclo[5.4.0]undec-7-ene (DBU)
applications, **42**, 91, 170; **45**, 31
basicity, **42**, 91
physicochemical properties, **42**, 91
reactions, **42**, 86, 169
synthesis, **42**, 85
use in chlorin synthesis, **43**, 108
uses, **42**, 91, 170, **45**, 31
13*H*-6,13a-Diazadinaphtho[2,3-*a;*2′,3′-*e*]pentalene, **45**, 203
1,3,2,4-Diazadiphosphorines, **43**, 25
1,3,4,6-Diazadiphosphorines, **43**, 26
1,4,2,5-Diazadiphosphorines, **43**, 27
1,3,2-Diazaphosphole, 4,5-dicyano-, **41**, 35
1,2,5-Diazaphospholes, ring-expansion, **43**, 16
1,2,3-Diazaphosphorines, **43**, 16
1,2,4-Diazaphosphorines, **43**, 17
1,3,4-Diazaphosphorines, **43**, 17
1,3,5-Diazaphosphorines, **43**, 18
1,4,2-Diazaphosphorines, **43**, 24
Diazaquinomycin, occurrence and structure, **45**, 92
1,3,2,4,6-Diazatriphosphorines, **43**, 28
6*H*-1,4-Diazepines, formation from DAMN, **41**, 34
Diazinium salts
addition of methanol to, **43**, 323
hydration, **43**, 321
polarographic reduction potentials, **43**, 321
reactions with dinucleophiles, **43**, 321
Diazoacetic ester, reaction with quinones, **45**, 76
Diazoalkyl heterocycles, ring-cleavage, **41**, 67, 70
1,4-Diazocane-5,8-dione, synthesis, **45**, 188
1,4-Diazocanes
synthesis, **45**, 187, 199
uses, **45**, 223
1,4-Diazocine dianion, **45**, 193
Diazocines, perhydro-, *see* Diazocanes
1,4-Diazocines (review), **45**, 185

1,4-Diazocines, 1,4-dihydro-
  conformation, **45**, 211
  spectra
    MCD, **45**, 210
    NMR, **45**, 210
  synthesis, **45**, 192
1*H*-[1,4]Diazocino[7,8,1-*hi*]indol-2(3*H*)-
  one, 4,5-dihydro-, **45**, 189
[1,4]Diazocino[2,1-*a*]isoindolones, **45**, 189
[1,4]Diazocino[8,1-*a*]isoindolones, **45**, 189
1,4-Diazocinones, formation by Schmidt
  reactions, **45**, 190
1,4-Diazocin-5(4*H*)-ones, 2,3,6,7-
  tetrahydro-
  synthesis, **45**, 207
  uses, **45**, 223
2-Diazo-4,5-dicyano-2*H*-imidazole, **41**, 9
Diazolium salts, adducts with dienes, **41**, 11
Diazomethane
  methylation
    of pyrimidinones, **43**, 140
    of 1,2,4,6-thiatriazine 1,1-dioxides and
      related species, **44**, 161
    of triazinediones, **43**, 146
  reaction with quinones, **45**, 74, 82
Dibenz[*b,e*]indolizine-8,11-diones, synthesis, **45**, 51
Dibenzo[*b,h*]carbazole-5,13-dione, 6-phe-
  nyl-, synthesis, **45**, 52, 118
Dibenzo[*b,h*]carbazole-5,7,12,13-tetrones,
  synthesis, **45**, 50, 52, 122
Dibenzo[*b,f*][1,4]diazocine-6,11-dione,
  5,12-dihydro-
  conformation, **45**, 215
  controversy surrounding structure, **45**, 197
  formation from a diazabenzobipheny-
    lene, **45**, 195
Dibenzo[*b,f*][1,4]diazocines and related
  compounds
  controversy surrounding, **45**, 202
  optical resolution, **45**, 205, 216
Dibenzo[*b,f*][1,4]diazocines,
  5,12-diacetyl-5,6,11,12-tetrahydro-, con-
    formation, **45**, 214
Dibenzo[*b,f*][1,4]diazocines, 6,11-diaryl-,
  diborane reduction with rearrangement, **45**, 219

Dibenzo[*e,g*][1,4]diazocines
  conformation, **45**, 212
  metal complexation, **45**, 214
  optical activity and racemization, **45**, 213
Dibenzo[*b,f*][1,4]diazocin-6(5*H*)-one, deriv-
  ative, synthesis, **45**, 187
Dibenzo[*b,f*][1,4]diazocin-6(5*H*)-one, 11,12-
  dihydro-12-tosyl-, synthesis, **45**, 188
Dibenzo[1,4]dioxin quinones, **45**, 114
Dibenzo[1,4]dithiin quinones, **45**, 115
Dibenzofuran-1,4-diones, **45**, 59, 60
Dibenzofuran-3,4-diones, **45**, 59
Dibenzofuran-1,4,5,8-tetrone, 3,6-dihy-
  droxy-2,7-dimethyl-, **45**, 60
Dibenzo[*e,g*]phenanthro[9,10-
  b][1,4]diazocine, 5,14-dihydro-, **45**, 196
Dibenzo[*c,e*][1,2,7]thiadiazepine 2,2-diox-
  ide, 1,3-dihydro-, synthesis, **44**, 184
Dibenzothiophene-1,4,6,9-tetrones, **45**, 115
2,3-Dichloro-4,5-dicyanobenzoquinone
  (DDQ), uses, **45**, 60, 103, 104
Dielectric effects and conformation study,
  41, 80
Diels-Alder reactions, inverse electron de-
  mand, **43**, 342
Diels-Alder reactions, hetero-, natural
  product synthesis by (review), **42**, 245
Dielsiquinone, structure and occurrence,
  45, 90
Difluorocarbene, azomethine ylid genera-
  tion using, **45**, 293
Diformyl heterocycles, conformation, **41**, 97
3,6-Dihydro-2*H*-1,2-thiazine 1-oxides, **42**, 297
Dihydromaritidine, synthesis, **42**, 290
Diimide reduction of tetraphenylporphyrin,
  43, 95
Diiminosuccinonitrile (DISN), **41**, 6, 15, 28
Diketene, condensation with sulfamide, **44**, 121
Dimers, proton-bound, dissociation, **41**, 198
o-Dimethoxy substitution, effect of on
  Chichibabin reactions, **44**, 19, 61
Dimethyl acetylenedicarboxylate (DMAD),
  see Acetylenedicarboxylic esters

Dimethyl sulfoxide, acid-base equilibria in, **41**, 217
Dimethylamides, rotational barriers, **43**, 246
Dimethylamino groups, rotational barriers for, **43**, 241
Dimethylamino-quinolines and isoquinolines, quaternization, **43**, 273
Dimroth rearrangement
  of a triazolo[4′,3′-1,2]pyrimido[4,5-*b*]azepine, **42**, 163
  of pyrazolo[3,4-*d*]pyrimidines, **41**, 335
  on alkylation
    of aminopyrazine, **43**, 144
    of 2-aminopyrimidine, **43**, 137
    of 3-amino-1,2,4-triazine, **43**, 145
    of amino-1,3,5-triazines, **43**, 147
Dinaphtho[1,2-*b*,2′,3′-*d*]furan-7,12-diones, **45**, 66
Dinaphtho[1,2-*b*,2′,1′-*d*]furan-5,6,7,8-tetrone, **45**, 65
Dinaphtho[1,2-*b*,2′,3′-*d*]furan-5,6,7,12-tetrone, **45**, 65
Dinaphtho[2,3-*b*,2′,3′-*d*]furan-5,7,12,13-tetrone, **45**, 65
Dinaphth[2,3-*b*,2′,3′-*e*][1,4]oxazine-5,7,12,14-tetrone, 13-phenyl-, synthesis, **45**, 118
Diodantunezone, occurrence and structure, **45**, 56
1,4-Dioxino-fused heterocycles, synthesis, **43**, 318
Dipole moments of 1,2,6-thiadiazine 1,1-dioxides, **44**, 85
2,2′-Dipyridylamine
  amination, **44**, 27
  formation in Chichibabin reaction, **44**, 31
Disaccharides
  FAB mass spectra, **42**, 375, 382
  nomenclature, **42**, 368
Disproportionation
  of acridinium pseudobases, **41**, 298
  of pyridinium pseudobases, **41**, 280
Dissociation of proton-bound dimers, **41**, 198
1,4,2,6-Dithiadiazine 1,1-dioxides, **44**, 163
Di[1,2,5]thiadiazolo[3,4-*b*;3′,4′-*e*]pyrazine 2,2,6,6-tetroxide, dihydro-
  acidity of, **44**, 148
  disodium salt of, **44**, 142
2*H*,6*H*-1,5,2,4,6,8-Dithiatetrazocine 1,1,5,5-tetroxides, tetrahydro-, synthesis, **44**, 184
1,3λ^4,2,4,6-Dithiatriazine 1,1-dioxide, 5-trimethylsilyloxy-
  mass spectrum, **44**, 158
  synthesis, **44**, 174
1,3λ^4,2,4,6-Dithiatriazine 1,1,3-trioxides, synthesis, **44**, 174
1,3λ^4,2,4,6-Dithiatriazin-5(6*H*)-one 1,1-dioxide, 3,3-dimethyl-
  acidity, **44**, 161
  synthesis, **44**, 175
1,3,2,4,8-Dithiatriazocine-5,7-dione, 1,1,3,3-tetroxide, synthesis, **44**, 186
1,4-Dithiino-fused heterocycles, synthesis, **43**, 314, 316
1,2-Dithiolium ion, 3-dimethylamino-, rotational barrier, **43**, 245
Dual substituent parameter (DSP) treatment, **43**, 179
Dunniones, occurrence and structure, **45**, 58
Dynamic stereochemistry, methods of study, **43**, 217

E

Elaeokanine synthesis, **42**, 261
Elbs oxidation of pyrimidines, **44**, 234
Electrochemical reduction
  of porphyrins and hydroporphyrins, **43**, 88, 91
  of pyrazolo[1,5-*a*]pyrimidines, **41**, 354
Electrochemistry
  of isoquinoline-5,8-diones, **45**, 97
  of quinoxaline-5,8-diones, **45**, 100
Electron acceptor substituents in Chichibabin reactions, **44**, 18
Electron donor substituents in Chichibabin reactions, **44**, 20
Electron impact (EI) mass spectrometry of sugar derivatives, **42**, 344
Electron spectroscopy for chemical analysis (ESCA) of uracil derivatives, **43**, 130

Electron spin resonance (ESR)
  and conformational study, **41**, 82
  of a benzobistriazole quinone, **45**, 83
Electron transfer in Chichibabin reactions, **44**, 6
Electronic effects
  of heteroaromatic groups (review), **42**, 1
  separation from steric, **43**, 175
Electrophiles, reactions of annular nitrogens of azines with (review), **43**, 127
Eleokanine, synthesis, **42**, 261
Eleutherins, occurrence and synthesis, **45**, 105
Elimination reactions, DBU in, **42**, 100
Ellipticine, synthesis, **42**, 308
Enantiomeric purity, determination of, **45**, 3
Ene reaction of formaldehyde, **42**, 285
Energetics of acyl group conformations, **41**, 77
Enzymes
  dehydrogenase, **45**, 118
  oxidation of heteroaromatic iminium salts using, **41**, 310
Epicorynoline, synthesis, **42**, 309
Epilupinine, synthesis, **42**, 263, 264, 306; **45**, 20
Epoxide deoxygenation by carbenes, **45**, 171
Erysopine I, rearrangement, **43**, 61
Erythrina alkaloids, **43**, 68; **45**, 341
Erythroidine, rearrangement, **43**, 61, 68
Eschenmoser's base, **45**, 32
Eserethole, synthesis, **45**, 341
Ethoxymethylenemalonic ester (EMME)
  condensation with 3-aminopyrazoles, **41**, 328
Ethylenediamines
  condensation with benzocyclobutenediones, **45**, 208
  in 1,4-diazocine synthesis, **45**, 188, 196, 199, 203, 206
Eupolauramine, synthesis, **42**, 316
Europium complexes as condensation mediators, **42**, 284
Evodiamine, synthesis, **42**, 249
Excited state pK_a values, **41**, 220
Experimental determination of pK values, **41**, 202

F

Fabianine, synthesis, **42**, 302
Fast atom bombardment (FAB) mass spectrometry, **42**, 357, 366, 370–398
Ferricyanide oxidation
  of pyridinium salts, **41**, 276
  of quinolinium salts, **41**, 279
Field desorption (FD) mass spectra
  of glycosides, **42**, 342, 354
  of oligosaccharides, **42**, 352
Five-membered heterocycles
  aromaticity of, **45**, 152
  transmission of electronic effects, **42**, 73
Flash desorption mass spectrometry, **42**, 357
Flavoskyrin, biosynthesis, **42**, 323
Flowing afterglow, **41**, 197
Fluorescence in porphyrins and hydroporphyrins, **43**, 82
Fluoride desilylation, **45**, 242
Foramycins, synthesis, **41**, 346
Formyl heterocycles, conformation, **41**, 83
Fourtneine, structure, **43**, 69
Free radicals, phosphorus-containing, **43**, 11
Frémy's salt, quinone synthesis using, **45**, 40, 42, 46, 55, 59, 69, 81, 85, 86, 96, 114, 120
Friedel-Crafts acylation reactions
  of dihydropyridines, **44**, 223
  of pyridines, **44**, 217
Friedel-Crafts cyclizations, forming azaazulene derivatives, **43**, 52
Friedel-Crafts triazination, **44**, 244
Fries rearrangement, anionic, in lithiopyridines, **44**, 219, 220
Frogs, poison-arrow, **42**, 293
Frondoside-A, mass spectrum, **42**, 397
Frontalin, synthesis, **42**, 321
Frontier molecular orbital (FMO) theory
  and azomethine ylid cycloadditions, **45**, 319
  and heterocyclic cycloadditions, **43**, 342
  and thiophene reactivity, **45**, 165
Fructosides (sucrose, turanose), FAB mass spectra, **42**, 384
Fucose, synthesis, **42**, 273
Furanoeremophilanes, synthesis, **42**, 318
Furan ring annelation, **43**, 311

SUBJECT INDEX

Furano-fused quinones, **45**, 54
Furans
 Claisen rearrangements in, **42**, 207
 cleavage of carbanions, **41**, 44
 synthesis using oxazoles, **42**, 308, 318, 320
Furans, acyl, conformation, **41**, 83
Furazano[3,4-c][1,2,6]thiadiazine 1,1-dioxides
 acidity, **44**, 110
 glycosylation, **44**, 114
 NMR spectra, **44**, 94
 synthesis, **44**, 112, 128
 tautomerism, **44**, 108
Furazano[3,4-c][1,2,6]thiadiazine 1,1-dioxides, glycosylated
 biological activity, **44**, 188
 synthesis, **44**, 114
Furazans, cleavage by base, **41**, 53
Furoxans, cleavage by base, **41**, 54
Furyl groups
 substituent constants, **42**, 43
 substituent effects, **42**, 61
Fusarubin and derivatives, structure, **45**, 106

G

Galactoside, p-nitrophenyl, FAB mass spectrum, **42**, 372
Galactosyl glycosides (lactose, melibiose), FAB mass spectra, **42**, 383
Gas-phase
 acidities and basicities of azoles in, **41**, 191, 195
 structures of protonated azoles in, **41**, 225
Gas-phase ion-molecule reactions, stereochemistry, **42**, 342
Gear effect, **43**, 233
Gear-clashed conformations in polymethylpyridines, **43**, 275
Gentiobiose, FAB mass spectrum, **42**, 385
Gephyrotoxin, synthesis, **42**, 293, 302
Gliotoxins, reactions of, **42**, 174
Glucosides, simple, FAB mass spectra, **42**, 370
Glutaric acid, meso-2,4-dimethyl-, asymmetric transformations, **45**, 24

Glutaric acid, 3-methyl-, asymmetric transformations, **45**, 22
Glycosides
 elimination reactions in, **42**, 103, 174
 mass spectra, **42**, 339, 342, 351, 354, 370
 nomenclature, **42**, 368
 of N-heterocycles, conformation and rotation barriers, **43**, 229
 of pyrazolopyrimidines, **41**, 340, 246, 356
 of 1,2,6-thiadiazine 1,1-dioxide, rotamerism, **44**, 102
Glycosylation
 of fused imidazoloquinones, **45**, 80
 of fused pyrazoloquinones, **45**, 77
 of 1,2,6-thiadiazine 1,1-dioxides, **44**, 112
Gnididione, synthesis, **42**, 320
Gold(III) complexes of organic bases, **43**, 207
Granaticin, occurrence and structure, **45**, 111
Grignard reagents, reactions
 copper-catalyzed, **44**, 211, 223, 225
 with pyridine oxides and derivatives, **44**, 214
 with pyridines, **44**, 200, 209, 224
 with pyrimidines, **44**, 233
Griseusin, occurrence and structure, **45**, 112

H

Halogenation
 of pyridazines, **44**, 231
 of pyridine N-oxides, **44**, 218
 of pyridines, **44**, 217
Halogens, complexation with pyridines, **43**, 208
Hammett substituent constants of heteroaromatic rings (review), **42**, 41
Hammick reactions of pyridinecarboxylic acids, **44**, 223
Helinudichromene, **45**, 101
Heliotridine, synthesis, **42**, 293
Herbicidal thiadiazine dioxides, **44**, 188
Heteroaromatic anions, ring-opening of five-membered (review), **42**, 41
Heteroaromatic rings as substituent groups, electronic effects (review), **42**, 1

Heterocycles, chiral induction using (review), **45**, 1
Heterocyclic chemistry, reviews in, **44**, 269
Heterocyclic iminium salts, oxidative transformations of (review), **41**, 275
Heterocyclic quinones (review), **45**, 37
Heterocyclic synthesis from hydrogen cyanide derivatives (review), **41**, 1
Hetero-Diels-Alder reactions, natural product synthesis by (review), **42**, 245
(-)-(9R)-Hexahydrocannabinol, synthesis, **42**, 326
Hexamethylenetrisulfohexamine, **44**, 186
High-pressure mass spectrometry (HPMS), **41**, 196
Hikosamine, synthesis, **42**, 284
Histamine H_2-receptor antagonists, **44**, 189
Hofmann degradation of an alkaloidal salt, **43**, 48
E-Homoeburnanes, 14-hydroxy-, stereochemistry, **43**, 39
Hückel M. O. parameters in Chichibabin reaction, **44**, 16
Hydantoin, 4-methyl-1-phenyl-, as imidazoledione precursor, **42**, 252
Hydrazine, as dinucleophile in reactions with heterocycles, **43**, 305
Hydride elimination in Chichibabin reaction, **44**, 7
Hydride transfer in Chichibabin reaction, **44**, 8
Hydroboration of zinc octaethylporphyrin, **43**, 95
Hydrogen, formation in Chichibabin reaction, **44**, 7, 23
Hydrogen-bonding association and azoles, **41**, 211
Hydrogen cyanide derivatives, heterocycles from (review), **41**, 1
Hydrogen peroxide oxidation
of acridinium salts, **41**, 299
of pyridinium salts, **41**, 294
Hydrogenated porphyrin derivatives (review), **43**, 73
Hydrogenation
catalytic, of porphyrins and hydroporphyrins, **43**, 94
photochemical, of porphyrins and hydroporphyrins, **43**, 92

Hydroporphyrins (review), **43**, 73
biochemical aspects, **43**, 116
crystal structure, **43**, 79
redox potentials, **43**, 88
synthesis, **43**, 90
Hydroxamic acid oxidation, **42**, 288
Hydroxyalkylpyridinium salts, oxidation, **41**, 289
Hydroxygalegine, synthesis, **42**, 288
4-Hydroxyimino-1,2,6-thiadiazine 1,1-dioxides, 3,5-diamino-
displacement of amino groups, **44**, 118
synthetic use, **44**, 128
4-Hydroxyimino-1,2,6-thiadiazin-3(2H)-one 1,1-dioxides, 5-amino-
isomerism of oxime group, **44**, 101
tautomerism, **44**, 105
Hydroxylamine, as dinucleophile in reactions with heterocycles, **43**, 305

I

Imidazole, 2-diazo-4,5-dicyano-, **41**, 9
Imidazoles
acidity and basicity data, **41**, 239
quaternization rates, **43**, 197
substituent effects in, **42**, 67
synthesis from HCN oligomers, **41**, 6
Imidazoles, N-acyl, conformation, **41**, 127, 133
Imidazoles, condensed, Chichibabin amination, **44**, 56
Imidazoline-2,5-dione, 4-methoxy-1-phenyl-, as aza-dienophile precursor, **42**, 252
Imidazoline-2-thiones, conformation of substituents, **43**, 231, 234
Imidazoline-2-thiones, 1-aryl-, rotational barriers, **43**, 267
Imidazolyl groups, substituent constants, **42**, 46
5H-Imidazo[2,1-a]isoindole, 2,3-dihydro-5-phenyl-, **45**, 203
Imidazo[1,2-a]pyridines, quaternization rates, **43**, 277
Imidazo[4,5-g]quinazoline-4,5,9-triones, **45**, 120
Imidazo[4,5-f]quinoline, 3-benzyl-, amination, **44**, 63

SUBJECT INDEX

4H-Imidazo[4,5,1-*ij*]quinoline, 5,6-dihydro-, Chichibabin amination, **44,** 65
Imidazo[1,2-*a*]quinoxaline, Chichibabin amination, **44,** 65
Imidazo[4,5-*c*][1,2,6]thiadiazine 2,2-dioxides, synthesis, **44,** 128
Imination, oxidative, of pyridinium salts, **41,** 282
Iminium salts, heterocyclic, oxidation of (review), **41,** 275
Indazole-4,5-diones
  biological activity, **45,** 75
  synthesis, **45,** 74
1H-Indazole-4,7-diones, synthesis, **45,** 74, 82
2H-Indazole-4,7-diones
  crystal structure, **45,** 76
  synthesis, **45,** 75
1H-Indazole-6,7-diones, synthesis, **45,** 74
Indazoles
  acidity and basicity data, **41,** 256
  cleavage by base, **41,** 51
Indazolyl groups, substituent constants, **42,** 51
Indole quinones, **45,** 39
Indole, 1-ethyl-4-hydroxy-2,6-dimethyl-, oxidation to quinones, **45,** 40
Indole-4,7-diones, synthesis
  from 2-azido-3-alkenylbenzoquinones, **45,** 43
  using benzocyclobutenones, **45,** 44
Indoles
  acidity and basicity data, **41,** 235
  Claisen rearrangements in, **42,** 205, 228
  cleavage of dilithio derivatives, **41,** 49
Indoles, N-acyl-
  conformation, **41,** 131, 134
  rotation barriers, **43,** 251
  synthesis from nitropyridinium salts, **43,** 340
Indoles, 1-isopropyl-, conformation, **43,** 235
Indoles, perhydro-, synthesis, **42,** 289
Indoles, 1-(trimethylsilylmethyl)-, azomethine ylids from, **45,** 248
Indolizidines, synthesis, **42,** 302, 305
Indolizines, hexahydro-, synthesis by intramolecular cycloaddition, **42,** 261, 269
Indolizino-fused quinones, **45,** 66

Indolo[3,2-*b*]carbazole-6,12-dione, 5,11-dimethyl-, synthesis, **45,** 51
Indolyl groups, substituent constants, **42,** 50, 56, 60
3-Indolylmalonic ester, synthesis, **45,** 171
Indoxazenes, cleavage by base, **41,** 50
Inductive substituent constants (σ_I) of azinyl groups, **42,** 16
Infrared spectra
  of 3a-azaazulenones, **43,** 39
  of porphyrins and hydroporphyrins, **43,** 84
  of 1,2,6-thiadiazine 1,1-dioxides, **44,** 98
  of 1,2,5-thiadiazole 1,1-dioxides, **44,** 135
Infrared spectroscopy in substituent constant determination, **42,** 5, 10
Inositolamine, hexaacetyl-, synthesis, **42,** 288
Intramolecular cycloaddition
  of carbonyl groups, **42,** 285
  of oxazoles, **42,** 316, 318, 320
  of pyrimidine-4,6-diones, **42,** 317
Inversion at trigonal sulfur
  in thiophene *S*-oxides, **45,** 153
  in thiophene *S*-ylids, **45,** 162
  in thiophenium salts (calculated), **45,** 158
Ion cyclotron resonance spectroscopy, **41,** 196
  and azole acidity and basicity, **41,** 214
Ionic strength correction of p*K* data, **41,** 203
Iridodial, synthesis, **42,** 322
Isatin, azomethine ylids from, **45,** 266
Isoalloxazine analogs, reaction with dinucleophiles, **43,** 307
Isobacteriochlorins
  structure, **43,** 74, 81
  synthesis, **43,** 110
Isobacteriophin, synthesis, **43,** 114
Isobenzofuran quinones, **45,** 67
Isobenzofuran-4,7-dione, **45,** 68
Isochroman-5,8-diones, **45,** 102
Isognididione, synthesis, **42,** 320
Isoindole, 2-(2-aminophenyl)-1,3-diphenyl-, formation, **45,** 219
Isoindole-4,7-dione, 5-methoxy-2,6-dimethyl- (reniera isoindole)
  occurrence, **45,** 53
  synthesis, **45,** 340

Isoindole-4,7-dione, 2-phenyl-, cycloaddition, **45**, 54
Isoindole-4,7-diones, synthesis, **45**, 53, 340
10*H*-Isoindolo[2,1-*a*]benzimidazole, formation, **45**, 202
10*H*-Isoindolo[2,1-*a*]benzimidazol-10-one, dichloro-, formation, **45**, 219
Isomerisations, DBU-induced, **42**, 93
Isoporphyrin, structure, **43**, 75
Isoprosopinine-B, synthesis, **42**, 258
Isoquinoline
  σ-adduct formation from, **44**, 10
  Chichibabin reaction of, **44**, 5, 20, 47
Isoquinoline quinones, **45**, 92
Isoquinoline-1,4-diones, synthesis and reactivity, **45**, 85
Isoquinoline-5,8-diones
  ^{13}C NMR, **45**, 97
  electrochemical reduction, **45**, 97
  reactivity, **45**, 96
  synthesis, **45**, 96
Isoquinolines, 3,4-dihydro-
  azomethine ylids from, **45**, 262, 267
  Diels-Alder reactions with, **42**, 246
Isoquinolines, halo-, amination, **44**, 48
Isoquinolinium pseudobases
  disproportionation, **41**, 300
  oxidation, **41**, 299
Isoquinolinium salts
  addition of nitroalkanes, **43**, 335
  cycloadditions, **42**, 309; **43**, 343
  steric effects on cycloadditions, **43**, 214
Isoselenazoles, lithiation and ring-cleavage, **41**, 52
Isotanshinone, structure, **45**, 123
Isothiazoles
  lithiation and ring-cleavage, **41**, 52
  quaternization rates, **43**, 186
Isotope effects, secondary (H/D), steric and electronic, **43**, 187
Isoxazoles
  cleavage by base, **41**, 49, 58
  rearrangement to pyrazoles, **42**, 172
Isoxazoles, acyl-, conformation, **41**, 105
Isoxazoles, 3-amino-, conversion into 1,2,6-thiadiazine 1,1-dioxides, **44**, 127
4-Isoxazolines, azomethine ylids from, **45**, 287
Isoxazolium salts, cleavage by base, **41**, 51

J

Janus groups, **43**, 234

K

Kalafungin and derivatives
  occurrence and structure, **45**, 106
  synthesis, **45**, 107
KDO (deoxyoctulopyranosate), synthesis, **42**, 132
Khellinquinone, structure, **45**, 120
Kigelinone, occurrence and structure, **45**, 57
Kinetic isotope effect (H/D) in base-cleavage reactions, **41**, 49, 43
Kinetics
  of acid-base equilibrium processes, **41**, 219
  of alkylation of monocyclic azines, **43**, 131
  of Chichibabin reaction, **44**, 4, 5, 6, 8
  of dequaternization of pyridines, **43**, 278
  of oxidation of pyridinium salts, **41**, 279
  of piperidinodechlorination of azines, **43**, 277
Knoevenagel condensation, using DBU, **42**, 132
Konduramin-F1 tetraacetate, **42**, 288

L

Lambertellin, occurrence and structure, **45**, 105
β-Lactam antibiotics containing thiadiazolidine dioxide rings, **44**, 189
β-Lactams
  cyclization using diazomalonic ester/Ph$_3$P, **45**, 12
  diastereoselective syntheses, **45**, 11, 13, 30
  see also Azetidinones, Cephams, Cephems, Carbapenems, Oxapenams, Penams
β-Lactams, 3-amino-, chiral recognition on acylation, **45**, 5
β-Lactones, reactions involving DBU, **42**, 104

SUBJECT INDEX

Lactose, FAB mass spectra, **42**, 383
β-Lactoside, phenyl, FAB mass spectra, **42**, 388
Lanthanide complexes, in condensation reactions, **42**, 277, 278, 284, 324
Lanthanide shift reagents, complexation with azines, **43**, 206
Lapachols, **45**, 62, 102
α-Lapachone, **45**, 102
  derivatives, **45**, 104
  rearrangement, **45**, 103
α-Lapachone, 3,4-didehydro-, **45**, 104
β-Lapachones
  basicity, **45**, 103
  occurrence, **45**, 102
  rearrangements, **45**, 62, 103
  structure, **45**, 102
Large rings containing the sulfamide moiety, **44**, 186
Laser desorption (LD) mass spectrometry, **42**, 356
Lavendamycin
  occurrence, **45**, 88
  synthesis, **42**, 312; **45**, 88, 123
Lawesson's reagent, reaction with acetophenone oxime, **43**, 4
Lead tetraacetate oxidation of 1-aminopyridinium salts, **41**, 296
Lewis acid complexation with pyridines and analogs, **43**, 203
Ligularone, synthesis, **42**, 318
Lincosamine, synthesis, **42**, 274
Linear free energy relationships (LFER), **43**, 175 and tautomerism **41**, 224
Literature of heterocyclic chemistry (review), **44**, 269
Lithiation
  of pyridines, **44**, 200, 218–222
  see also Carbanion cleavage
Lithiomethyl derivatives of heterocycles
  cleavage of, **41**, 65
  addition to pyridine ring, **44**, 212
Lithium chelates in azomethine ylid generation, **45**, 284, 313, 332
Lithium organometallics, addition to pyridines, **44**, 200–203, 212
Lone-pair (electron) on nitrogen, ring-cleavage promoted by, **41**, 55
epi-Lupinine, synthesis, **42**, 263, 264

2,6-Lutidine, 4-dialkylamination, **44**, 34
3,4-Lutidine, Chichibabin amination, **44**, 25, 36
Lutidines, amination, **44**, 36
Lycorane alkaloid synthesis, **45**, 342
Lysergic acid, synthesis, **42**, 259

M

Macromolecules, syntheses using DBU, **42**, 140, 178
Magnetic circular dichroism (MCD) spectroscopy
  of 1,4-dihydro-1,4-diazocine, **45**, 210
  of porphyrins and hydroporphyrins, **43**, 86
Maltose, FAB mass spectrum, **42**, 385
Maltose acetate, FAB mass spectrum, **42**, 389
Manganese(III)-induced oxoalkylation of pyridines, **44**, 229
Mannich reactions
  of pyridazines, **44**, 231
  of pyrimidines, **44**, 234
Mansonones, structure and synthesis, **45**, 109
Mass spectra
  of S-alkylthiophenium salts, **45**, 156
  of a $1,3\lambda^4,2,4,6\lambda^4$dithiatriazine 1,1-dioxide, **44**, 158
  of 3-dimethylaminocholestane derivatives, **42**, 339
  of fused quinolizidines, **42**, 338
  of hydroporphyrins, **43**, 86
  of 1,2,6-thiadiazine 1,1-dioxides, **44**, 99
  of 1,2,5-thiadiazole 1,1-dioxides, **44**, 136
  of 1,2,4,6-thiatriazine 1,1-dioxides, **44**, 158
  see also Mass spectrometry
Mass spectral techniques
  chemical ionization (CI), **42**, 349
  electron impact (EI), **42**, 344
  fact atom bombardment (FAB), **42**, 357
  flash desorption, **42**, 356
  field desorption (FD), **42**, 351
  field ionization (FI), **42**, 351
  laser desorption (LD), **42**, 356
  matrix support, **42**, 362

Mass spectral techniques (cont.)
  molecular beam for solid analysis (MBSA), **42**, 360
  plasma desorption (PD), **42**, 356
  secondary ion mass spectra (SIMS), **42**, 357
Mass spectrometry of carbohydrates and other oxygen heterocycles (review), **42**, 385
Matrix support mass spectra, **42**, 364
Maturinone
  occurrence and structure, **45**, 56
  synthesis, **45**, 62
Maturone, occurrence and structure, **45**, 57
Medium effects
  in acid-base equilibria of azoles, **41**, 213
  in conformation of acyl groups, **41**, 65
  in substituent constants for pyridyl groups, **42**, 10
Meisenheimer complexes in Chichibabin reaction, **44**, 3 see also σ-Adducts; σ-Complexes
Melezitose, FAB mass spectrum, **42**, 391
Melibiose, FAB mass spectrum, **42**, 383
Melicope alkaloids, quinones from **45**, 90
(-)-Menthol, mesophase induction in **45**, 214
Mesoionic oxazoles, cycloaddition, **45**, 53
Mesomeric betaines, fused pyrimidine, **42**, 146, 155, 179
Mesophases (cholesteric, nematic) in (-)-menthol solutions, **45**, 214
Metal coordination species in azomethine ylid generation, **45**, 280
Metal ion complexation
  of dibenzo[e,g][1,4]diazocines, **45**, 214
  of pyridine and analogs, **43**, 203
Methoxatin
  biological activity, **45**, 119
  occurrence, **45**, 118
  reactions, structure and synthesis, **45**, 119
O-Methylarnottiamide, synthesis, **42**, 309
Methylphosphonate esters, reaction with azines, **43**, 209
Microwave spectroscopy
  and molecular conformation, **41**, 79
  in stereochemical studies, **43**, 217
Micellar effects, **44**, 214

Mimosamycin
  occurrence and structure, **45**, 93
  synthesis, **45**, 95
Mitiromycin, structure, **45**, 45
Mitomycins
  structure, **45**, 45
  synthesis, **45**, 122
Mitomycins, quinones related to, **45**, 43
Mitosanes, structure, **45**, 45
Mitosenes, structure, **45**, 45
Molecular beam for solid analysis (MBSA mass spectra), **42**, 360
Molecular mechanics calculations
  general considerations, **43**, 219
  on poly-t-butyl heterocycles, **43**, 238
Molecular orbitals
  of 3a-azaazulenes, **43**, 37
  of porphyrins and hydroporphyrins, **43**, 77
  usefulness in explaining reactions, **42**, 226
Molecular spectra, see Spectra
Molecular structure
  of porphyrins, **43**, 79
  see also X-ray crystallography
Molybdenum carbonyl complexes, reactions of, **42**, 179
Mouse androgen, synthesis, **42**, 270
Munchnones, cycloaddition, **45**, 53
Murrayaquinones, **45**, 49, 122

N

Nanaomycins, structure, **45**, 106
Naphthimidazole quinones, **45**, 79
Naphthimidazoles, N-methyl-, Chichibabin amination, **44**, 17, 63
Naphth[2,3-b]indolizine-6,11-dione, synthesis, **45**, 51, 122
4λ^5-Naphtho[2,1-c][1,2]azaphosphorine, 4-chloro-1,2,3,4-tetrahydro-4-thioxo-, **43**, 2
Naphthocyclinones, structure, **45**, 112
Naphtho[2,3-**d**]-1,3-dithiole quinones, **45**, 82
Naphthofuran quinones fused to other heterocyclic systems, **45**, 67
Naphtho[1,2-b]furan-4,5-diones
  natural occurrence, **45**, 58

synthesis, **45**, 63
Naphtho[1,2-*b*]furan-4,5-diones, 2,3-dihydro-, occurrence and synthesis, **45**, 59, 61
Naphtho[2,3-*b*]furan-4,9-diones
  natural occurrence, **45**, 56
  synthesis, **45**, 61, 63
Naphtho[2,3-*b*]pyran-5,10-dione, 3,4-dihydro-, synthesis, **45**, 103
Naphtho[1,8-*bc*]pyran-7,8-diones, occurrence and synthesis, **45**, 109
Naphtho[2,3-*b*]pyran-5,10-diones, occurrence, **45**, 104
Naphtho[2,3-*c*]pyran-5,10-diones, 3,4-dihydro-, occurrence, **45**, 105
Naphtho[2,3-*c*]pyrazole-4,9-diones
  formation, **45**, 76
  reactions, **45**, 77
1*H*,3*H*-Naphtho[1,8-*cd*][1,2,6]thiadiazine 2,2-dioxide, synthesis, **44**, 131
Naphtho[2,3-*c*][1,2,5]thiadiazole-4,9-dione, synthesis, **45**, 84
Naphtho[2,3-*d*]thiazole-4,9-dione, synthesis, **45**, 81
Naphtho[2,3-*b*]thiophene-4,9-diones, synthesis, **45**, 71
Naphth[2,3-*d*]oxazole-4,9-dione, 2-methyl-, synthesis, **45**, 80
1,8-Naphthyridine, hexachloro-, **42**, 145
1,8-Naphthyridine, 1,2,3,4,4a,5,6,7-octahydro-2,2,4a,7,7-pentamethyl (Eschenmoser's base), use, **45**, 32
Naphthyridines
  σ-adducts with amide ion, **44**, 10, 52–55
  Chichibabin amination, **44**, 53
  quaternisation rates in, **43**, 197
Naphthyridinomycin, biological activity, structure and synthesis, **45**, 95
Narceine imide derivatives, reactions, **43**, 51
Natural product synthesis by hetero-Diels-Alder cycloadditions (review), **42**, 245
Nectriafurone, structure, **45**, 67
Negative ion mass spectra of phenol glucosides, **42**, 373
Nickel-catalysed coupling reactions, **44**, 208, 233
Nicotinamide, Chichibabin amination, **44**, 27
Nicotinamide, *N,N*-diethyl-2,4-dimethyl-, methiodide, optical stability of, **43**, 271

Nicotine and derivatives
  conformation, **43**, 223
  free radical alkylation, **44**, 227
  quaternization, **43**, 201
Nicotinic acid, Chichibabin amination, **44**, 27
Ninhydrin, reaction with α-aminoacids, **45**, 273
Nitration
  of pyridazines, **44**, 231
  of pyridine *N*-oxides, **44**, 218
  of pyridines, **44**, 218
  of 1,2,4-triazines, **44**, 242
Nitrene cyclisations, intramolecular, forming aza-benzazulenes, **43**, 54
Nitrenes, heterocyclic, ring-cleavage of, **41**, 67
Nitrile ylids, cycloaddition to quinones, **45**, 53
Nitrogen-15 labelling in S_N(ANRORC) mechanism study, **44**, 12
Nitrones, cycloaddition to acetylenic esters, **45**, 288, 335
Nitroso compounds as dienophiles, **42**, 286
Nitrosobenzene, cycloaddition to vinylporphyrins, **43**, 104
Nitrosoformate, benzyl, cycloaddition, **42**, 291
Nonbonding interactions and stereochemistry, **43**, 216
  see also Steric effects
Norephedrine, in synthesis of chiral oxazolidine-2-thiones, **45**, 2
Norporphyrins
  structure, **43**, 77
  ultraviolet spectra, **43**, 82
Nuclear magnetic resonance (NMR) spectra, proton and general
  of 3a-azaazulenes, **43**, 38
  of 3a-azaazulen-4-ones, **43**, 42
  of 1,4-azaphosphorinane derivatives, **43**, 9
  of 3a-azoniaazulene salts, **43**, 42
  of 3-bromo-1-ethylbenzothiophenium salt, **45**, 158
  of bridgehead phosphorus compounds, **43**, 16, 23
  of 1,4,3,5-oxathiadiazine 4,4-dioxides, **44**, 155
  of porphyrins and hydroporphyrins, **43**, 84, 117, 121

Nuclear magnetic resonance (*cont.*)
  of 1,2,6-thiadiazine 1,1-dioxides, **44**, 87
  of thiadiaziridine 1,1-dioxides, **44**, 177
  of 4*H*-1,2,4,6-thiatriazine 1,1-dioxides, **44**, 155
  of thiophenium salts, **45**, 156, 158
  of thiophenium *S*-ylids, **45**, 163
Nuclear magnetic resonance (NMR) spectra, carbon-13,
  of 3,5-dimethylbenzofuran-4,7-dione, **45**, 56
  of *S*-methylthiophenium salts, **45**, 158
  of naphthocyclinones, **45**, 112
  of porphyrins, **43**, 86
  of 1,2,6-thiadiazine 1,1-dioxides, **44**, 85, 89, 107, 156
  of thiadiaziridine 1,1-dioxides, **44**, 177
  of 1,2,4,6-thiatriazine 1,1-dioxides, **44**, 156
  of thiophenium salts and ylids, **45**, 158, 163
Nuclear magnetic resonane (NMR) spectra, fluorine-19, of 1,2,4,6-thiatriazine 1,1-dioxides, **44**, 156
Nuclear magnetic resonance (NMR) spectra, nitrogen-15
  of porphyrins, **43**, 86
  of 1,2,6-thiadiazine 1,1-dioxides, **44**, 94, 107
  of 1,2,4,6-thiatriazine 1,1-dioxides, **44**, 157
Nuclear magnetic resonance (NMR) spectra, oxygen-17, of azine oxides, **43**, 150
Nuclear magnetic resonance (NMR) spectroscopy
  and proton transfer reactions, **41**, 219
  in conformation study
    of acyl groups, **41**, 81
    of dibenzo[1,4]diazocines, **45**, 214–217
    of heterocycles with steric effects, **43**, 218
  in determination of enantiomeric purity, **45**, 3
  in investigation
    into azine protonation sites, **43**, 130
    into triazinone alkylation site, **43**, 146
  in substituent constant evaluation, **42**, 4
Nuclear magnetic resonance (NMR) spectroscopy, carbon-13, and dinucleophile-diazine reactions, **43**, 325, 329

Nuclear magnetic resonance spectroscopy, dynamic (DNMR),
  of 2,5-disubstituted thiophene *S*-oxides, **45**, 153
  see also NMR in study of conformation and steric effects
Nucleophilic attack
  on thiadiaziridine 1,1-dioxides, **44**, 179
  on 1,2,6-thiadiazine 1,1-dioxides, **44**, 117
  on 1,2,5-thiadiazole 1,1-dioxides, **44**, 141
  on 1,2,4,6-thiatriazine 1,1-dioxides, **44**, 162
Nucleophilic substitution
  in pyrazines, **44**, 236
  in pyridazines, **44**, 230
  in pyridines, **44**, 200
  in pyrimidines, **44**, 232
  in 1,2,6-thiadiazine 1,1-dioxides, **44**, 118
  in 1,2,4,6-thiatriazine 1,1-dioxides, **44**, 163
  steric effects on
    in pyridines and pyrimidines, **43**, 277
    in thiophenes, **43**, 212
  see also Amination
Nupharidine, deoxy- and epideoxy-, synthesis, **42**, 302

O

Oligomers of hydrogen cyanide, heterocycles from **41**, 3
Oligosaccharides and derivatives
  mass spectra
    chemical ionization, **42**, 350
    electron impact, **42**, 345, 347
    fast atom bombardment, **42**, 358, 366
    field desorption, **42**, 352
    nomenclature, **42**, 368
    optical rotation, **42**, 380
Olivose, synthesis, **42**, 325
Optical resolution
  of diaryldibenzo[*b,f*][1,4]diazocines, **45**, 205, 216
  of racemic acids, **45**, 6
Optical rotation and oligosaccharide conformation, **42**, 380
Optical rotatory dispersion (ORD) spectroscopy and conformation study, **42**, 380; **43**, 218

Oxacephem, chiral recognition on acylation, **45**, 5
1,2,4-Oxadiazoles, cleavage by base, **41**, 54, 61
1,2,5-Oxadiazoles, cleavage by base, **41**, 53
1,2,5-Oxadiazoles, fused, thermal cleavage, **41**, 65
1-Oxa-diene systems, cycloadditions, **42**, 321
Oxalyl cyanide, **41**, 18
Oxapenam, chiral recognition on acylation, **45**, 5
1,4,3,5-Oxathiadiazine 4,4-dioxide, 2-methoxy-6-methyl-, ring interconversion, **44**, 127
1,4,3,5-Oxathiadiazine 4,4-dioxides, synthesis, **44**, 171
1,4-Oxathiino-fused heterocycles, synthesis, **43**, 318
1,4-Oxathiocin
 formation from thiophenium ylids, **45**, 169
 rearrangement to phenol derivative, **45**, 170
2H-1,2-Oxazines, 3,6-dihydro-, synthesis, **42**, 286
1,4-Oxazino-fused heterocycles, synthesis, **43**, 317, 318
Oxazoles
 cycloaddition
  intermolecular, **42**, 307
  intramolecular, **42**, 316, 318, 320
 lithiation and cleavage, **41**, 53, 57, 61
Oxazoles, mesoionic, cycloaddition, **45**, 53
Oxazolidine-2-thione, (4R and 4S)-4-ethyl-, synthesis, **45**, 2
Oxazolidine-2-thione, (4S)-4-isopropyl-, synthesis, **45**, 2
Oxazolidines, formation by azomethine ylid cycloaddition, **45**, 269
Oxazolidine-2-thione, (4R,5S)-4-methyl-5-phenyl-, synthesis, **45**, 2
Oxazolidine-2-thiones, 3-acyl-, chiral, use in diastereoselective aldol synthesis, **45**, 7
Oxazolidin-5-ones, cycloreversion to azomethine ylid, **45**, 271, 272
2-Oxazolines, pyridyl-, nucleophilic substitution, **44**, 200

4-Oxazolines, rearrangement to azomethine ylids, **45**, 287
Oxazolium salts, reduction to 4-oxazolines, **45**, 290
Oxidation
 by lead tetraacetate, of α-hydroxyketones, **43**, 44
 by mercuric acetate, of octahydro-3a-azaazulenes, **43**, 44
 by osmium tetroxide, of porphyrins, **43**, 99, 100
 of hydroporphyrins, **43**, 88
 of porphyrins, **43**, 98
 of 1,2,6-thiadiazine 1,1-dioxides, **44**, 116
 of thiadiaziridine 1,1-dioxides, **44**, 178
N-Oxidation of monocyclic azines, **43**, 149
Oxidative imination of pyridinium salts, **41**, 282
Oxidative ring-contraction of pyridinium salts, **41**, 294
Oxidative transformations of heterocyclic iminium salts (review), **41**, 275
N-Oxides, heterocyclic, ring-opening, **41**, 67
Oxidizing agents
 in Chichibabin amination, **44**, 7, 10, 43, 49, 52–55, 65, 66
 producing quinones, **45**, 42, 44
Oxime stereoisomerism, **44**, 101
Oxiranes, acyl, conformation, **41**, 119
Oxy-Cope rearrangement, **42**, 320
Oxygen, singlet, reactions
 with cyclic 1,4-dipoles, **41**, 303
 with pyrazolium-4-olates, **41**, 309
Oxygen heterocycles, mass spectrometry of (review), **42**, 335
Oxygen-17 NMR of azine oxides, **43**, 150
Ozonolysis of an aza-benzazulenone, **43**, 44

P

Palladium-catalyzed substitution
 of pyrazines, **44**, 238
 of pyridines, **44**, 206, 208, 224
Paniculide-A, synthesis, **42**, 318
Penams
 chiral recognition on acylation, **45**, 5
 synthetic approaches to, **45**, 30

Penems, synthesis, **45**, 31
Pentaarylpyridines, restricted rotation in, **43**, 271
Pentacyclic and higher fused 1,4-dioxin quinones, **45**, 114
Pentacyclic and higher fused furan quinones, **45**, 65
Pentacyclic and higher fused phenothiazine quinones, **45**, 118
Pentacyclic and higher fused pyran quinones, **45**, 112, 114
Pentacyclic and higher fused pyridazine quinones, **45**, 98
Pentacyclic and higher fused pyridine quinones, **45**, 92
Pentacyclic and higher fused pyrrole quinones, **45**, 50
Pentazole, acidity and basicity (estimated), **41**, 265
Peptides, conformation in, **41**, 142
Perfluoroalkylation (radical) of pyridine, **44**, 229
Perimidine, 1-methyl-, Chichibabin amination, **44**, 5, 16, 17
Perimidines, Chichibabin amination, **44**, 69
Perimidinyl group, substituent constants, **42**, 36
Permanganate-promoted amination
  of naphthyridines, **44**, 52
  of pyridazines, **44**, 48
  of pyrimidines, **44**, 49
  of quinazoline, **44**, 66
  of quinoline, **44**, 43
  of quinoxaline, **44**, 65
  of 1,2,4-triazines, **44**, 241
Petaselbine, synthesis, **42**, 318
Pharmacological properties of 1,2,5-thiadiazole 1,1-dioxides, **44**, 189
Phenanthridines, Chichibabin amination, **44**, 45
Phenanthro[1,2-*a*]furan-10,11-diones, **45**, 63
Phenanthroline-9,10-diones, **45**, 91
Phenanthrolines, quaternization rates, **43**, 197
Phenanthro[9,10-*c*]thiadiazole 2,2-dioxide, pyrolysis, **44**, 138
Phenazine-1,4-dione, 2,3-dihydroxy-, **45**, 99

Phenophosphazines, 5,10-dihydro-, **43**, 12
  synthesis, **43**, 13
  uses, **43**, 13, 15
Phenothiazines, fused, quinones from, **45**, 118, 120
Phenoxazines, 1,4-diazocino-fused, **45**, 191
Phenyl groups, substituted, electronic effects of, **42**, 56
Phlorin, structure, **43**, 75
Phomazarin
  structure and occurrence, **45**, 90
  synthesis, **45**, 123
Phosphole oxide derivatives, synthesis, **42**, 106
Phosphorus-nitrogen heterocycles (review), **43**, 1
Phosphorus trichloride, cyclizations using, **43**, 3, 12
Photochemical reduction of porphyrins and hydroporphyrins, **43**, 92
Photochemistry of hexahydropyrimido[1,2-*a*]azepin-4-ones, **42**, 150
Photocyclization of a (chloroacetyl)amine, **45**, 188
Photocyclisation reactions, forming azaazulenes, **43**, 57
Photodehydrogenation (cyclization)
  of 1,2-diarylpyridinium salts, **41**, 294
  of 1,2-diarylquinolinium salts, **41**, 298
(2+2)Photodimerization of 11b-azadibenz[*e*, *ij*]azulene, **43**, 48
Photoelectron spectroscopy
  and proton affinities, **41**, 199
  of azines, **43**, 147
  see also ESCA
Photolysis of 1-azatriptycene, **43**, 55
Photo-oxidation
  of phenylpyridinium betaines, **41**, 294
  of porphyrins, **43**, 101, 102
Photo-stimulated radical substitution ($S_{RN}1$), **44**, 227
Phthalazine-5,8-diones, **45**, 97
Phthalazines, quaternization rates, **43**, 197
1-(*N*-Phthalimido)aziridines, ring-opening to azomethine ylids, **45**, 238
Physostigmine alkaloids, synthesis, **45**, 340
2-Picoline
  alkylamination, **44**, 33
  amination with ANRORC rearrangement, **44**, 27

Chichibabin amination, **44**, 2, 27
3-Picoline
  alkylamination, **44**, 33
  Chichibabin amination, **44**, 4, 22
4-Picoline, alkylamination, **44**, 33
Piperazines, N-acyl, conformations of, **41**, 151
Δ^1-Piperideine, Diels-Alder reactions with, **42**, 246
Piperidines, synthesis by cycloaddition, **42**, 257
Piperidines, acyl, conformations of, **41**, 123, 129, 148
Piperidin-2-one, 6-acetoxy-, chiral induction in use as electrophile, **45**, 18
Plasma desorption (PD) mass spectrometry, **42**, 356
Polarographic reduction of N-alkyldiazinium salts, **43**, 321
Polarography
  in substituent constant determination, **42**, 5, 7
  of phenophosphazines, **43**, 15
  see also Electrochemistry
Polonovski reaction of mitosane N-oxide analogs, **45**, 46
Polymer syntheses using DBU, **42**, 140, 178
Polymer-supported DBU, **42**, 170
Polymerization, cationic, **43**, 283
Polysaccharides
  inter-residue linking and hydrogen-bonding, **42**, 379
  mass spectra (review), **42**, 335
Porfiromycin
  structure, **45**, 45
  synthesis, **45**, 121
Porphodimethene, structure, **43**, 75
Porphomethene, structure, **43**, 75
Porphyrin derivatives, hydrogenated (review), **43**, 73
Porphyrinogen, structure, **43**, 75
Porphyrinogen, substituted, rearrangements of, **43**, 96
Porphyrins
  biochemical aspects, **43**, 116
  crystal structure, **43**, 81
  oxidation, **43**, 98
  reduction, **43**, 90
Praziquantel, synthesis, **42**, 267

Prelog-Djerassi lactone, synthesis, **42**, 270
Prelog-Djerassi lactonic acid, methyl ester, chiral synthesis, **45**, 25
Pressure effects
  on alkylation of sterically hindered bases, **43**, 181
  on Chichibabin reaction, **44**, 22
Proline, N-acyl, conformations of, **41**, 138
Propargyl ethers, Claisen rearrangements of, **42**, 208, 218, 224, 228, 231, 236, 237
Propargyl thioethers, Claisen rearrangements of, **42**, 209, 212, 223
Prosopis alkaloid, synthesis, **42**, 257
Prostaglandin analogs, synthesis, **45**, 27
Protoberberine, synthesis, **42**, 246
Proton affinities (gas phase)
  and photoelectron spectroscopy, **41**, 199
  of azoles, **41**, 214
Protonation
  of azines at ring N, **43**, 128
  of hydroporphyrins, **43**, 87
Pseudobases, disproportionation of, **41**, 280
Pseudomonic acid, synthesis, **42**, 285
Pseudotropine, synthesis, **42**, 293
Pseudouridine, conversion into pseudoisocytidine, **43**, 339
Psoralene-derived quinones, **45**, 120
Pteridine, reaction with β-dicarbonyl compounds, **43**, 319
Pteridine-4-carboxylic ester, reaction
  with 1,4-dinucleophiles, **43**, 306
  with ethanol, **43**, 321
Pteridinium salts
  addition of methanol, **43**, 323
  reaction with acetoacetamides, **43**, 329
  reaction with thioureas, **43**, 328
Purines
  σ-adduct formation with amide ion, **44**, 10
  Chichibabin amination, **44**, 70
  Claisen rearrangements in, **42**, 205
  formation from HCN oligomers, **41**, 33
Puromycin analog, synthesis, **42**, 252
Push–pull ethylenes, rotation barriers in, **43**, 255
Pyrayaquinone-B, **45**, 49
Pyrazine
  σ-adduct formation with amide ion, **44**, 9, 10, 48
  Chichibabin amination, **44**, 48

Pyrazine, 2.5-bis(di-*t*-butylmethylene)-1,4-diethyl-, **45**, 260
Pyrazine, 2,3-dichloro-, reaction with dinucleophiles, **43**, 314
Pyrazine, tetrachloro-, reaction with β-dicarbonyl compounds, **43**, 311
Pyrazine 1-oxides, chlorodeoxygenation, **44**, 237
Pyrazine quinones, **45**, 99
1*H*,4*H*-Pyrazine-2,6-diones, dipolar, **45**, 237
Pyrazines
  N-amination, **43**, 161
  formation in reactions of DAMN and DISN, **41**, 16, 22, 28
  N-oxidation, **43**, 156
  quaternization, **43**, 143, 197
  rates of quaternization, **43**, 197
  substitution
    electrophilic, **44**, 239
    free-radical, **44**, 239
    nucleophilic, **44**, 236
Pyrazines, dichloro-, reaction with nucleophiles, **43**, 318
Pyrazinium salts
  addition
    of methanol, **43**, 323
    of nitromethane anion, **43**, 304
  reaction
    with acetoacetamides, **43**, 329
    with 1,4-diamines, **43**, 330
    with thioureas, **43**, 328
Pyrazinium salts, fused, addition of methanol, **43**, 324
Pyrazinones, alkylation, **43**, 143
Pyrazino[2,3-*c*][1,2,6]thiadiazine 2,2-dioxides
  glycosylation, **44**, 114
  spectra, **44**, 94, 97
  synthesis, **44**, 130
  tautomerism, **44**, 107
Pyrazoles
  acidity and basicity, **41**, 234
  cleavage by base, **41**, 51
  comparison of ^{13}C NMR data with 1,2,6-thiadiazine1,1-dioxides, **44**, 90, 93
  pK_a data, **41**, 250
Pyrazoles, *N*-acyl, conformations of, **41**, 127
Pyrazolium-4-olates, oxidation, **41**, 309

Pyrazolo-bipyrimidines, formation, **41**, 353, 354
Pyrazolo[4,3-*f*]indazole-3,5-dicarboxylic acid, 4,8-dioxo-, synthesis, **45**, 76
Pyrazolopyrimidines
  chemistry of (review), **41**, 319
  spectra, **41**, 362
  tautomerism, **41**, 363
  theoretical studies, **41**, 361
Pyrazolo[1,5-*a*]pyrimidines
  biological activity, **41**, 366
  reactions
    with electrophiles, **41**, 349
    with nucleophiles, **41**, 351
  reduction, **41**, 354
  synthesis, **41**, 321
  tautomerism, **41**, 364
Pyrazolo[1,5-*c*]pyrimidines
  biological activity, **41**, 367
  synthesis, **41**, 348
  tautomerism, **41**, 364
Pyrazolo[3,4-*d*]pyrimidines
  alkylation, **41**, 356
  biological activity, **41**, 367
  reactions
    with electrophiles, **41**, 355
    with nucleophiles, **41**, 357
  rearrangements, **41**, 359
  synthesis, **41**, 333
  tautomerism, **41**, 363
Pyrazolo[4,3-*d*]pyrimidines
  biological activity, **41**, 367
  synthesis, **41**, 345
Pyrazolyl groups, substituent constants, **42**, 46
Pyridazine
  σ-adduct formation with amide ion, **44**, 9, 10, 48
  Chichibabin amination, **44**, 48
Pyridazine, 3,4,6-trichloro-, reaction with dinucleophiles, **43**, 318
Pyridazine, 3-methoxy-, Chichibabin amination, **44**, 50
Pyridazine, 3-phenyl-, Chichibabin amination, **44**, 48
Pyridazine 1-oxides, nucleophilic attack, **44**, 230
Pyridazine quinones, **45**, 97
Pyridazines
  acylation rates, **43**, 147

N-amination, **43**, 161
basicity, **43**, 129
N-oxidation, **43**, 151
quaternisation, **43**, 132, 197, 201
rates of quaternisation, **43**, 197, 201
substitution
  electrophilic, **44**, 231
  free-radical, **44**, 232
  nucleophilic, **44**, 230
Pyridazines, dichloro-, reaction with dinucleophiles, **43**, 318
Pyridazino[1,2-*a*]pyridazine-1,4,6,9-tetrone, **45**, 97
Pyridazinyl group, substituent constants, **42**, 14
Pyridine
  substitution
    free-radical, **44**, 227
    nucleophilic, **44**, 3
Pyridine, 2-benzyl-, Chichibabin amination, **44**, 27
Pyridine, 3-*t*-butyl-, amination, **44**, 37
Pyridine, 4-butyl- (sec, tert), amination, **44**, 36
Pyridine, 2-chloro-5-nitro-
  amination, **44**, 18
  ANRORC reaction with, **43**, 303
Pyridine, 2,6-di-*t*-butyl-
  catalytic uses of, **43**, 283
  protonation, **43**, 177
  quaternization, **43**, 182
  sulfonation, **43**, 211
  uses, **43**, 283
Pyridine, 2,3-dichloro-, reaction with dinucleophiles, **43**, 314
Pyridine, 2,6-dimethyl-, trifluoromethanesulfinyloxylation, **43**, 283
Pyridine, dimethyl-, *see also* Lutidine
Pyridine, 2-dimethylamino-
  Chichibabin amination, **44**, 39
  rotation barrier, **43**, 243
Pyridine, 3-dimethylamino-, Chichibabin amination, **44**, 39
Pyridine, 4-dimethylamino-, Chichibabin amination, **44**, 15, 27, 39
Pyridine, methyl-, *see* Picoline
Pyridine, pentafluoro-, reaction with nucleophiles, **43**, 309
Pyridine, 3-phenylthio-, Chichibabin amination, **44**, 37

Pyridine, 2,3,4,5-tetrahydro-, Diels-Alder reactions with, **42**, 246
Pyridine, 2,3,5-trichloro-, reaction with nucleophiles, **43**, 308
Pyridine 1-oxides
  substitution
    electrophilic, **44**, 218
    free-radical, **44**, 230
    nucleophilic, **44**, 214
  uses in synthesis of substituted pyridines, **44**, 214, 218
Pyridine quinones, **45**, 84
Pyridine-2-carboxaldehyde, azomethine ylids from, **45**, 266
Pyridine-3-carboxaldehyde, azomethine ylids from, **45**, 274
Pyridine-2,5-dione, 3(4)-*t*-butyl-6-cyano-, synthesis, **45**, 84
Pyridine-2,5-diones, 6-hydroxy-, **45**, 84
Pyridines
  N-acylation, **43**, 207
  basicity
    compared with azoles, **41**, 230
    steric effects on, **43**, 177, 184, 202
  complexation
    with halogens, **43**, 208
    with metal ions, **43**, 203
    with Lewis acids, **43**, 203
  deactivation effects in, **42**, 62
  N-oxidation, kinetics, **43**, 208
  quaternisation kinetics, **43**, 191
  reaction with methylphosphonate esters, **43**, 209
  substitution
    electrophilic, **44**, 217
    free-radical, **44**, 227
    nucleophilic, **44**, 1, 200
  synthesis by hetero-Diels-Alder reactions, **42**, 307, 312, 316
Pyridines, sterically-hindered, basicity, **43**, 177
Pyridines, acetyl-, rotational barriers in, **41**, 112
Pyridines, acyl-, conformations of, **41**, 106
Pyridines, alkyl-, Chichibabin amination and dimerisation, **44**, 29
Pyridines, alkynyl-, synthesis, **44**, 206
Pyridines, amino-
  Chichibabin amination, **44**, 39
  quaternisation, **43**, 201

Pyridines, azido-, nucleophilic attack at ring, **44,** 206
Pyridines, dihydro-, synthetic uses, **44,** 223
Pyridines, halo-
  nucleophilic substitution of halide, **44,** 203–206
  quaternisation rates, **43,** 186
Pyridines, 2-isopropyl-
  basicity, **43,** 274
  quaternization, **43,** 274
Pyridines, lithiated, synthetic uses, **44,** 218
Pyridines, methylnitrosoamino-, nucleophilic displacement of substituent, **44,** 206
Pyridines, 3-(2-oxazolin-2-yl)-, nucleophilic substitution in, **44,** 200
Pyridines, pentaaryl-, restricted rotation in, **43,** 271
Pyridines, polymethyl-, gear-clashed conformation and quaternisation, **43,** 275
Pyridines, 3-substituted, pressure amination, **44,** 25
Pyridines, 4-substituted, amination, **44,** 27
Pyridines, trialkylstannyl-, synthetic uses, **44,** 219, 222, 224 see also Lutidine, Picoline
Pyridinio-groups, substituent constants, **42,** 11, 14, 37, 42
Pyridinium salts
  dealkylation kinetics, **43,** 189
  ring contraction (oxidative), **41,** 294
  oxidation, **41,** 276, 291, 294
Pyridinium salts, 1-acyl-, nucleophilic attack on, **44,** 208
Pyridinium salts, 2-acyl-, oxidation, **41,** 291
Pyridinium salts, 1-alkoxycarbonyl-, nucleophilic attack on, **44,** 211
Pyridinium salts, 1,2-diaryl-, photodehydrogenation, **41,** 294
Pyridinium salts, 1-fluoro-, nucleophilic attack on, **44,** 212
Pyridinium salts, 2-hydroxyalkyl-, oxidation, **41,** 291
Pyridinium salts, isopropyl- and polymethylisopropyl-, rotamer populations, **43,** 235
Pyridinium salts, 2-carboxy-, oxidation, **41,** 291

Pyridinium salts, 1-(9-fluorenyl)-, rotamer isolation, **43,** 227
Pyridinium salts, 1-phenoxycarbonyl-, nucleophilic attack on, **44,** 209, 211
Pyridiniumyl groups, substituent constants, **42,** 11
Pyridinones, see Pyridones
Pyrido[1,2-a][1,3]diazepines, formation, **41,** 295
Pyrido[2,3-b][1,4]diazocine-6,9-diones, 5,8,9,10-tetrahydro-, **45,** 188
4-Pyridones, N-phenyl-, chiral, **43,** 265
2-Pyridones, formation from pyridinium salts, **41,** 276
Pyrido[2,3-b]pyrazine
  σ-complexes from, **44,** 52
  quaternisation rates, **43,** 198
Pyrido[2,3-d]pyridazine
  σ-adduct formation with amide ion, **44,** 11
  Chichibabin amination, **44,** 51
Pyrido[2,3- and 3,4-d]pyridazinediones, generation, **45,** 85
Pyrido[3,2- and 3,4-g]quinoline-5,8-diones, synthesis, **45,** 91
Pyrido[2,3-g]quinoxaline-5,10-dione derivative, **45,** 121
Pyrido[2,3-c][1,2,6]thiadiazine 2,2-dioxides
  acidity and basicity, **44,** 110
  spectra, **44,** 94, 97
  synthesis, **44,** 130
  tautomerism, **44,** 107
Pyridoxal-derived azomethine ylids, importance of, **45,** 254
Pyridoxine, synthesis, **42,** 308
Pyridyl groups, substituent constants, **42,** 7
Pyrimidine
  σ-adduct formation with amide ion, **44,** 9, 10, 48
  Chichibabin amination, **44,** 48
Pyrimidine mesomeric betaines, fused, **42,** 146, 155, 179
Pyrimidine quinones, **45,** 98
Pyrimidine, 5-bromo-, amination, **44,** 51
Pyrimidine, 4-t-butyl-, amination, **44,** 51
Pyrimidine, 5-methyl-, amination, 29
Pyrimidine, 5-nitro-
  cycloaddition to, **43,** 343, 346

nucleophilic addition (ANRORC), **43**, 346
reaction with amidines, **43**, 336, 338
Pyrimidine, 4-phenyl-
 alkylamination, **44**, 33
 Chichibabin amination, **44**, 12, 35, 50
Pyrimidine, 5-phenyl-, amination, **44**, 50
Pyrimidine-2,4-dione, 1,3-dimethyl-, reaction with guanidine, **43**, 337
Pyrimidine-2,4(1*H*,3*H*)-dione, 1,3-dimethyl-5-nitro-, reaction with 1,3-dinucleophiles, **43**, 339
Pyrimidines
 N-amination, **43**, 161
 basicity, **43**, 130
 Chichibabin amination, **44**, 29, 33, 48, 232
 deactivation effects in, **42**, 62
 formation from HCN oligomers, **41**, 32
 N-oxidation, **43**, 153
 protonation site, **43**, 130
 quaternization, **43**, 136, 197
 rates of quaternization, **43**, 197
 substitution
  electrophilic, **44**, 234
  free-radical, **44**, 235
  nucleophilic, **44**, 232
 transmission of substituent offects in, **42**, 71
Pyrimidines, dimethylamino-, rotation barriers, **43**, 241, 244
Pyrimidinethiones, alkylation, **43**, 141
Pyrimidine-2-thiones, 1-phenyl-, chiral, **43**, 267
Pyrimidin-2-one, 5-nitro-, reaction with diethyl acetonedicarboxylate, **43**, 336
Pyrimidinones
 acylation, **43**, 149
 alkylation, **43**, 137
Pyrimidin-2-ones, 1-phenyl-, chiral, **43**, 267
Pyrimidinyl groups, substituent constants, **42**, 20, 31
Pyrimidoazepines (review), **42**, 83
Pyrimido[1,2-*a*]azepine, octachloro-, **42**, 145
Pyrimido[1,2-*a*]azepine, octahydro-, *see* 1,8-Diazabicyclo[5.4.0]undec-7-ene (DBU)

Pyrimido[1,2-*a*]azepines
 spectra, **42**, 155
 synthesis, **42**, 145, 179
 uses, **42**, 156, 180
Pyrimido[1,6-*a*]azepines
 reactions, **42**, 158
 spectra, **42**, 158
 synthesis, **42**, 157
 uses, **42**, 159
Pyrimido[4,5-*b*]azepines, synthesis, **42**, 159, 180
Pyrimido[4,5-*c*]azepines, synthesis, **42**, 164
Pyrimido[4,5-*d*]azepines
 reactions, **42**, 165
 rearrangements, **42**, 166
 synthesis, **42**, 164
Pyrimido[5,4-*b*]azepines, **42**, 169
Pyrimido[1,2-*a*]azepin-4-ones, hexahydro-, photochemistry, **42**, 150
Pyrimidyl groups, substituent constants, **42**, 14, 58
Pyrones, acyl, conformations of, **41**, 115
4-Pyrones, 2,3-dihydro-, synthesis by cycloaddition, **42**, 270
Pyronyl groups, substituent constants, **42**, 42
Pyrrocorphin, octaethyl-, formation from octaethylporphyrinogen, **43**, 96
Pyrrocorphins
 structure, **43**, 75
 synthesis, **43**, 111
Pyrroles
 acidity and basicity, **41**, 234
 Claisen rearrangements in, **42**, 205
 pK_a data, **41**, 234
 synthesis from azomethine ylids, **45**, 235, 239, 289, 290, 293, 336
Pyrroles, acyl, conformations of, **41**, 94, 127
Pyrroles, 2,5-dihydro- and tetrahydro-, synthesis from azomethine ylids, **45**, 231–344
Pyrroles, *N*-phenyl-, chiral, **43**, 263
Pyrrolidines, *N*-acyl, conformations of, **41**, 129, 138
Pyrrolidino group, rotation barrier of, **43**, 241, 246
Pyrrolidin-2-one, 5-acetoxy-, chiral induction in use as electrophile, **45**, 18

Pyrrolizidine alkaloids
  chiral synthesis, **45**, 20
  synthesis by cycloaddition reactions, **45**, 338
Pyrrolizine, 3,3-dimethyl-, cycloaddition to, **43**, 59
Pyrrolo[1,2-*a*]azepines (review), **43**, 35 *for detailed index see* 4-Azaazulenes
Pyrrolo[1,2-*a*]azepinium salts, **43**, 42
Pyrrolo[1,2-*b*][2,5]benzodiazocines, 6,11-dihydro-, synthesis, **45**, 190
Pyrrolo-fused quinones, **45**, 39
Pyrrolo[3,2-*e*]indole-4,5-dione, 6-benzenesulfonyl-1-*t*-butoxycarbonyl-5-methyl-, **45**, 42
Pyrrolo[1,2-*a*]indolediones, **45**, 45–48
Pyrroloisoquinolines, Chichibabin amination, **44**, 45
1*H*-Pyrrolo[2,3-*f*]quinoline-2,7,9-tricarboxylic acid, 4,5-dioxo- (PQQ), *see* Methoxatin
Pyrroloquinolinequinone (PQQ) derivatives, **45**, 123
Pyrroloquinolines, Chichibabin amination, **44**, 44
Pyrrolo[2,3-*c*][1,2,6]thiadiazin-4 (3*H*)-one 2,2-dioxide, 1,7-dihydro-5,6,7-triphenyl-, synthesis, **44**, 133
Pyrrolyl groups, substituent constants, **42**, 45
Pyrylium salts
  reactions with primary amines, **43**, 283
  use in aminoheterocycle substitution, **44**, 203
Pyrylium salts, dimethylamino-, rotation barriers, **43**, 241, 245

Q

Quantitative analysis of steric effects in heteroaromatics (review), **43**, *173*
Quaternization
  of azines
    rates of, **43**, 131
    steric effects on, **43**, 180, 273
  of pyridazines, **43**, 132, 273
  of ring nitrogen
    mechanism, **43**, 180
    steric effects, **43**, 180, 197

Quinazoline
  Chichibabin amination, **44**, 66
  quaternization, **43**, 198
Quinazoline-5,6-diones, **45**, 98
Quinazoline-5,8-diones, synthesis and reactivity, **45**, 98
Quinazolines and analogs, 4-dimethylamino-, rotation barriers, **43**, 244, 245
Quinazolinium methiodide, cycloaddition of acetimidates, **43**, 345
Quinazolin-4-one, synthesis by cycloaddition, **42**, 248
Quinazolinyl groups, substituent effects, **42**, 60
Quinoline
  Chichibabin amination of, **44**, 41
  σ-adducts in, **44**, 10, 42
    in presence of oxidant, **44**, 43
  pressure effects on, **44**, 29
  steric effect of adjacent ring on *N*-alkylation in, **43**, 185
Quinoline quinones, **45**, 85
  biological activity, **45**, 88
Quinoline, 3-nitro-
  addition of amidines, **43**, 334
  amination, **44**, 43
Quinoline, 4-nitro-, amination, **44**, 43
Quinoline, trifluoromethyl-, amination, **44**, 43
Quinoline-3,4-dione, **45**, 85
Quinoline-5,6-dione, **45**, 85, 88
Quinoline-5,8-diones, **45**, 86, 87
  ^{13}C NMR, **45**, 88
  reactivity, **45**, 87
  synthesis, **45**, 86, 92, 121, 123
Quinolines, Claisen rearrangements in, **42**, 230
Quinolines, methyl-, amination, **44**, 41
Quinolines, 5,6,7,8-tetrahydro-, synthesis by cycloaddition, **42**, 302
Quinolinium salts, oxidation by ferricyanide, **41**, 296
Quinolinium salts, 1-methoxy-, addition of enamines, **43**, 335
Quinolinium salts, 3-nitro-, addition of acetone, **43**, 341
Quinolizidine alkaloids, chiral synthesis, **45**, 21
Quinolizidines
  mass spectra, **42**, 338

synthesis, **42**, 302, 305; **45**, 21
Quinones, cycloaddition
    to an azomethine ylid, **45**, 340
    to benzonitrile oxide, **45**, 77
    to diazoalkanes, **45**, 74
Quinones, [2.3.4]cyclazine, **43**, 44
Quinones, heterocyclic (review), **45**, 37
Quinones, pyridazine, **45**, 97
Quinones, pyridine, **45**, 84
Quinones
    1,4-dioxano-fused, **45**, 114
    1,4-dithiino-fused, **45**, 115
    1,4-oxathiino-fused, **45**, 115
    1,4-oxazino-fused, **45**, 117
    furano-fused, **45**, 54
    imidazo-fused, **45**, 79
    isothiazolo-fused, **45**, 78
    isoxazolo-fused, **45**, 77
    pyrazolo-fused, **45**, 74
    pyridazino-fused, **45**, 97
    pyrido-fused, **45**, 85
    pyrimidino-fused, **45**, 98
    pyrrolo-fused, **45**, 39
    thieno-fused, **45**, 69
    triazolo-fused, **45**, 82
Quinoproteins, biological activity, **45**, 118
Quinoxaline, Chichibabin amination with oxidant, **44**, 65
Quinoxaline, 2-chloro-, reaction with nucleophiles, **43**, 310, 312
Quinoxaline, 2,3-dichloro-, reaction
    with amidines, **43**, 313
    with 1,4-dinucleophiles, **43**, 316, 318
    with dithiocarboxylates, **43**, 313
    with enolates, **43**, 310
    with sulfur dinucleophiles, **43**, 312
Quinoxaline-5,8-diones
    reactivity, **45**, 123
    synthesis, **45**, 99
Quinoxalinium salts
    addition
        of acetoacetamides, **43**, 329
        of β-diketones, **43**, 326
        of 1,4-dinucleophiles, **43**, 330, 333
        of dithiocarbamates, **43**, 326
        of enamines, **43**, 306
        of imidate anions, **43**, 327
        of methanol, **43**, 323
        of nitromethane anion, **43**, 304
        of thioamides, **43**, 326, 327

        of thioureas, **43**, 328
    oxidation by ferricyanide, **41**, 304

R

Rabbit livers, **41**, 310
Racemisation rates of 6,11-diaryldibenzo[*b*,*f*][1,4]diazocines, **45**, 216
Radical anion reduction of porphyrins, **43**, 91
Radical chain substitution mechanism, **44**, 227
Raffinose, FAB mass spectra, **42**, 391
Raman spectroscopy of iron chlorin derivatives, **43**, 84
Reactions of annular nitrogens of azines with electrophiles (review), **43**, 127
Rearrangements
    aza-Cope, **42**, 213
    by reversible ring-opening, **41**, 67
    Claisen, in heterocyclic systems (review), **42**, 203
    DBU-induced, **42**, 93
    Dimroth, *see* Dimroth rearrangements
    forming pyrazolopyrimidines, **41**, 332
    of 4-acyloxazoles (Cornforth), **41**, 57
    of an octahydroquinoline hydroperoxide, **43**, 41
    of hydroporphyrins, **43**, 96
    of pyrazolopyrimidines (Dimroth), **41**, 335, 337
    of pyrazolo[1,5-*a*]pyrimidines, **41**, 351, 352
    of pyrazolo[3,4-*d*]pyrimidines, **41**, 359
    of spiro-3*H*-pyrazoles, **41**, 349
    of triphosphazines, **43**, 28
Redox potentials
    of hydroporphyrins, **43**, 88
    see also Polarography
Reduced porphyrins (review), **43**, 73
Reduction
    of porphyrins and hydroporphyrins
        analytical, **43**, 88
        synthetic, **43**, 90
    of thiadiaziridines, **44**, 179
Reduction, electrochemical, of pyrazolo[1,5-*a*]pyrimidines, **41**, 354
Reduction potentials, polarographic, of diazinium salts, **43**, 321

Reissert analog, pyridine, **44**, 225
Reissert compounds, quinoline, azomethine ylids from, **45**, 268
Relaxation techniques in acid-base equilibra investigation, **41**, 219
Reniera isoindole, *see* Isoindole-4,7-dione
Renierone and derivatives
  occurrence and structure, **45**, 92
  synthesis, **45**, 95
Resonance, steric inhibition of, **43**, 276
Resonance constant (σ_R)
  definition, **42**, 3
  of azinyl groups, **42**, 18
Restricted rotation
  in biaryls and analogs, **43**, 256
  in pentaarylpyridines, **43**, 271
  see also Rotation barriers
Retro-Diels-Alder fragmentation in mass spectrometry of cyclic 1,4-diones, **42**, 337
Retro-mass-spectral synthesis, **42**, 246
Retronecine, synthesis, **42**, 293; **45**, 339
Rhodin g_7, trimethyl ester, synthesis, **43**, 91
Rhodium complexes, use in synthesis, **45**, 73
Rhodium(II) acetate-catalysed diazoester reactions, **45**, 18, 160
Ring-chain isomerisation in berberine derivatives, **43**, 41
Ring-chain tautomerism of an azidothiatriazine, **44**, 159
Ring-cleavage of fused pyrazoles, **41**, 360
Ring-contraction of pyrylium salts, oxidative, **41**, 294
Ring-expansion of fused pyrazole *N*-oxide, **41**, 359
Ring-opening of five-membered heteroaromatic anions (review), **41**, 41
Rotation barriers
  about amide C-N bonds, **43**, 247
  about C=C double bonds, **43**, 238, 252
  of alkyl groups, **43**, 190
  of biaryls and analogs, **43**, 256
  of dimethylamino groups, **43**, 242
  of isopropyl groups in heterocycles, **43**, 224
  of ring alkyl groups, **43**, 220
  past methylthio and thione groups, **43**, 221

Rubradirins, occurrence and structure, **45**, 116
Ruhemann's purple, formation, **45**, 273
Rutaecarpine, synthesis, **42**, 249

S

Safracins
  structure, **45**, 93
  synthesis, **45**, 95
Saframycins, structure, **45**, 93, 123
Saponins, FD mass spectra, **42**, 354
Sarubicin, occurrence and structure, **45**, 109
Scabequinone, occurrence, structure, synthesis, **45**, 120
Sceletium alkaloid synthesis, **45**, 341
Schmidt reaction, forming diazocinones, **45**, 190
Secondary ion mass spectrometry (SIMS), **42**, 357
Secondary steric effects, **43**, 211, 275
1,2,5-Selenadiazole, 3,4-dicyano-, **41**, 35
1,2,3-Selenadiazoles, base cleavage, **41**, 63
Selenienyl groups, substituent constants, **42**, 43
Selenophenes, cleavage on lithiation, **41**, 48
Selenophenes, acyl, conformation, **41**, 84
[3.3]-Sigmatropic rearrangements, *see* Aza-Cope and Claisen rearrangements
Silylation, using DBU, **42**, 128
Sirohydrochlorins, **43**, 111, 118
Slaflamine, synthesis, **42**, 263
Solid state, structure of charged azoles in, **41**, 226
Solution, structure of charged azoles in, **41**, 225
Solvent effects
  on acyl group conformation, **41**, 165
  on Chichibabin reaction, **44**, 21
  on quaternization reactions, **43**, 182
Soret bands in porphyrins and hydroporphyrins, **43**, 83
Spectra, see the individual spectroscopic techniques
Spectroscopic properties of pyrimidoazepines, **42**, 158, 163, 180

Sphingosine, *erythro* and *threo*, synthesis, **42**, 297
"Spin-lattice" relaxation times of methyl protons, **43**, 237
10-Spirobiphenophosphazines, **43**, 13
Spirophosphonium salts, **43**, 14, 19
Spirophosphoranes, **43**, 16, 21
Sputtering, in mass spectral ionisation, **42**, 363
Stachyose, FAB mass spectrum, **42**, 393
Standardisation of thermodynamic pK_a data, **41**, 212
Statistical corrections in acid-base ionizations, **41**, 193
Stauroporine, synthesis, **42**, 297
Stemine, structure, **43**, 69
Stemphone, occurrence and structure, **45**, 110
Stenocarpoquinone-A, rearrangement, **45**, 62
Stereochemical considerations in mass spectroscopy of carbohydrates, etc. (review), **42**, 335
Steric effects
  in dequaternization of pyridinium salts, **43**, 278
  in heteroaromatics, quantitative analysis of (review), **43**, 173
  of heteroaromatic rings, **42**, 3; **43**, 173
  separation from electronic, **43**, 175
Steric effects, secondary, **43**, 211, 275
Steric inhibition of resonance, effect
  on amino and nitro groups, **43**, 276
  on reactivity of π-system, **43**, 211
Steric parameters (E_s)
  and quaternization
    of pyridines, **43**, 185
    of thiazoles, **43**, 187
  modification of, **43**, 189
Steric parameters, *ortho*-, (S^o), **43**, 216
Stevens rearrangement of spiro-ammonium salt, **43**, 61
Strained homomorphs, **43**, 185
Streptonigrin
  biological activity, **45**, 88
  biosynthesis, **45**, 89
  occurrence, **45**, 89
  synthesis, **42**, 252, 311; **45**, 88, 123
Structural determination of ionic azoles, **41**, 224

Structure, *see also* Molecular structure, X-ray crystallography
Structure of charged azoles
  in solid state, **41**, 226
  in solution, **41**, 225
Substituent constants (σ), definition, **42**, 2
Substituent effects
  on azole acidity and basicity, **41**, 228
  on Chichibabin reaction, **44**, 18
Substituent rate factors, **43**, 199
Substituents, acidity and basicity of,
  in benzimidazoles, **41**, 250
  in carbazoles, **41**, 238
  in imidazoles, **41**, 243
  in indoles, **41**, 237
  in pyrazoles, **41**, 256
  in pyrroles, **41**, 235
  in tetrazoles, **41**, 264
  in 1,2,3-triazoles, **41**, 262
  in 1,2,4-triazoles, **41**, 259
Substitution
  radical ($S_{RN}1$)
    of pyridines, **44**, 227
    of pyrimidines, **44**, 235
  regioselective, in six-membered aromatic N-heterocycles (review), **44**, 199
  for other types of substitution see the relevant ring system
Substitution reactions, using DBU, **42**, 122
Sucrose, FAB mass spectrum, **42**, 384
Sucrose acetate, FAB mass spectrum, **42**, 389

Sugars, reduced, synthesis, **42**, 324
Sulbactam analog synthesis, **45**, 32
N-Sulfinyl compounds as dienophiles, **42**, 297
Sulfamide, condensation forming heterocycles, **44**, 121–124, 131, 149, 184, 185, 186
Sulfamide moiety, heterocycles containing the (review), **44**, 81
Sulfur extrusion, **45**, 115, 118, 180

T

Tabtoxin, synthesis, **42**, 291
Tanshinones, occurrence and structure, **45**, 63

Tautomerism
  of amino-1,2,5-thiadiazole 1,1-dioxides, **44**, 137
  of azepino[1,2-*a*]indoles, **43**, 40
  of hydroporphyrins, **43**, 87
  of *meso*-hydroxyporphyrins, **43**, 87
  of pyrazolopyrimidines, **41**, 363
  of 1,2,6-thiadiazine 1,1-dioxides, **44**, 103–106
  of 1,2,5-thiadiazolones and -diones, **44**, 136
  of 1,2,4,6-thiatriazine 1,1-dioxides, **44**, 158
Tautomerism, ring-chain
  of actinorhodin, **45**, 108
  of chromenediones, **45**, 101
  of lapachones, **45**, 102
Tecomaquinone-II, structure and synthesis, **45**, 103
Tellurophenes, cleavage on lithiation, **41**, 48
Temperature correction of pK data, **41**, 212
Temperature effects on Chichibabin reaction, **44**, 21
Tetraaza[14]annulenes, 1,8-dihydro-, **41**, 34
Tetracyanoethylene, reaction
  with an aza-dibenzazulene, **43**, 55
  with protoporphyrin systems, **43**, 102
Tetracyanoethylene oxide, reaction with azines, **43**, 209
Tetrahydroalstonine, synthesis, **42**, 326
5,6,7,8-Tetrahydroquinolines, synthesis by Diels-Alder reactions, **42**, 302
Tetramethylenedisulfotetramine, **44**, 186
Tetrasaccharides, FAB mass spectra, **42**, 393
Tetrasulfimide ($S_4N_4H_4O_8$), **44**, 186
Tetrasulfur tetranitride S,S-dioxide, **44**, 186
1,2,4,5-Tetrazine, 3,6-di(methoxycarbonyl)-, in cycloadditions, **42**, 252
1,2,4,5-Tetrazines
  nucleophilic substitution, **44**, 246
  use in synthesis for bridge removal, **45**, 54, 68
Tetrazoles
  acidity and basicity, **41**, 262
  Claisen rearrangements in, **42**, 212
  cleavage by base, **41**, 63

Tetrazoles, *N*-acyl, conformation, **41**, 133
Tetrazolium salts, cleavage by base, **41**, 63
Tetrazolo-fused diazocines, **45**, 191
Tetrazolyl groups, substituent constants, **42**, 47
Thelephoric acid, occurrence and structure, **45**, 65
Theoretical calculations on/studies of
  acidity and basicity of azoles, **41**, 200
  acyl conformations in heterocycles, **41**, 161
  3a-azaazulenes, **43**, 37, 40
  charge densities in 1,2,6-thiadiazine 1,1-dioxides, **44**, 85
  1,4-diazocines, **45**, 209, 212
  molecular conformation, **43**, 219, 235
  pyrazolopyrimidines, **41**, 361
  reactivity of isoindole-4,7-diones, **45**, 54
  thiophene S-oxide inversion barrier, **45**, 153
  thiophene reactivity, **45**, 182
  thiophenium ylids, **45**, 164
  see also Molecular mechanics
Thermodynamic aspects of 1,2,6-thiadiazine 1,1-dioxides, **44**, 100
Thermodynamic cycle of imidazole and pyrazole, **41**, 226
Thermodynamic parameters
  of acidity data of azolium ions, **41**, 205
  standardisation of data on, **41**, 212
2-Thiabicyclo[3.1.0]hex-3-enes, formation, **45**, 168
1,2,4,3-Thiadiazaboretidine 1,1-dioxides, **44**, 182
1,2,4,3-Thiadiazaphosphetidine 1,1-dioxides, **44**, 180
1,2,7-Thiadiazepine 1,1-dioxide, hexahydro-, synthesis, **44**, 184
1,2,6-Thiadiazine 1,1-dioxide glycosides, rotamerism, **44**, 101
1,2,6-Thiadiazine 1,1-dioxide, 3,5-diamino-4-hydroxyimino-
  electrophilic attack, **44**, 115
  isomerism of oxime group, **44**, 101
  tautomerism, **44**, 105
1,2,6-Thiadiazine 1,1-dioxide, 3,4,5-triamino-, fused systems from, **44**, 128
1,2,6-Thiadiazine 1,1-dioxides
  aromaticity, **44**, 100
  reaction with carbenes, **44**, 117

reduction, **44**, 117
spectra
  infrared, **44**, 98
  mass, **44**, 99
  nuclear magnetic resonance, **44**, 87
  ultraviolet, **44**, 94
structure, **44**, 84
synthesis, **44**, 120
tautomerism, **44**, 103
theoretical studies, **44**, 84
thermodynamic aspects of, **44**, 100
1,2,6-Thiadiazine 1,1-dioxides, amino-
  basicity, **44**, 109
  tautomerism, **44**, 104
1,2,6-Thiadiazine 1,1-dioxides, 3,5-
  diamino-
  biological activity, **44**, 188
  hydrolysis, **44**, 118
  synthesis, **44**, 124
  uses, **44**, 188
1,2,6-Thiadiazine 1,1-dioxides, 3,5-diami-
  no-4-oxyimino, nucleophilic substitution of amino groups, **44**, 118
1,2,6-Thiadiazine 1,1-dioxides, 3,5-dioxo-
  acidity, **44**, 109
  biological activity, **44**, 188
  carbon-13 nmr, **44**, 93
  conformation, **44**, 86
  electrophilic attack on, **44**, 114, 115
  isolation as salts, **44**, 111
  synthesis, **44**, 123
  tautomerism, **44**, 106
  uses, **44**, 188
1,2,6-Thiadiazine 1,1-dioxides, 3-oxo-
  carbon-13 nmr, **44**, 92
  electrophilic substitution, **44**, 115
  reaction with PCl$_5$, **44**, 118
  synthesis, **44**, 124
1,2,6-Thiadiazine-3,5-dione 1,1-dioxides,
  see 1,2,6-Thiadiazine 1,1-dioxides,
  3,5-dioxo-
1,2,6-Thiadiazine-3,5(2*H*,6*H*)-dione 1,1-di-
  oxide, 4-oximino-, ring contraction,
  44, 153
1,2,6-thiadiazin-3-one 1,1-dioxides, *see*
  1,2,6-Thiadiazine 1,1-dioxides, 3-oxo-
10*H*-[1,2,6]Thiadiazino[3,4-*b*]quinoxalin-
  4 (3*H*)-one 2,2-dioxides, synthesis, **44**,
  131
Thiadiaziridine 1,1-dioxides

nucleophilic attack
oxidation, **44**, 178
reactivity, **44**, 177
spectra, **44**, 177
structure, **44**, 176
2*H*-1,2,8-Thiadiazocine 1,1-dioxide, hexa-
  hydro-, synthesis, **44**, 185
1,2,5-Thiadiazole, chlorocyano-, **41**, 34
1,2,5-Thiadiazole 1,1-dioxide, 4-amino-2,3-
  dihydro-3,3-dimethyl-, basicity, **44**, 148
1,2,5-Thiadiazole 1,1-dioxide, 3,4-dichlo-
  ro-, reaction with nucleophiles, **44**,
  142
1,2,5-Thiadiazole 1,1-dioxide, 3,4-dimeth-
  oxy-, methyl migration in, **44**, 144
1,2,5-Thiadiazole 1,1-dioxide, tetrahydro-
  4,4-dimethyl-3-methylene-, synthesis
  and rearrangement, **44**, 147, 153
1,2,5-Thiadiazole 1,1-dioxides
  biological activity, **44**, 189
  pyrolysis, **44**, 138
  reactions of substituents, **44**, 141
  spectra
    carbon-13 and proton nmr, **44**, 134
    infrared, **44**, 135
    mass, **44**, 136
    ultraviolet, **44**, 135
  structure, **44**, 133
  synthesis, **44**, 148
  thermolysis, **44**, 138
  uses, **44**, 189
  x-ray diffraction, **44**, 133
1,2,5-Thiadiazole 1,1-dioxides, amino-
  acidity, **44**, 147
  biological activity, **44**, 189
  tautomerism, **44**, 137
1,2,5-Thiadiazole 1,1-dioxides, 2,3-
  dihydro-
  reaction of nucleophiles, **44**, 145
  reduction, **44**, 140
1,2,3-Thiadiazoles, cleavage by base, **41**, 63
1,3,4-Thiadiazoles, quaternisation rates,
  43, 197, 201
1,2,5-Thiadiazolidine 1,1-dioxides, **44**, 140,
  150, 151
1,2,5-Thiadiazolidine 1,1-dioxides, 3,4-
  dioxo-
  acidity, **44**, 147
  dialkylation, **44**, 138
  hydrolysis, **44**, 141

1,2,5-Thiadiazolidine 1 (*cont.*)
  infrared spectra, **44**, 135
  tautomerism, **44**, 136
1,2,5-Thiadiazol-3(2*H*)-one 1,1-dioxide, 4-amino-, hydrolysis, **44**, 142
1,2,5-Thiadiazol-3(2*H*)-one 1,1-dioxide, 4-piperidino-, **44**, 137
1,2,5-Thiadiazol-3(2*H*)-one 1,1-dioxides
  acidity, **44**, 147
  crystal structure, **44**, 133
  spectra
    carbon-13 nmr, **44**, 134
    ultraviolet, **44**, 135
  synthesis, **44**, 149
  tautomerism, **44**, 136
  x-ray diffraction, **44**, 133
[1,2,5]Thiadiazolo[3,4-*b*]quinoxaline 2,2-dioxide, 1,3-dihydro-, **44**, 137, 144
[1,2,5]Thiadiazolo[3,4-*c*][1,2,3]thiadiazine 5,5-dioxide, 7-amino-, synthesis, **44**, 128
[1λ^4,2,5]Thiadiazolo[3,4-*c*][1,2,5]thiadiazole, **44**, 142
1,2,4,6-Thiatriazine 1,1-dioxides
  acidity, **44**, 160
  biological activity, **44**, 189
  nucleophilic substitution in, **44**, 163
  ring interconversion of, **44**, 126
  spectra
    mass, **44**, 158
    nuclear magnetic resonance, **44**, 155
  synthesis, **44**, 163, 184
  tautomerism, **44**, 158
  uses, **44**, 189
1,2,4,6-Thiatriazine 1,1-dioxides, fused derivatives
  hydrolysis, **44**, 162
  ultraviolet spectra, **44**, 157
1,2,4,6-Thiatriazine 1,1-dioxides, saturated derivatives with adamantane skeleton, **44**, 164
1,2,4,6-Thiatriazine 1,1-dioxides, 3,4-dihydro-
  hydrolysis, **44**, 161
  synthesis, **44**, 165
1,2,4,6-Thiatriazine 1,1-dioxides, tetrahydro-, synthesis **44**, 164
1,2,4,6-Thiatriazin-3-one 1,1-dioxides
  carbon-13 nmr, **44**, 156
  synthesis, **44**, 167
  x-ray diffraction, **44**, 155

1,2,3,5-Thiatriazole 1,1-dioxides, 2,3-dihydro-, **44**, 183
1,2,3,5-Thiatriazole 1,1-dioxides, 2,5-dihydro-, **44**, 182, 183
2*H*-1,2-Thiazine 1-oxide, 3,6-dihydro-, synthesis and uses, **42**, 297
1,4-Thiazino-fused heterocycles, synthesis, **43**, 314, 315
1,2-Thiazocine, 2,3,4-trialkoxycarbonyltetrachloro-, formation, **45**, 178
Thiazoles
  quaternization rates, **43**, 186, 187
  substituent constants, **42**, 66
Thiazolidine-2-thione, (4*S*)-4-ethyl-, synthesis, **45**, 3
Thiazolidine-2-thione, (4*S*)-4-isopropyl-, synthesis, **45**, 3
Thiazolidine-2-thione, (4*R*)-4-methoxycarbonyl-, synthesis, **45**, 2
  use, **45**, 4
Thiazolidine-2-thiones, 3-acyl-, chiral recognition
  in aldol reactions, **45**, 7
  in aminolysis, **45**, 4
  in asymmetric dicarboxylic acid synthesis, **45**, 22, 24, 26, 27
  in β-lactam synthesis, **45**, 13, 16
Thiazolines, in antibiotic β-lactam synthesis, **45**, 30
Thiazoline-2-thiones, 3-alkyl-, rotational barriers, **43**, 221, 229, 232
Thiazoline-2-thiones, 3-aryl-, rotational barriers, **43**, 267
Thieno-fused quinones, **45**, 69
Thieno[3',4'-4,5]dibenzo[*b,d*]thiophenes, synthesis, **45**, 73
Thieno[3,2-*b*]furan-3-carboxylic acid, 5-chloro-2-methoxy-, methyl ester, formation, **45**, 171
Thienyl groups, substituent effects, **42**, 43
Thioacyl heterocycles, conformation, **41**, 158
Thiocarbonyl compounds, cycloaddition to azomethine ylids, **45**, 304
Thio-Claisen rearrangements
  in benzimidazoles, **42**, 212
  in benzothiazoles, **42**, 213
  in furans, **42**, 207
  in indoles, **42**, 206

in pyrimidines, **42**, 223
in quinolines, **42**, 220
in thiophenes, **42**, 208
in triazines, **42**, 223
Thioisatin, syntheses from, **45**, 73
Thiones, amination under Chichibabin conditions, **44**, 66
Thionitroso compounds, formation and reactivity, **45**, 178
Thiophene, S-alkylation, **45**, 153
Thiophene 1,1-dioxides, reactivity, **45**, 153
Thiophene 1,1-dioxides, 2,5-dihydro-, diene precursors, **42**, 263
Thiophene 1-oxide, 2,5-di-t-butyl-, shape, **45**, 153
Thiophene 1-oxide, 2,5-bis(1,1,3,3-tetramethylbutyl)-, ^{1}H NMR and inversion at S, **45**, 153
Thiophenes
  acylation kinetics, **43**, 211
  Claisen rearrangements in, **42**, 208
  steric inhibition of resonance in, **43**, 212
Thiophenes, acyl, conformation, **41**, 84, 94
Thiophenium S-imines
  S-N bond cleavage, **45**, 177
  cycloadditions, **45**, 177
  formation as intermediates, **45**, 179
  synthesis, **45**, 176
Thiophenium salts (review), **45**, 151
Thiophenium salts, 1-alkyl-
  photochemical alkyl migration, **45**, 159
  preparation, **45**, 154
  reactivity, **45**, 159
  spectra
    mass, **45**, 156
    nuclear magnetic resonance, **45**, 156
    ultraviolet, **45**, 155
  structure, **45**, 158
Thiophenium ylids (review), **45**, 151
Thiophenium S-ylids
  cycloadditions, **45**, 171
  fragmentation to thiophene and carbenes, **45**, 170
  reactivity, **45**, 165
  rearrangements
    to 1,4-oxathiocins, **45**, 169
    to 2H-thiopyrans, **45**, 167, 173
    to 2-substituted thiophenes, **45**, 167
    to 3-substituted thiophenes, **45**, 168, 170
  to 2-thiabicyclo[3.1.0]hex-3-enes, **45**, 168, 174
  spectra, **45**, 163
  structure, **45**, 165
  synthesis, **45**, 159
Thiophenium 1-di(alkoxycarbonyl)methylides, 2-benzyl-, dynamic ^{1}H NMR spectra, **45**, 162
Thiopyrylium ion, 4-benzylmethylamino-, rotational barriers, **43**, 245
Thiourea, reaction with trichloropyridine, **43**, 308
Three-membered rings, acyl, conformation, **41**, 118
Tin(II) triflate, enolates from, **45**, 8, 13, 18, 26
Tirandamycin, synthesis, **42**, 278
Titrations
  calorimetric, **41**, 203
  nuclear magnetic resonance, **41**, 204
  potentiometric, **41**, 202
  spectrophotometric, **41**, 203, 208
α-Tocopherol, oxidation by $FeCl_3$, **45**, 103
α-Tocopurple, structure, **45**, 100
α-Tocored, structure, **45**, 100
Toluene, amination, **44**, 73
Tosyl cyanide, as dienophile, **42**, 252
Trachelanthamidine, synthesis, **45**, 20, 338
Transmission of electronic effects of substituents, **42**, 68
Trehalose, FAB mass spectrum, **42**, 385
1,2,4,5-Triazaphosphorines, **43**, 24
1,3,5-Triazine in heterocyclic synthesis, **41**, 3
1,3,5-Triazine, diphenyl-, Chichibabin amination, **44**, 70
1,3,5-Triazine, phenyl-, Chichibabin amination, **44**, 11
Triazines, σ-adducts with amide ion, **44**, 10
1,2,3-Triazines
  in Diels-Alder reactions, **42**, 302
  nucleophilic substitution, **44**, 239
  N-oxidation, **43**, 159
  quaternization, **43**, 144
1,2,4-Triazines
  in Diels-Alder reactions, **42**, 310
  substitution
    electrophilic, **44**, 242
    free-radical, **44**, 242

1,2,4 Triozines (*cont.*)
  nucleophilic, **44**, 240
  N-oxidation, **43**, 160
  quaternization, **43**, 144
1,3,5-Triazines
  nucleophilic substitution, **44**, 243
  N-oxidation, **43**, 161
  quaternization, **43**, 147
  transmission of electronic effects in, **42**, 72
1,2,4-Triazinium salts, reaction
  with acetoacetamides, **43**, 329
  with 1,4-dinucleophiles, **43**, 333
1,2,4-Triazinones, alkylation, **43**, 145
1,3,5-Triazinones, alkylation, **43**, 147
1,3,5-Triazinyl groups, substituent effects, **42**, 16, 34
1,2,3-Triazole, 4,5-dicyano-, **41**, 36
Triazoles, *N*-acyl, conformation, **41**, 133
1,2,3-Triazoles
  acidity and basicity, **41**, 261
  ring-cleavage of anions, **41**, 54, 62
1,2,4-Triazoles
  acidity and basicity, **41**, 258
  cleavage on lithiation, **41**, 61
1,2,4-Triazolo[4,3-*c*]pyrazolo[4,3-*e*]pyrimidines, **41**, 361
1,2,3-Triazolo[4,5-*c*][1,2,3]thiadiazine 5,5-dioxide, 7-amino-2,4-dihydro-, synthesis, **44**, 128
Triazolyl groups, substituent constants for, **42**, 46
Tribenzodi[1,4]thiazine-6,12-dione, **45**, 118
Trichloromethyl groups, nucleophilic displacement of, **41**, 344
Trichothecene, mass spectrum, **42**, 394
Trifluoromethyl azomethine ylids, **45**, 238
Triptycene, phosphazine analogs, **43**, 15
Trisaccharides, FAB mass spectra, **42**, 391
2*H*,4*H*-1,3,5,2,4-Trithiadiazepine 3,3-dioxides, 6,7-dihydro-, synthesis, **44**, 184
2*H*-1,3,5λ^4,2,4,6-Trithiatriazine 1,1-dioxide, 2-trimethylstannyl-
  synthesis, **44**, 174
  x-ray crystallography, **44**, 155
Tropacocaine, synthesis, **42**, 293
Trypethelone, occurrence and structure, **45**, 59
Tuberostermonine, structure, **43**, 69
Tunicaminyl uracil, synthesis, **42**, 282
Turanose, FAB mass spectrum, **42**, 384
Tylophorine, synthesis, **42**, 261, 306

U

Ultraviolet/visible spectra
  of 3a-azaazulenones, **43**, 40, 43
  of fused 1,2,4,6-thiatriazine 1,1-dioxides, **44**, 157
  of oxazolidine-3-thiones, **45**, 3
  of porphyrins and hydroporphyrins, **43**, 82, 117, 121
  of 1,2,6-thiadiazine 1,1-dioxides, **44**, 94
  of 1,2,5-thiadiazole 1,1-dioxides, **44**, 135
  of thiazolidine-3-thiones, **45**, 3
  of thiophenes and thiophenium salts, **45**, 155
Uracil, alkylation, **43**, 140
Uracil, 5-fluoro-, synthesis from a triazinedione, **43**, 340
Uracils, Claisen rearrangements in, **42**, 222

V

Valerianine, synthesis, **42**, 321
Ventilones, occurrence and structure, **45**, 68
Ventiloquinones, occurrence and structure, **45**, 108
Vesparion, structure, **45**, 110
Vibrational spectroscopy and molecular conformations, **41**, 78
Vicarious nucleophilic substitution
  in pyridines, **44**, 202
  in 1,2,4-triazines, **44**, 240
Vilsmeier formylation/reactions
  of 3a-azaazulene derivatives, **43**, 44
  of dihydropyridines, **44**, 225
  of pyrimidines, **44**, 232
Vindoline, synthesis, **42**, 249
Vindrosine, synthesis, **42**, 249
Vinemycinone-B$_2$, synthesis, **42**, 277
Viopurpurin, occurrence and structure, **45**, 67
Virginiamycin M1, synthesis, **45**, 11
Visible spectra, *see* Ultraviolet
Vitamin B$_{12}$ synthesis, **43**, 118

SUBJECT INDEX

W

Water
   as solvent for acid-base equilibria, **41,** 216
   reaction with sulfuryl diisocyanate, **44,** 173
Wilsonine, structure, **43,** 69

X

Xanthones, Claisen rearrangements in, **42,** 238
X-ray crystallography/structure determination
   in stereochemical studies, **43,** 217
   of 5-azido-2H-1,2,4,6-thiatriazin-3(4H)-one, 1,1-dioxide, **44,** 155
   of a bridged azaazulene, **43,** 68
   of cyanocycline-A, **45,** 95
   of 1,3,2,4,6-diazatriphosphorines, **43,** 28
   of diazine-dinucleophile adducts, **43,** 329
   of 1,4-diazocine derivatives, **45,** 210, 211
   of an indazole-4,7-dione, **45,** 76
   of methoxatin, **45,** 119
   of 5-methoxy-2-methyl-2H-1,2,4,6-thiatriazin-3(4H)-one 1,1-dioxide, **44,** 155
   of S-methyldibenzothiophenium tetrafluoroborate, **45,** 158
   of S-methylnaphtho[2,3-b]thiophenium tetrafluoroborate, **45,** 158
   of mitomycin C, **45,** 46
   of naphthyridinomycin, **45,** 95
   of phenophosphazines, **43,** 14
   of porphyrins, **43,** 79
   of sarubicin, **45,** 109
   of 1,2,6-thiadiazine 1,1-dioxides, **44,** 86, 113
   of thiadiaziridine 1,1-dioxides, **44,** 176
   of 1,2,5-thiadiazole 1,1-dioxides, **44,** 133
   of thiophenium S-di (methoxycarbonyl)-methylide, **45,** 162
   of thiophenium S-tosylimine, **45,** 176
   of 2-trimethylstannyl-2H-1,3,5λ^4,2,4,6-trithiatriazine 1,1-dioxide, **44,** 155
Xyloidine, reactivity, **45,** 123
Xylopinine, synthesis, **42,** 248
o-Xylylenes, as Diels-Alder dienes, **42,** 247, 248

Y

Ylid(e)s
   azomethine (review), **45,** 231
   thiophenium (review), **45,** 159
Ytterbium complexes in condensation catalysis, **42,** 277, 278, 324

Z

Zeatin, synthesis, **42,** 287
Zinc chelates in azomethine ylid generation, **45,** 281, 282, 313

DEC 1 9 1989